‘Even as the climate warms, so too does the debate about thermal comfort in buildings. *Adaptive Thermal Comfort: At the Extremes* is appropriately hard-hitting and insightful – a true guide for hotter times ahead.’

Jonathon Porritt, *writer and campaigner, president of The Conservation Volunteers and Population Matters*

‘We can’t air-condition our way out of the climate crisis – this book is an essential manifesto for designing a resilient future, showing how our bodies, buildings, and cultures must adapt to a heating world.’

Prashant Kapoor, *chief industry specialist, Green Buildings and Cities, Climate Business Department, IFC*

‘This is a truly stupendous piece of work covering all dimensions of the question of thermal comfort in buildings – historical time, location, built space and the challenge of Climate Change. A must read for designers and researchers for an urgently needed “reset” for theory and practice.’

Ashok Lall, *principal of Ashok B Lall Architects, Delhi, India*

‘How wonderful to read a book on science that is written so clearly: for in a time of adjustment we need plain speaking… As Sue Roaf says: “the first step to escaping from an echo chamber is to realise that you are in one.” May this book assist you to thrust the window of your enclosed space wide open and brave this world which created you – if you can find one that opens!’

Phil Harris, *Troppo Architects*

‘Many congratulations on this new book. It’s brimming with ideas related to radically cutting carbon in the face of the drastic rates of climate change we are now caught in.’

Aubrey Meyer, *The Global Commons Institute, UK*

‘*Adaptive Thermal Comfort: At the Extremes* is a provocative reading of how comfort has been defined, captured, codified, exported and enforced in our built environment – with highly uncomfortable results for our heating world. We are honored that our projects have been identified as exemplars for localized, adaptable and culturally attuned comfort. This book is a necessary critique of how we got here and offers actionable suggestions for designers, professionals and policy-makers moving forward.’

Richard Hassell, *WOHA, Singapore*

ADAPTIVE THERMAL COMFORT

To remain comfortable in a world of ever more extreme weather events and climate trends, we need a building revolution. Building designers, owners, managers, and occupants must prepare now for future climates with new ways to stay comfortable indoors. This book is a compendium of information on comfort that provides an overview of the complexity of the many ways that comfort is achieved in buildings. It outlines the impacts and implications of current design practices on greenhouse gas emissions from buildings and the health and well-being of people in them. In reality, many modern buildings, and particularly homes, are already failing in various ways. During extreme weather events, they overheat. During power outages, many buildings do not even remain habitable. During the COVID pandemic, cross-infections between occupants were rife in buildings like hospitals and hotels without opening windows. As energy prices soared and globally economies flatlined, many found themselves unable to pay for high-cost comfort solutions and so either had to change their lifestyles and expectations or learn to live with discomfort. Underlying many of the growing global problems is the trend towards an overdependence on mechanical systems to produce comfort, coupled with a decrease in the passive climatic performance of the buildings themselves. Both factors are resulting in a generation of increasingly un-resilient buildings.

The theory of *adaptive thermal comfort* states that people adapt to those temperatures they normally occupy, and if they become uncomfortable, they tend to change themselves or their surroundings to return to comfort, if they are able or can afford to. This is the third of three volumes, which builds on the practical and theoretical foundations of the subject laid out in the first two volumes. It builds on their premises to shape a new and better roadmap going forward for imagining, designing, and constructing adaptable buildings, and for the behavioural lifestyle changes needed to prepare humanity to survive and thrive comfortably in the very different weather and climates ahead.

Susan Roaf is Emeritus Professor of Architectural Engineering at Heriot Watt University. Raised in Malaysia and the Australian bush and educated in Britain, she has lived and worked as an architect, anthropologist, and archaeologist in Iran, Iraq, Pakistan, California, and Antarctica, experiences that colour her unique understanding of buildings and comfort in different climates and cultures and that inspired her work on adapting buildings and cities to a heating world. She pioneered UK building–integrated solar technologies and eco-design and, with Nicol and Humphreys, has promoted adaptive thermal comfort globally. Her expertise in ancient technologies informed some of her 23 books and other publications, all aimed at better understanding building performance in the past, present, and future.

Fergus Nicol is an award-winning leader in the field of adaptive thermal comfort, having started as a physicist at the Building Research Establishment in the 1960s. He moved on to work with the UK Medical Research Council and into teaching before leaving both to start the radical book shop Bookmarks. Returning to research in 1992, he is now Emeritus Professor in a number of universities and a top cited scholar across his many publications. He led influential pan-European and Pakistan studies on comfort, and he leads NCEUB, the Network for Comfort and Energy Use in Buildings. He co-founded and ran the Windsor Conferences on Comfort and is internationally respected for his support of fellow researchers and students.

Michael Humphreys is known for his pioneering work on the adaptive approach to comfort. He was Head of Human Factors at the Building Research Establishment and has been Research Professor at Oxford Brookes University. His scientific interests are the methodology of field studies of environmental comfort, the structure and statistical modelling of human adaptive behaviour, and the interactions between the several aspects of the indoor environment.

This is the third volume in the *Adaptive Thermal Comfort* trilogy. See also:

Adaptive Thermal Comfort
Principles and Practice
Fergus Nicol, Michael Humphreys and Susan Roaf

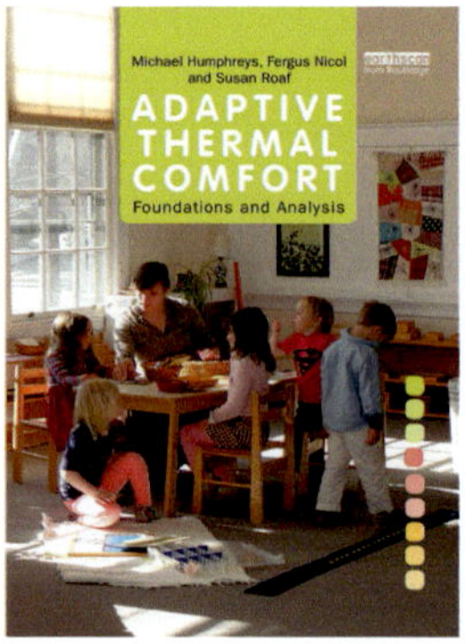

Adaptive Thermal Comfort
Foundations and Analysis
Michael Humphreys, Fergus Nicol and Susan Roaf

ADAPTIVE THERMAL COMFORT

At the Extremes

Susan Roaf, Fergus Nicol and Michael Humphreys

Routledge
Taylor & Francis Group
LONDON AND NEW YORK

earthscan
from Routledge

Designed cover image: Susan Roaf, Fergus Nicol, Richard Roaf and Luisa Brotas

First published 2026
by Routledge
4 Park Square, Milton Park, Abingdon, Oxon OX14 4RN

and by Routledge
605 Third Avenue, New York, NY 10158

Routledge is an imprint of the Taylor & Francis Group, an informa business

British Library Cataloguing-in-Publication Data
A catalogue record for this book is available from the British Library

ISBN: 978-0-367-13769-4 (hbk)
ISBN: 978-0-415-69163-5 (pbk)
ISBN: 978-1-315-64507-0 (ebk)

DOI: 10.1201/9781315645070

Typeset in Times New Roman
by Apex CoVantage, LLC

This volume is dedicated to Michael Humphreys and his pioneering work on adaptive thermal comfort.

It is also dedicated to all those pioneering designers and developers brave enough to break free of the shackles of 20th-century thinking on comfort in buildings to create buildings at all scales, that can run for as much of the year as possible on free, local, renewable energy.

Pioneers, we salute you for the future-facing buildings you create and inform. On your shoulders stand the best hopes for humanity of radically reducing the impact of the built environment on our shared atmosphere and planet. By showing the world how to design genuinely low energy and low emissions buildings that can keep occupants thermally safe and healthy even at the extremes you draw into focus the dangerous folly of those who continue to build and occupy the Super-Polluters, *the all glass buildings in cities around the world that are a major and growing driver of our dangerously heating climate.*

A pioneering building: the hybrid, largely naturally ventilated[1] BRAC University in Dhaka, Bangladesh, designed by WOHA Architects,[2] Singapore, and Engineered by Transsolar, Germany.[3]

Notes

1 https://transsolar.com/de/projects/dhaka-brac-university
2 https://woha.net/
3 https://transsolar.com/de/

CONTENTS

PREFACE

This book is a compendium on comfort. It has been written to explain and educate on the many and varied factors and forces that contribute to people feeling comfortable in and around buildings. Comfort plays such a huge part driving greenhouse gas (GHG) emissions from the built environment that we also touch on some of the multiple consequences of the human quest for comfort. Understanding comfort and acting on that knowledge is one of the most effective levers societies have to hand to batten down the hatches of buildings and cities to ensure that we can weather the coming climate storms with our health, wealth, and well-being intact.

The book is built on the foundations of a combined total of over 170 years of research, discussion, and development by the three authors on the subject of *adaptive thermal comfort*! We have usefully concluded that this hugely complex subject can, in fact, be summed up by one simple idea:

> *People adapt to the thermal conditions they normally occupy, and if they are uncomfortable, they change themselves, or their environment, to restore their own comfort where possible.*

This volume has a focus on the design of buildings. They are our third protective layer against the climate, after skin and clothing. Buildings should ideally be designed to ensure the thermal conditions people occupy are (1) generally safe and acceptable to occupy and (2) that people have sufficient usable tools to hand in them to adapt both themselves and their surroundings to be able to return to comfort if and as conditions change. Any actions that deny people the possibility of adapting themselves, or their buildings, retard the ability of occupants to cope with and adapt to those changes. This presents a practical and an existential problem in the rapidly changing world around us as the buildings of wealthier populations are, in practice, being increasingly detached from the world around them as machines take over the control of internal environments.

The following chapters show clearly that humanity is facing a growing comfort crisis in the built environment. Many modern buildings are already overheating, driving up GHG emissions from the energy needed to cool them and making 'comfort' less affordable and accessible for the many. These are symptoms of a deep malaise in late 20th-century design thinking that must urgently be tackled at all stages of the building's life cycle. Three issues that influence these failings are the increasing over-mechanisation of buildings at the expense of their passive climatic performance; the overbearing role of mechanical services professionals in setting and applying building regulations, codes, and standards; and thirdly, the almost-universal use of limited building simulation tools at all stages of the design. Bundling decisions about construction, form, and materials together with mechanical services in simulations means that designers never get a clear view of how the actual building itself will perform without its services, neither in current nor in future climates.

The false 20th-century assumption that all people, everywhere, are comfortable in the same range of temperatures has needlessly resulted in the waste of the vast amounts of energy that continues to be a major driver of climate change. Many designers today have little understanding of the actual conditions necessary in a building to enable people to make themselves comfortable. A common preconception for designers now is that comfort is a product measured in temperatures, air velocity, and humidity and sold in measures like kilowatts or kilowatt-hours. This book attempts to shine a light on such flawed thinking.

The following chapters cover many different aspects of how people experience and perceive comfort and the impacts comfort has on the health and well-being of people and populations in the built environment. The book covers a wide range of psycho-physical, social, and economic factors that affect comfort, but it is important to remember that every building, client, climate, and culture is different. One size does not fit all. You cannot transmute design rules from one side of a deep valley to the other, let alone from one latitude to another. Comfort in a particular building is achieved through the interaction of the local climate and the environment, its design and construction, and the customary habits and lifestyles of its occupants.

The icing on the comfort cake is that if we take a step further and look at what comfort means to all of us in deeper senses, then we can use that knowledge not just to improve physical health but also to substantially improve the mental and spiritual quality of life and well-being of people in the buildings we design and occupy.

This is the third book in a trilogy. The first volume, led by Fergus Nicol, covered the principles and application practices of the adaptive thermal comfort model. The second volume, written largely by Michael Humphreys, expounded on the theoretical foundations and analysis methods of the science of adaptive thermal comfort. This third volume is largely written by Susan Roaf and sets out to explain why adaptive comfort thinking is so important for building designers and for humanity in its broadest senses. The chapters set out a very flexible framework on which next-generation design thinkers can create more climate-safe buildings that perform well in a rapidly changing world of more extreme weather events, pandemics, population demographics, and economics.

The first two chapters of the book explain some of the growing challenges around creating low-energy, climate-safe, and comfortable buildings, then explain how some modern building types evolved over a century to become ever more un-resilient, and in some cases even super-polluters. Six further sections and 26 chapters follow, dealing with:

1. *People and comfort.* On the relationship between people, comfort, climate, and buildings.
2. *Comfort and cultures.* Including the colonisation of comfort by building engineers.
3. *Thermal heartbeats of buildings.* Including when they become lethal at the extremes.
4. *Ecology and comfort.* A major theme of the book, dealing with the reconnection of the climates and ecologies within buildings to those around them. Eight chapters explore ways in which the flows of energy between spaces within buildings and nature outside can be managed, harvested, stored, and controlled to provide comfort indoors for much of the year.

5. *Design tools and a three-step design process for buildings.* Adaptive opportunities and delight.
6. *Well-being and comfort.* Chapters cover physical, mental, emotional, and spiritual well-being.

The scope of the book is enormous, and the text had to be limited in places to fit into one book. We have tried to provide key references that expand issues on individual subjects for the really interested to follow. Forgive us for glaring emissions or mistakes and we may be able to rectify them in future editions. To make the book easier to dip in and out of, we subdivided the chapters into digestible sub-sections. Many of the ideas dealt with in the book have been triggered by contributions from the delegates to the 30 years of our *Windsor Conferences* on Adaptive Thermal Comfort[1] and more recent ones dealing with *comfort at the extremes*.[2] Detailed papers on many subjects included here can be found in their proceedings, which are free online, and from books that have resulted from the Conferences.[3] It was in those meetings with their open and constructive discussions that the foundations were laid for our broader understanding of what actually constitutes comfort in buildings. Those important collaborations informed our thinking around what the next steps might be to creating climate-safe and largely natural-energy buildings that are needed to protect both people and the planet in the evermore extreme weather and climates we will all share in our overlapping futures in this heating world.

Delegates at the 2016 Windsor Conferences on Thermal Comfort.

Source: Ashak Nathwani.

Notes

1 https://windsorconference.com/history/
2 https://comfortattheextremes.com/past-conferences/
3 Nicol, F., H. Rijal and S. Roaf (Eds) (2022). Resilient comfort standards, in *The Routledge Handbook of Resilient Thermal Comfort*, London: Routledge, ISBN: 978-1-032-15597-5

ATC3 ACKNOWLEDGEMENTS

This is the third book of our trilogy on adaptive thermal comfort. The lead author on Volume 3 is Susan Roaf. This book builds on the themes of both previous volumes. Volume 1, led by Fergus Nicol, covered the basic theory of adaptive thermal comfort and described related methods. Volume 2, led by Michael Humphreys, took those themes to a more advanced level and included a detailed history of the development of the Adaptive Model in Britain.

Volume 3 has expanded the scope of the original adaptive comfort corpus of research interests, not least in response to the growing challenges faced by designers in a heating world. This book is built on the ideas shared by the authors over more than 30 years and the ever-evolving themes and discussions they have navigated to better understand the complex subject of comfort.

The following 28 chapters bear testament to the expanding reach of the subject and the importance of also understanding *comfort* in its broader senses to the health and well-being of future generations.

All three volumes acknowledge both the pioneers whose work we have built on and also the importance to our work of the growing international fellowship of adaptive comfort researchers, many of whom we came to know and value through the 11 *Windsor Conferences* we organised from 1994 to 2020 and the *Comfort at the Extremes* conferences that succeeded them. Our greatest thanks go to the hundreds of friends and related researchers we have made around the world over the past decades through those conferences who have helped us build a broader and deeper understanding of how people become comfortable and what makes them comfortable. This book is, in a way, a tribute to all of you, even if your name is not listed in what follows. You have all mattered to our endeavours in one way or another. Thank you all.

This book was ten years in the writing because every time it approached a closing chapter, a new comfort disaster occurred. In 2020, it was almost ready for the press, and then COVID happened. That was followed by the record-breaking extreme hot and cold weather events and their reverberations on both on the mental health and well-being of a generation so further chapters needed to be added.

Much pioneering work on adaptive thermal comfort was done notably by two luminaries, Fergus Nicol and Michael Humphreys. Their foundational ideas and research also underpin this volume. A main conclusion of theirs is that there is no such thing as a single comfort temperature or comfort zone, because people adapt to those temperatures in the climates that they normally occupy. This idea was originally brought home by Michael Humphreys in the late 1970s in a single graph that related indoor comfort temperatures to those outside, showing their strong interdependence. A second idea developed here is that comfort in different regions and cultures has a shape that inevitably changes over time, an idea expounded on in Chapter 6 by Fergus Nicol. The thinking of all three authors of our three volumes has evolved over the last decades of working together, often radically, in the face of changing circumstances.

There are new lateral themes in this volume, and I would like to thank David Campbell of Heriot Watt University in particular for the course he handed over to me to teach from 2007 until my retirement at Heriot Watt University in 2016, on ecology and the built environment. His notes and thinking set me off on a path to seeing buildings as larger parts of ecosystems, which is reflected in eight of the chapters ahead, and help in fine-tuning them was given me by Brian Orland of the University of Georgia.

Another key theme concerns the role of physiology in adaptive comfort at the extremes, which was triggered not least by a chance meeting in Groningen with Wouter van Marken Lichtenbelt from the University of Maastricht. He and his group, including Hannah Pallubinsky and Boris Kingma, provided new insights and evidence into the workings of the human body at high and low temperatures. Thanks also to physiologist George Havenith and building physicist Dennis Loveday from Loughborough, who have also provided unstinting support to us.

The importance of seeing energy as a dynamic player in the shape of comfort in buildings and landscapes was initially stimulated by the team from Delft University, nurtured under the wing of the brilliant late Jon Kristinsson and including Arjan van Timmeren, Andy van Dobbelsteen, and Rob Roggema.

As we waded our way through the lessons learnt during COVID, the importance of the link between health and comfort became clearer, and for helping explain and making me understand many of the complex issues involved, I am extremely grateful to Philomena Bluyssen of Delft University and Stephanie Dance of the Lanarkshire NHS Trust in Scotland for their research that shed light on these complex issues, highlighting the overriding benefit of natural ventilation in purging spaces of pathogens. Over the years, Ulrike Passe of Iowa State University has supported and informed our work with her research and publications on ventilation in buildings.

Michael Humphreys contributed a core narrative for the book that people find too much change thermally uncomfortable in buildings. This idea that the building, the weather, and the inhabitants create their own unique daily rhythms of indoor temperature was evidenced and reinforced by the work of Geshe Huebner and her colleagues at UCL and Abdul Rahman Al Shaikh in Damam Saudi Arabia. They helped us visualise daily thermal 'heartbeats' of buildings, the temperature pulses that people move through over the day and year to enable them to make themselves comfortable. In the book we promote to building designers the benefits of well-behaved buildings with acceptable customary, or at least steady, indoor climates that require minimal energy input over a year to keep occupants thermally satisfied.

But another key lesson is that there is never ONLY ONE solution to a design challenge. We are grateful to Dolaana Kholvaag, who provided another narrative with her study of how Siberian herdsmen stay thermally safe in thin tents, yurts, in the freezing winters on Russian steppes. She and van Marken Lichtenbelt showed another way to create comfort at the extremes, by fine-tuning when and where a person is heated, at least enough to stay healthy and happy, with very limited fuel in an extreme climate.

Thanks to Bjarne Olesen for the light bulb moment he gave to Sue Roaf in Nice, France, in 2016 about comfort in a resource-constrained future – that buildings will only be heated and cooled when absolutely necessary and for the rest of the year will be naturally ventilated. Thank you, Bjarne. Thanks also to DAIKIN Europe for the annual think tank gatherings that she was honoured to be invited to join for over a decade. Those events helped her understand and admire the immense efforts gone to by the HVAC industry to continually improve their products and to provide low-cost and low-impact comfort.

Many of the subjects covered in this compendium on comfort are not our core fields of expertise, and thanks go to those who have contributed to informing and improving sections in the book, including Bill Bordass, Robyn Pender, Tim Ingold, Lauri Nummenmaa, Victor Olgyay, Harvey Bryan, Federico Tartarini, Stefano Schiavon, Tim de Toldi, Isaac Meir, Peter van den Engel, Karel Kabele, Rajan Rawal, Jessica Fernández-Agüera, Samuel Domínguez-Amarillo, Veronica Soebarto, Terry Williamson, Abdul Rahman Al Shaikh, Rajat Gupta, Ali Chesmehzangi, David Cox, Kevin Lomas, Catherine Simmons, Andrew Sullivan, Brian Ford, Wolfgang Kessling, Fan Wang, Bo Han, Daniel Barber, Hanan Al Khatri, Madhavi Indraganti, Elizabeth Shove, and also to Glen T. Remington, Prashant Kapoor, Valpy Fitzgerald, Manuel Fuentes, Joe Saady, Abraham Yezioro and statisticians Jane and Rex Galbraith.

It is not only the authors of papers who are important in the studies that underpin this book but also those who enable the research, the subjects who provide the data on which our results should be based and the research workers who carefully collect together and analyse the results. For collaborations on fieldwork, thanks go to friends and colleagues working in:

Antarctica. Manuel Corriea Guedes, Joao-Pinelo, Gonçalo Roque Araújo, and Gonçalo Vieira of the Portuguese PROPOLAR team and Marcella Andrades Soto and César A. Cárdenas at the Chilean Professor Julio Escudero Base on King George island, Antarctica.

Australia. Geelong: Jane Matthews and Glen Rogers; Adelaide: Monica Oliphant and John Fargher; Western Australia: Robert Toland, Victory Hunter, Esther Kok and, Alan Yuille; Heron Island: Helen Appleton;

Pakistan. Gul Najam Jamy, Hamid Afridi, and Yasmin Lari.

Much of the data used in our studies has been collected from office workers in Britain, Europe, Pakistan, and beyond and stored and analysed by the researchers in the Oxford Brookes Thermal Comfort Unit, especially Hom Rijal, Iftikhar Raja, Kate McCartney, Maita Kessler, Mary Hancock, Gary Clark, and Ollie Sykes.

A special thanks to Murray Milne, Jane Matthews, Nick Baker, and the late John Harrington-Lyn for their Appendices, and to Hom Bahadur Rijal for his tireless assistance and unstinting support for us in the preparation of the book and its images. Thanks also to Oliver Ball, who helped in assembling the final manuscript in a useful form. Many thanks to all those who have supplied or helped us with the illustrations for the book, including Jim Eveson and Freya Morgan for their graphics. A special thanks in particular also goes to all those we should have thanked but have omitted from these acknowledgements.

Finally, the greatest thanks must go to our editor at Routledge, Ed Needle, who has patiently allowed us to develop the manuscript year after year, in the hope that we would eventually produce a volume that is worth waiting for. Thank you, Ed, and all in the great Routledge team including Chris Mathews, who have helped bring this book to fruition. We hope all three volumes will prove useful to many who are involved in helping humanity adapt in the built environment to the much more extreme conditions we all face in in our rapidly changing world.

Needless to say, responsibility for errors and omissions remain our own.

Susan Roaf and Fergus Nicol

Comfort in Buildings

1

DESIGNING FOR COMFORT AT THE EXTREMES

Why Comfort Matters

We spend over 90% of our lives in buildings. Buildings can protect us from the climate or expose us to it. They can be our friends or enemies. They are the largest investment most of us will ever make. Buildings are usually described according to how they look or by the functions and spaces they contain. Often, however, the most important thing about a building is how *thermally comfortable* it is, not least at *temperature extremes*.

But the very subject of comfort now has a greater role to play in our shared global future. Today, the choices we make around how we design buildings to be comfortable profoundly affect the ability of many individuals and even whole populations to adapt to a rapidly heating climate. Buildings are our third skin, our third basic defence against the elements after our own skin and clothing. The shelters that protect us from sun and wind, heat and cold, and rain evolved as *Homo sapiens* emerged from Africa over 300,000 years ago to inhabit ecosystems across the planet, from the hot deserts to the frozen wastelands of the Arctic. Human adaptation involved changing physiologies, clothing, shelters, habits, and aptitudes, utilising local resources to build with in vastly different climates and landscapes. Such adaptations were possible because humans are clever and can move, and because of the adaptive thermal comfort principle:

> *People adapt to the thermal conditions they normally occupy, and if they are uncomfortable, they change themselves, or their environments, to restore their own comfort where possible.*

People in hot climates will make themselves comfortable at higher temperatures, and people in cold climates find ways to make themselves comfortable in much cooler conditions. This is the simple fact that underlies the adaptive thermal comfort principle, evidenced by data from comfort surveys around the world in which real people were simply asked what temperatures they feel comfortable at in the buildings they usually occupy. Figure 1.1 clearly shows that the reported comfort (neutral) temperatures on the left-hand axis vary according to the occupied or

DOI: 10.1201/9781315645070-2

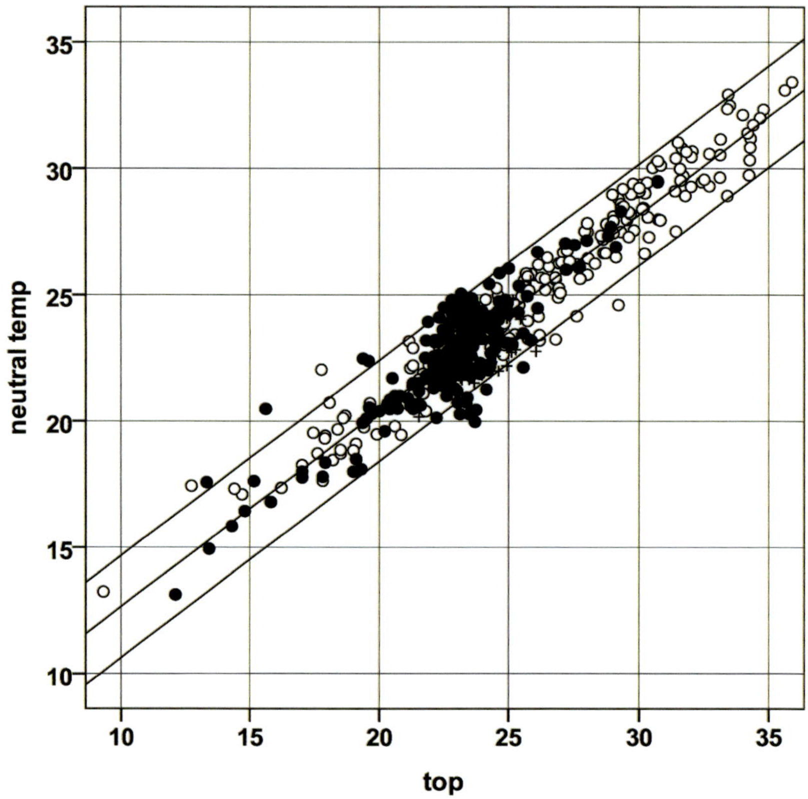

FIGURE 1.1 Results from comfort surveys of subjects in their normal, everyday environment showing the dependence of temperature for comfort (neutral) temperature on the mean indoor operative temperature (*top*). The type of mark shows whether at the time of the surveys the buildings were not heated or cooled (white dot), heated or cooled (black dot), or cross (both).

operative indoor temperatures shown on the horizontal axis. It is patently obvious that indoor comfort temperatures are related to outdoor comfort temperatures, so:

1. There is no such thing as a single universal comfort temperature because
2. People adapt to the temperatures they normally occupy.[1]

Before buildings were connected to gas and electric grid supplies that provided heat and coolth at the flick of a switch, they were basically designed to minimise discomfort for occupants and, at best, to create conditions in which people could experience feelings of thermal delight. Most of us might have felt pleasurably warm in front of a fire in winter or deliciously cool in a fresh breeze from an open window in summer. Pre-grid architects and builders shaped buildings to use locally available energy – solar, wind, and shade – to heat and cool occupants while protecting them from weather extremes. These were the *slow buildings* that evolved over centuries and millennia to form bonds between cultures, climates, buildings, and their landscapes to keep people inside them as comfortable as possible given the available opportunities.

FIGURE 1.2 The comfort crossroads. The front cover of Gail Cooper's classic study of the evolution on air conditioning in the USA.[2]

Source: Johns Hopkins University.

The Comfort Crossroads

Over a hundred years ago, research on 'comfort' started with people like Thomas Bedford in England systematically investigating the effects of heating and ventilation on workers in factories and offices, informing legislation aimed at minimising ill-health, fatalities, and injuries in the workplace.[3] By the early decades of the 20th century, thinking on thermal comfort was already being largely moulded by the growing heating, ventilating, and air conditioning (HVAC) industry. The late Ole Fanger wrote of the engineer's interpretation of what comfort is in 1970[4]:

> Creating thermal comfort for man is a primary purpose of the heating and air conditioning industry, and this has had a radical influence . . . on the whole building industry. . . . thermal comfort is the 'product' which is produced and sold to the customer . . .

If comfort is a 'product' of machines, then it needed to be defined in terms of what machines could produce in order to be managed, codified, and certified. Measures like volumes of air at certain temperatures and humidities within 'comfort zones', moving at certain rates, powered by energy measured in units like BTUs (British thermal units) or kWhs (kilowatt-hours) per floor area or person, were developed. Laboratory experiments in the middle decades of the 20th century based on small-scale studies with limited subject cohorts in constant and controlled environments provided the data from laboratories across northern Europe and America. By 1970, Ole Fanger, the late doyen of the HVAC industry, had produced a *steady, steady* comfort zone prediction method that assumed that the occupied conditions were not dynamic. The consequence became a universally and internationally applied system that defined comfort temperatures as falling within a fairly narrow band of temperatures, from roughly between around 20°C (68°F) and 25.5°C (78°F), regardless of what climate, culture, or economy the building was located in. This engineer's method and its assumptions became hard baked into the language of comfort that is still used in all national and international building codes and standards, regardless of the construction of the building or its mode of operation, whether naturally or mechanically conditioned.

As far back as the 1920s, the HVAC profession pushed the idea that old-fashioned approaches to producing comfort had reached a crossroads. There was a simple binary choice in the direction of travel for the provision of comfort: between *natural ventilation with opening windows* and *mechanical air conditioning solely reliant on machines.* Gradually, over the decades, window opening ceased to be necessary to comply with codes, and legislative strangleholds were ratcheted up by those who wrote the codes, resulting in evermore machines being sold and engineers employed to meet mandated comfort standards.[5] It is extraordinary to think that, nowadays, across the world, some buildings are even constructed with no opening windows at all, completely shut off from nature and the free energy around them, regardless of the impact of that simple decision on the people inside or the environments around them.

That *age of mechanical comfort* was a 20th-century dream, nurtured at a time when people thought that energy would become 'too cheap to meter'.[6] It also came from an era when the impacts of fossil fuel burning on the global climate were unheard of and the threat of the grid's failing was not as universal a problem as it is now. An unfortunate side effect of the mechanical comfort dream is that it has resulted in buildings that are no longer able to provide comfort passively without machines and, consequently, often consume unnecessarily high levels of energy to remain habitable. The consequential costs to society are enormous. Even many in the middle classes in developed economies are finding it increasingly hard to afford the energy needed to keep their homes thermally safe, let alone comfortable, even before the onset of the evermore extreme weather events we all face in the future.

We are now at another crossroads. One direction points to continuing to design buildings that are shut off from the climate around them in order to be conditioned by *efficient machines*. The other direction points to the seemingly revolutionary idea that we urgently need to reconnect indoor climates to those outside them to create buildings in which people are enabled to use natural energy from local sun, wind, and planting to stay comfortable indoors for as much of the year as possible, free of cost to themselves and with minimal impacts on the planet.

The central fallacy in the crossroads paradigm is that it is a binary choice. It is not. Inevitably, in evermore extreme weather, most people will have to heat or cool at some times of the year, and as the climate worsens, those periods may increase. But the key point is that in most climates, at some times of the day or year, the temperatures outside a building will be perfectly adequate to keep people indoors comfortable with natural ventilation where ambient air pollution levels permit.

Fast Architecture – Soaring Impacts

Fast architecture occurs when the evolution of building types goes into overdrive as money floods into societies from rapacious empires, booming commodity markets, or investments from conquering nations. Buildings become the crowning glories of transient economic booms. Britain between 1700 and 1900 boasted bulging coffers from the exchange of slaves, sugar, and other spoils from its empire. Spices from the Far East in the 17th and 18th centuries inflated the Portuguese economy. In the mid-19th century, ornate wind towers soared above the roofs of Yazd in central Persia, paid for by the British need to feed opium into China out of the new port of Hong Kong.[7] Similarly, oil-funded glass and steel skyscrapers now tower over once modest cities in the petrostates of America, the Gulf, and Asia. In 1992, the three modest Hilton apartment blocks were the only towers on the Sheikh Zayed Road in Dubai, and now, 30 years later, it is one of the tallest cities in the world.

It is that scale and pace of change in the buildings of this *fast architecture* that are now driving runaway climate change, not the modest homes of the billions of low earners around the world. Yet it is typically people in their ordinary homes who are the first to feel the rising heat. Estimates over the past couple of decades claim that around 40% of global energy consumption and 34% of greenhouse gas (GHG) emissions result from the building sector. The burning of coal, natural gas, and oil for electricity and heat is the largest single source of GHG emissions. In some countries in the Middle East, and also in parts of the United States, space cooling can represent more than 70% of peak residential electrical demand on extremely hot days, but year on year, the peak demands are rising as buildings become more dependent on air conditioning and global temperatures soar. Cooling accounted for an estimated 37% of the increase in US electricity demand during April–September 2024 over the demand in April–September 2023. Cooling power demand doubled the year-on-year increase of China's electricity demand between August and September 2024, as shown by Rangelova et al. in Figure 1.3.[8] The scale of such year-on-year increases are truly alarming.

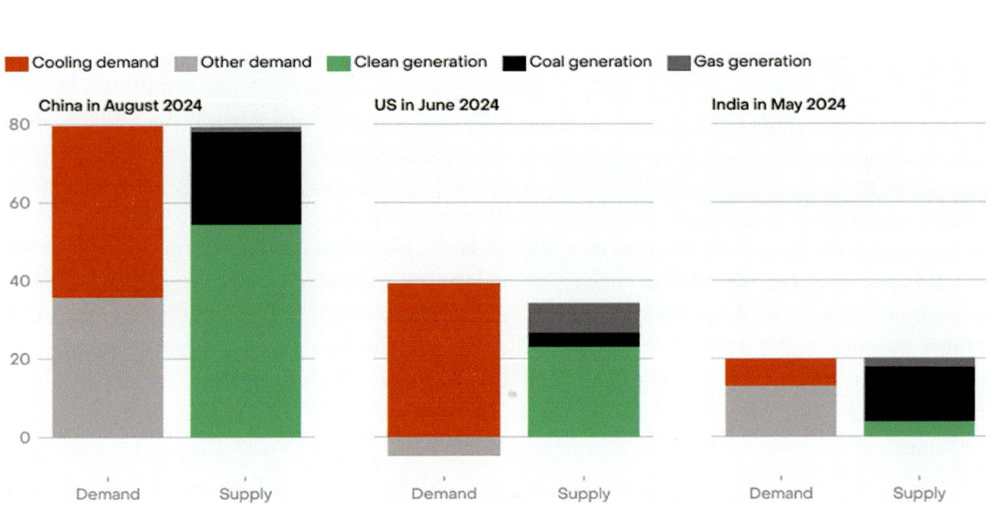

FIGURE 1.3 The chart shows only the *higher cooling demand* in TWh in ONE YEAR, the heatwave year of 2024 over that in 2023, in China, the USA, and India, that drove soaring fossil fuel consumption

Source: EMBER. Published under a Creative Commons ShareAlike Attribution Licence (CC BY-SA 4.0).

Such soaring GHG loadings now actually threaten global climate breakdown as our oceans and atmosphere choke on emissions from the burning of fossil fuels. Thorough and honest reviews of the changes that are urgently needed must now be made for humanity to survive decently with runaway climate change. A good starting place should be in the way we design and legislate for thermal comfort in buildings and societies.

Hi-, Low-, or No-Tech Buildings?[9]

To design a reasonably sized building that can act as the primary modifier of the climate is not difficult. After all, it has been done for millennia by the *slow builders* of the world. Just as the growers of grapes now need to change their vine stock in a timely fashion with the changing climate, designers need now to adapt local designs for a much hotter climate. The biggest barrier to evolving the next generation of *climate-safe buildings* has been the de-skilling over the past decades of architects and building engineers, who are seldom taught the basic tenets of good climatic design, like how to naturally ventilate a building or use thermal mass to store heat and coolth in it as and when needed. Both these simple functions improve the thermal character of a building and minimise its need for imported grid energy to keep occupants comfortable. Students are taught instead how to use computer programmes that do not encourage non-expert designers to undertake these two simple design exercises. The world has spent billions in *energy-efficiency* programmes to reduce CO_2 emissions over the last decades, with the results that emissions have been steadily ratcheting upwards because the very concept of 'efficiency' requires a machine at its heart, and 'more efficient' is typically interpreted in the codes as meaning more machines and/or more expensive machines that all consume imported energy.

Transforming markets in the built environment in time to enable most populations to survive in a world of accelerating climate heating takes vision and leadership. Are we really heading for a world of smart, high-tech structures increasingly reliant on AI to manage their systems ever more efficiently? Or are we headed towards a world of lo-tech or no-tech simpler, smaller, climatically designed buildings. Intellectuals and industries have very different views on this dilemma from either side of that fence.[10] What is sure is that if we do not change the way we design buildings for comfort, things will get a whole lot worse very quickly, not least when grids fail.

Can We Evolve in Time?

The whole landscape of current building design could change tomorrow with the right legislation. It is not as it was for the European settlers landing in North America in the mid-17th century, when it took around a hundred years to transform the simple building blueprints the Pilgrim Fathers brought with them into an extraordinary new palette of American building types. The compact, centralised, close-clipped house of New England differed radically from the Palladian plantation homes of the Seaboard colonies, the airy pavilions of the Louisiana French, and the mud masonry court-yarded haciendas of the South-west, as shown in Figure 1.4 examples.[11] This rapid evolution of American architecture is a tribute to the enormous skills and ingenuity of their builders, transferring lessons learned in other world climates, to novel challenges.

In evolutionary terms, the original European population moved rapidly into drastically different 'adaptive zones' across America from those they were familiar with. They had to learn how to adapt their design strategies for the new climates they settled in or not survive at all in them. What is happening today is different. Populations are encountering drastic shifts in the adaptive zones they already occupy, and many have done so for centuries. They must now evolve their

FIGURE 1.4 American builders had established new climatic design archetypes by the late 18th century, ranging from the compact, centralised, close-clipped, boarded buildings like the 1727 Old Town House in Marblehead, New England, on the left to the heavyweight mud masonry haciendas of the South-west that are still built today.

Source: Susan Roaf.

existing dwellings to cope with the new conditions of a changing climate. This transition for advanced and wealthy populations is being currently masked by the adoption of mechanical conditioning systems. However, the narratives around the benefits of these technological solutions need to be accompanied by their dark sides as well, so decision-makers can weigh up the implications of different pathways forward for the majority of the global populations. We learnt, and should remember, the COVID mantra that ‘no-one is safe until everyone is safe’.

Natural Energy Buildings

The slow builders in almost every ecological niche on the planet knew well how to harness the elements to make themselves more comfortable within their own local, well-understood climates. Those climates are changing, and we need to speed up our understanding of adaptive architecture to keep up with those changes. Fortunately, humanity has been thrown a benign lifeline in its struggle to stay thermally safe and comfortable in more extreme climates. Durable solar and wind technologies can generate electric energy with minimal impacts on the climate. The sun can produce heat but can also produce electricity by the use of photovoltaics (PVs). Almost miraculously, the problem of cooling building occupants during periods of high solar gain coincides with maximum generation from PVs. During winter, winds often pick up to cover heating loads.

Home and business owners around the world are now finding out that with enough photovoltaics on the roof of a building, much of the comfort cooling loads can be met with local solar gain, except for during late afternoons and evenings. These technologies are making the idea of running buildings for as much of the year as possible on local natural energy a potential reality in most climates, with two provisos, that (1) the building itself be climatically well-designed to require modest amounts of energy to keep people indoors comfortable and that (2) the building contain the potential to store excess heat, cold, or electricity within the mass of its structure, hot water in tanks, and electricity in batteries (Figure 1.5).

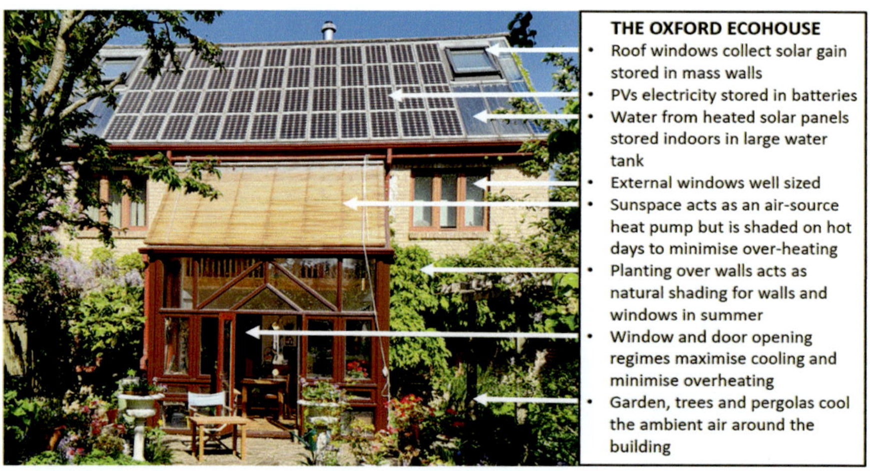

FIGURE 1.5 Adapting to future climates does not require rocket science; it requires careful design to minimise overheating while maximising the contribution of local natural energy to providing comfort indoors as happens in the Oxford Ecohouse shown here.

Source: Susan Roaf.[12]

Brittle Buildings in Extreme Conditions

Natural energy buildings and naturally comfortable buildings can only work if the building itself is not climatically 'brittle', if it does not break quickly during weather or climate extremes. The scale of the challenge of surviving in unadaptable buildings became apparent during 2003 European heatwave that killed over 72,000 people, many in their own homes. In that same summer in the USA, 50 million people were plunged into darkness, some for days, when the entire grid of the Eastern Seaboard failed. People had less than an hour to pour out of their buildings onto the pavements and into heatwave temperatures. Many slept on the streets of New York because they had no way of getting home. Many of the city's high towers where people lived and worked had insufficient back-up generation to cover the air conditioning loads, their glass skins had no opening windows, and within minutes the indoor air became superheated, and in places putrid, where there was no water to flush toilets. Blackouts in high-tech districts of global cities are now happening regularly.

Just how brittle buildings are when they are closed off completely from the outside world also became even clearer when, from 2020 to 2022, the COVID-19 virus hit us all. Tall buildings with lift access only and no opening windows became infectious workplaces. Offices were abandoned in high-rise city centres for months by all but the hardiest. Hospitals built with no opening windows became sources of infection to their own patients, and their design investigated over whether such buildings were ever 'fit for purpose' in protecting the health of both patients and staff, many of whom actually caught the lethal infection when inside a hospital.[13]

In a post-COVID world, Mark Zuckerberg's now-famous motto of 'move fast and break things' seems to reflect another possibility of 'move too fast and get broken'. We hope this book is a wake-up call for humanity in a rapidly heating climate. The way we currently measure and design buildings for the climate is not working. It is leading us to a dead-end generation of unadaptable and thermally dangerous buildings that require so much energy to remain habitable, that only the very wealthiest will be able to afford to occupy them. Even then they will only be safe when supplied with increasingly unreliable grid energy. In addition, the often-exorbitant amounts of energy over-engineered buildings use for heating, cooling, and ventilation are major drivers of the heating atmosphere that lead inevitably to climate instability.

We hope that readers will learn in the book ahead about the simple reality of what makes people comfortable in and around buildings in different cultures and climates. As we prepare for different climates in the future, there are sections on when buildings themselves become thermally dangerous to occupy. Advice offered for designers suggests ways in which to reconnect the ecosystems inside and outside buildings to enable comfort to be sourced from local energy resources to reduce the impact of buildings on the climate. We also touch on the many different ways in which the issue of comfort influences the physical, mental, social, and spiritual health and well-being of individuals and populations. We are not concerned here with the comfort of the richest people in the world in high-cost, mechanically conditioned buildings but, rather, with the health, well-being, and occupied thermal conditions of the many, the billions who will need help to survive in the increasingly hot and challenging decades ahead.

Overheating Buildings – The Biggest Challenge

This book is also aimed in part at the designers of buildings. Chapter 2 concerns the main interface between ordinary people and climate change in the buildings they spend over 90% of their lives in. Over the 20th century, standards of climatic design deteriorated as more reliance was placed on mechanical systems, leading to the generation of modern buildings that even now are overheating. Even in the moderate UK climate, where 40°C was recorded for the first time in July 2022, a 2025 study showed that the prevalence of indoor overheating reported in UK dwellings has increased from 20% in 2011 to 82% in 2022, driving a rapid increase in installed domestic air conditioning.[14] The middle classes in already-hot climates have, for decades, had to spend much of their incomes in the hottest months on cooling to keep families thermally safe indoors. How sustainable these climatically brittle buildings in fragile economies are, and in evermore extreme climates, will become clearer soon enough.

What is clear now is that humanity is currently unprepared for the scale of adaptation that will be required to survive in the extremes of heat and weather in the coming decades. Policymakers urgently need to re-think their strategies for keeping populations thermally safe, as 20th-century approaches to the provision of comfort in buildings can be seen now as part of the problem around timely adaptation – not their solutions.

Notes

1 Humphreys, M.A., H.B. Rijal and J.F. Nicol (2013). Updating the adaptive relation between climate and comfort indoors; new insights and an extended database, *Building and Environment*, 63(May), pp. 40–55. DOI: 10.1016/j.buildenv.2013.01.024

2 Cooper, G. (1998). *Air-Conditioning America: Engineers and the Controlled Environment, 1900–1960*, Baltimore: The John Hopkins University Press.

3 Bedford, T. (1936). *The Warmth Factor in Comfort at Work. A Physiological Study of Heating and Ventilation*, Industrial Health Research Board Report. Medical Research Council, No. 76, iv+102 pp.

4 Fanger, O. (1970). *Thermal Comfort*, Copenhagen: Danish Technical University Press.

5 Cass, N., and E. Shove (2018). Standards, whose standards?, *Architectural Science Review*, 61(5), pp. 272–279, Published online: 27 July 2018,

6 Statement was made by Atomic Energy Commission Chairman Lewis Strauss in a 1954 address to science writers. The full statement is available on page 9 at: https://www.nrc.gov/docs/ML1613/ML16131A120.pdf

7 Roaf, S. (2008). The traditional technology trap (2): More lessons from the Windcatchers of Yazd, *Proceedings of the PLEA Conference*, Dublin, October, Accessed on 14/05/2025 at: http://web5.arch.cuhk.edu.hk/server1/staff1/edward/www/plea2018/plea/2008/content/papers/oral/PLEA_FinalPaper_ref_739.pdf

8 Rangelova, K., D. Jones and N. Fulghum (2025). Powering through the heat: How 2024 heatwaves reshaped electricity demand, Published by Ember, Accessed on 13/05/2025 at: https://ember-energy.org/latest-insights/powering-through-the-heat-how-2024-heatwaves-reshaped-electricity-demand/

9 Roaf, S. (Ed) (2012). Innovative approaches to the natural ventilation of buildings, *Architectural Science Review*, 55(1), pp. 1–3. http://dx.doi.org/10.1080/00038628.2012.642190

10 Roaf, S. (Ed) (2010). *Transforming Markets in the Built Environment: Adapting for Climate Change*, Earthscan Publications Ltd. April.

11 Fitch, J.M. (1966). *American Building: The Forces That Shaped It*, Cambridge, MA: The Riverside Press, p. 5.

12 Roaf, S., M. Fuentes and S. Thomas (2013). *Ecohouse: A Design Guide*, 4th Edition, London: Routledge.

13 Roaf, S. (2022). COVID-19: Trust, windows and the psychology of resilience, in *The Routledge Handbook of Resilient Thermal Comfort*, Edited by F. Nicol, H. Bahadur Rijal and S. Roaf, London: Routledge, ISBN: 978-1-032-15597-5

14 Khosravi, F., C. Demski, L. King, et al. (2025). A nation unprepared: Extreme heat and the need for adaptation in the United Kingdom, *Energy Research & Social Science*, 124(104065). https://doi.org/10.1016/j.erss.2025.104065

2

DANGEROUS CURVES[1]

The Overheating Buildings Problem

Introduction: The Greenhouse Effect

A chronic problem with many modern buildings is that they overheat, not because the climate is warming, but because they are poorly designed for their climates. Rising temperatures globally largely result from an accelerating *greenhouse effect*, in which greenhouse gasses (GHGs) become trapped as heat inside the protective envelope of Earth's atmosphere rather like sunlight is trapped as heat inside a garden greenhouse, as shown in Figure 2.1.

Light travels from the sun as *shortwave radiation* in the ultraviolet, visible light, and near-infrared spectrums. Once in the Earth's atmosphere, it is either reflected back out into space (c. 30%) or absorbed in the atmosphere or at the Earth's surface. The Earth and its atmosphere then remit that absorbed energy as *longwave radiation*, also known as *thermal infrared radiation* or heat. The *outgoing longwave radiation* passes back out towards space through the atmosphere, but some is trapped by the thin, high barrier of GHGs that include water vapour, carbon dioxide (c. 75%), methane, nitrous oxide, and ozone. As the GHG emissions increase, the outer layer thickens and locks more heat inside the atmosphere, driving global warming.

Overheating in buildings is caused in a similar way. Once the radiation is through the outer glass wall of a building, it is absorbed at a surface, changes from shortwave light into long-wave heat, and can no longer pass easily back out of the space through the glass. This is why external shading keeps the heat out of the building but internal shading like blinds do not. Where there are no opening windows or doors in the envelope to purge indoor heat, it is trapped in the materials and air of a space, causing them to heat up fast in strong sunlight and more gradually in lower daylight levels. What is critical for the rate of heat gain are the size and orientation of glazed areas and the angle at which the sunlight hits the glass (Figure 2.2). The angle at which incoming sunlight strikes glass (angle of incidence) significantly affects the amount of light that enters the space through the glass, and that's how much is reflected. When sunlight hits the glass at a 90° angle perpendicular to the surface, a small amount, around 4–6% for ordinary glass, of light is reflected. The rest is transmitted through the glass or absorbed in it. As the angle of incidence increases, until the light hits the glass almost parallel to the surface, the amount of reflection increases dramatically. For example, at a grazing angle of 5°,

DOI: 10.1201/9781315645070-3

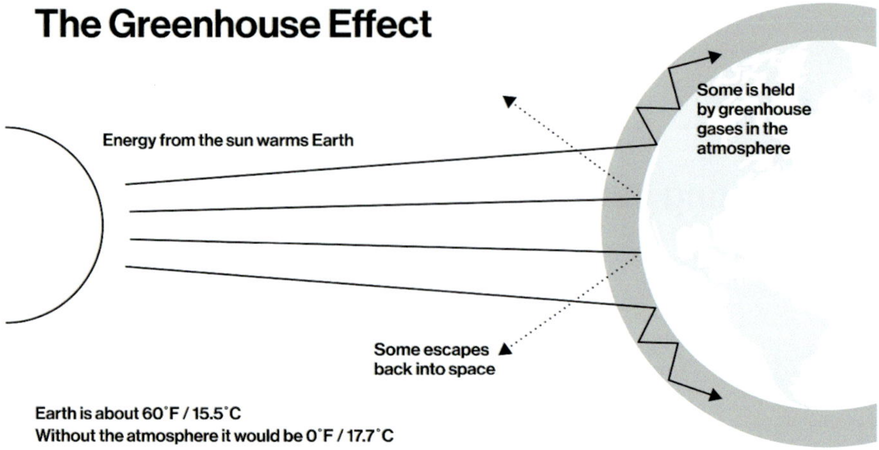

FIGURE 2.1 The greenhouse effect. Greenhouse gases (also known as GHGs) form a denser layer of gasses at the outer edge of the Earth's atmosphere that prevent some heat from escaping into outer space.

Source: Freya Morgan.

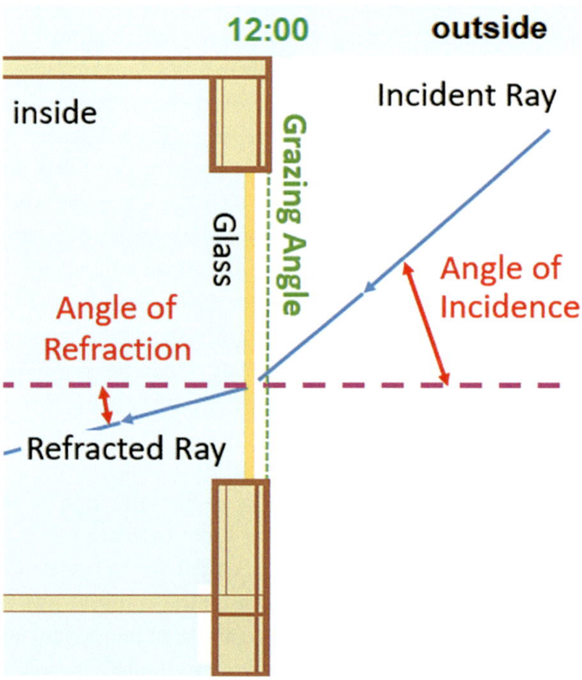

FIGURE 2.2 The smaller the angle of incidence of the solar gain hitting a glass window, the lesser the light reflected from its surface and the more it penetrates through the glass into the internal space.

Source: Glenn T. Remington.

glass reflects around 60–62% of the light. The same is true for radiation coming in through the atmosphere where incoming rays are closer to the perpendicular from the sun. Over the tropics, more light and heat get through the atmosphere to the Earth's surface, creating hotter climates around the equator. Lower angles of incidence of afternoon solar gain on a west-facing façade can drive serious overheating because not only is it difficult to shade against externally but also the lower angle incidence at which the afternoon sun penetrates the glass means less is lost in reflection at the surface. This is also why buildings in northerly latitudes can overheat badly too, because although they receive less solar gain than low latitudes, the angle of incidence they receive it at on the walls means more light can penetrate the glass and, on long summer days, for far longer.

In traditional buildings, windows were sized and oriented to keep occupants well-ventilated and adequately lit in the local climate while avoiding overheating. Mechanical conditioning systems have enabled an 'international style' of buildings, where the internal climates of a building are decoupled from the surrounding climate by the glass walls used that are now common in cities from the Gulf States to Siberia, at a devastating cost for both people, who pay the energy bills, and our planet. This chapter outlines the evolution of what may turn out to be *humanity's nemesis* and is the main engine driving our heating atmosphere – the glass building.

Glass

People have made glass for at least 3,600 years, evidenced by glass objects found in Mesopotamia and Egypt. The Romans were thought to be the first to use glass for windows, around AD 100, and by the 13th century, the dazzling stained glass windows of Venice were a wonder to a world where oiled paper, alabaster, and layers of animal horns had previously provided translucent weatherproofing for building windows.

The explosion of large glass building windows dates from the great palaces of the Tudor court in England, like Hardwick Hall, completed in 1597, which was said to be 'more glass than wall'. Rooms were flooded with light, and wealth was flaunted, but in the Little European Ice Age from the 16th to the 19th centuries, this probably led more to freezing problems indoors than overheating ones. Louis XIV was appropriately called the Sun King, not least for his glass- and mirror-filled Palace of Versailles, designed to dazzle rival nations with their brilliant light.[2] Developments in lead glass, coal furnaces, and glass production lines paved the way for more ordinary buildings to benefit from advances in affordable wall and roof windows, and the widespread availability of gas and electricity supplies in the 20th century liberated vernacular buildings from the need to rely on sun and wind to power and light them.

Design in the 20th Century

Everything changed in the 20th century. The old world became the modern world, peopled by societies that see-sawed between decades of wars, economic turmoil, and unprecedented economic growth. Two European wars bookended the Great Depression of the 1920s and 1930s, which saw the collapse of national economies and many traditional industries. A key political fiscal lever used to repair both societies and economies after those wars involved the stimulation of building markets. After the First World War in 1918, the threat of Bolshevism was countered by the promise of *homes fit for heroes*. After the Second World War (WW2), men returning from the battlefronts to the UK were lured away from the attractions of communism with promises of council housing for them all.[3] Social mobility was evidenced by the transition of some of the working classes into decent homes in middle-class suburbs and garden cities.[4]

The global population rose from 1.6 to 6 billion people over the 20th century, in which technology and cheap energy promised new solutions to provide ever-higher living standards for the masses. Unfortunately, the idea of cheap and endless energy swept designers away from centuries of continuously evolving vernacular design wisdom into the dreamscape of effortless comfort offered by high-energy, fossil-fuelled, and machine-dependant buildings. In the rush for change, old tried-and-tested ways of constructing and operating buildings were abandoned. Traditional construction methods were sidelined after the First World War with the demise of the masters and apprentice system for passing on complex knowledge and experience and feeding experiences back into working practices.

The 19th-century Industrial Revolution saw the emergence of the industrial production of faster, cheaper, and bigger buildings serviced by air conditioning and elevators powered by abundant, affordable, and accessible energy supplies. This period is well documented by J. M. Fitch in his classic books on American buildings and the historic and environmental forces that shaped them.[5] Between 1850 and 1950, the rate and scale of growth in such new methods, technologies, and materials resulted in many being widely adopted before they were sufficiently tried, tested, and understood. An obsession with 'innovation', radical changes in materials and designs and their marketing, and the siloing of expertise resulted in many mistakes made on the way, including, perhaps, the worst impacts today of *fast architecture* in modern buildings: the problem of overheating.

The Bauhaus: Out With Visual Clutter

The most influential architectural movement of the 20th century was the Bauhaus school, founded in Weimar by Walter Gropius in 1919 and led in Berlin from 1930 till its closure in 1933 by Mies van der Rohe. In the aftermath of the horrific First World War, outbreaks of Spanish flu and tuberculosis, and the Great Depression, a brilliant, liberal cohort of teachers and students came together under his leadership to create a strong design ethos combining the principles of mass production with a unique, artistic, modernist design style featuring simple geometric shapes like rectangles and spheres, without elaborate decorations.[6]

Bauhaus designers sought a clean break from tradition using an abundance of light, fresh air, and sunshine, not least to promote the health of occupants. Clean lines swept away traditional visual clutter from façades, like ornament, shutters, shades, and ornate balconies. The Bauhaus mantra of 'form follows function' was to be echoed globally as Hitler, a declared proponent of '*Volks Architektur*', the people's architecture, drove many of the Bauhaus adherents out of Germany in the late 1930s. The clean lines of their buildings influenced many emerging styles, including the post–Art Deco fashion for Streamline Moderne and the modernist movements, culminating in the international style which has now almost universally wiped out the last vestiges of vernacular forms in recent larger buildings around the world.

Modernism Sweeps Away Passive Climate Controls

Out with the visual clutter went many features that protected buildings from the climate, like cornices, overhanging eaves, pitched roofs that also kept the water out, and external shading devices to keep the buildings cool as walls were reduced to intersecting two-dimensional planes.

FIGURE 2.3 Villa Malaparte on the island of Capri, Italy. An iconic modern movement house of geometric forms with no visual clutter of shades and shutters, although security grilles have had to be installed.

Source: Susan Roaf.

The Villa Malaparte on the Italian island of Capri was an iconic example of the Bauhaus style. Built in 1930 by Curzio Malaparte, the three-story house nestled onto a rock bluff above the sea. A 1998 study of the thermal performance of five different houses built in the Naples region over 2,000 years explored the evolution of domestic architecture on that ancient Mediterranean coastline.[7] The oldest house surveyed, the Villa Giulio Polibio in Pompeii, was destroyed in the eruption of Vesuvius in AD 73 and had an astonishing 20 different climate-adaptive features in it, including thermal buffer rooms, verandas, a hypocaust, external shading, colonnades, ornamental evaporative pools, an internal garden, natural ventilation systems, and so on.

The Villa Malaparte, shown in Figure 2.3, had only six adaptive features used to manage the internal climate, including an insulated roof, 'smart windows', and earth sheltering around parts of the land-side rooms that the owner resorted to in winter to be near his ceramic stove. The external geometric forms and surfaces, including large fixed picture windows, were unencumbered by visual clutter. By contrast, the Villa Ranzo, a modern vernacular house on the other side of Capri, had 17 traditional climatic design features, including shutters, shades, and a rainwater storage system. Useful climatic design features in and around modern buildings gradually became an anathema to many purist modern architects. Interestingly, even within the Bauhaus pioneers, a schism existed between those who promoted climatic awareness in design, like Le Corbusier, and those who handed responsibility for the thermal performance of their designs to the engineers, like Mies Van de Rohe.

Le Corbusier

Charles-Édouard Jeanneret (1887–1965), known as Le Corbusier, was a Swiss French architect, polymath, and global influencer. Liberated by new materials and construction methods, in his

1923 book *Vers une architecture*,[8] he laid out his *five points of a new architecture*,[9] which were the use of the following:

- *Pilotis.* Pylons on which a building is raised to allow for the free circulation of air below, eliminating dark and damp parts of the house.
- *Roof terraces*. Flat roofs are used as gardens, for walks or sports or swimming pools.
- *Free plans*. Load-bearing walls are replaced by internal steel or reinforced concrete columns supporting open floors, on which walls can be put anywhere or left out entirely.
- *Ribbon Windows*. Since the walls do not support the house, the windows can run the entire length of the house, so all rooms can get equal light.
- *Free façades*. Non-load-bearing external walls mean façades can be lighter, use more glass, and eliminate lintels or other window structures.

Corbusier never taught at the Bauhaus schools, but they were well-versed in his thinking. Corbusier's early travels around the warmer Mediterranean countries inevitably informed his later emphasis on designs appropriate to local climates, his empathy towards solar design and understanding of the problems of overheating in certain climates, and his interest in solar cycles.[10]

Corbusier's concern over the potential for overheating in buildings was evidenced in his objections to the proposed new designs for the UN Headquarters in New York. Before work had begun on planning the interiors of the buildings, he wrote to Senator Warren Austin, president of the Headquarters Advisory Committee, in November 1947, overseeing the construction of the New York Headquarters of the United Nations:[11]

> *My strong belief, Mr. Senator, is that it is senseless to build in New York, where the weather is terrible in the summer, large glass areas which are not equipped with a 'brise-soleil.' I say this is dangerous, very seriously dangerous!*

He received no response and wrote again in high dudgeon, asking that the senator 'please . . . not consider this problem beyond [his] comprehension because it appear[ed] to be too technical'. He painted a grim picture of the thermal sufferings of an occupant of the building, who would be, he claimed, 'in an intolerable position, surrounded by hot air but feeling intense cold' radiating from the glass, resulting in them being 'not able to fill their jobs and the entire Secretariat of the United nations will be incapacitated. . . . This is extremely grave. It is decisive. One has no right to admit even such an alternative'. His two solutions were:

1. *Neutralising wall.* The double glass wall he invented in 1929 with panes a foot apart, in which solar-warmed air would circulate in winter, and cool air in summer, preventing excessive heat gain in summer and providing free solar heating in winter.
2. *Brise-soleil.* Sun breakers – a form of 'sun control' he invented in 1929 for North African climates.

He ended his letter by explaining:

> [O]ne cannot make window walls without a 'neutralizing wall' and a 'brise-soleil' (just as one cannot make an automobile with a powerful motor without brakes, etc.). . . . An error made by the United Nations in this direction could have serious consequences.

FIGURE 2.4 Frank Lloyd Wright designed his studio at Taliesin West in Arizona to provide shading for windows at all levels and from all directions.

Source: Susan Roaf.

Harvey Bryan pointed out that while the brise-soleil is undoubtedly a good idea for buildings in many warmer climates, Corbusier had misunderstood the basic physical principles that were at work in the *mur neutralisant*[12] that would inevitably result in very high energy consumption to fill the cavity between the glass panes with air at the right temperature to provide indoor comfort. Corbusier was much wedded to the glass double wall and proposed in his skyscraper residential towers that they should have 'a soundproof, well-lit living room and a double-façade of glass to control the effects of outside temperature'. They were very different times in a world where the motor car dominated city planning and separation from the heavily polluted air of cities was a benefit.

Again in the words of Le Corbusier to Senator Austin: 'Please do not consider this problem beyond your comprehension because it appears to be too technical'.

Corbusier, like Frank Lloyd Wright (1867–1956), was an architect who was not content to simply let engineers or manufacturers solve his technical problems but was an astute climatic and environmental designer for many of his projects, despite this ignorance of his own ignorance.

Rayner Banham wrote in 1969:[13]

> *Le Corbusier might admonish in 1925 that 'an engineer should stay fixed and remain a calculator, for his particular justification to remain within the confines of pure reason'. . . but many of Le Corbusier's own buildings would have been unbuildable or uninhabitable had engineers ever heeded his advice . . .*[14]

Mies van der Rohe

Ludwig Mies van der Rohe (1886–1969) was a contemporary of Corbusier from an equally humble background. A fellow design colossus and towering ego, he was the last director of the Bauhaus in the 1930s and a pioneer of the international style of architecture. In 1921, his competition proposal for the faceted all-glass Friedrichstrasse skyscraper was followed by a taller, curved version in 1922 named the Glass Skyscraper. After Nazism's rise to power, with its strong opposition to modernism, Mies immigrated to the United States in 1938, becoming head of the architecture school at the Illinois Institute of Technology (IIT). While continuing his traditional neoclassical design practice, he published a series of visionary projects, of which few were built, but that did emphasise his role as a leading modernist architect.[15]

A master of form and glass, he passed the responsibility for the thermal performance of his buildings to others. His work was not popular with all. In *House Beautiful* in 1956, its editor, Elizabeth Gordon, a great fan of Frank Lloyd Wright (Figure 2.4), whom she called the true American modernist, divided the architecture world with an editorial titled 'The Threat to the Next America'. She accused Ludwig Mies van der Rohe of moving America towards totalitarianism and communism by way of visual austerity.[16] On behalf of American consumers, she fuelled a highly visible legal feud between Mies and his client Dr Edith Farnsworth, whom she described as 'a highly intelligent, now disillusioned, woman who spent more than $70,000 building a 1-room house that is nothing but a glass cage on stilts'. In 1966, Robert Venturi coined the postmodern motto 'less is a bore' as counter-vision epithet to Mies's motto 'less is more'.

Architects around the world, however, continued to copy his lead, with glass houses, often 'floating' romantically in charming landscapes, appearing around the world. The much-vaunted quality of building 'transparency' they espoused has been painfully shoehorned into excuses for all sorts of glass buildings, from parliaments to offices, schools, and even public libraries, but was never popular. Issues around privacy, modesty, and emotional security seem irrelevant to designers promoting a hundred-year-old 'modern-feeling' building despite the fact that a good glass pane may have a U-value of 1.5, but a good wall can provide ten times more insulation.

People in Glass Houses

Philip Johnson (1906–2005) was a lifelong friend and collaborator of Mies. His first famous building was the Glass House in New Canaan, Connecticut (USA), in which he allegedly lived and worked for 58 years (Figure 2.5a)

The building was so controversial that a police officer was posted nearby to keep trespassers out. Johnson put up a sign near the street stating: 'This House Is *Now Occupied.* Please Respect the Privacy of the Owner'.[17] In fact, Johnson actually slept, worked, and lived in the adjacent brick guest house and in other buildings on the estate, described once as an 'architectural theme park'. The brick house offered privacy and protection from the elements and much more comfort

FIGURE 2.5 The Glass House by Philip Johnson, built in 1949. (a) The glass structure which he owned and occasionally occupied for 58 years. (b) The adjacent brick guest house, where Johnson actually slept.[18]

most times of the year than the glass-and-steel structure of his glass folly home. His glass house proved almost impossible to heat comfortably, let alone affordably, relying eventually on radiant heated floors that had to be run at such high temperatures that they were impossible to walk on without shoes. He learnt early on that people in glass houses pay very high bills.

Modernism in Britain

Modernist window designs in solid concrete walls that were popular in Britain in the middle of the 20th century never really worked well in cooler, damper northern climates. Fergus Nicol explains the thinking of his own uncle, the pioneering British modernist Colin Lucas:

> *Their idea was to have as much sunshine, daylight and fresh air as possible in the house for the mental and physical well-being of their occupants which is why they designed homes with large opening windows.*

The Connell, Ward, and Lucas firm created some of the most iconic 'modern' British buildings between 1930 and 1937, using Corbusier's design tenets and materials, including glass, reinforced concrete, and steel. In 1931, he designed 'Sunlight House' for Frances Lucas, Nicol's grandfather, at Chelwood Gate, East Sussex.[19] One can imagine how uncomfortable it could become simply by looking at it in Figure 2.6.

Concrete proved to be a particular Achilles heel in following mid-century styles, for instance, as applied in many materially flawed Brutalist buildings and town centres that were rapidly thrown up after WW2, many of which are now being demolished. Few will be missed with their

FIGURE 2.6 The Sunlight House, designed by Colin Lucas in 1930–1931, showing that early modern movement design embraced large opening windows for healthy ventilation. Demolished in 2016.

Source: Fergus Nicol.

grim, damp concrete walls and single-glazed, metal-framed windows. Overuse of untested materials was a problem with the rush to be innovative and lead architectural fashions, in particular, the universal use of toxic ones, including all forms of asbestos, which were finally banned for import, export, use, or sale in the UK as late as 1999, and flawed economical structural products. In 2023, in the UK, an overnight order was issued to close all buildings roofed in reinforced autoclaved aerated concrete (RAAC), for fear of collapses, including school, college, hospital, residential, and office buildings. The repair of over-glazed buildings will be more difficult.

The rapid move to higher glazing ratios on external walls did drive overheating in commercial buildings in the pre–air conditioning era.[20] A 1961 study showed that office workers were already less comfortable, reporting increasing problems of overheating in structurally lighter buildings with ever more glass.[21] In pre–Second World War offices, the window areas averaged around 20% of the floor area, and 85% of occupants wanted more sunshine in their offices, while only 9% were concerned about the potential for overheating in summer in their workspaces.[22] Windows were over 50% larger in the 1961 survey than in a 1948 survey. Moreover, 80% of the internal partitions on post-war buildings were lightweight, more flexible, but warmed up and cooled down much more quickly than in the pre-war buildings with heavy internal walls. In 1961, as many as 40% of the occupants were sometimes too hot in the summer.[23] This study provided the data from which the 'admittance method' (see Appendix 1) was developed to calculate exactly how much thermal mass was optimal in a particular building.[24]

Few air-conditioned buildings existed in London in 1966 when the Building Research Station looked at the 'experience of air-conditioned buildings' in London.[25] Their surveys found that air-conditioned offices were perceived to be cleaner and quieter in very hot weather, when they were preferred to naturally ventilated buildings. However, in less extreme weather, that is, most of the year, the thermal environment in the air-conditioned buildings was less favoured than that in offices with opening windows. The studies found that for offices, the heavier, more conventional construction was more comfortable over a year than the lightweight building with large areas of glazing.

The change from pre-WW2 to post-war commercial and residential building types was largely driven by the switch to concrete construction from masonry walls with timber floors. It was determined then that English buildings in the 1960s of heavy construction could have up to 50% of the external wall glazed and still provide comfort, but that unless sun controls were provided, glass areas in buildings with lightweight internal walls and ceiling should be restricted to 20% of the wall area if they faced south. If temperatures of 27°C (80.6°F) were considered acceptable, then this area could be extended to 30% of the wall area in noisy areas and 50% in quiet areas for lightweight buildings where windows could be opened;[26] 100% glazed facades were not even considered as a possibility then.

Lightweight, demountable internal wall partitions were increasingly favoured, which could be replaced or moved around if necessary to re-shape spaces. This also maximised the internal office area to the advantage of the occupying firm and the landlord, in a move characterised by the shift to 'open plan' offices. Rather like factory farming, the drive was to get more people per floor area to tick the cost–benefit boxes in annual performance reviews. Individual offices with their own operable windows largely disappeared, replaced by floors of warm workers and their warm machines, for whom

> *[i]n many instances, 'copresence' (working side by side) via an open office or a digital channel does not result in productive collaboration.*[27]

TABLE 2.1 Typical temperatures in 259 UK houses with three grades of insulation from the 1949–1950 study recorded with a mean outside temperature of 7.2°C (45°F)

	Grade of insulation		
	High	*Medium*	*Low*
Mean living room temperature	16.6°C (62°F)	16°C (60.9°F)	15.5°C (59.9°F)
8:00 p.m. living room temperature	18.4°C (65.2°F)	17.7°C (63.9°F)	17.2°C (62.9°F)
Mean bedroom temperature	12.8°C (55°F)	11.6°C (52°F)	10.7°C (51.2°F)
Mean house temperature	14.2°C (57.6°F)	12.2°C (55.8°F)	12.5°C (54.4°F)

Source: Danter (1951).

Common symptoms of overheating are visible in cities, campuses, or estates where buildings have often have most of their internal blinds down much of the year to reduce heat/cold gain and glare, or in buildings around the world where occupants even try to paper over the windows from the inside to make indoor conditions more bearable.

But of course, it is not just overheating that mattered. People lived in homes in the UK with levels of cold indoors many of us would not tolerate today. After the hardships of the two wars, many incomes soared, and along with the rising desire for a higher quality of life came the call for more comfort, not least because the buildings were very often uncomfortable. In the UK, central heating only began to become widespread in the 1980s, associated with government support for better boilers and insulation in homes. As building fabric was incrementally improved from the 1960s onwards, by the 1990s, emphasis had shifted to increasing the energy efficiency of the machines needed to heat and cool buildings.[28] Extraordinary to think that in Britain itself, we have gone from extremely cold indoor comfort temperatures to wide-scale overheating of buildings in half a century[29] as buildings modernised. Table 2.1 provides temperatures from a UK survey done in 1949 to 1950, just after WW2, showing that mean winter living room temperatures in 259 studied homes were much lower than is typical today, at around 15.5°C to 16.6°C, even in what were then considered highly insulated homes.[30]

Trends in Late 20th-Century Architecture

By the mid-1950s, architecture was heading in three different directions:

1. *The vernacular* continued everywhere, with traditional building forms in common use and production by, and for, the masses, with house-building at its height in the post-WW2 period, when traditional design elements were latterly copied in new materials and structures that often performed less well than the originals.
2. *The mechanisation of buildings* was embraced by designers, who revelled in being liberated from climate design constraints by affordable air conditioning systems that also liberated designers to create evermore imaginative sculptural forms.
3. *The solar building movement* was born, focussed on using solar energy to heat buildings in winter, combined with sensible strategies to prevent overheating in summer.

Frank Lloyd Wright embraced the usefulness of machines, writing in 1930:[31]

> *The machine does not write the doom of Liberty, but is waiting at man's hand as a peerless tool, for him to use to put foundations beneath a genuine Democracy. Then the machine*

may conquer human drudgery to some purpose, taking it upon himself to broaden, lengthen, strengthen and deepen the life of the simplest man.

Who has not benefitted from machines to clean our clothes, dishes, and homes; rationalise our working practices; and entertain and inform us around the clock? But it is actually relatively recently that the most ordinary people have benefitted from the mechanical revolution in their own lives.

The Solar Building Movement

In the 1950s, pioneering architects like the Olgyay brothers in Hungary and the USA strove to develop ever more scientific methods to enable buildings to be run in winter and the shoulder seasons on free, clean solar heat gain. To do that while avoiding problems of overheating in warmer seasons needed appropriate solar building forms, openings, and shading systems.[32] The debate over the role of the building itself in providing the climatic filter between outdoor and indoor climates raged across the 20th century[33] as some ignored and some embraced the climate in their design philosophies:

> *The Climate is the Architect's true adversary . . . as the climate varies so should shelter.*
>
> (Victor Olgyay, 1951)

Curbing the Curves

The sun powers the seasons, drives day and night, and influences weather in climates. In the 1920s and 1930s, the use of concrete and glass buildings was proliferating, and overheating became a serious concern to architects. The climatic designer and teacher Victor Olgyay proposed four essential steps to flatten the temperature curve with buildings using the following properties (Figure 2.7):

1. Climate
2. Site
3. Building as a modifier of the amplitude of the solar thermal impact
4. Role of mechanical heating and cooling

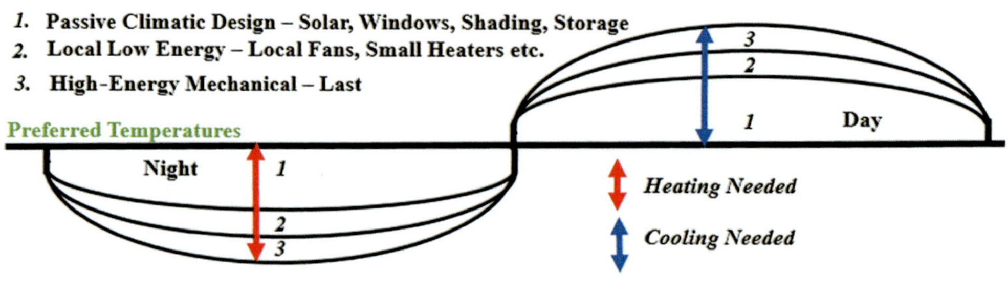

FIGURE 2.7 Victor Olgyay saw the importance of reducing the amplitude of the swings in temperature indoors in his 1951 article 'The Temperate House'.[34]

He saw that all these four steps were needed to enable designers to provide his preferred mean indoor temperature of 70°F (21°C).

Increasingly concerned with the problem of the overheating of modern US buildings, the Olgyay brothers' solutions centred on solar control. Victor proposed as necessary that

> *interception of energy happens at the right place, at the façade, before it enters the building.*

He understood the fundamental importance of reducing the damage that could be done by indoor temperature swings resulting from uncontrolled solar gain. Once solar heat has entered the building through the glass, whatever its g-value, it then changes wavelength and remains in the space. External solar control design would need to change with different climates (Figure 2.8), and in his influential 15-page article 'The Temperate House', he provided clear and useful methods for accurately assessing how large the overhangs should be on different orientations at each season.[35]

The work and teachings of Olgyay and his contemporaries in the USA were paralleled in the UK by that of experts at the Building Research Station and authors like Baruch Givoni[36] in Israel and Otto Koenigsberger[37] working in India, later at the School of Tropical Architecture at the Architectural Association in London. Until the first Oxford Conference on Architectural

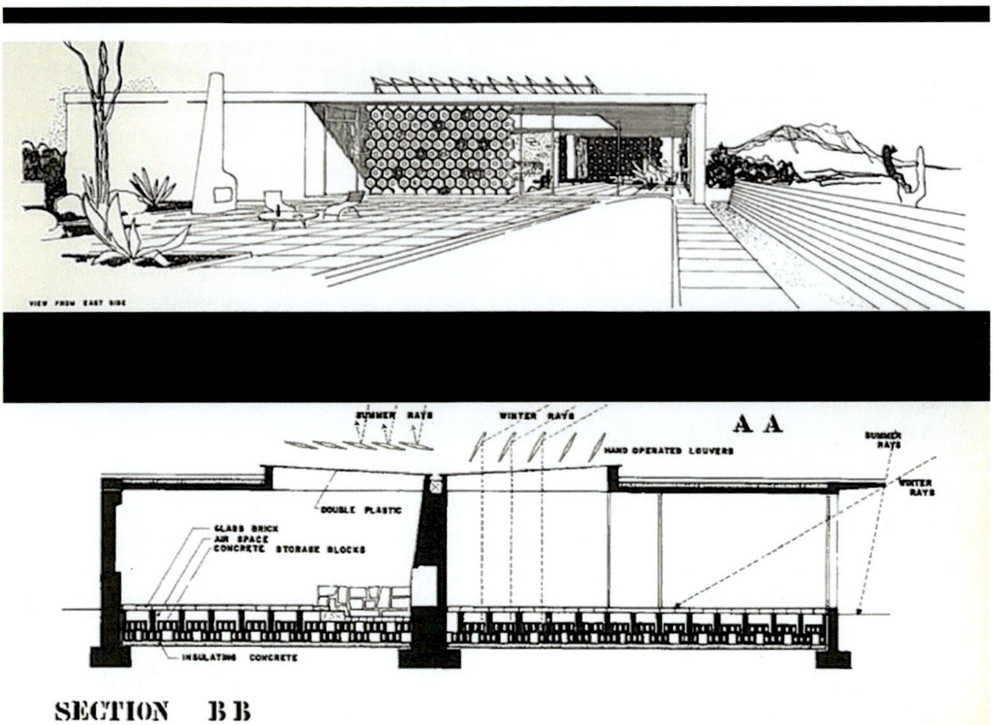

FIGURE 2.8 Elevation and section of a solar house designed by Victor Olgyay in 1957 emphasising the importance of energy storage in the massive core feature and the hypocaust floor, using well-designed shading to moderate overheating.

Source: Olgyay.[40]

Education in 1958, most professional architectural training was done in offices, on live projects mentored by qualified architects. In the 1960s, architectural training moved largely into university-based schools of architecture, where gradually professional academics took over the teaching roles of practice-based architects. Studio design teachers taught more theory-based subjects, and some schools evolved to teach very little at all on the more practical aspects of design.[38] Some schools still continued to carry forward 'the golden thread' of solar energy[39] by championing good climatic design using appropriate shading, ventilation, and orientation.

In the 1980–1990s, many schools of architecture had building science laboratories, a number of which were dismantled by the turn of the 20th century as design became increasingly focussed on creating buildings using simulation packages in which the lines between passive and active thermal buildings were blurred and few really noticed as buildings slipped from being passively heated and cooled where possible to relying entirely on active mechanical conditioning systems.

Buildings Slip Unnoticed From Passive to Active Modes

St George's School, Wallasey, near Liverpool, UK, was designed by Emslie Alexander Morgan and opened 1962 as part of an innovative movement amongst local authorities to create low-energy, comfortable schools in England.[41] It immediately attracted attention because of its distinctive 'solar section' with impressive fully glazed south-facing elevation, shown in Figure 2.9. Morgan took out a number of patents on his design system, with the final sentence in one claiming it was '[a] building capable of being maintained at a habitable temperature solely or mainly from the sun and electric lighting lamps . . .'[42]

Its 'genius' designer harnessed key features to manage energy flow through the building and reduce energy consumption. He built a large double-skin glass solar wall and included high levels of thermal mass in walls, floors, and roof. His stated aims were to replace conventional heating in classrooms in winter, to use daylight to reduce electrical lighting loads, to utilise thermal mass to absorb solar gain in the summer to reduce overheating and provide steady stored solar heat over the day in the winter, to advance ventilation control with opening windows to provide cooling in the summer and distribute heat during winter, and to use insulation to conserve the small amounts of heat that are input in winter.

Michael Davies of Liverpool University studied the performance of the school in the 1970s and the 1980s and concluded that the building was often uncomfortable, with classrooms getting very cold in the winter and overheating in the summer. Lessons had to move outside on very hot days, with classrooms recorded as reaching 33.5°C (92.3°F) on at least one occasion, while others claimed even higher temperatures. The glass window-wall led to high heat loss in the winter, and Davies concluded that the amount of insulation used around other elements of the structure was secondary to the overall energy heat loss and gain performance of the building as a whole. Davies studied it extensively for an internationally interested audience.[43] The building finally proved uninhabitable and was abandoned in 1996, ostensibly due to falling school roles.[44]

Thirty years later, Wolfgang Feist and Bo Adamson designed the world's first Passive House in Darmstadt-Kranichstein, North Germany. The section echoed that of the Wallasey School with a large south-facing solar window-wall, and the energy assumptions were similar. In winter, solar gain would replace the need for imported energy for heating that would overall replace imported energy for heating. In summer the 'high-performance' triple-glazed low-e glass would reduce the overheating, and some cooling would be provided by the sub-soil heat exchanger. Heat losses and gains from the walls, floor, and roof would be reduced by insulation, and the

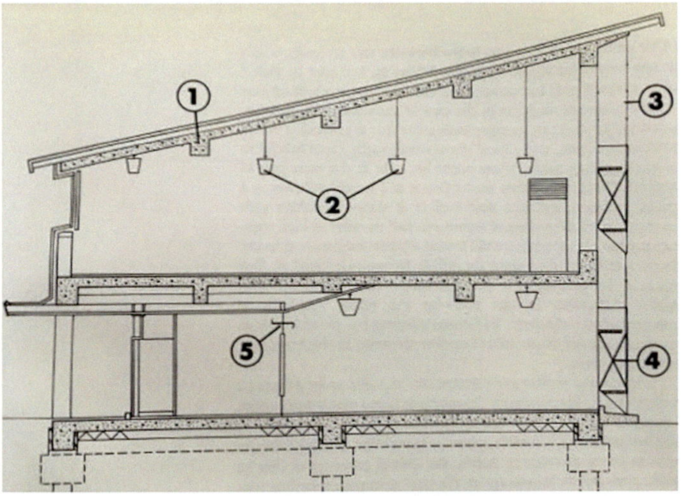

FIGURE 2.9 (a) South-facing wall of the main wing of the Wallasey School building in 1961. (b) Section through the main wing with (1) insulated roof, (2) light fitting, (3) double-skin solar wall, (4) adjustable ventilating windows, and (5) ventilating windows at the back of the classroom.

Source: Elain Harwood.

need for opening windows would be replaced by the use of mechanical ventilation, heat recovery, and distribution systems for the whole house.

The logic in the Passive House design balancing heat gains and losses over the year, as shown in Figure 2.10(b), was akin to that of Emslie Morgan's thinking:

> If during a 24-hr period the incoming heat is greater than the loss, the window is thermally profitable for that day. Furthermore, if during a year the incoming heat is greater than the loss, the window is thermally desirable, If overall the window gives a net positive gain, its size should be as large as possible. Conversely, if the long-term window net gain is negative, its best size is zero. These results are independent of all other thermal aspects of the building.[45]

There are two basic flaws in this logic evidenced in such buildings over time. The first arises possibly from the fact that the designers came from northerly cooler latitudes and did not

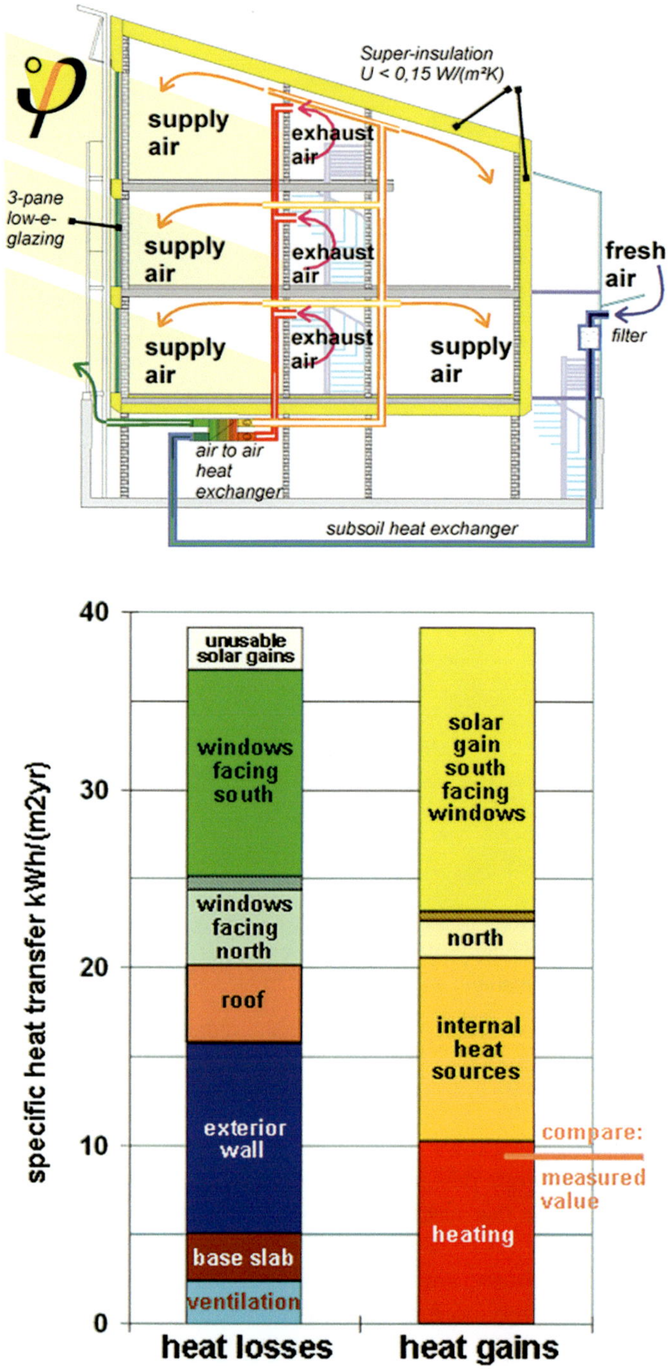

FIGURE 2.10 (a) Section through the first Passive House in Darmstadt-Kranichstein, Germany, showing the systems employed to provide comfort. (b) The annual heat losses and gains from the design were calculated to balance out over the year.

Source: Passivhaus Institute.

believe the rooms behind the solar window/walls would overheat. Second is the fact that solar gain occurs mainly in the summer, when additional heat indoors is not needed, rather than in the winter, when it is. Both buildings installed heating systems in the spaces, and in Wallasey, it was eventually calculated that the electric heating systems meant the running costs of the schools were similar to those of others in the area.

The important fact about these two buildings that look similar in aspiration and form is that they represent a *late 20th-century paradigm shift in design*. The Wallasey school was designed to run largely in *passive mode* using thermal mass and natural ventilation to temper the indoor climate both in summer and winter.

The Darmstadt-Kranichstein house designed out thermal mass and sidelined natural ventilation to shift to an overweening reliance on *active systems* with mechanical heat recovery and ventilation to provide comfort and breathing air indoors. Ironically, that move to *active systems* kick-started the *Passive House* movement. Both those versions of solar buildings did suffer from overheating in key spaces from the start due to excessive south-facing glazing in the summer and high levels of heat loss in the winter, but that was largely the fault of the form of the building and its openings.

As Davies said of the Wallasey school, it would also not matter how much insulation you put around the Darmstadt-Kranichstein house; it would always be intermittently overheated as designed. It would, however, have benefitted from some active thermal mass indoors to absorb excess solar gain on hot days. It did fortunately have good opening doors onto the balconies to dump heat out of in the summer. Could it have performed significantly better if it had been more passive and less reliant on active systems?

Many discussions on building performance today focus on secondary issues, like thickness of insulation, the performance of the glass, air infiltration rates, or the energy efficiency and controls of the 'smart' mechanical systems.[46] What the Wallasey school and the Darmstadt-Kranichstein house clearly show is that it is the design of the basic form of the building itself that holds the key to its climatic performance and its ability to also work in 'passive mode'.

That surreptitious late 20th-century *modal shift from passive to active* operation for buildings has been masked by marketing tropes and the outputs of black box simulation models. The advice of the US Resilient Design Institute to mandate that all buildings be modelled at the design stage with the HVAC function off might go some way to weeding out the thermal nightmares before they are built.

The next surreptitious modal shift in design is already happening from active/passive buildings to fully air-conditioned structures. Therein lies the real danger for the planet and the populations, as Margaret Ackerman hinted at in her 2002 book on America's romance with air conditioning:

> *Air-conditioning (that) has made it possible to erect structures that must be evacuated when the power fails, to make buildings in which people get sick. It gulps electricity . . . makes urban heat islands even hotter with the exhausts of a million air-conditioned cars and thousands of sealed buildings.*[47]

Time for the Next Major Modal Shift in Design

The journey from Bauhaus to Passivhaus has shown the importance of thinking about the big picture in terms of thermal performance in building design that inevitably involves model shifts in building architypes.

To thrive in a different future, buildings must be able to keep occupants comfortable, healthy, and alive, even during more extreme weather events and climate trends. To create such *climate-safe* buildings will require a good knowledge of what thermal comfort is, how it is achieved, and what adaptive opportunities can be provided in buildings to enable people to make, and keep, themselves comfortable indoors, even during the extremes. We hope the following chapters will help readers gain some insights they can build on to create the next generation of resilient buildings.

Notes

1 Cheyney, P. (1939). *Dangerous Curves*, London: Collins. Films of this title were also made in 1929, 1961, and 1988.
2 Hamon, M. (1999). *From Sun to Earth 1665–1999: A History of Saint-Gobain*, Paris: JC Lattès
3 Swenarton, M. (2021). *Homes Fit for Heroes: The Politics and Architecture of Early State Housing in Britain*, London: Routledge, ISBN: 978-1-138-36027-3
4 NHBC (2915). *Homes Through the Decades: The Making of Modern Housing*, London: Published by the NHBC Foundation, Accessed on 30/07/2024 at: https://www.nhbc.co.uk/foundation/the-making-of-modern-housing
5 Fitch, J.M. (1966). *American Building: The Forces That Shaped It*, Cambridge, MA: The Riverside Press, p. 218. Quoting a pamphlet written by Sullivan between 1906–1908. This was followed by: Fitch, J.M. (1975). *American Building: The Environmental Forces That Shaped It*, New York: Schocken Books.
6 Wingler, H.M. (1969). *Bauhaus: Weimar, Dessau, Berlin*, Chicago: MIT Press.
7 Roaf, S., D. Crichton and F. Nicol (2009). *Adapting Buildings and Cities for Climate Change*, Oxford: Architectural Press, pp. 192–201.
8 Le Corbusier (1931). *Towards a New Architecture*, unabridged translation of the 13th French Edition, London: Published by John Rodker.
9 Oechslin, W., and W. Wang (1987). Les Cinq Points d'une Architecture Nouvelle, *Assemblage, No. 4*, pp. 83–93. https://doi.org/10.2307/3171037
10 Le Corbusier, and P. Jeanneret (1945). *Oeuvre Complete 1934–1938* 2nd Edition, Zurich: Dr. H. Girsberger.
11 Le Corbusier (1947). Letter to Senator Warren Austin, Dated December 4th. Reprinted in *Architectural Record*, July 1950, pp. 69–71. Letter accessed in Columbia University's Abramovitz archive (Box:15 Folder:33): https://www.columbia.edu/cu/lweb/eresources/archives/avery/abramovitz/subs/d0e9935.html
12 Bryan, H. (1991). Le Corbusier and the 'Mur Neutralisant': An early experiment in double envelope construction, *Proceedings of the Ninth International PLEA Conference*, Kluwer Academic Publishers, Dordrecht, Netherlands, pp. 257–262.
13 Le Corbusier (1925). *Urbanisme*, Paris: Les Editions G. Crès & Cie, p. 48.
14 Banham, R. (1969). *The Architecture of the Well-Tempered Environment*, Chicago: The University of Chicago Press, p. 16.
15 Mertins, D. (2024), *Mies*, London: Phaidon Press, ISBN: 978-1838668174
16 Allotto, D. (2017). Elizabeth Gordon's international style, *Curbed.com*, Accessed on 17/07/2024 at: https://archive.curbed.com/2017/5/10/15592658/elizabeth-gordon-house-beautiful-frank-lloyd-wright
17 Accessed on 27/05/2025 at: https://en.wikipedia.org/wiki/Glass_House
18 Accessed on 27/05/2025 at: https://commons.wikimedia.org/wiki/File:Guest-house-philip-johnson-glass-house.jpg
19 Sharp, D., and S. Rendel (2008). *Connell Ward & Lucas: Modern Movement Architects in England 1929–1939*, London: Frances Lincoln Ltd.
20 This term was coined in the early 1970s, see: Baird, G., M. Donnne, F. Pool, W. Brander and C. Aun (1984). *Energy Performance of Buildings*, Florida: CRC Press Ltd., Boca Raton, p. 7.
21 Loudon, A.G. (1968). Window design criteria to avoid overheating by excessive solar gains, *Building Research Station Current Paper*, CP4/68, BRS.
22 Gray, P.G., and T. Corlett (1952). A survey of lighting in offices, *Postwar Building Research*, 30, appendix 1, London: HMSO.

23 Loudon, F.J., and E.C. Keighley (1964). User research in office design, *Architect's Journal*, 139(6), pp. 333–339.
24 The evolution of the admittance method is illustrated in: Danter, E. (1965). Periodic heat flow characteristics of simple walls and roofs, *IHVE Journal*, 28, pp. 136–146.
25 Black, F., and E. Milroy (1966). Experience of air-conditioning in offices, *Journal of the Institute of Heating and Ventilating Engineers*, (September), pp. 188–196.
26 Loudon, F.J., and E. Danter (1965). Investigations of summer overheating, *Building Science*, 1(1), pp. 89–94.
27 Bernstein, E., and B. Waber (2019). The truth about open plan offices, *Harvard Business Review*, (November–December), Accessed on 27/05/2025 at: https://hbr.org/2019/11/the-truth-about-open-offices
28 Roaf, S., and M. Hancock (1992). *Energy Efficient Buildings*, Oxford: Blackwell Publishers.
29 Lomas, K.J., and S.M. Porritt (2017). Overheating in buildings: Lessons from research, *Building Research & Information*, 45(1–2), pp. 1–18. https://doi.org/10.1080/09613218.2017.1256136
30 Danter, E. (1951). A survey of temperatures in houses, *The Heating and Ventilating Engineer*, 25(290), pp. 57–63.
31 Wright, F.L. (1931). *Modern Architecture: Being the Kahn Lectures for 1930*, Princeton: Princeton University Press, p. 23.
32 Barber, D.A. (2020). *Modern Architecture and Climate Design before Air Conditioning*, Princeton: Princetown University Press.
33 Mumford, L. (1952). *Roots of Contemporary American Architecture: A Series of Thirty-Seven Essays Dating from the Mid-Nineteenth Century to the Present*, New York: Reinhold Publishing Corporation.
34 Olgyay, V. (1951). The temperate house, *Architectural Forum*, (March), pp. 179–194.
35 Olgyay, V. (1963). *Design with Climate: Bioclimatic Approach to Architectural Regionalism*, Princeton: Princeton University Press, p. 14.
36 Givoni, B. (1969). *Man, Climate and Architecture*, London: Elsevier.
37 Koenigsberger, O., T.G. Ingersoll, A. Mayhew and S. Szokolay (1974). *Manual of Tropical Housing and Building: Climatic Design Pt. 1*, London: Longmans.
38 Roaf, S., and A. Bairstow (Eds) (2008). *The Oxford Conference: A Re-Evaluation of Education in Architecture*, Southampton: WIT Press.
39 Butti, K., and J. Perlin (1980). *A Golden Thread: 2500 Years of Solar Architecture and Technology*, New York/London: Van Nostrand Reinhold Company.
40 Olgyay, V. (1958). *Living with the Sun Volume 1: 60 Plans Selected from the Entries in the 1957 International Architectural Competition to Design a Solar Heated Residence*, Phoenix Arizona: Published by the Association for Applied Solar Energy.
41 Davies, M.G. (1979). The solar technology of St. George's school, Wallasey, England, *Architectural Science Review*, 22(4), pp. 89–93. DOI: 10.1080/00038628.1979.9696424
42 McKenna, G.T. (1985). St. George's school Wallasey, the visual environment, *Building and Environment*, 20(1), pp. 61–71. https://doi.org/10.1016/0360-1323(85)90033-2
43 Perry, J.E. (1976). The Wallasey school, *Proceedings of the Passive Solar Heating and Cooling Conference*, Albuquerque, USA, Accessed on 05/07/2025 at: https://www.osti.gov/servlets/purl/7345962
44 Harwood, E. (2023). *The Solar Campus, former St Georges School, Wallasey Merseyside*, Published by the 20th century Society. Accesses on 05/07/2025 at: https://c20society.org.uk/building-of-the-month/the-solar-campus-former-st-georges-school-wallasey-merseyside
45 Davies, M. (1987). The passive solar-heated school in Wallasey. V. Energy requirements for A passive school building, *International Journal of Energy Research*, 11(1), pp. 1–20. DOI: 10.1002/er.4440110102
46 Katzenbach, R., C. Frithjof and Z. Jie (2014). Efficiency for sustainability – smart energy management in social housing, *Journal of Power and Energy Engineering*, 2, pp. 1–6. https://doi.org/10.4236/jpee.2014.212001
47 Ackermann, M. (2002). *Cool Comfort: America's Romance with Air-Conditioning*, Washington: Smithsonian Institute Press, p. 184.

People and Comfort

3

HOW BODIES ADAPT

We All Know What Comfort Is: Our Own Comfort

We are all familiar with our own body and the conditions in which it is comfortable. There are many interpretations of comfort, but in this book, we are concerned with thermal comfort in terms of our relationship with the thermal world we live in. Over days and seasons, we may occupy many different temperatures at home and at work. We roughly know when and where we will feel comfortable in each place at different times of day and year. At home, temperatures may vary a lot, being cooler in winter than in summer, for instance, or being in the bedroom rather than the kitchen. In the office, they may be more constant. The different ways we enjoy temperatures also change with the seasons.

In winter, we may like a walk in bracingly cold air, wrapped up warmly in draught-proof and insulating clothing. On a summer holiday, we might delight in the feel of the warm sun on the skin, or of a cool breeze on a hot day. We can be very comfortable at a lot of different temperatures. Our own comfort temperatures are unique to each of us and depend on our attitudes, our bodies, our economic status, and our thermal histories: whether we were brought up, for instance, in Sweden or the Caribbean. Our personal comfort preferences may vary considerably from those of the people around us, as some will know from squabbles over a thermostat setting or an energy bill. The temperatures we actually occupy with other people inevitably require compromises. Many of us would suffer at the temperatures accepted as normal by people in tents in the Arctic, in the desert, or even possibly the house next door, because we have not learnt how to live in their temperatures.

There are many different ways of seeing comfort: It is not a product sold in a bottle or by the kWh from a plug in the wall or the volume of air pumped from a ceiling duct. Some maintain that 'comfort is a state of mind',[1] unique to the individual who feels it. That is true but perhaps should be qualified by adding 'at a particular time', because fundamental to the ability to thermally adapt is a dependence on our ability to continually adjust to the temperatures where we are. But there are some core truths about what temperatures people find comfortable, and they relate to the 'thermo-physiology' of the body (Figure 3.1). In this chapter we look at the

DOI: 10.1201/9781315645070-5

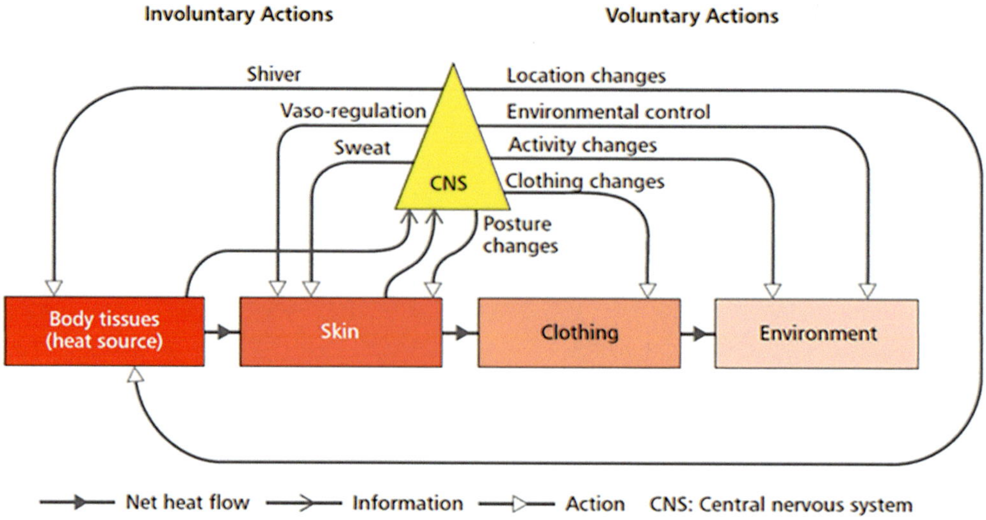

FIGURE 3.1 The thermoregulatory system of the body is controlled via the central nervous system and is continually monitoring local temperatures via sensors in the skin to inform involuntary and voluntary actions to maintain safe body core temperatures at around 37°C.

Source: Diagram by Fergus Nicol, redrawn by Ken Butcher.

physiological systems of the body and the central nervous system and the involuntary ways they impact on our comfort.

The DNA of Thermal Comfort

Homo sapiens, thinking man, has walked this planet for at least 300,000 years. Individual humans share 99.9% of their individual DNA with each other. If our DNA sequence were a book, it would have 262,000 pages, of which only 500 relate to us as unique human beings, like our eye, skin, and hair colour or body form (Figure 3.2).[2] All of us have roughly the same core temperatures, similarly wired thermoregulatory systems, and the same driving need to ensure that our core body temperature is kept at around 37°C. It is an extraordinary system, considering how many variables are involved. The two systems of both conscious and unconscious responses are linked by the brain, which constantly fine-tunes their various functions to keep the body thermally safe and comfortable.

People Adapt

Despite our similarities, humans have adapted themselves to live in almost every climate on this planet, both culturally and physiologically, shaped by those forces we are exposed to on a regular basis. In environments ranging from burning deserts to Arctic white-lands, humans have created structures they can survive in occupying with acceptable indoor conditions.[3] This they can do in accordance with the *adaptive principle*:[4]

> *If change occurs that produces discomfort, people react in ways that tend to restore their comfort.*

FIGURE 3.2 Despite great variations in personal attributes, like skin and eye colour and body shape, people around the world share the same basic thermoregulatory systems and core body temperatures that are individually fine-tuned by many factors, like those outlined in what follows.

The temperature a person experiences in a building results from the three-way interaction between that person and their customary habits and lifestyle, the thermal performance of a building on its site, and the climate and weather around it (Figure 3.3). The nature and form of that building in different climates is largely dictated by the difference between temperatures experienced around it and those needed inside it to maintain the core body temperatures of occupants within safe limits. The wider the gap between the two temperatures, the better the building needs to be, and the more important the habits and opportunities available to its occupants need to become to adapt successfully to the climate. The traditional building really is the third skin of our relationship with the outdoor climate, after the skin itself and, secondly, our clothing.

Many modern buildings, and their 'envelopes', are incapable of being adjusted and adapting to changing weather and consequently rely mainly on machines to cope with external weather conditions. In this comfort relationship, people cease to be major players, resorting to imported energy to provide acceptable indoor conditions at a huge environmental cost to purse and planet. As we desperately try to reduce energy use and related greenhouse gas emissions globally, we will have to re-learn our thermal limits, and the first place to look is at those temperatures the body needs to function safely.

FIGURE 3.3 Skin temperatures can be similar in the Arctic Circle or on a summer holiday in Spain with the right behaviour, clothing, posture, and shelter. They may also be very different depending on the physical properties of each environment at that time.

Involuntary Responses: Homeostasis

Within our human skin, the human thermoregulatory system is immensely efficient at adapting itself to survive using the self-regulating process of *homeostasis*. This works to steady internal temperatures, keep the core temperature close to 37°C, and let the skin temperatures be roughly where they need to be to ensure that steady state is maintained. Skin temperatures can normally and comfortably vary a lot over a day by controlling blood flow to the skin, to change it from being an insulation layer (blood vessels closed) to a thermal radiator (blood vessels opened), depending on what is needed to keep the core stable (Figure 3.4).

Homeostasis is achieved using both involuntary and conscious actions. *Involuntary physiological actions* are enacted largely via the central nervous system. The heart pumps blood around the body at a safe temperature, heated on its path through arteries and veins by energy generated by the activity of muscles and organs. The heart and organs are in themselves small heat engines working to keep blood pumping around the body at useful temperatures. At the skin interface between the body and the external environment, the body is then further cooled, or heated by radiation, convection, conduction, and evaporation (Figure 3.5). These processes are well explained in volume 1 of this trilogy on adaptive thermal comfort.

If body temperatures begin to rise, driven by heat inside or outside the body and approaching levels that start to overheat the core body temperatures above 37°C, a range of *involuntary physiological actions* kick in to cool the body, including *vasodilation*. Blood vessels near the body surface expand to move more blood closer to the skin, so heat can be lost more quickly from it to the ambient environment. If body temperatures continue to increase, either because the body starts to gain heat from the surroundings or because of increased exercise, this is compensated for by the evaporation of sweat from the skin, particularly at ambient temperatures higher than around 30°C, depending, as always, on the dynamic interaction of many different factors, including the thermal history of the individual.

Humans are one of the few animals with sweat glands all over their bodies. Moisture produced by sweat glands in the epidermis cools the body as it evaporates from the skin's surface due to the latent heat of vaporisation, a process for which energy is absorbed when water is turned into vapour.

When conditions get colder, the blood vessels then contract, reducing blood flow to the skin and extremities, conserving heat within the trunk of the body and around the heart. The hands and the feet are the first to get cold in cold weather because of this *vasoconstriction*. Continued

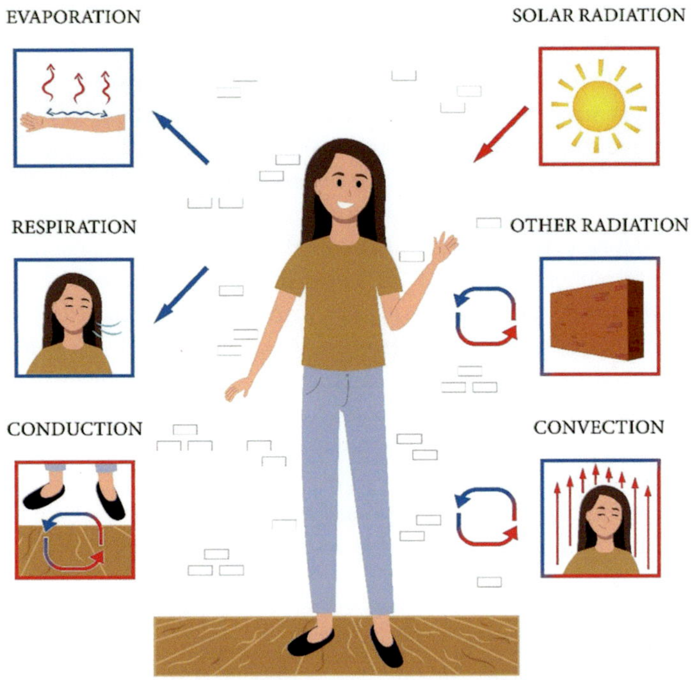

FIGURE 3.4 The body interacts with its thermal environment radiantly, absorbing thermal radiation and/or solar radiation. It also exchanges heat with its environment, by convection, which adds or removes heat from the body via air movement over the skin. Heat is also lost by evaporation of sweat from the skin, or moisture exhaled from the lungs. Heat can also be exchanged directly by conduction through contact between the body and surrounding surfaces.[5]

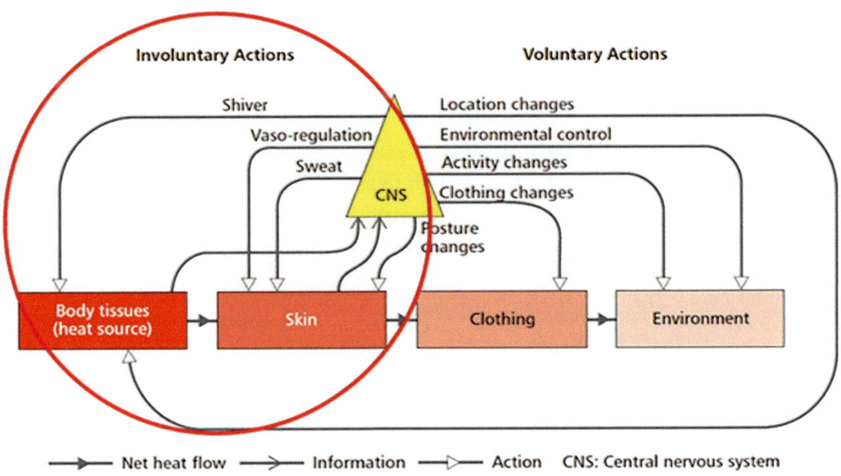

FIGURE 3.5 People may be occasionally aware of excess shivering or sweating, but the body is continuously harmonising heat losses and gains via the skin and body tissues to maintain a core body temperature of around 37°C through involuntary thermoregulatory actions.

Source: Diagram by Fergus Nicol, redrawn by Ken Butcher.

bodily cooling can give rise to an increase in muscular tension and shivering as the body attempts to increase heat production in the muscles to compensate.

Habitual use of this dynamic cardiovascular system has been shown to be important for health as it exercises the systems needed to adapt to thermal excursions outside the accepted comfort zones. People also feel that moderate changes of temperatures can be pleasant.[6] By contrast, those who inhabit constant temperatures in, for instance, care homes for the elderly can lose that vital ability to adapt to, and survive in, a wider range of temperatures.

The thermoregulatory system ensures that a body is continually, unconsciously managing its own thermal environments to prevent a person from becoming *hypothermic* in extremely cold conditions (when the body loses heat faster than it can produce heat), or *hyperthermic* in extreme heat, when the reverse happens. Both states cause thermal stress and, if severe, even death.

The whole system is powered by the *metabolism,* a word suitably derived from the Greek *metabolē*, meaning 'change'. This work engine operates a set of chemical reactions that:

- Convert food (calories) into energy to power the cellular processes of the body
- Convert food into the body's building blocks, including carbohydrates, fat, proteins, and amino acids
- Eliminate metabolic wastes

The amount of work being done by a metabolism is measured by its metabolic rate (MR), which at rest is defined as its basal metabolic rate (BMR). Many factors impact on an individual's BMR, as many of us are aware as we actively count our calories to lose or gain weight. The number of calories we need are related to an individual's BMR, which is calculated in online tools using declared age, gender, height, weight, and activity level. There is no single BMR for 'a person', just as there is no single comfort temperature for everyone. We are all different.

The Body at Rest: The Thermoneutral Zone (TNZ)

Some readers, particularly those in Europe and the USA, may be surprised to learn what temperatures the unclothed body is universally comfortable at. They may have been brought up believing that mechanical comfort zones developed by the HVAC industry relate to the actual human experience of comfort. They do not. The desire to be comfortable goes way back to the origins of our species, and its need to survive in different places and changing climates.

Humans have evolved over millennia to use as little energy and water reserves as possible to stay safely comfortable without having to either shiver to get warmer or sweat to cool down. Early hominids roamed Africa over 5 million years ago and started making tools around 2 and 1/2 million years ago. By 1.7 million years ago, the *Homo erectus* were known to have spread from China to the Caucasus in Georgia, but our ancestors, the modern *Homo sapiens*, only appeared in the excavated records from around 300 to 200,000 years ago. By then, the species was already physiologically adapted to the climate conditions they lived in, like the nice, warm steppes of East Africa, even while Neanderthals were adapting to the colder northern climates of Europe.[7] When the *Homo sapiens* started to migrate across the Middle East, Europe, and beyond around 200,000 years ago, they took with them the human *thermoneutral zone (TNZ).*

The body is actually very efficient at making itself comfortable. The TNZ is a physical and physiological state in which a naked, resting body requires no additional energy to be able to

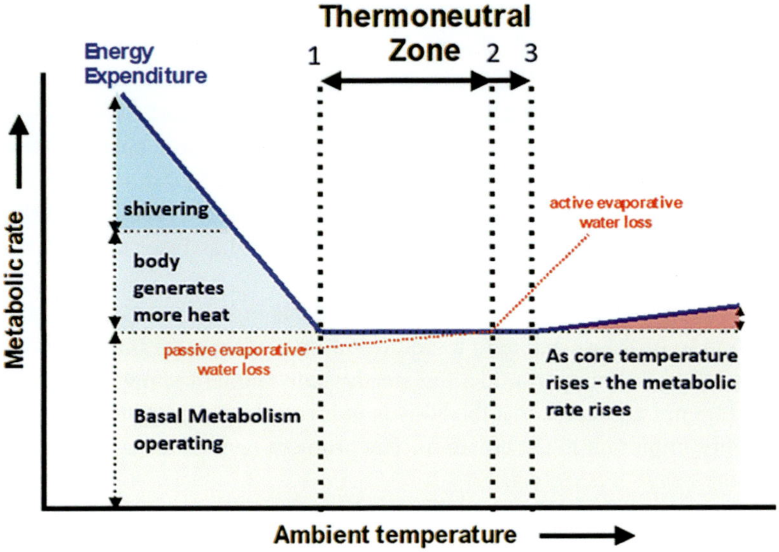

FIGURE 3.6 The thermoneutral zone is bounded by the (1) lower critical temperature (LCT), below which metabolism works harder to generate heat, and two upper critical temperatures. (2) At upper critical temperatures evaporative (UCTe), the sweat rate increases. (3) At upper critical temperature metabolic (UCTm), the metabolic rate increases as the body temperature rises.

Source: Hannah Pallubinsky.[8]

maintain its core temperature of 37°C. It roughly equates to the periods in which the body is just ticking over, running on basal metabolism alone (Figure 3.6).

Thermal discomfort is the signal the body uses to register the need to change ambient conditions to minimise the physiological effort required to maintain body temperature within the thermoneutral zone. Temperature sensor systems are triggered in the skin and body as it becomes too hot or too cold,[9] signalling that the body needs to take action.

Feeling too warm may cause a person to take off a coat, step into the shade, or take another sip from a cool drink. But well before that, the body starts taking involuntary actions to maintain thermally unthreatening core temperatures. At the initial upper critical temperatures evaporative (UCTe), the sweat rate increases to cool the skin. Then, as the body warms even more, a second upper critical temperature metabolic (UCTm) will be passed, at which heart rate and breathing rates increase to lower body temperatures to maintain stable thermal conditions internally.

A body learns over time what combinations of different conditions will result in comfort or discomfort for itself and acts accordingly to ensure discomfort is avoided. The body continually balances signals from multiple sensors (skin, core, eyes, etc.) to stay safe, including external environmental conditions, like airspeed, solar gain, and skin-wettedness, and body-related factors, ranging from clothing to activity, with mental state, health conditions, body form, and thermal history.

It is impossible to predict a precise TNZ temperature bandwidth for a particular person at any one time. A very broad brushstroke generalisation of upper and lower critical temperatures for a naked, resting, fasted adult TNZ in the laboratory suggests they may lie in the region of 28°C to 32–33°C. Roughly below the LTC, the body starts non-shivering thermogenesis, and above the UTCe, sweating becomes the major cooling process for the body. Obviously, these limits are very different to those experienced by clothed people sitting at a desk, doing office work. Studies of the human TNZ are scarce because they are difficult and costly to do. Pioneers of the field Hardy and Dubois arrived at a figure for the TNZ of 'normal' men of LCT of 28.5°C and UTC of 32°C.[10] Since then, results on different subjects have demonstrated little consistency, ranging from LCTs of as low as 26.5°C to UTCs of as high as 35.5°C.

Kingma et al. proposed a model that loosely calculates heat production and transport within a body in relation to heat loss and gain to and from the environment that result in a theoretical TNZ.[11] He concluded that simply modelling steady-state ambient temperature range associated with the TNZ does not guarantee that the body is in thermal balance, but core and skin temperatures are also very important in the equation. The problem here is, as usual, with comfort – it is complex.

Sleep Temperatures

Around about the centre of ranges posited for the TNZ of a body is the environmental temperature of 29–30°C that Nicol proposes as optimal thermal environment for a nude sleeping body to be comfortable and get adequate sleep.[12] The choice of sleepwear, bedding, and mattress type allows the sleeping person to achieve this temperature in their immediate personal space within the bed related to the bedroom temperatures at the time.

Research in the UK on acceptable bedroom temperatures for sleeping during warm periods has proposed that during heatwaves, bedroom temperatures can be tolerated between 26°C for people wearing summer nightwear and 29°C bedding for near-naked, uncovered persons. For persons sleeping on a typical mattress, a threshold of 30°C was deemed probably too high, but if a fan was used to cool the body, this threshold might be elevated by around 1 K.[13] This points to the need to start designing in cooling features to buildings, such as higher ceiling. into which fan can be fitted as and when needed for low-energy cooling.

Countries around the world are experiencing ever-higher summer temperatures and heatwaves, driving the rising use of air conditioning. One study in Osaka, Japan, found that sleep quality and related habits were strongly affected by the attitudes and customary lifestyles adopted by individuals in their own homes, with younger adults moving to air conditioning while their older parents preferred traditional, naturally ventilated sleeping rooms. It was clear that the latter led to a wider tolerance of higher overnight temperatures, better sleep patterns, and lower energy bills in homes where mean indoor temperatures in naturally ventilated rooms averaged 28.2°C. The quality of the building in which people lived, and their location within that building in terms of floor height and orientation, often affected the quality of the sleep they experienced in it, with homes with more insulation and shading that lowered temperatures indoors passively faring well. Feelings of personal security were also found to be important where natural ventilation was used.[14]

A fairly universal response is for air-conditioned bedrooms to be very cold. One study reported that some young Singaporeans suffer from real cold discomfort at night when they set their air conditioners down to 20°C even when they have duvets on their beds. As in Osaka, their elders slept better with the windows open in outdoor temperatures of 25°C to 32°C.[15] This really highlights that the air conditioning industry needs to step up and:

1. Provide sensible guidance on the night-time temperature settings for sleep
2. Provide more nuanced sleeping modes where indoor and outdoor temperatures can be better synchronised to reduce energy use at night-time

It is impossible to report being comfortable while asleep, but being very comfortable when going to sleep is often part of a thermal lifestyle in different cultures. The British love a hot water bottle in the bed to make it cosy on entry. The habit of having a hot bath before bed was traditionally common in Japan, helping improve the quality of sleep between the mattress and the covers, meaning room air temperatures were less important.[16] It is very common to hear people say they get a much better night's sleep in a cold bedroom, and in former generations, it was common to sleep with bedroom windows open even in winter.

The psychology of sleep is also complex, as anxieties can interfere with sleep patterns, and perhaps more attention should be paid to enhancing the affordability of sleeping condition improvements to help people sleep better as domestic energy bills soar.

Alliesthesia: The Pleasure Pump

The body, over time, has developed an effective mechanism to maintain homeostasis, employing a 'sensory siren system' to lure the conscious mind into taking action to return the body to safe temperatures by rewarding it for doing so with feelings of *thermal delight.*[17] Cabanac originally researched this effect and called it *alliesthesia*. He found in the 1960s that *thermal pleasure* is *useful* – it makes the body do what it needs to in order to survive.

The pulse of pleasure produced as a result of the alliesthesia response has been likened to a chemical sugar rush, one that steers the body in the direction of homeostasis and comfort.[18] This pleasure pump pushes the body into action when it is in danger of becoming too hot or too cold, and interestingly, the temperatures at which the subjects record neither cold or heat discomfort nor pleasure in associated studies overlap well over the temperatures of the thermoneutral zone, well above the often-accepted level of 20°C–26°C of the air conditioning industry comfort zones, because people are more or less clothed, as we deal with in the next chapter.[19] More on the importance of alliesthesia in our experiences of buildings in Chapter 23.

What Is Physiological Comfort?

At its most basic level, comfort is that state in which a body is not signalling to itself that it is uncomfortable. Physiologically, the human body is in thermal equilibrium at environmental temperatures that can be surprisingly high or low depending on what we are experiencing and what we have been used to.

Over the course of this book, it will become ever clearer that there is no such thing as 'a comfort temperature'. The thermal images in Figure 3.7 show how men, women, and children are very different in the thermal landscapes of their bodies at a relatively low air temperature. What we outline in the next chapter are three different ways in which humans act consciously to adapt themselves to their thermal environments and/or their environment to them in their everyday lives.

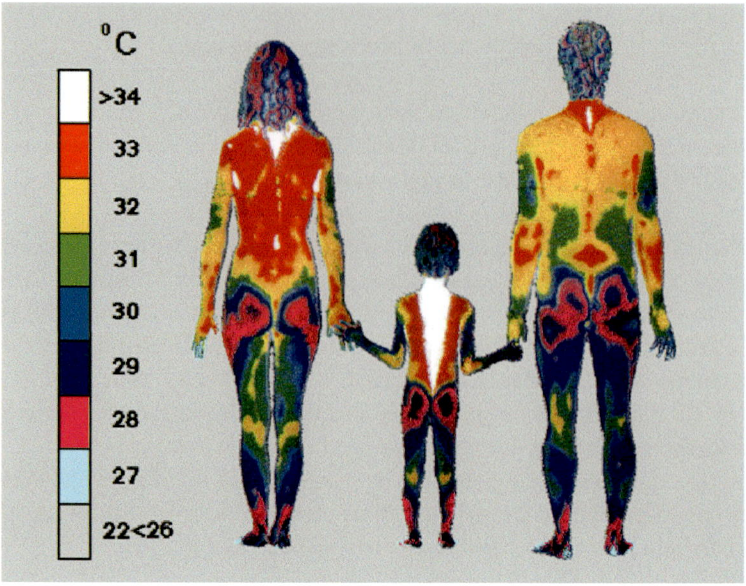

FIGURE 3.7 Thermograms showing the skin temperatures of a naked man, a woman, and a 7-year-old boy in an ambient air temperature of 22°C. Temperatures range from around 28°C to 35°C. Note that in general, the upper part of the body is hotter than the lower part. The boy has a typical 'V'-shaped warm area seen in childhood that becomes a 'T' or 'Y' in adults, more pronounced in the woman.

Source: Clark and Edholm (1985, Plate I).[20]

Why Is Physiological Comfort Important?

Every degree centigrade an indoor thermostat is set away from the outdoor ambient temperature can result in an increase of between 7% and 10% in the energy used to make people in a building comfortable. Current statistically and conceptually flawed 'steady-state' comfort standards force people to set thermostats to temperatures that are too high in winter and too low in summer, with dire consequences for the planet and every other species that lives on it. Overly narrow comfort standards are key drivers for climate change and have huge economic costs for fragile economies and countries.

It has been estimated that overcooling of buildings in the USA alone in 2012 wasted around 104,000 GWh of electricity, resulting in some 50,000 kt of CO_2 emissions, at a cost of over 10 billion US dollars.[21] The next chapter deals with the ways in which people consciously and continuously fine-tune their thermal environments using postures, clothing, and activities to both make themselves more comfortable and avoid discomfort.

From now on, architects and engineers need to understand the fundamentals of what temperatures real people in real buildings can and do find acceptable not only to maintain thermal neutrality but also to live decent, comfortable lives in buildings in a changing climate.

Notes

1 Nicol, F., M. Humphreys and S. Roaf (2012). *Adaptive Thermal Comfort: Principles and Practice*, London: Earthscan/Routledge, p. 71.

2 Sabatini, R. (2016). How to read the genome and build a human being, *TED Talk*, Accessed on 15/06/2023 at: https://www.ted.com/speakers/riccardo_sabatini
3 Oliver, P. (1997). *Encyclopaedia of the Vernacular Architecture of the World*, Cambridge, UK: Cambridge University Press.
4 Humphreys, M.A., and J.F. Nicol (1998). Understanding the adaptive approach to thermal comfort, *ASHRAE Transactions 1998*, 104, Part 1B: Symposium Papers, pf. 1162, Accessed on 15/06/2023 at: https://www.osti.gov/biblio/649430
5 Humphreys, M.A., and J.F. Nicol (1998). Understanding the adaptive approach to thermal comfort, *ASHRAE Transactions 1998*, 104, Part 1B: Symposium Papers, pf. 1162, Accessed on 15/06/2023 at: https://www.osti.gov/biblio/649430. These processes are explained in more detail in volume 1 of this trilogy: Nicol, F., M. Humphreys and S. Roaf (2012). *Adaptive Thermal Comfort: Principles and Practice*, London: Earthscan/Routledge, p. 71.
6 van Marken Lichtenbelt, W.D., M. Hanssen, H. Pallubinsky, B. Kingma and L. Schellen (2017). Healthy excursions outside the thermal comfort zone, *Building Research & Information*, 45(7), pp. 819–827. https://doi.org/10.1080/09613218.2017.1307647
7 Gee, H. (2021). *A (Very) Short History of Life on Earth*, London: Picador, pp. 175–193.
8 Pallubinsky, H., L. Schellen and W.D. van Marken Lichtenbelt (2019). Exploring the human thermoneutral zone – A dynamic approach, *Journal of Thermal Biology*, 79, pp. 199–208. https://doi.org/10.1016/j.jtherbio.2018.12.014
9 Parsons, K. (2002). *Human Thermal Environments: The Effects of Hot, Moderate, and Cold Environments on Human Health, Comfort, and Performance*, 3rd Edition. https://doi.org/10.1201/b16750
10 Hardy, J.D., and E.F. DuBois (1938). Basal metabolism, radiation, convection and vaporization at temperatures of 22 to 35°C, *Journal of Nutrition*, 15, pp. 477–497.
11 Kingma, B.R., N.P. Frijns, L. Schellen. and W.D. van Marken Lichtenbelt (2014). Beyond the classic thermoneutral zone: Including thermal comfort, *Temperature (Austin)*, 1(2), pp. 142–149. https://doi.org/10.4161/temp.29702
12 Nicol, F. (2019). Temperature and sleep, *Energy & Buildings*, 204, pp. 1–5. https://doi.org/10.1016/j.enbuild.2019.109516
13 Lomas, K.J., and M. Li (2023). An overheating criterion for bedrooms in temperate climates: Derivation and application. *Building Services Engineering Research & Technology*, 44(5), pp. 485–517. https://doi.org/10.1177/01436244231183113
14 Umemiya, N., and Y. Chen (2022). Causes and effects of partial cooling during sleep, in *The Routledge Handbook of Resilient Thermal Comfort*, Edited by F. Nicol, H. Bahadur Rijal and S. Roaf, London: Routledge.
15 Yang, B., and T. Olofsson (2017). A questionnaire survey on sleep environment conditioned by different cooling modes in multistorey residential buildings of Singapore, *Indoor and Built Environment*, 26(1), pp. 21–23. https://doi.org/10.1177/1420326X15604206
16 Harding, E.C., N.P. Franks and W. Wisden (2019). The temperature dependence of sleep, *Frontiers in Neuroscience*, 13. https://doi.org/10.3389/fnins.2019.00336
17 Parkinson, T., and R. de Dear (2015). Thermal pleasure in built environments: Physiology of alliesthesia, *Building Research & Information*, 43(3), pp. 288–301. https://doi.org/10.1080/09613218.2015.989662
18 M. Cabanac (1971). Physiological role of pleasure, *Science*, 173, pp. 1103–1107.
19 Mower, G.D. (1976). Perceived intensity of peripheral thermal stimuli is independent of internal body temperature, *Journal of Comparative and Physiological Psychology*, 90(12), pp. 1152–1155.
20 Clark, R.P., and O.G. Edholm (1985). *Man and His Thermal Environment*, London: Edward Arnold.
21 Alnuaimi, A., S. Natajan and T. Kershaw (2022). The comfort and energy impact of overcooled buildings in warm climates, *Energy and Buildings*, 260, pf. 111938, ISSN: 0378-7788

4

HOW PEOPLE ADAPT

Introduction

The thermoregulatory system of a body has similar components to those in many buildings today. Both have two heat/cold sources that matter: ambient temperatures around the building/body and those within the building/body. Both are influenced by the:

- *Building/body shape.* Its form, mass, and size.
- *Insulation.* The degree of protection provided between the outside and the inside climates to maintain homeostasis.
- *Internal heat source.* Wood or coal in older buildings, mechanical in modern ones, and metabolism in the body.
- *Storage capacity of the building/body.* Thermal mass in the former, and body mass and brown fat in the latter, charged and discharged with energy as needed or available.
- *Thermal supply networks of services or vascular systems.*
- *Control system.* In the body operates using signals passed via the central nervous system, and in a building either electrically via thermostats or energy management systems or by the voluntary or involuntary actions of its occupants.

Following are the main ways in which a person's body can adapt to a local environment indicated in Figure 4.1:

- Regulating the rate of heat generation (e.g. by activity, by eating, or with 'emotions')
- Regulating the rate of the body's heat loss (e.g. with clothing insulation)
- Changing the thermal environment around the body (windows, fans, shades, heaters, etc.)
- Changing thermal environment (moving rooms, buildings, or regions)
- Modifying the physiological comfort conditions of the body (acclimatisation)

Human beings evolved over time to know what complex combinations of these capabilities could be effective in maintaining homeostasis in well-understood local micro-climates.

DOI: 10.1201/9781315645070-6

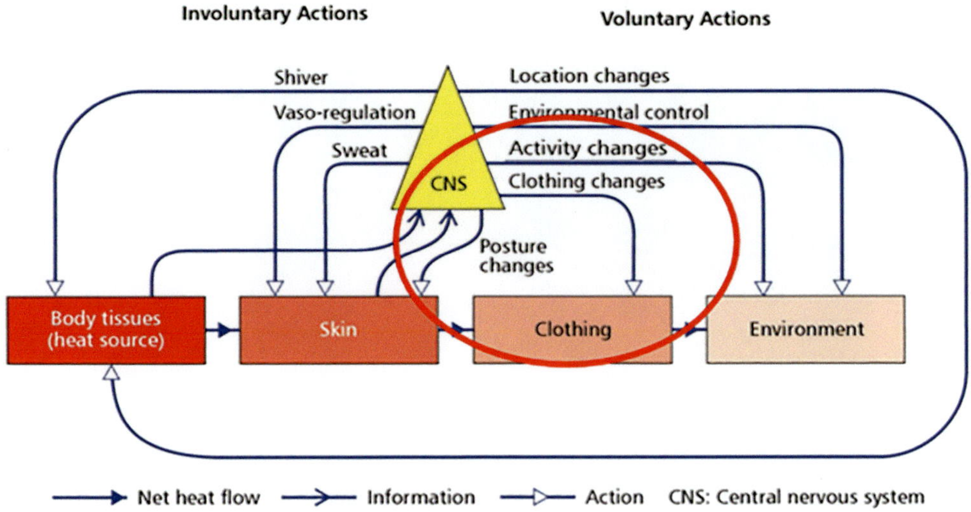

FIGURE 4.1 The mechanisms to keep core body temperature constant at about 37°C are controlled by the nervous system and include involuntary physiological actions, including shivering, sweating, and vasoregulatory responses,[1] and voluntary conscious or sub-conscious actions, which are expressed through behaviour.

Source: Fergus Nicol, redrawn by Ken Butcher at CIBSE.

In traditional buildings, local populations have also evolved complex combinations of physiological and cultural behaviours and material cultures to ensure discomfort is avoided as far as possible, at an acceptable, affordable, and achievable energy costs. Here we look at three key voluntary actions undertaken by people to remain comfortable: posture, clothing, and activities.

Posture

If involuntary actions like shivering or sweating are not enough to manage internal temperatures, then a range of semi-conscious actions is taken by the body, starting with body posture. Across a day or year, people remain largely unaware of the choices made by their own bodies about the posture it adopts. Look at the people in any audience or classroom. Some will be splayed out in a T-shirt and jeans to lose more heat, while others in the same room may be huddled up, with arms folded, in a warm jumper to keep warm.[2] Neither will have made an overtly conscious decision to adopt that posture, but their body has decided, based on thermal preferences and experience, that a particular posture will minimise their risk of discomfort.

Postural actions become conscious ones when people start to feel uncomfortable and consequently re-engage consciously with their ability to control their own body temperature using posture. For a naked body, simply changing postures has been shown to alter the surface area of the body available for heat exchange by up to a third,[3] and mean steady state body temperature by up to 4°C.[4] This explains why the semi-naked Selk'nam natives of Patagonia, shown in Figure 4.2, use a huddled posture beneath their skin cloaks and snuggle into their neighbours to maintain homeostasis.

FIGURE 4.2 A closed posture and snuggling up close together help conserve body heat when people are cold, especially with minimal clothing beneath their fur cloaks and no shoes. The Selk'nam people of Terra del Fuego at the southern tip of Chile were photographed by Alberto de Agostini in 1917.[5] Mean local winter temperatures there average close to 0°C (32°F), and summer temperatures at around 10°C (50°F).

Conscious Voluntary Reponses

Table 4.1 lists some of the opportunities available in most buildings for people to take voluntary, conscious actions to raise or lower the local temperatures, from opening a window to turning on a fan or a heater. People can move about in a room, in or out of the sun, or change rooms, have a hot or cold drink, or simply adjust their psychological expectations of what temperatures are acceptable or desirable. Many of us have done just that, while lying on the beach on a hot day on holiday, knowing that once we become too hot, a cool plunge in the sea beckons to pleasurably solve the discomfort problem.

TABLE 4.1 List of some of the opportunities available for people to take voluntary, conscious actions to improve comfort

Feeling cold?	*Feeling hot?*
• Increase activity	• Reduce activity
• Increase clothing	• Reduce clothing
• Close posture	• Open posture
• Cuddle up	• Separate
• Heat the room	• Cool the room
• Find a warmer place	• Find a cooler place
• Have a hot drink/food	• Have a cold drink/food
• Avoid draughts	• Use a fan
• Modify the room/building	• Modify the room/building
• Emigrate	• Emigrate

A mark of a good comfort designer is that they strive to provide a wide and useable range of comfort action options to the occupants of their buildings. Adaptive comfort opportunities provide a 'win/win' situation because not only can they offer people with a richer, more sensually pleasurable and thermally secure palette of thermal experiences, but also micro-managing ways to keep comfortable manually can save energy and GHG emissions by negating the need to turn on a machine.[6] A common riposte to this suggestion is that occupants opening windows, etc., will interfere with the efficient running of HVAC systems; however, simple devices to turn off air conditioning systems when a window is opened are widely available.

Less occupant-friendly designers might provide no way for an occupant to interact with their heat/cold source, controls, or supply system, leading them to become 'the problem' when they turn off a machine, tamper with a thermostat, tape up noisy air vents, or complain about the temperatures they occupy to the management or a partner.[7]

Clothing

Clothing is the first line of defence against discomfort as people take off, or put on, some clothes. We saw in Chapter 2 that humans' physiological and anatomical systems are designed for a life in moderate and warm environments. We are tropical animals, and while it is culturally impossible for those wearing a sarong in Malaysia, or a loincloth in India, to take off that final modesty wrap, humanity has been able to expand to almost every climate on the planet by first putting on more clothes and then building stronger buildings.

When the ambient temperature falls below 20°C, homeostasis in a naked body cannot be sustained over time without taking voluntary actions to help heat it back up. The colder it gets, the more clothes we put on, but how conscious are the decisions we make to do so?

Every morning we choose an ensemble of underclothes and overclothes to wear, depending on the weather. Most often, this is a habitual response, informed by short- or long-term experience, tempered by attitudes and opportunities that result in our individual clothing preferences and actions on a day-to-day basis. It is always as well to remember that clothing has other functions than thermal comfort.

Nightwear provides a good example of barely reasoned change in clothing choices: nights spent naked between the sheets can merge into ones spent in light cotton, then in medium- or heavyweight nighties or pyjamas, adjusted according to changing bedroom temperatures to maintain that important resting mean skin temperature of around 30°C between the body and its bedding. Only when sudden step changes in the outdoor climate occur, like a strong weather front passing over or a sudden seasonal shift, will the mind consciously engage in the definite decision that change is needed to avoid discomfort. We do not say that on the 1st of October, we will replace our cotton nightie with a flannelette one or put on another blanket or higher-tog duvet, and yet building thermostats were traditionally set to change from winter to summer mode and vice versa on a set date every year, at a high cost in terms of comfort and the energy needed to run the systems. Building occupants can adapt to most conditions, an over-cooled air-conditioned building to an overheated one, by simply changing their clothing.

CLO Units

The insulation value of clothing is measured in CLO units. A CLO value of 1 is equal to the amount of clothing required by a resting human to maintain thermal comfort at a room temperature of 21°C (71°F). Figure 4.3 shows that 1 CLO is roughly equivalent to a medium-weight, traditional business suit, reflecting perhaps that when the unit was invented in 1941, it was men in suits who were key clients of comfort equations.

This very simplistic vision of what the people we design for wear, shown in Figure 4.3, was useful for the development of basic steady-state comfort equations and suited a White middle-class male society. But not only can one apparently uniform ensemble have a range of different CLO ratings; the world has also changed enormously in the last eight decades. For example, Indraganti demonstrates that the sari of an Indian lady might provide a wide range of different CLO units depending on the material used; the actual type of sari ensemble; the personal habits, preferences, and cultures of the wearers; and so on. Similarly, a suit can have different CLO values if the buttons are undone or if it is made of light cotton or heavy hebridean worsted!

FIGURE 4.3 CLO units for different ensembles of clothing, ranging from (L–R) beachwear to the ordinary office suit, heavy winter suit, winter suit and overcoat, winter suit and shoes with heavy winter overcoat, to the outfits worn by natives in arctic climates.

Source: Based on Auliciems (2007).[8]

FIGURE 4.4 In a modern office in the Gulf, many workers are ladies wearing a wide range of Western and local ensembles in the same office spaces, each of which will have different thermal insulation values and also thermal environments within and beneath them.[9] Characterising them with a single (assumed) CLO value to calculate thermostat setting temperature inevitably leads to discomfort for some.

Source: Madhavi Indraganti.

Comfort and Clothing

The effect a piece of clothing has on the thermoregulatory system and feelings of comfort depends, first, on the material it is made of, which will provide thermal insulation as well as resistance against vapour transfer in or out. But often the greater impact on heat flows results from air pockets and layers contained within the garment. The material fibres mainly affect radiative heat transfer, which is less important than the thickness of the material, which typically dictates its insulation value. People are often told to wear more layers when it is cold, and it is the layers of air between and outside the material that act to enhance the thermal resistance of an ensemble. A detailed database of the CLO values of different clothing ensembles is available online.[10]

Metabolic Rate and Clothing: The Sad Case of Scott in the Antarctic

The metabolic rate of a body can be affected by the weight of clothing, which can be quite efficiently carried on the torso or head, but weight carried on the hands and feet has a much higher impact per unit of mass on the energy needed to support it. In Figure 4.6, the Sherpa men of

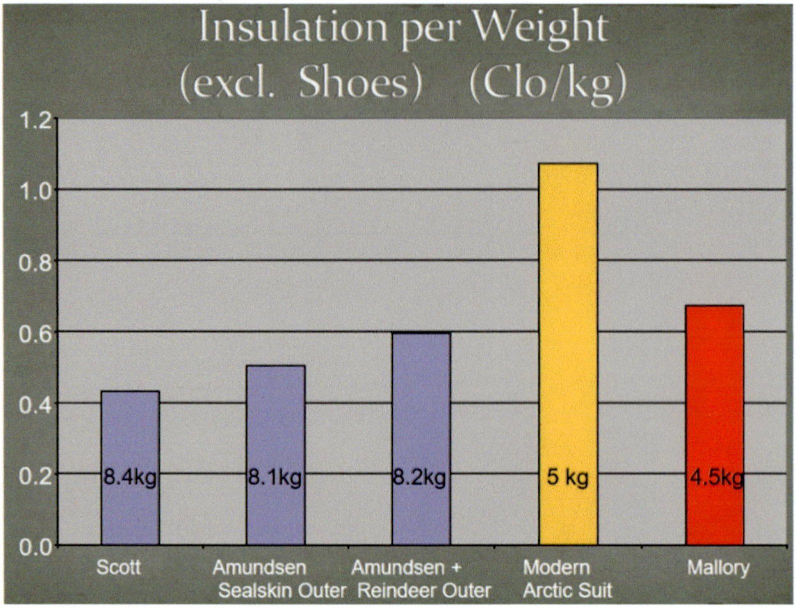

FIGURE 4.5 Insulation values (CLO) of clothing ensembles worn by Scott and Amundsen in the Antarctic and in Mallory on Everest and a modern arctic suit.

Source: George Havenith (2010).

Nepal, who are out and about, working with the flocks all day, wear much less weight of clothing than did the women, who were home-based and less active. Different cultures develop their own 'metabolic rate vs. clothing weight' balance to keep adequately warm over the day while doing the tasks at hand.

Clothing weight can be important in day-to-day living and can also be critical for survival. This was sadly borne out by a study of the outfits worn in Antarctica by great explorers Roald Amundsen and Robert Falcon Scott as they raced to the South Pole. Between December 1911 and January 1912, both Amundsen and Scott reached the South Pole within five weeks of each other. But while Scott and his four companions died on the return journey, Amundsen's party managed to reach the geographic South Pole first and subsequently return to their base camp at Framheim without loss of human life, not least, Havenith (2010) posits, because of the clothing they were wearing.

Based on his reconstructions of the outfits both men wore, and the clothes worn by George Mallory, who died on the North Face of Everest in 1924 shown in Figure 4.5, Havenith concluded that Scott's were inferior to Amundsen's fur-based clothing and had a worse insulation-to-mass ratio. Amundsen's expedition used dogs to haul loads right to the South Pole, while Scott had to abandon his ponies and had the sledges hauled by his team, so needing less insulation while active but causing a major problem while inactive outdoors. Scott's clothing was also heavier than Amundsen's extra and, because of its weight, bulkiness, internal friction, and stiffness, required more energy to be expended for every step taken on the fateful journey, greater metabolic activity, and more calories to keep going. Scott's team died only

FIGURE 4.6 In the remote Nepalese walled town of Lo Manthang, located in the Mustang district, this Sherpa population lives comfortably in low indoor temperatures by wearing high levels of clothing insulation, drinking large quantities of hot butter tea, and sitting close to and around the fire. Men wear, on average, 2.9 CLO, and women nearly double that, as they are less active. The residents' measured mean comfort temperature was 10.7°C, which 59% of the people asked deemed to be comfortable.

Source: Hom Rijal.[12]

18 km south of the final food and fuel depot on their journey. Havenith's simulated race to the Pole showed that had their clothes been lighter and less stiff and caused less friction, they may just have had enough calories in reserve to reach that final depot.[11] The failing with Mallory's far lighter outfit on Everest was that it was useful to −35°C, and on his descent, the weather turned, plunging the mountain temperatures to well below that with the wind chill factor.

Both these narratives demonstrate how the impact of clothing on the working of the metabolism and the thermal health of the body does not simply rely on their insulation value but also on many other aspects of the physical and hydroscopic performance of the materials in a clothing ensemble and the airspaces within and around them.

Populations in the high Himalayas are adept at fine-tuning their outfits to stay alive and comfortable in their daily lives. The Sherpas of Mustang, who have had centuries of practice coexisting with extremely low temperatures, temper their ambitions and balance their activities, food intake, and clothing to maintain their thermal equilibrium in temperatures as low as −25°C in winter.

Humidity and Comfort

People often ask about what the influence of humidity is on comfort. While water content has a substantial effect on the enthalpy of air, at low temperatures, many agree that there is, in fact, 'nothing to worry about'.[13] As it gets hotter, people sweat more, as we saw in Chapter 2. The evaporative capacity of air is a function both of how much water it holds (its humidity or vapour pressure) and how fast the air is moving over the skin. Dry heat loss from the skin and lungs enables people to be comfortable and maintain the heat balance in the body while exerting little energy to do so in the thermoneutral zone (TNZ). When temperatures rise above the TNZ, usually above skin temperatures of c. 30–35°C, sweating increases to compensate for the extra heat gain. Dry heat exchange is a function of air and radiant temperatures and wind speed, but once the need for evaporative cooling from the skin begins, then the humidity of the air affects the rate at which heat is lost from the skin.[14]

At the other end of the scale, very low humidity affects health rather than comfort, exacerbating Sick Building Syndrome (SBS) symptoms to proliferate, such as dry, cracking, or itching skin and lip, eye, nose, or throat irritation and dry coughs. For historical reasons, many mechanical conditioning systems use de-humidification in their cooling and heating processes, and low resulting humidity levels may explain to some extent why SBS symptom prevalence is significantly higher in air-conditioned buildings than in buildings with simple mechanical ventilation and no humidification.[15] This feeling of ill health can increase sickness and absenteeism and causes a decrease in productivity of workers, and while it has become less of an issue recently, it remains a major occupational hazard, and care should be made in comfortable buildings to manage and prevent its causes, a key one being overly dry air.[16]

One myth about why humidity impacts comfort at lower temperatures is that the wetter the air is, the heavier it is, and therefore, the better it is at conducting heat away from the skin, thereby making a subject less comfortable. In reality, the moister the air is, the worse it is at conducting heat away from the skin, so that theory is not valid (Figure 4.7).

Humidity discomfort is mainly felt at high temperatures because with high humidity, the evaporative cooling potential of air decreases as air becomes more saturated with water. Moderate sweat levels seldom cause discomfort as moisture is evaporated efficiently from the skin, but as more sweat pools on the skin and air speeds are not enough to remove it, then the skin 'wettedness' of a person increases and is felt in the following manner:

> Sensors in the skin register (1) temperature and (2) the rate of change in temperatures at the skin and then send that information to the brain, resulting in a bi-modal temperature sensation. There are no sensors in the skin that measure moisture, or skin wettedness, so the main way in which moisture is associated with comfort is by the rate of cooling of the skin when sweat is evaporated from it.

Touch sensors in the skin do report drops of sweat trickling across the skin, which the brain interprets as a moisture sensation. Havenith (2002) explains that clothing can amplify the sensation of 'wetness' dramatically, because when the skin is wet, it dampens the local clothing, which registers as a 'tactile sensation' that can be interpreted as a sticky or clammy, uncomfortable feeling.

Wet clothing may cool more quickly than the skin and is pulled away from the skin during normal movement. When it touches the skin again, it can induce a strong cold sensation that registers as wetness, a 'cold chill'. If consisting of natural fibres or rayon, it absorbs and desorbs

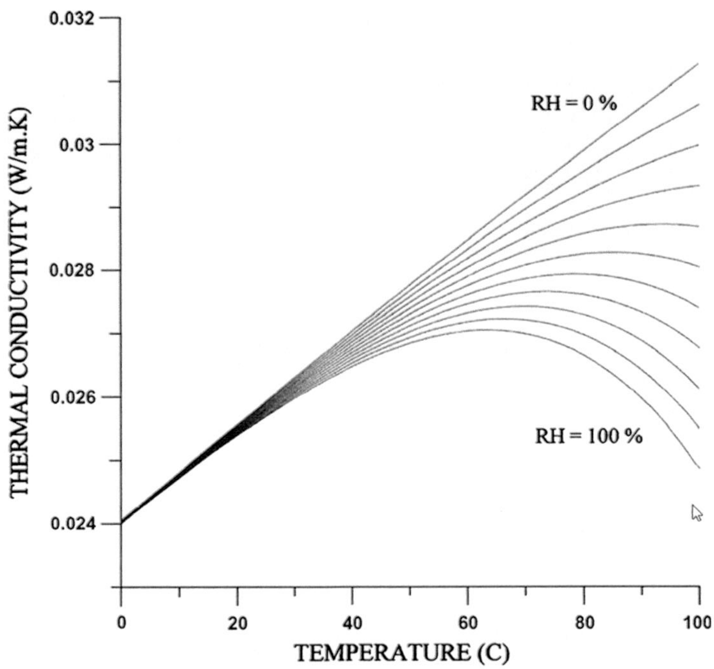

FIGURE 4.7 The thermal conductivity of moist air as a function of temperature, with the relative humidity (RH) ranging from dry air (0% RH) to saturated conditions (100%RH) in 10% steps.

Source: Tsilingiris (2008).[17]

moisture to come into equilibrium with the new environment. Desorption is the same as evaporation; it cools the fabric and pulls heat from the adjacent skin. For people who have been active outdoors, building up sweat on their skin or getting damp clothes, who then sit down, still with a warm skin, this effect can cause 'after-chill'. If you want to know more on this, read Havenith's excellent paper on the subject.[18]

Humidity and Comfort at High Temperatures

Baruch Givoni (1998) did extensive research on the impact of humidity on comfort and found that at temperatures around and below 25°C, people had little idea of what humidity they occupied between 30% and 80% relative humidity. Nicol reinforced this finding with data from the extensive European SCATs database and with data from many subsequent studies. He found that the dependence of the comfort vote on the global temperature (operative temperature) of vapour pressure is insignificant below and above 25°C, until one gets further south in France and down into Portugal, where summer temperatures edge over 30°C. Even at higher temperatures and humidities, like those found in the Gulf, people still have little clue as to what actual relative humidity levels they occupy in temperatures over 35°C.[19]

Enhanced air movement helps cool the body by dry heat exchange, when the surrounding air is lower than upper skin temperature of around 35°C (95°F). When air temperatures rise above skin temperature, then that dry heat loss is supplanted by the cooling effect of sweat evaporating on the skin. *Evaporation* is an endothermic process that absorbs heat, in this case the latent heat of vaporisation, which is the heat required to break the bonds between water molecules, allowing them to transition from liquid to gas. This heat is drawn from the surrounding environment, including the body, which results in a cooling effect. While evaporative efficiency is already high at a low ambient humidity, the benefits of faster air movement, for example, from a fan, in reducing thermal and cardiac strain become more pronounced in humid conditions as air movement from natural ventilation pathways or from fans enhance evaporative efficiency compared to still air.

Cooling people with air movement generated by the simple opening of well-designed windows and using desk, floor, or ceiling fans is obviously the lowest-energy and lowest-cost way to enhance comfort at higher temperatures, and especially where the humidity is also high. Many national and international health guidelines actively advise people not to use fans when indoor air temperatures exceed temperatures between 32°C and 35°C (95°F), but there is little supporting evidence to back up this advice, and there is no evidence at all to support or refute the notion of using electric fans during a heatwave. Research from Berkeley demonstrated that electric fans can be used safely even if the indoor dry bulb temperature exceeds 35°C (95°F), and even with elevated air speeds of 0.8 m/s and above, the critical operative temperature at which heat strain is expected to occur can be reduced by a remarkable average of 14°C, enabling comfort even at temperatures above 40°C, depending also on the relative humidity values of the air.[20]

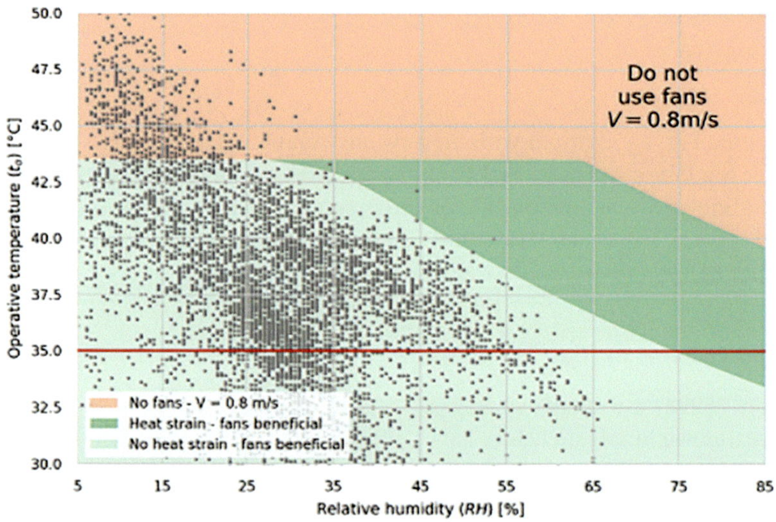

FIGURE 4.8 Environmental conditions under which the use of electric fans is beneficial. The dots show the maximum extreme climate conditions recorded over the last 20 years in more than 5,000 locations worldwide. The green area shows the combinations of temperatures and relative humidities in which the use of fans is beneficial, and they should be kept on and running to avoid heat strain. The image does not mean that people will necessarily be comfortable at 42°C and 10% RH.

Source: Tartarini et al.[21]

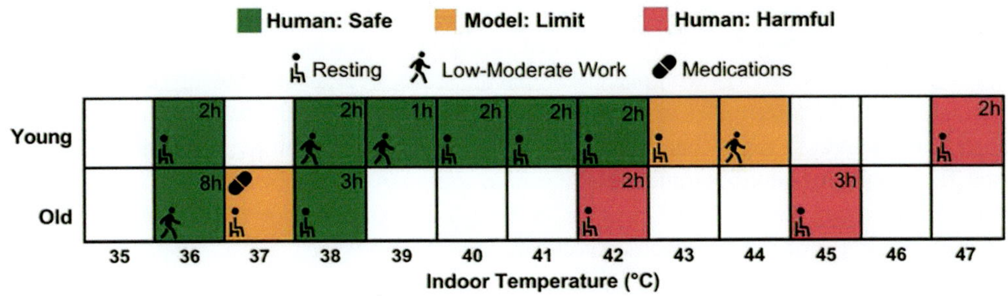

FIGURE 4.9 The current state of evidence on fan use in indoor temperatures exceeding public health guidance in young (*top*) and older (*bottom*) adults. The colours represent where a fan has been proven to be safe and/or beneficial in humans (green), proposed to be safe based on modelling (orange), or has been proven to be harmful at this temperature (red), at any humidity.

Source: Mihalcin et. al. (2025).

Humidity and Health

In more temperate climates, most of the adverse health effects associated with high humidity involve respiratory diseases and skin irritation symptoms resulting from damp and mould in buildings, recurrent respiratory infections, and unspecific, neurological, or general symptoms.[22] These ailments are caused by health-related agents that are affected by indoor humidity, including dust mites, fungi, bacteria, viruses, and non-biological pollutants. All these affect human health, primarily through their inhalation from the air, although some of them have lesser effects through the skin. Most of the identified biological health agents grow on or within surfaces of the building, its systems, and its furnishings, or in standing water within or outside the building.[23],[24] Some of the most damaging health impacts are from the condensation and/or puddling of water, where cold bridging occurs through the construction, often around windows and through walls.

Humidity can be beneficial to comfort at high temperatures, as higher air speeds enhance cooling rates and feelings of thermal relief at high temperatures, as indicated in Figure 4.9. While evaporative efficiency is already high above skin temperatures at a low ambient humidity, the benefits of a good breeze from a fan for reducing thermal and cardiac strain become more pronounced in humid conditions as fans enhance evaporative efficiency compared to still air. Recent work from the College of Environmental Design at Berkeley has shown that fans are safe, up to a point. For healthy young people, that was found to be true until room temperatures hit and exceeded 43°C (109°F). That's 3–11°C (5–20°F) higher than suggested by various heat health guidance globally. In terms of health effects, heart stress came first, as the body's cardiovascular system begins to show signs of strain with fan use at 44°C (111°F) and above before core body temperatures began rising at 45–46°C (113–115°F). The health impacts of air movement at high temperatures also depend on many factors, including age, the amount of activity undertaken, and the medications taken.[25]

Are Women Hotter Than Men?

Men and women have been found to have similar thermal responses from similar contexts. Harimi undertook a meta-analysis of many different studies on the relationship between thermal comfort and gender.[26] Her results showed no significant differences between male and female thermal comfort perceptions. Studies have shown that women are more likely to complain about the thermal conditions they occupy than men, and this can be explained in a number of ways, not least by the differences in dress worn in a workplace between the sexes, with women having been shown to use often uncomfortable ensembles in the workplace to enhance their attractiveness and, in turn, their career opportunities, a factor that in the USA has been shown to be more important to older workers.[27] The amount of control women feel in shared spaces over their own thermal environments is also critical,[28] as are issues of who pays for the energy used to condition the occupied spaces.

Comfort and gender comparisons involve many different issues, and as always, some dimensions can compensate for others; for example, more clothing insulation compensates for a lower metabolic rate, as we saw in the Nepalese group, or higher activity levels compensate for lower resting metabolic rate, and so on. Some widely reported papers purporting to show men and women feel comfort differently cover only limited dimensions of the issue, such as that of heat production, which is easily isolated and studied with a model but very hard to do in practice.[29] When looking at the drivers of what makes people more or less comfortable, the majority of studies conclude that if you take into account clothing and activity, there really is not a significant gender difference in neutral thermal effect. A further meta-analysis found that there is also a 75% overlap in the thermoneutral zone of females vs males when using their estimated resting metabolic rate based on age, gender, height, and weight.[30] So women are usually not hotter or cooler than men, but they often traditionally wear different clothes, do different work, and have different daily routines than men; but in many parts of the world, that is changing.

This issue highlights a big challenge for those attempting to understand 'comfort'. people tend to extrapolate from their own experiences to come to general conclusions around how, when, where, and why people are comfortable. Simple factors in a home may trigger a pre-conception on the comfort/gender issue. This may be as simple as who pays the energy bills at the end of the month, as evidenced by the oft-heard phrase 'My wife likes the room much warmer than I do, and I go 'round turning down the thermostat after her!' Who pays the energy bills?

For women, the issue is complicated by the fact that many women can feel hotter and experience hot flashes during pregnancy, the post-partum period, and on menopause due to hormonal fluctuations. During pregnancy, increased blood flow and metabolic rate contribute to feeling warmer, while menopause involves hormonal changes that affect the body's temperature regulation. More is included about hormonal impacts on comfort in Chapter 23. However, this does not invalidate the findings of meta-analyses with thousands of data points from around the world that conclude that under comparable conditions, men's and women's perceptions of comfortable temperatures are roughly similar.

The Political Power of Clothing

After the 2011 Fukushima nuclear disaster in Japan, all small commercial offices and homes were required to reduce their energy consumption by 15% to compensate for taking all Japanese nuclear plants offline overnight. Many government buildings were mandated to turn on their air-cooling (AC) systems only when the temperatures indoors reached 28°C. Many suffered from discomfort as a result because modern offices often had windows fixed closed, and the practice of utilising small desk fans was not prevalent. Studies of the comfort response at that time in offices found that actually setting the maximum acceptable temperature at 1°C lower, at 27°C, was more effective in practice because worker productivity could be better maintained at that temperature, minimising the financial impacts of the newly imposed regulation temperatures.

The reason these seemingly high temperatures were considered acceptable was the radical *Cool Biz* programme that was started in 2005 by the Japanese Ministry of the Environment. To reduce energy consumption, traditional business suits were replaced in offices by casual summer wear, loose cotton shirts, jeans, and sandals, making it much easier to tolerate higher temperatures. In 2011, as a result of the Fukushima disaster, *Super Cool Biz* was introduced. Some, of course, found it more difficult than others to change their business wear habits of a lifetime, to exchange their dark suits for Hawaiian shirts and board shorts. This fitted well with simple calculations of the impact of CLO on comfort temperatures. Havenith (2002) tells us that long trousers with a short-sleeved shirt will push the comfort temperature down 3°C (0.2 CLO per 1°C), so a median, thermally neutral temperature of 30°C translates into a comfort temperature of 27°C for those doing sedentary office work. A clothing ensemble might also be worn in many different ways to make the wearer more or less warm, such as an Indian sari, which can consist of many different parts, layers, and materials, and is also draped around the body in many different ways, as is clear from Figure 4.4 earlier.[31]

Flexibility of clothing protocols at work is a powerful economic and political tool to reduce carbon emissions from the built environment as a result of reduced energy. A sure way to enable a rapid widening of the temperature settings in offices at different times of day or year, potentially saving a large percentage of office conditioning costs at the same time. What is increasingly clear is that it is impossible to predict what clothing ensembles will be worn in any particular building in the future. Which 1950s be-suited businessman could have foreseen the *Cool Biz* or *Normcore* T-shirt, jeans, and trainer aesthetic popular in so many modern offices today?

The Pumping Action of Clothing

In extreme climates, exposure of the limbs to cold or heat can be fatal, and interestingly, at both ends of the thermal spectrum, people adopt a portable clothing tent approach. In Figure 4.2, the Selk'nam people of Southern Patagonia carry around, over what clothes they are wearing, their own individual fur skin tent/cloak. For them, the secret is to stay in a sedentary position so that a warm, insulating micro-climate can build up between their skin and the fur so that any heat lost from their bodies can be reused to buffer them from the external climate. Traditionally, the cloak or shawl around the shoulders performed a similar function.

In extremely hot climates, people also adopt tent-like outer clothing, but it functions differently. Notice that some women in Figure 4.3 are wearing the local long black flowing garments in Qatar. These robes protect the skin from not only direct sunlight when out and about but also, in the heat, the evaporation of sweat from the skin, which cools the air, and the movement of limbs pumps more air over the skin, so increasing the rate of its cooling. When energy is being exerted in activity, then it can be compensated for by this accelerated process. Men in the Middle East who spend much more of their lives outdoors also wear long bright white robes that also reflect more sunlight away from the body, while also benefitting from this micro-climatic cooling process around the skin.[32]

Activity, Energy, and Customary Lifestyles

In many comfort textbooks, a diagram is presented rather like that in Figure 4.10, showing the results of laboratory studies whose outputs provide a theoretical mean figure for the metabolic energy expended for the activity levels of subjects from reclining to running fast.

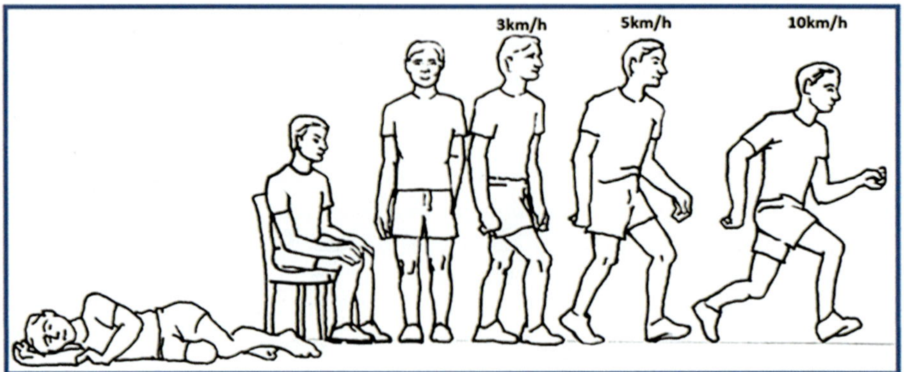

	Reclining	Seated Quiet	Filing Standing	Cleaning Walking	Fast Walking	Running
Activity Level (W/Person)	81	108	144	207	175	466
Metabolic rate (Met)	0.8	1.0	1.4	2.0	3.0	8.0

FIGURE 4.10 Showing very general levels of heat generated (W/person) by the body and the metabolic rates (Met) of a person doing six different levels of activity.

We have already seen that many different variables that can affect the thermal responses of an individual are only very crude approximations to what any single person will be generating in terms of heat and metabolic activity. The comfort experienced while active is, of course, heavily affected by the clothes they are wearing, amongst many other factors.

Just as the way we use clothing over a day or year as part of the thermal strategies and habits of our lifestyles, so too can activity be used to top up or reduce the heat generated and stored in the body. An increase in the metabolic rate, the amount of work being done in the body, by 20 Watts pushes the comfort range for a person down by around 1°C, as does an increase of 0.2 CLO. In warm conditions, an increase in air speed can push the comfort range up by 10°C for 0.2 ms^{-1} (Havenith, 2002).

The intake of hot and cold food also affects body temperature, often simply by warming tissues rather than impacting on heat generated metabolically. Effects can be counter-intuitive; for instance, in hot conditions, when the body is sweating, the intake of very hot drink will produce a sharp increase in the sweat rate, which paradoxically leaves the skin temperature lower than it was and the person feeling cooler. Thus, a hot drink on a hot day can be more effective at cooling the body than a cool drink.[33]

Humidity Impacts on Thermal Stress During Activity

One problem with comfort research is that there is an inevitable disconnect between results recorded in a laboratory and those in field studies. A great example of this is a recent study designed to characterise the thermoregulatory and performance responses of elite road-race athletes competing in hot, humid, night-time conditions during the 2019 IAAF World Athletic Championships in Doha. Very innovative methods were used to collect dynamic skin and core temperatures from 108 elite athletes, as shown in Figure 4.11, during outdoor events at the World Athletics

FIGURE 4.11 Infrared thermography (IRT) images were taken with a cooled camera programmed to record mean skin temperatures in regions of interest (ROIs) over the body to help understand the relationship between core and skin temperatures during an extreme road race in Doha. They show the very different thermal responses to the heat in different racers and their clothing.

Source: George Havenith[32].

Championships.[34] Exposed mean skin temperatures were collected with infrared thermography, and core temperatures with an ingestible telemetry pill. The athletes ran in air temperatures of 29.3–32.7°C and relative humidities of 46–81%. These measurements showed that key issue for the runners was not the humidity of the air but the speed at which it flowed over their skin. The *core temperatures* of the athletes rose with the duration of the exercise, whereas their *skin temperature* showed a decline over the duration of the race, as the skin was continuously cooled by evaporation in the brisk wind movement over the skin as the runners rapidly cut through the air. In this case, the impact of the elevated air flow worked positively to reduce the excessive heating of the core temperature. This finding contradicts the commonly assumed pattern of a recorded rise and plateau of skin temperatures resulting from laboratory studies at similar ambient temperatures, but without realistic air movement rates being applied. The lesson, as always, in comfort studies is that it really does all depend on the context of the person in question as to whether they consider themselves to be comfortable or not.

There Is No Such Thing as a Single Comfort Temperature

A person's body is constantly balancing the heat gains and losses to and from it with voluntary and involuntary actions to maintain basic core temperature as efficiently as possible. Thermal

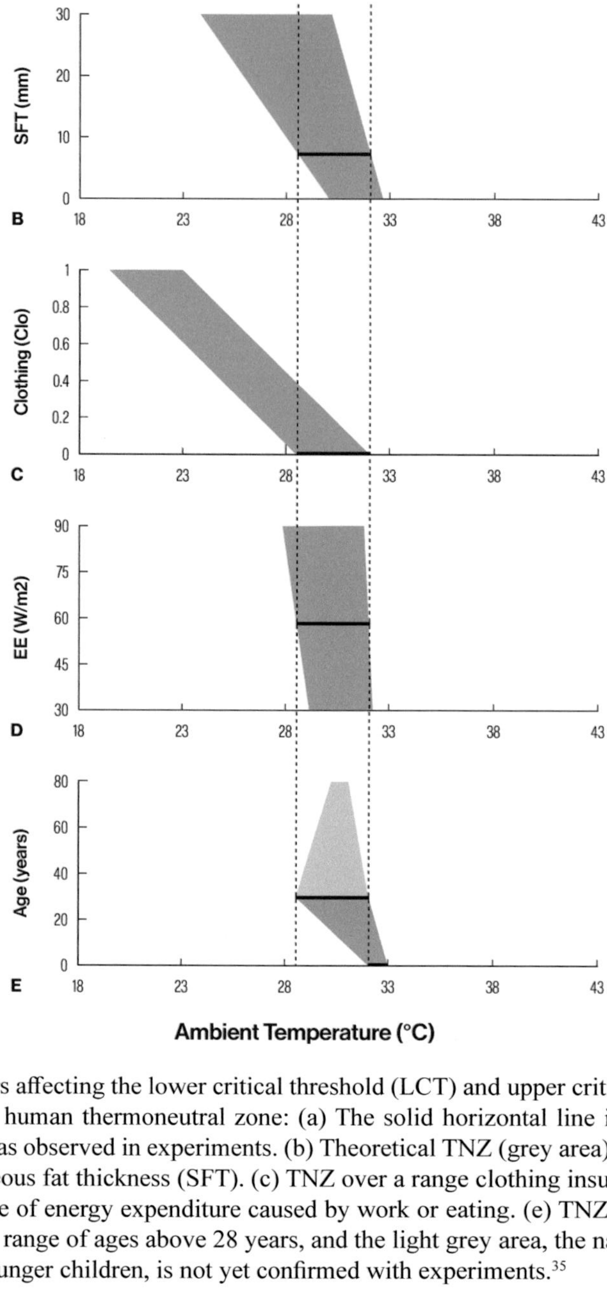

FIGURE 4.12 Factors affecting the lower critical threshold (LCT) and upper critical threshold (UCT) of the human thermoneutral zone: (a) The solid horizontal line indicates a reference TNZ, as observed in experiments. (b) Theoretical TNZ (grey area) over a range of subcutaneous fat thickness (SFT). (c) TNZ over a range clothing insulation. (d) TNZ over a range of energy expenditure caused by work or eating. (e) TNZ (the dark grey area) over a range of ages above 28 years, and the light grey area, the narrowing of the TNZ for younger children, is not yet confirmed with experiments.[35]

neutrality for a person is hugely influenced by a whole range of body-centric factors, such as age, activity, clothing, and how much fat they carry, as shown in Figure 4.12. Subcutaneous fat functions well in a cooler environment, giving an advantage over people who have less fat insulation. Then add to that their intake of hot or cold drinks and food, emotional states, the passage of time, their personal thermal histories, and the evolution of cultures and expectations etc. and the achieving of comfort must be understood as a complex process. In the next chapter, we look at the role of environmental controls and location changes on the nature and timing of very different comfort cultures occupied in the wide and diverse world around us.

There is not, and cannot ever be, a single 'comfort temperature' that applies to everyone in a room, building, city, or region, and certainly not one that can be reduced to one temperature defined to two decimal places!

Notes

1 Nicol, J.F., and M.A. Humphreys (1973). Thermal comfort as part of a self-regulatory systems, *Building Research and Practice* (J.CIB), 6(3), pp. 191–197.

2 Day, J. (2022). Tools and rules for behavioural agency in buildings: Minimising energy use while maintaining comfort, in *The Routledge Handbook of Resilient Thermal Comfort*, Edited by F. Nicol, H. Bahadur Rijal and S. Roaf, London: Routledge.

3 Raja, I.A., and J.F. Nicol (1997). A technique for postural recording and analysis for thermal comfort research, *Applied Ergonomics*, 28(3), pp. 221–225.

4 Tikuisis, P., and M.B. Ducharme (1996). The effect of postural changes on body temperatures and heat balance, *European Journal of Applied Physiology and Occupational Physiology*, 72(5–6), pp. 451–459. https://doi.org/10.1007/BF00242275

5 See: Accessed on 15/06/2023 at: https://imgur.com/gallery/F4YKAfG

6 Heschong, L., and J. Day (2022). Why occupants need a role in building operation: A framework for resilient design, in *The Routledge Handbook of Resilient Thermal Comfort*, Edited by F. Nicol, H. Bahadur Rijal and S. Roaf, London: Routledge.

7 Nicol, F., H. Rijal and S. Roaf (2022). Resilient comfort standards, in *The Routledge Handbook of Resilient Thermal Comfort*, London: Routledge.

8 Auliciems, A., and S. Szokolay (2007). *Thermal Comfort*, 2nd Edition, Published by PLEA: Passive and Low Energy Architecture International in association with Department of Architecture, The University of Queensland, Accessed on 15/06/2023 at: https://www.healthyheating.com/HH_Integrated_Design/Week%201/Thermal%20Comfort.pdf.

9 Indraganti, M. (2010). Thermal comfort in naturally ventilated apartments in summer: Findings from a field study in Hyderabad, India, *Applied Energy*, Elsevier, 87(3), pp. 866–883.

10 Smallcombe, J.W., S. Hodder, K. Kuklane, M. Młynarczyk, D. Loveday, et al. (2021). Updated database of clothing thermal insulation and vapor permeability values of Western ensembles for use in ASHRAE Standard 55, ISO 7730, and ISO 992, *ASHRAE Transactions*, 127, pp. 773–799.

11 Havenith, G. (2010). Benchmarking functionality of historical cold weather clothing: Robert F. Scott, Roald Amundsen, George Mallory, *Journal of Fiber Bioengineering and Informatics*, 3(3), pp. 121–129. https://doi.org/10.3993/jfbi12201001

12 Rijal, H. (2021). Thermal adaptation of buildings and people for energy saving in extreme cold climate of Nepal, *Energy and Buildings*, 230. https://doi.org/110551.10.1016/j.enbuild.2020.11

13 Lasance, C. (2003). The thermal conductivity of moist air, *Electronics-Cooling*, Accessed on 15/06/2023 at: https://www.electronics-cooling.com/2003/11/the-thermal-conductivity-of-moist-air/

14 Givoni, B. (1998). *Climate Considerations in Building and Urban Design*, New York: John Wiley and Sons.

15 Seppänen, O., and W.J. Fisk (2002). Association of ventilation system type with SBS symptoms in office workers, *Indoor Air*, 12(2), pp. 98–112.

16 Joshi, S.M. (2008). The sick building syndrome, *Indian Journal of Occupational and Environmental Medicine*, 12, pp. 61–64.

17 Tsilingiris, P.T. (2008). Thermophysical and transport properties of humid air at temperature range between 0 and 100°C, *Energy Conversion and Management*, 49(5), pp. 1098–1110. https://doi.org/10.1016/j.enconman.2007.09.015

18 Havenith G. (2002). Interaction of clothing and thermoregulation, *Exogenus Dermatology*, 1(5), pp. 221–230. https://doi.org/10.1159/000068802

19 Alshaikh, A.M.A. (2016). *Design Principles for Thermally Comfortable and Low Energy Homes in the Extreme Hot-Humid Climatic Gulf Region, with Reference to Dammam, Saudi Arabia*, PhD Thesis at Heriot Watt University, Edinburgh, Accessed on 15/06/2023 at: https://ethos.bl.uk/OrderDetails.do?uin=uk.bl.ethos.712727

20 Schiavon, S., and P. Raftery (2023). Fans for cooling people guidebook, published by the centre for the Built Environment, University of California, Berkeley. https://cbe-berkeley.gitbook.io/fans-guidebook/

21 Tartarini, F., S. Schiavon, O. Jay, E. Arens and C. Huizenga (2022). Application of Gagge's energy balance model to determine humidity-dependent temperature thresholds for healthy adults using electric fans during heatwaves, *Building and Environment*, 207, Part B, 108437, ISSN: 0360-1323.

22 Nevalainen, A., and M. Seuri (2005). Of microbes and men, *Indoor Air*, 15(supplement 9), pp. 58–64. https://doi.org/10.1111/j.1600-0668.2005.00344.x

23 Baughman, A.V., and E.A. Arens (1996). Indoor humidity and human health – Part I: Literature review of health effects of humidity influenced indoor pollutants, *ASHRAE Transactions*, 102(1), pp. 193–211.

24 Arens, E.A., and A.V. Baughman (1996). Indoor humidity and human health: Part II – Buildings and their systems, *ASHRAE Transactions*, 102(1), pp. 212–221.

25 Mihalcin, E., S. Schiavon and N. Ravanelli (2025). Examining the physiological strain with electric fans during high indoor heat stress. *Building and Environment*, 113261. DOI: 10.1016/j.buildenv.2025.113261

26 Harimi, D. (2019). A meta-analysis on gender thermal comfort, *Proceedings of the 1st Conference of Comfort at the Extremes*, Dubai, Accessed on 15/06/2023 at: https://windsorconference.com/wp-content/uploads/2019/04/CATE2019_Proceedings_v3_web.pdf

27 Engeln, R., and A. Zola (2021). These boots weren't made for walking: Gendered discrepancies in wearing painful, restricting, or distracting clothing, *Sex Roles*, 85, pp. 463–480. https://doi.org/10.1007/s11199-021-01230-9

28 Heschong, L., and J. Day (2022). Why occupants need a role in building operation: A framework for resilient design, in *The Routledge Handbook of Resilient Thermal Comfort*, Edited by F. Nicol, H. Bahadur Rijal and S. Roaf, London: Routledge.

29 Kingma, B. (2019). Personal communication.

30 Schweiker, M., G.M. Huebner, B. Kingma, R. Kramer and H. Pallubinsky (2018). Drivers of diversity in human thermal perception – A review for holistic comfort models, *Temperature*, 5(4), pp. 308–342. https://doi.org/10.1080/23328940.2018.1534490

31 Indraganti, M. (2010). Thermal comfort in naturally ventilated apartments in summer: Findings from a field study in Hyderabad, India, *Applied Energy*, 87(3), pp. 866–883.

32 Berger, X. (1988). The pumping effect of clothing, *International Journal of Ambient Energy*, 9(1), pp. 37–46. DOI: 10.1080/01430750.1988.9675909

33 Clark, R.P., M.R. Goff and J.B. Mullam (1977). Skin temperatures during sunbathing and some observations on the effect of hot and cold drinks on these temperatures, *Journal of Physiology*, 267, pp. 8–9.

34 Aylwin, P., G. Havenith, M. Cardinale, et al. (2023). Thermoregulatory responses during road races in hot-humid conditions at the 2019 Athletics World Championships, *Journal of Applied Physiology*. DOI: 10.1152/japplphysiol.00348.2022

35 Kingma, B., A. Frijns and W. van Marken Lichtenbelt (2012). The thermoneutral zone: implications for metabolic studies, *Frontiers in Bioscience*, 1(4), pp. 1975–1985. https://doi.org/10.2741/518

Buildings and Comfort

5

HOW PEOPLE ADAPT IN BUILDINGS

Introduction

Nicol and Humphreys joined the UK Building Research Station in the 1960s working in the thermal comfort section. It was there that Charles Webb had conducted thermal comfort field surveys in Singapore,[3] North India, and Iraq[4] and then worked in the UK using the then new-fangled electronic data-logging methods to conduct thermal comfort surveys using the Bedford scale, shown in Figure 5.1, while asking people to record their comfort feelings. Environmental temperatures were also recorded, and people reported whether they were comfortable and neither too warm nor too cool. The surveys were completed at widely different room temperatures in different countries, from the high teens or the low 20s Celsius in the UK to the high thirties in summer in Baghdad.[5]

Webb preferred to use Thomas Bedford's scale because it asks respondents how comfortable they actually feel. Ole Fanger later used the ASHRAE scale because he maintained that a person selecting 'Hot', 'Warm', 'Cool', or 'Cold' meant that the people were uncomfortable, regardless of the fact that subjects may have wanted to be cool or warm. Issues around gathering comfort votes are dealt with in detail in volume 2 of this trilogy on adaptive thermal comfort.

By the time Webb's projects ended in the 1960s, it had become clear that the achievement of thermal comfort could not just be primarily concerned with the physiology of heat exchange or the science of clothing and activity. Although these remained important, Figure 5.2 shows the wide spread of Webb's comfort data that confirmed that powerful feedback systems are at work that enable people to be comfortable at approximately the room temperatures they had become accustomed to. Each point on the graph is the mean value of the comfort vote on the Bedford scale from a single person over a period of time plotted against the mean indoor temperature at the times of the survey.

People obviously achieved comfort in many different ways in the different climate surveys, by either sub-consciously or consciously adapting in their environments, or by changing them physically using the tools at hand, some of which are shown in Figure 5.3. In more extreme conditions, they often remain comfortable by moving between locations inside and outside the building to avoid discomfort.

DOI: 10.1201/9781315645070-8

ASHRAE SCALE		BEDFORD SCALE	
Hot	3	Much Too Warm	3
Warm	2	Too Warm	2
Slightly Warm	1	Comfortably Warm	1
Neutral	0	Comfortable	0
Slightly Cool	-1	Comfortably Cool	-1
Cool	-2	Too Cool	-2
Cold	-3	Much Too Cool	-3

FIGURE 5.1 Two of the most commonly used comfort scales with a seven-point comfort vote in each: the ASHRAE (USA) and Bedford (UK) scales. NB: The Bedford scale is also numbered from 1 to 7, from Much Too Cold to Much Too Hot.

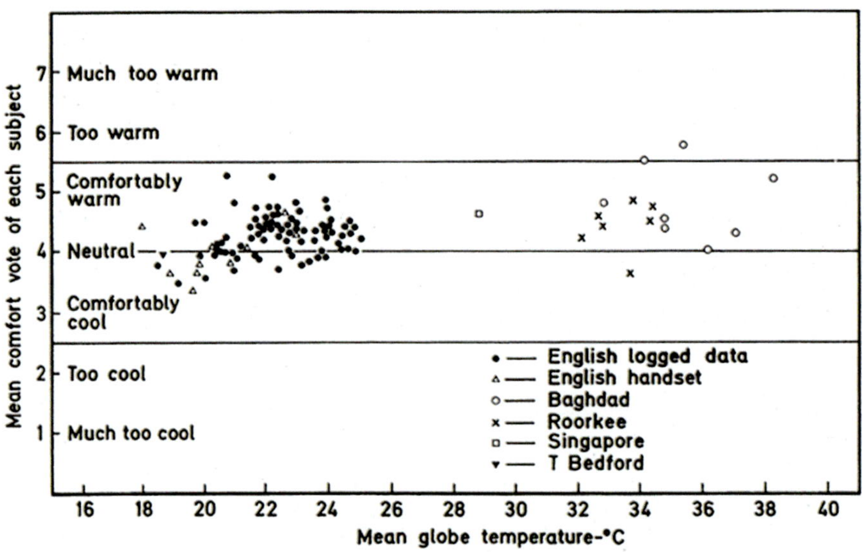

FIGURE 5.2 This influential diagram published by Nicol and Humphreys in 1971[6] shows a wide spread of comfort votes from BRS surveys ranging from 18°C to 38°C.[7] The Bedford scale is numbered 1–7 here.

Many scientists at the BRS in the 1960s did not consider comfort surveys to be 'proper science' and queried the funding of them. After all, why undertake field surveys when you can conduct controlled scientific experiments in laboratories to answer questions around comfort? The obvious response was that animal behaviour is best studied in the wild, not in the zoo, and

FIGURE 5.3 There are a multitude of choices we make, and take for granted to remain comfortable over time, relating to our activities, environmental controls, clothing, food and drink and lifestyles, some of which can be enabled, or disabled through design

Source: Drawn by Jim Eveson.

people are best studied in their everyday lives. After all, even laboratory studies need to be validated in everyday conditions (although even today many are not), so why not start with the field study?

Furthermore, it is impossible to study in a laboratory with a handful of subjects and limited resources, the richness of how different individuals, in different buildings and cultures, use the very different spaces they occupy and the opportunities they have at hand to make themselves more comfortable. The truth of this has been born out in 50 years of research since then, as Olesen and Parsons pointed out in 2002: 'Laboratory studies have often supported the validity of Standard 7730 (PMV method) whereas field studies have not'.[8] The simple fact of the matter is that comfort standards derived from laboratory studies do not well reflect comfort as experienced in most buildings in the real world. Comfort is achieved by harnessing all the available resources to hand in the dynamic feedback system that underpins the thermoregulatory system of the body, both internally with involuntary actions and outside the body using a range of voluntary actions to modify the real world around it, as shown in Figure 5.4.

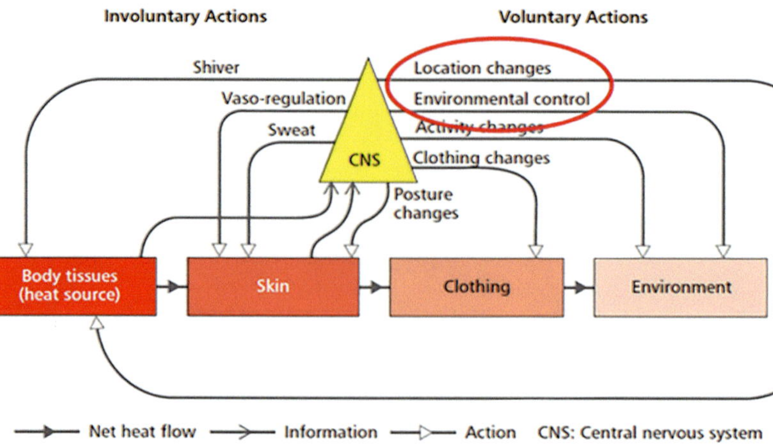

FIGURE 5.4 In the dynamic comfort feedback systems, prominent amongst the voluntary actions people take to make themselves comfortable are the environmental controls and the possibility to move around a room, building, or beyond to stay as comfortable as possible.

Source: Fergus Nicol, redrawn by Ken Butcher at CIBSE.

Voluntary Actions: Changing Positions

In the 20th century, cooling was largely done by moving air around with *punkha wallahs*, fans, or opening windows systems designed to manage cooling breeze flows through spaces. Moving a chair nearer an open window or door could provide a delightful feeling of cooling down. Heating was done initially by fires, so moving towards or away from a fire might result from how hot the source was or how great the need for heat was by the seated person. As heating systems moved to boilers with water-filled radiators, people might also move close to the radiant heat source to warm up. The move since the 1950s in the USA, and later the rest of the world, to fan-driven air systems, to heat and cool spaces the temptation to cosy up to a hot air vent rather disappeared, not least because of the speed with which the air exits the vent but also for the associated noise and actual thermal discomfort often caused by those air streams moving over the skin. For some indefinable reason, moving natural outdoor air from an open window often appears much more acceptable than the experience of air from a mechanical system vent. Perhaps because occupants feel they have power to open and close the window and less so over the machine, particularly in a public building.

Voluntary Actions: Changing Rooms

Interactions between the spaces within a building and the needs of people seeking comfort in them depend largely on the availability, and usability, of opportunities for occupants to modify their local environments and move around in them.

Location changes occur when a particular spot in a building becomes uncomfortable, causing people to move to a different part of a space or to a different room, building, region, or even a different country. Many societies traditionally moved inter-seasonally to cooler or warmer regions to exploit seasonally available resources and enjoy more comfortable climates. Nomadic and transhumant societies depend on such migrations to seasonally occupy peripheral territories between barren and permanently cultivated lands to simply survive.

FIGURE 5.5 Tourists from around the world flock to experience the novel thermal thrill of bathing in hot spring water in Iceland, here shown at Ylstrondin Geothermal Beach at Nauthólsvík, near Reykjavik, Iceland.

Source: Guðmundur Ingi Guðmundsson.

People in more affluent societies can move to the seaside or mountains in summer to enjoy thermal pleasure on holidays or to avoid excessive heat and polluted environments, in order to enrich their quality of life over a year. They may seek winter thermal thrills on skiing holidays, invigorated by moving from the freezing snow slopes to the warm fireside for their Après Ski fun, or mix the two with a hot dip outdoors in a cold country like Iceland, as shown in Figure 5.5. The search for novel thermal experiences can offer luxurious sensual adventures for those who can afford them.

Climate change and soaring energy prices are driving the shift back towards intra-mural migration in search of low-cost comfort in extreme weather. For example, occupying thermally diverse spaces in a building to lower bills and carbon emissions from buildings while avoiding discomfort. Many now choose to sleep downstairs, or in a cooler room in their homes during heatwaves, not least city dwellers suffering from the high night-time temperatures in urban heat islands. It can be cheaper, and possibly more comfortable, to move to a cooler space than to turn on an air conditioner. There is an increasing move to look at existing and new buildings in terms of the thermal safe havens they can provide for people during extreme weather events. Of course, larger homes have more opportunities for cool or cosy rooms, whereas they are less common in small homes and flats.[9] The modern fashion to 'knock rooms through' to provide more spacious 'open plan' living areas and to make whole living floors open plan make it much more difficult to adequately heat or cool those larger spaces when needed.

Grand 19th-century homes in Britain had rich thermal landscapes with breakfast rooms warmed by the morning sun, winter gardens oriented to take advantage of the low mid-winter sun, and cool summer houses for hot afternoons. Modern homes cannot run to such extravagances but do have warmer and cooler spaces in them between which people can move over a day, and between seasons, to benefit from as much 'free-energy comfort' as possible by exploiting the thermal landscapes around their homes.

Voluntary Actions: Environmental Controls

Environmental controls to improve the thermal conditions in a room can be passive and manually adjusted, like external sunshades, openable windows, blinds, curtains, screens, etc. Active systems might range from a small desk fan or lamp, a ceiling fan, or a heater to room- and building-level heating and cooling systems either manually managed locally using switches and thermostats or remotely controlled using pre-set control systems.

How can we design in appropriate adaptive opportunities into a particular building when each culture and climate is so different and each individual within that culture is unique? Might the answer to such questions be helped by the fact that human beings are one species, with similar thermoregulatory systems, from Siberia to the Sahara? How is it possible to know the temperatures at which people will want to open windows, for instance, or turn on a fan?

The simple answer is that you ask people how they feel and then how they would prefer to feel. If you ask enough people, you can get an idea of what thermal comfort today in a particular location involves, enabling you to design with that knowledge in mind.[10] Here we use examples from Pakistan to Japan to show how field studies have shed light on what comfort is for populations in buildings in local climates. The analysis of huge comfort survey data sets has employed statistics to show trends in the data on which more generalised comfort standards can be based.

The Adaptive Algorithm

A breakthrough in the understanding of human comfort was published in 1975, when Humphreys plotted the neutral or comfort temperatures against the mean indoor temperatures, as shown in Figure 5.6. Suddenly, what had been implied in the diagrams like that in Figure 5.2

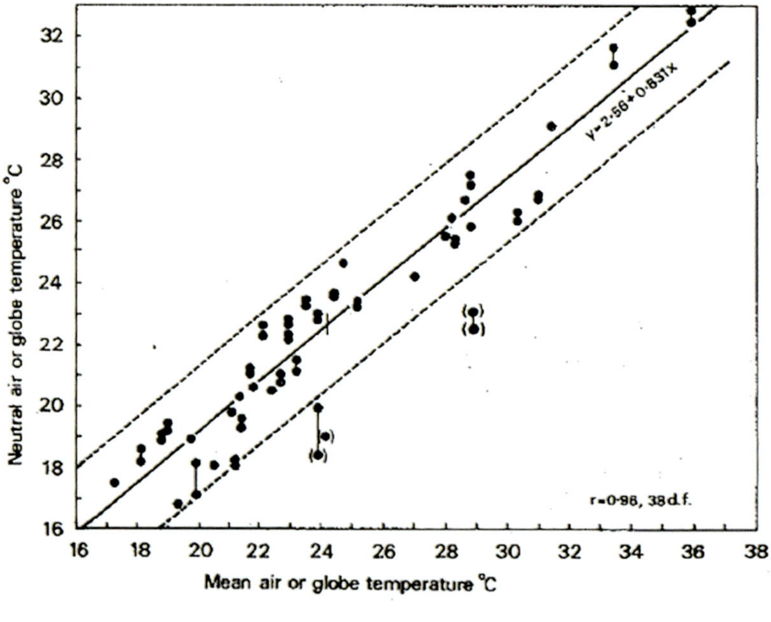

FIGURE 5.6 Scatter diagram of indoor neutral temperatures plotted against monthly mean outdoor temperatures. The free-running state means that there was no heating or cooling of the building during the period of the field study.

Source: Humphreys (1971).[11]

became crystal clear: people obviously adapt to those temperatures they normally experience, in hotter climates they are comfortable at higher temperatures, and vice versa.

Using statistics, it was possible to show that this was a very strong relationship, with a correlation coefficient of 0.96 for the data then presented, and importantly, it was also possible to use the line of best fit for the data to write an equation to predict what the neutral temperature indoors (T_n) might be for any given indoor temperature (T_m). For the data he plotted then, that equation was:

$$T_n - 2.6 + 0.831\ T_m$$

The closeness of the relationship between comfort and indoor, and, in turn, outdoor climates, revealed in Figure 5.6, and its algorithm underpinned much of the future adaptive thermal comfort research and standards and clearly demonstrated how powerful the impact of the opportunities used by people must be to adopt them, to adapt to a huge range of different indoor and outdoor temperatures.

Two previous volumes of this trilogy on adaptive thermal comfort give a simple introduction to the principles and practice of the adaptive model in volume 1 and a detailed account of the foundations and analyses on which the model is based. Stanley Kurvers and Joe Leyten have also produced an excellent book on designing with the adaptive model that anyone interested the subject of designing for comfort should read.[12]

Temperatures at Which People Act to Restore Comfort

From that large thermal range of temperatures at which people report being comfortable, can we extract some rules of thumb about when people might act to take advantage of environmental technologies to hand? A key thing designers need to know is what are the temperatures at which people can use no- or low-energy cooling solutions, like simply opening windows for fresh air and for heating or cooling. Those temperatures roughly also indicate acceptable temperatures for the use of local fans that simply move air around to enhance sweat rates from the skin, thereby cooling people, rather than changing the actual temperature of the air. We are happy to use rather loose estimates of what might constitute comfortable temperatures for individuals. Previous chapters have shown that people differ in their temperature preferences for many different reasons. Anyone who tries to persuade us that temperatures required for comfort can be predicted accurately to one or two decimal points either knows very little about comfort or is trying to use faux science to blind their audience.

Between 1993 and 1996, the Oxford thermal comfort unit worked in Pakistan undertaking comfort surveys in cities in the five very different climatic zones of the country, as described in more detail in the following chapter. Nicol's analysis of that data sheds light on the temperature ranges that are suitable for using local unheated or uncooled moving air to enhance comfort, like opening windows, for example.

Figure 5.7 shows that as temperatures warmed up above around 15°C, the locally adapted Pakistani populations started to shed clothes (green), and by 30°C, they can take no more off decently, being reduced to wearing the minimum of 'Shalwar and Kameez', loose cotton bottoms and long shirts. Laboratory studies estimate every 0.2 CLO unit of clothing that is removed equates to 1°C increase in comfort temperature.[13] Obviously, this is only a small part of the adaption opportunities exploited there, as the 0.8 CLO units shed would imply an increase of 4°C in acceptable comfort temperatures, but people were occupying up to 20°C differences in temperatures in the survey, in apparently acceptable conditions. Sweating increased rapidly (blue) above the thermoneutral zone of clothed subjects, noticeably at temperatures above 30°C.

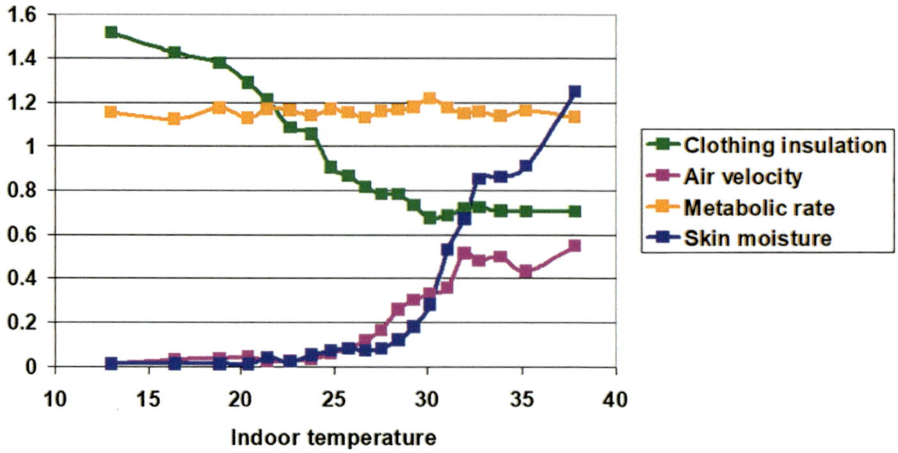

FIGURE 5.7 As the ambient temperature rises, human beings everywhere tend to respond roughly similarly, trying to maintain a stable metabolic rate so the body does not get overheated from within, while they take off more clothes, sweat more, and increase the velocity of the air around themselves.

Source: Nicol. Data from surveys in Pakistan (see Chapter 6).

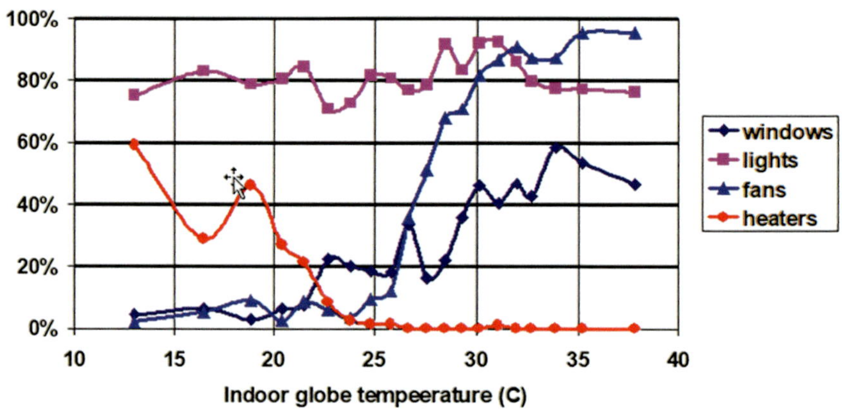

FIGURE 5.8 Adaptive opportunities: As it gets warmer, people increase local air velocity by opening windows and turning fans and reduce the room temperature by turning off heaters, but in offices, light levels are kept fairly constant as a feature of the working environment.

Source: Nicol. (See Chapter 6, following.)

A major factor in their lived comfort experience becomes air velocity (pink) at higher temperatures, and the increased velocities around the subjects are important in removing sweat and heat from the body by evaporation and convection. At around mean upper skin temperature of c. 35°C, the increased speed trend in ventilation then falters as too much hot air starts to heat the body, and skin moisture goes into overdrive to speed up heat loss from the surface of the skin.

The data also shown in Figure 5.8 highlights the key adaptive opportunities that the surveyed office workers resorted to as temperatures rose. Lighting levels remained fairly constant, not least because most of the buildings surveyed were traditional in style, with smaller windows that would require supplementary artificial light sources during working hours. Heaters were sparingly used, and when room temperatures rose above 20°C, they were largely turned off, and by 24°C, fully so. At around 20°C, windows were opened to increase air velocity, but at most, only 60% of them were in use, possibly because of the often-extreme noise and pollution in many of the Pakistani cities that were surveyed. At 25°C, most workers in the buildings had manually switched on their mechanical fans, and by 30°C, 80% had them on, and by 35°C, they were almost universally used. This data showed clearly that for adapted populations in the different climatic regions of Pakistan, the combination of clothing and environmental controls like heaters, windows, and fans can be effectively used to provide acceptable comfort levels in fairly traditional buildings.[14]

In the heat, Figure 5.9 shows the importance of air movement at higher temperatures for improving the comfort of subjects in India and Iraq. It demonstrates that in warm and hot

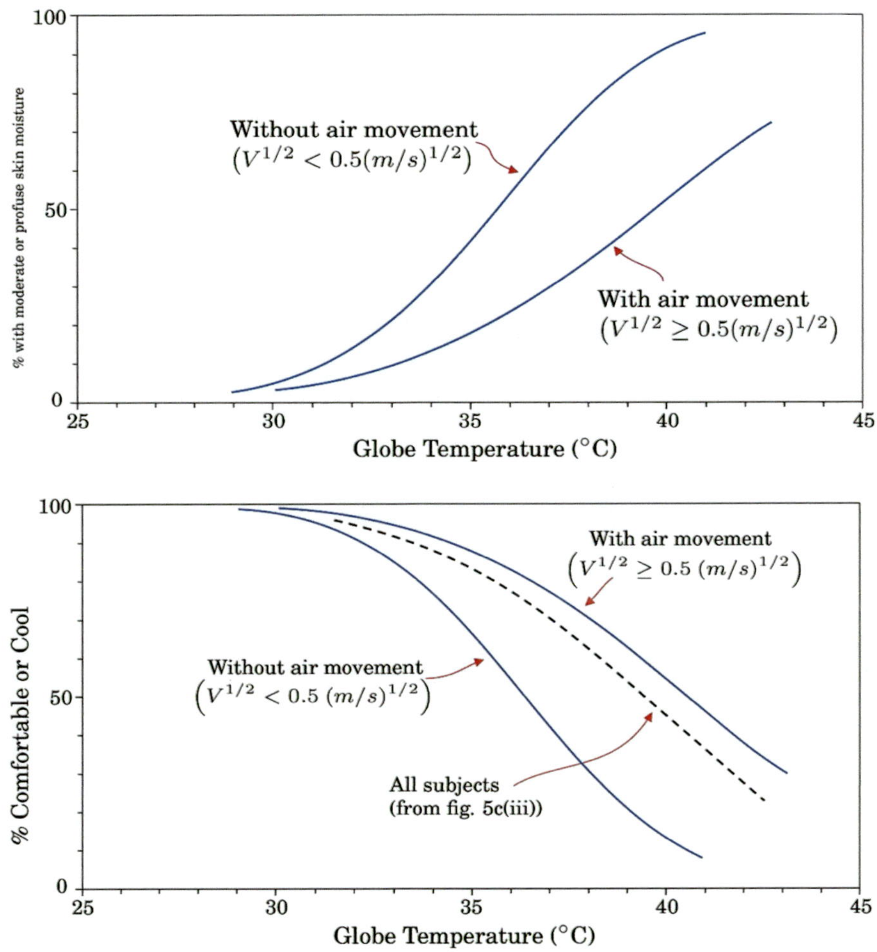

FIGURE 5.9 From field measurements collected by C. G. Webb and reported in Nicol (1974),[16] the data show clearly that (a) between 30°C and 40°C, the effect of air movement over the skin reducing skin moisture (sweat) is significant; in (b) improving comfort. The mid-20th-century data that underpins these data indicates the extent to which populations have become adapted to the lower-temperature conditioning provided by air conditioning.

climates, changes in the sweating processes of the body are one way that locally acclimatised populations can counter rising temperatures to remain comfortable, particularly at temperatures above 30°C. Physiological mechanisms like delays in the beginning of sweating, increases in sweat rates, and the decline of sweat rates after an hour or two due to fatigue in the sweat glands and their controlling mechanisms can all play a role in the acclimatisation of people and populations to higher temperatures.[15]

In the cold, clothing is a critical factor, and in the heat, air movement becomes more important. These are simple underlying principles of the physics and physiology of comfort, but they only go so far in explaining how people make themselves comfortable. Laboratory studies can be used to explain part of the comfort equation but can very seldom be used to validate or explain the outputs of survey studies, except in tightly controlled thermal environments, as in air-conditioned spaces.

How can we then make useful conclusions from comfort surveys that we can apply to the design of low-emission buildings in climates that we are unfamiliar with? One way is to use statistical methods using data from a number of studies to provide a rough approximation of what an appropriate local comfort temperature might be in a particular season, as explained in the next chapter. Another way is to do a detailed adaptive thermal comfort study of comfort responses in a city on which one can base accurate building simulations using real comfort data. To ensure that the outputs of studies by different teams in different locations around the world are scientifically robust, rigour must be instilled in the experimental design of the survey, the collection of data, and the analysis of its findings.

When Do People in Japan Open Windows?

Building on the work of Humphreys and Nicol, Hom Rijal looked at window opening behaviour and thermal comfort over a four-year period in the living rooms and the bedrooms of dwellings in the Kanto region of Japan. He based his conclusions on 36,144 sets of physical and subjective data collected from 243 residents of 120 dwellings. He found that people opened windows differently if the building was being either actively heated or cooled or naturally ventilated by opening the windows in what researchers call 'free-running mode' (FR), when a building is neither heated nor cooled (Figure 5.10).

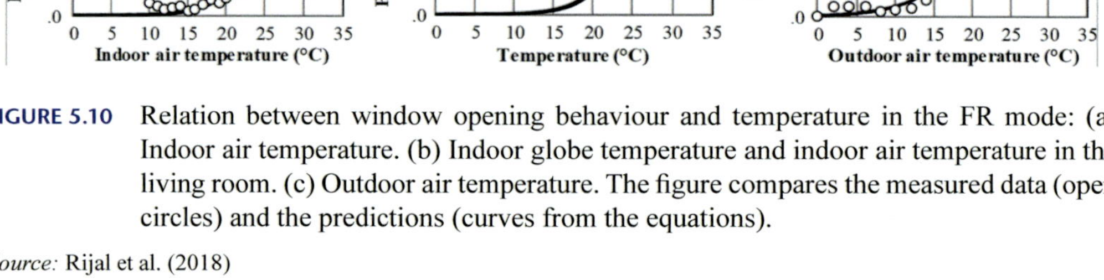

FIGURE 5.10 Relation between window opening behaviour and temperature in the FR mode: (a) Indoor air temperature. (b) Indoor globe temperature and indoor air temperature in the living room. (c) Outdoor air temperature. The figure compares the measured data (open circles) and the predictions (curves from the equations).

Source: Rijal et al. (2018)

The likelihood of a window being opened was found to be dependent on both the indoor and the outdoor air temperatures in the FR mode, when neither heating or cooling are used. Indoor comfort temperature correlated well with the running mean of the outdoor temperature as people were comfortable at higher temperatures as it got warmer. He used logistic regression analysis to develop a model for window opening behaviours in the city that agreed closely with the measured data. Thus, he could develop validated equations to enable window opening and comfort temperature to be predicted from mean values of the outdoor temperature.[17] However, in reality, the wide scatter of data on which those means are based showed that some people opened their windows at indoor temperatures ranging from 5°C to 35°C.

No data exists to show what happens above 35°C from this study, but it is usual when people are feeling too hot at such high temperatures and the outdoor temperature is higher than the indoor temperature that they may close windows to prevent themselves from overheating, as indicated in Figure 5.8.

Windows that may be kept open are those at the tops of buildings that perform the vital task of dumping air out of the upper stories of the building, drawn upwards in the thermal stack of air rising up through the building, driven by buoyancy. This often-powerful engine is effective in drawing cooler air from basements or adjacent cool gardens to ventilate and cool lower rooms in the building that are occupied on hot days, when upper rooms are abandoned as being too hot for comfort. Such air flows could be expertly designed into sophisticated naturally ventilated buildings today and were widely used even in simple vernacular buildings in hot regions for millennia. Different windows can perform very different tasks in the aeolian landscapes of a building.

Temperature Triggers for Comfort Behaviours

During the 1973 oil crisis in Britain, the UK government proposed mandating an upper thermostat temperature for government offices at the Building Research Establishment (BRE) outside London of 17°C. Humphreys, who then headed up the Human Factors Section at the BRE, argued against the proposal because it would require winter outdoor clothing to be worn all day in the offices. Those arguing for the proposal referred to a report they had obtained from BRS that said 17°C would cause no problems. It turned out to be the report written by Humphreys himself! This is a good example of not generalising from particular studies because, in the words of Steven Szokolay when asked about a design challenge, ‘it always all depends’. It is obvious that the temperatures that some adapted populations find perfectly acceptable indoors at a particular time of year may not be acceptable to other populations in different climates, or even the same population in a different building or at a different time of year.

For a country that wants to accurately develop real low-carbon buildings that can run on local natural energy for as much of a year as possible, it is important to know what temperatures are considered acceptable for building occupants locally and at what temperatures are triggers for them to activate environmental controls there. It must be madness to try to apply temperature standards developed in Sweden to inform design models for southern India, and yet that is how it is done for building design, HVAC system sizing, and regulation development today around the world.

It is much better to do climate-region-by-climate-region field studies to accurately define thermal variables and keep building energy wastage to a minimum. That is exactly what Hom Rijal is doing for Japan. In an initial large-scale study of 23 offices that were all naturally ventilated (free running) as well as heated and cooled in Tokyo and Kanagawa, 7,297 comfort votes

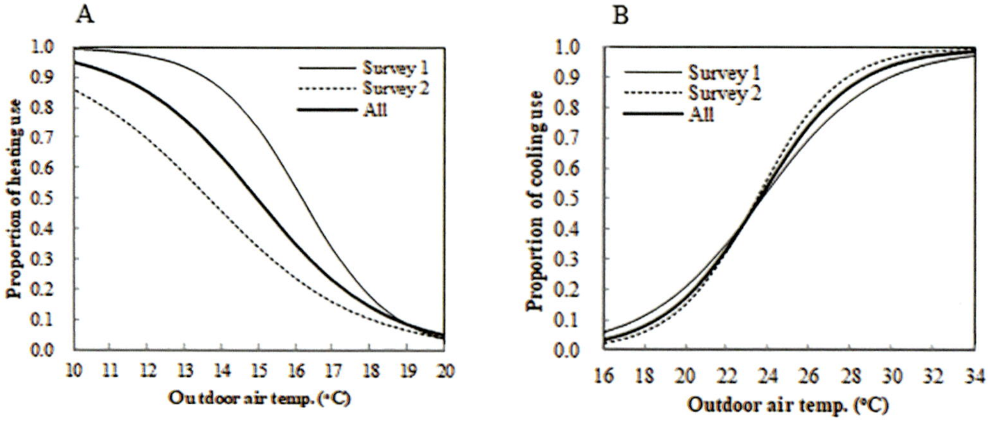

FIGURE 5.11 Relation between heating use or cooling use and outdoor air temperature in a mixed-mode building in Kanagawa: (a) Heating use. (b) Cooling use. Cooling and window opening started to be used when the outdoor temperature rose above about 20°C, and heating when the outdoor temperature dropped below 17°C.

Source: Rijal et al. (2022).

were collected.[18] The occupants reported being highly satisfied with the thermal environment of their offices, in which average comfort temperature was found to be 25.6°C when cooled, 24.3°C when heated, and 24.9°C when neither heating nor cooling were used. The comfort temperature for heating mode is surprisingly high in terms of energy consumed in the office, where the seasonal difference in the mean comfort temperature was only 2.1 K. Contrast this with 9.4 K in the seasonal comfort temperature differences in Japanese dwellings and the impact of who pays for the heating and cooling energy becomes clear.[19]

There is obviously plenty of leeway to permit more seasonal variation in indoor temperature in office buildings in Japan, as was found in dwellings where behavioural adaptations like window opening, clothing adjustments, and heating/cooling use are related to both the outdoor air temperature and the affordability of the energy involved (Figure 5.11). They underlie the adaptive models which are important for resilient thermal comfort and energy saving. Rijal suggests that in offices in those two cities, safe limits could range between 21.3°C and 28.5°C, provided there are enough adaptive opportunities to hand for their occupants.

Again, we see the complexity of factors involved, where some people may open a window for a cooling breeze or to purge the room for health and freshness, to deal with excess indoor humidity, or simply to connect with the world outside for mental refreshment. As with the Pakistan data, around 20°C is an indoor temperature trigger point to start opening windows, and hereby, 35°C indoors, nearly all the windows are open to reduce discomfort by maximising heat loss through sweating. But then again, tell that to the Bavarian housewife who daily throws open the bedroom windows to enhance the healthiness of the indoor environment and who knows a thing or two about deep cleaning from experience. Or to the cleaner at Escudero, the Chilean base in Antarctica, where Roaf worked in 2019 and 2020, who threw open the windows of public rooms regularly as he cleaned and freshened them, despite sub-zero temperatures outside. It is not just comfort issues that drive such behaviours but also those of hygiene, cleanliness, convenience, and health.[20]

Personal Comfort Systems

There are many different ways to enhance local environments around people with and without involving moving air. The University of California has done much work on what they call personal comfort systems (PCSs). These they define as thermal systems that heat and cool individuals without affecting the environments surrounding occupants, and that are under the individual's control. They tend to be devices locally positioned on, or incorporated in, workstation furniture, such as chairs, desktops, and near the feet and legs (Figure 5.12). Some are readily available, such as small DC-powered desk fans, while some are mostly still in the prototype stage, like heated and cooled chairs or efficient foot and bed warmers. Anything that can be turned on or off locally and moved to be as near as suits a person is a PCS. Passive adaptive opportunities may also overlap somewhat on PCS, such as relaxing office dress codes and providing access to operable windows. PCSs are particularly important in warmer climates and where people simply cannot afford full air conditioning systems, and are also very popular as they respond to the needs of individuals and produce welcome feelings of thermal relief.[21]

Turning on a mechanical heating or cooling system in a room may not produce an appreciable sensation of alliesthesia, or a thermal pleasure buzz, due to its slow rate of changing the room temperature, although walking into a conditioned space from colder or hotter outdoors may initially do so. By contrast, in an office situation, a PCS like a foot warmer, heated seat cushion, neck warmer, contact hand warmer, and radiant hand warmers were reported as producing immediate, thermally pleasant sensations across the whole body, and then in local body parts being warmed. Such systems have huge potential to lower energy consumption in a building by using lower or higher background temperatures across a space and only using PCSs to heat and cool those who feel the need for them. Every individual will have their own trigger temperatures

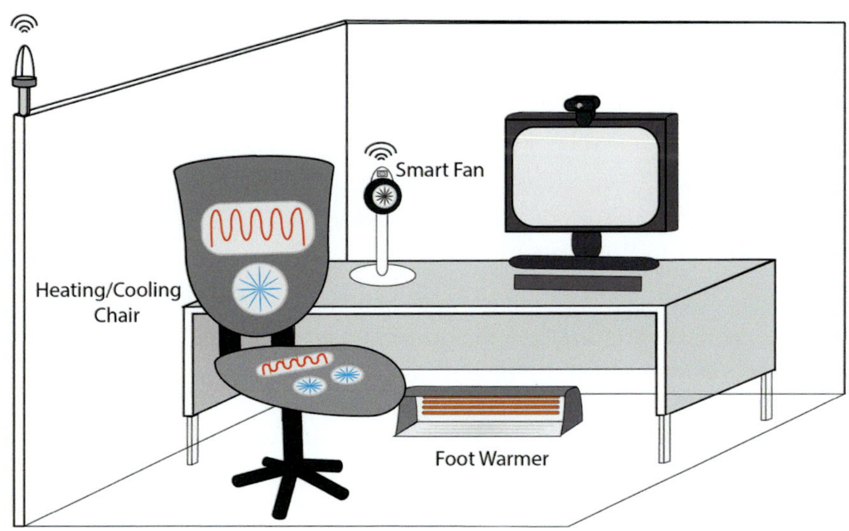

FIGURE 5.12 Personal comfort systems (PCSs), such as those designed for the California Office Environment here, are increasingly seen as enhancing individuals' personal comfort preferences while enabling low-energy use in occupied spaces of buildings.

Source: Ghahramani et al. (2020).[22]

to act to improve their personal comfort, but there will also be a tendency to adapt to the ambient room temperatures with clothing and activity as well.[23]

Multi-Modal Comfort in Buildings: Hardware

PCSs give occupants control of their own thermal environments, potentially enhancing their satisfaction with them. What is significant in a world of increasing extreme weather events and related power outages is that if a building is designed in such a way that building occupants retain the competence and information needed to change their surroundings to become or stay comfortable, then not only are they likely to be safer in that building, but they will also feel so.[24] People like to be in control of their own environments,[25] and there can be enormous energy, cost, and emissions advantages from them being so. From more pleasurable environments to safer and more resilient buildings, the availability and usability of local environmental controls, including opening windows and the use of PCS, point the way to a new thinking on the conditioning of buildings.[26] Already, the rise of adaptive thermal comfort standards, codes, and related regulations applied in larger buildings has enabled the shift towards mixed-mode buildings that can be naturally ventilated for as much of the year as possible and heated and cooled only when necessary.

The move from the full mechanical heating and cooling of buildings to enabling people to make and keep themselves comfortable in a variety of different space-specific ways requires a new type of building. Hybrid buildings, or mixed-mode buildings, allow occupants and management to run the building with and without mechanical systems. This requires the usable and accessible hardware necessary for occupants to action the adaptive opportunities that enable them to operate comfort systems flexibly as required by changing conditions in local microclimates. Comfort systems in use may vary, as shown in Table 5.1, in buildings that are:

1. Fully heated and cooled when necessary and affordable
2. Heated and cooled locally as required and desired with natural ventilation, passive adaptive hardware, and PCSs where useful
3. Naturally ventilated with passive adaptive fixtures only employed
4. Naturally ventilated only

Multi-Modal Comfort in Buildings: Behaviours

The actual systems used locally in a mixed-mode building will depend on the thermal design of the building, its comfort hardware, and the personal temperature preferences and locations of occupants. Workers in the summer in a south-facing office may prefer to use the external shading with the windows opened and fans on, while in the north-facing office, open windows may be sufficient for comfort.[27] Multi-modal buildings can benefit from both fitted and portable equipment to modify local conditions because it is almost impossible to predict what certain people will require in terms of comfort in a particular weather, day, or season. What one can do is set general background to avoid thermal discomfort for the majority in a space, but even that is often best left to local operative to determine, as influenced by many different factors, and within safe limits. Many readers will, to a greater or lesser extent, agree or disagree with the temperature thresholds proposed in Table 5.2, but this only emphasises how different all our understandings are of what comfort temperatures might be under different environmental conditions.

TABLE 5.1 How people adapt in buildings depends on the available usable hardware at hand that enables them to either operate spaces in a fully HVAC mode or flexibly in a variety of different comfort modes using both passive and active systems

HVAC BUILDINGS	MIXED MODE
MECHANICALLY COOLED	FULL MECHANICAL COOLING PERSONAL COMFORT SYSTEMS PASSIVE ADAPTIVE ACTIONS SHUTTERS, SHADES, AWNINGS NATURAL VENTILATION
MECHANICALLY HEATED	NATURAL VENTILATION PASSIVE ADAPTIVE ACTIONS CURTAINS. BLINDS, CLOTHING PERSONAL COMFORT SYSTEMS HEATERS, HEATED FURNITURE FULL MECHANICAL HEATING

Key: High Energy · Medium Energy · Low Energy · No Energy

Source: Susan Roaf.

People can be comfortable in a very wide range of temperatures, as becomes clearer in Chapter 6. Comfort is also a seasonally dynamic experience. An air temperature at the end of a hot summer may feel very different to the same temperature at the end of a cold winter, when the once sun-warmed walls of a building have chilled, providing a very different radiant landscape in the space. Given accessible and usable adaptive hardware in the thermal fixtures and fitting of a building and local access to heaters, cooler, and fans, people can 'sail' the thermal conditions of the spaces they occupy. They may open or close windows according to weather and winds, change clothing or location in a space, and use just enough local heat and coolth to become and remain comfortable according to their own personal preferences and how much effort and money they want to throw at the task.

Multi-modal buildings require multi-modal occupants and management to enable available opportunities to be actioned so people can adapt their own buildings to be comfortable. Clever design will mean that this is achieved without wasting energy but, rather, radically reducing its use for keeping occupants thermally comfortable.

Looking at data from the actual temperatures occupied by real people in real buildings around the world in Chapter 6 shows the very strong influence of climate on acceptable indoor thermal conditions. This can be explained in part by the extraordinary range of different ways in which people adapt to the myriad thermal conditions people normally occupy across very different cultures over time, as outlined in the following chapters.

TABLE 5.2 Multi-modal buildings can enable occupants to tailor local temperatures to their own comfort requirements if the adaptive comfort hardware is available to them. The actual temperatures that might trigger such model shifts can only be guessed at, as is done here based on global comfort studies

Temperature from	Temperature to	HYBRID STRATEGY
VERY COLD	24^0C (75^{0}F)	FULL BUILDING HEATING
VERY COLD	24^0C (75^{0}F)	HEATING IN OCCUPIED ROOM ONLY
VERY COLD	24^0C (75^{0}F)	LOCAL HEATING OF PERSON WITH PCS
10^0C (35^{0}F)	35^0C (95^{0}F)	NATURAL VENTILATION
24^0C (75^{0}F)	35^0C (95^{0}F)	NATURAL VENTILATION + PERSONAL FAN
24^0C (75^{0}F)	35^0C (95^{0}F)	NATURAL VENTILATION + ROOM FAN
24^0C (75^{0}F)	35^0C (95^{0}F)c	ROOM FAN ONLY
26^0C (80^{0}F)	50^0C+ (122^{0}F+)	PERSONAL / LOCAL OR ROOM FAN + AC
26^0C (80^{0}F)	50^0C+ (122^{0}F+)	LOCAL AC IN OCCUPIED ROOM ONLY
26^0C (80^{0}F)	50^0C+ (122^{0}F+)	FULL BUILDING AIR-CONDITIONING

Key: High Energy Medium Energy

Low Energy No Energy

Source: Susan Roaf.

Notes

1 Webb, C.G. (1959). An analysis of some observations of thermal comfort in an equatorial climate, *British Journal of Industrial Medicine*, 16(3), pp. 297–310.

2 Nicol, F.J. (1974). An analysis of some observations of thermal comfort in Roorkee, India, and Baghdad, Iraq, *Annals of Human Biology*, 1(4), pp. 411–426.

3 Webb, C.G. (1959). An analysis of some observations of thermal comfort in an equatorial climate, *British Journal of Industrial Medicine*, 16(3), pp. 297–310.

4 Nicol, F.J. (1974). An analysis of some observations of thermal comfort in Roorkee, India, and Baghdad, Iraq, *Annals of Human Biology*, 1(4), pp. 411–426.

5 Humphreys, M.A., F. Nicol and S. Roaf (2016). *Adaptive Thermal Comfort: Foundations and Analysis*, London: Earthscan, pp. 260–270.

6 Humphreys, M.A., and J.F. Nicol (1971). Theoretical and practical aspects of thermal comfort, *Building Research Station Current Paper*, CP14/71.

7 Humphreys, M.A. (1970). A simple theoretical derivation of thermal comfort conditions, *Journal of the Institute of Heating and Ventilating Engineers*, 38, pp. 95–98.

8 Olesen, B.W., and K.C. Parsons (2002). Introduction to thermal comfort standards and to the proposed new version of EN ISO 7730, *Energy and Buildings*, 34(6), pp. 537–548.

9 Drury, P., S. Watson and K. Lomas (2021). Summertime overheating in UK homes: Is there a safe haven?, *Buildings and Cities*, 2(1), pp. 970–990. https://doi.org/10.5334/bc.152

10 Detailed explanations of the development and methodologies of adaptive thermal comfort can be found in volumes 1 and 2 of this trilogy on the subject by the present authors.

11 Humphreys, M.A. (1975). Field studies of thermal comfort compared and applied, *Building Research Establishment Current Paper*, Garston, UK.

12 Kurvers, S., and J. Leyten (2022). Indoor climate and adaptive thermal comfort, Published by Stichting Kennisbank Bouwfysica/Delft Digital Press.

13 Havenith, G. (2002). Interaction of clothing and thermoregulation, *Exogenus Dermatology*, 1(5), pp. 221–230. https://doi.org/10.1159/000068802

14 The methodologies and related publications for the Pakistan project are described in Humphreys et al. (2016).

15 Parsons, K. (2003). *Human Thermal Environments*, 2nd Edition, London: Taylor and Francis.

16 Nicol, J.F. (1974). An analysis of some observations of thermal comfort in Roorkee, India, and Baghdad, Iraq, *Annals of Human Biology*, 1(4), pp. 411–426.
17 Rijal, H.B., M.A. Humphreys and J.F. Nicol (2018). Development of a window opening algorithm based on adaptive thermal comfort to predict occupant behavior in Japanese dwellings, *Japan Architectural Review*, 1(3), pp. 310–321. https://doi.org/10.1002/2475-8876.12043
18 Rijal, H.B., M.A. Humphreys and F. Nicol (2022). Adaptive approaches to enhancing resilient thermal comfort in Japanese offices, in *The Routledge Handbook of Resilient Thermal Comfort*, Edited by F. Nicol, H. Bahadur Rijal and S. Roaf, London: Routledge.
19 Rijal, H.B., M.A. Humphreys and J.F. Nicol (2019a). Adaptive model and the adaptive mechanisms for thermal comfort in Japanese dwellings, *Energy and Buildings*, 202. https://doi.org/10.1016/j.enbuild.2019.109371
20 Shove, E. (2003). *Comfort, Cleanliness and Convenience*, Oxford: Berg Publishers.
21 Rawal, R., M. Schweiker, O.B. Kazanci, et al. (2020). Personal comfort systems: A review on comfort, energy, and economics, *Energy and Buildings*, 214, p. 109858.
22 Ghahramani, A., P. Galicia, D. Lehrer, Z. Varghese, Z. Wang and Y. Pandit (2020). Artificial intelligence for efficient thermal comfort systems: Requirements, current applications and future directions, *Frontiers in Built Environment*, 6(49). https://doi.org/10.3389./fbuil.2020.00049
23 He, Y., T. Parkinson, E. Arens, H. Zhang, N. Li, J. Elson and C. Maranville (2021). Creating alliesthesia in cool environments using personal comfort systems, *Building and Environment*, 209, p. 108642. https://doi.org/10.1016/j.buildenv.2021.108642
24 Heschong, L., and J. Day (2022). Why occupants need a role in building operation: A framework for resilient design, in *The Routledge Handbook of Resilient Thermal Comfort*, Edited by F. Nicol, H. Bahadur Rijal and S. Roaf, London: Routledge, pp. 39–52.
25 Day, J. (2022). Tools and rules for behavioural agency in buildings: Minimising energy use while maintaining comfort, in *The Routledge Handbook of Resilient Thermal Comfort*, Edited by F. Nicol, H. Bahadur Rijal and S. Roaf, London: Routledge, pp. 312–329.
26 Brager, G., and L. Baker (2009). Occupant satisfaction in mixed-mode buildings, *Building Research & Information*, 37(4), pp. 369–380. https://doi.org/10.1080/09613210902899785
27 Parkinson, T., R. de Dear and G. Brager (2020). Nudging the adaptive thermal comfort model, *Energy and Buildings*, 206. https://doi.org/10.1016/j.enbuild.2019.109559

6

COMFORT CLOUDS

Introduction: Finding a Shape for Comfort

The 20th-century engineering approach to deciding how people will feel about their physical environment concentrated on creating a single 'comfort index' which brings together different factors, such as temperature, humidity, and the movement of the air, all of which are taken to be the main contributors to subjective comfort. But the relationship between people and their surroundings is usually more complex than is assumed by comfort indices such as PMV, the Predicted Mean Vote index favoured by the air-conditioning industry.[1] People do not interact passively with their thermal environment. They respond dynamically to their surroundings, and if they find themselves in an uncomfortable or thermally dangerous situation, they try to find ways to make themselves more comfortable and safer. Their responses to the thermal conditions are not random but directed to this essential goal.

To help understand the way an internal climate actually feels to people inhabiting a particular building, an approach used in comfort research is the **comfort survey**. The key elements in this approach are illustrated in Figure 6.1 from a study in Japan by Hom Rijal.[2] A variety of physical and environmental factors are observed, measured, and noted down. In a longitudinal survey, the data are collected on several occasions, preferably over a period of a year, from a small number of people so that the full variety of weather conditions experienced by the building and its occupants will be covered. The outdoor conditions are also of interest and can be measured on-site or from a nearby meteorological site. The data collected can be analysed using statistical and other scientific techniques. In a transverse survey, a larger number of people are polled just once within a limited survey period.

Earlier chapters deal with different aspects of the thermal relationship between people and the buildings they inhabit using knowledge of physiology, psychology, behaviour, and building physics. Another driving force in the thermal comfort of occupants is the relationship between indoor and outdoor temperatures.

To design a comfortable building, an architect or engineer needs to know how the *building itself*, our 'third skin', performs thermally in relation to the outdoor climate before any mechanical or passive appliances are added to enable people indoors to be more comfortable. To find out how good the building is at modifying the outdoor climate and weather, Nicol suggested in 2019

DOI: 10.1201/9781315645070-9

FIGURE 6.1 The conduct of a comfort survey in Japan involves the collection of a wide variety of data about the response of the subject as well as their physical environment and their sensory and physiological response to it.

that the physical relationship between them could be made visible by plotting the temperatures of the indoor environment against the outdoor environment at the same time[3] in what he called 'temperature clouds'.

Temperature Clouds for Buildings With and Without Controls

Figure 6.2 is a graph showing the indoor temperature (vertical axis) plotted against the outdoor temperature at times when the subject is being questioned by the researcher. The data illustrated is taken from a number of office buildings in Southern France, taken from the Pan-European SCATS comfort study.[4]

Figure 6.2 shows a temperature cloud from France with two distinct groups of data: a summer group of about above 12°C outdoor air temperature and a winter group below 12°C outdoor air temperature. In the winter, indoor temperatures are all greater than the outdoor temperature, suggesting that the buildings all needed some form of winter heating.[5]

Such a cluster of points with a definite but not fixed shape can be called a 'temperature cloud'. The general 'thickness' of the winter cluster of indoor temperatures lies between about 20°C and 26°C, with occasional excursions to lower and higher temperatures. In winter, different occupants may prefer different temperatures, or different spaces may require different levels of heating, both factors affecting temperature control settings.

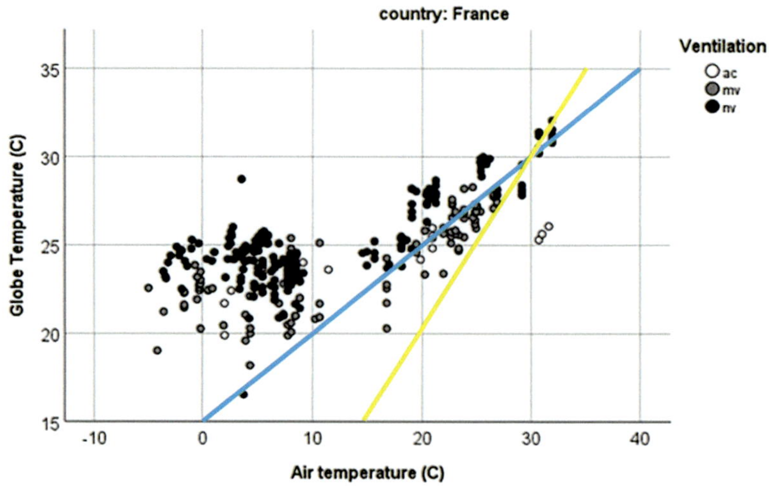

FIGURE 6.2 Showing the change of indoor globe temperature with concurrent outdoor air temperature in a group of buildings in Southern France. The yellow line shows when the indoor temperature is equal to that outdoors; the blue line shows the approximate line of slope for the summer cloud. Shading of the points on the graph indicates if the buildings is fully air-conditioned (*ac*), mechanically ventilated (*mv*), or naturally ventilated without mechanical ventilation (*NV*).

Source: Fergus Nicol.

The summer cloud is typical of naturally ventilated buildings without conditioning (heating or cooling), with the exception of three readings from one air-conditioned building, which cooled down to 25°C with outdoor temperatures of around 30°C. Generally, as outdoor temperatures increase, so do those indoors.

The yellow line shows when the indoor and the outdoor temperatures are the same. Some poorly performing buildings fall below that line, showing that the summer indoor temperatures are actually hotter than those outdoors, and vice versa. In this French survey, the indoor temperatures change by roughly half a degree for every degree change of outdoor temperature in buildings without conditioning. Many European buildings are heated in winter but not mechanically cooled in summer in the way demonstrated by this survey.

Temperature Clouds for Buildings With Year-Round Controls

Figure 6.3 shows such a 'cloud' of measurements in buildings which are both heated in the winter and cooled in the summer (fully air-conditioned). This data was collected at monthly intervals in five non-domestic buildings in the Gothenburg/Malmo region of south-west Sweden during the SCATS multinational comfort survey in the 1990s.[6] Each point on the graph presents the (globe) temperature measured near a survey subject plotted against the outdoor temperature at the same time from local meteorological data.

The limits of measured indoor temperatures in buildings with temperature controls typically range from 10°C at the lower end to 35°C as the upper limit for the wide variety of buildings and climates measured in different parts of the world, shown in Figure 6.4.[7] The temperature clouds

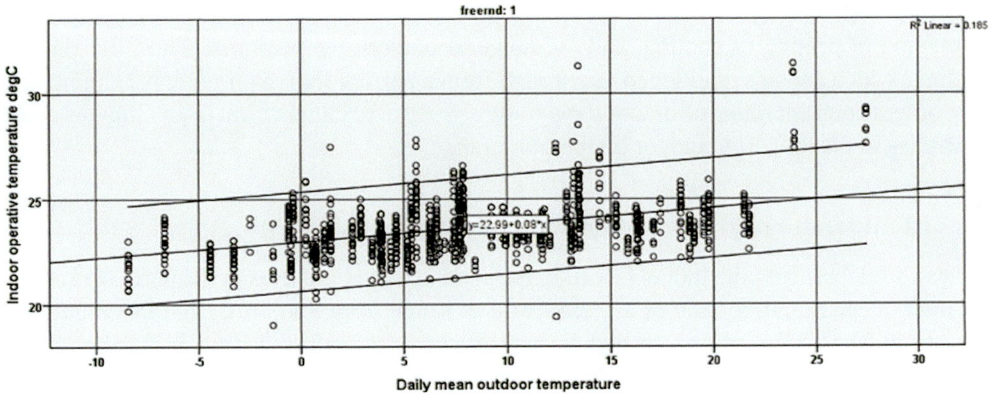

FIGURE 6.3 If we draw an approximate line around the points in the Swedish cloud, the top and bottom 95% lines are roughly parallel and about 4.5°C apart, with a low slope of about 0.08, meaning that for every change of 1°C in the outdoor temperature, there will be, on average, a change in indoor temperature of only 0.08°C.

Source: Fergus Nicol.

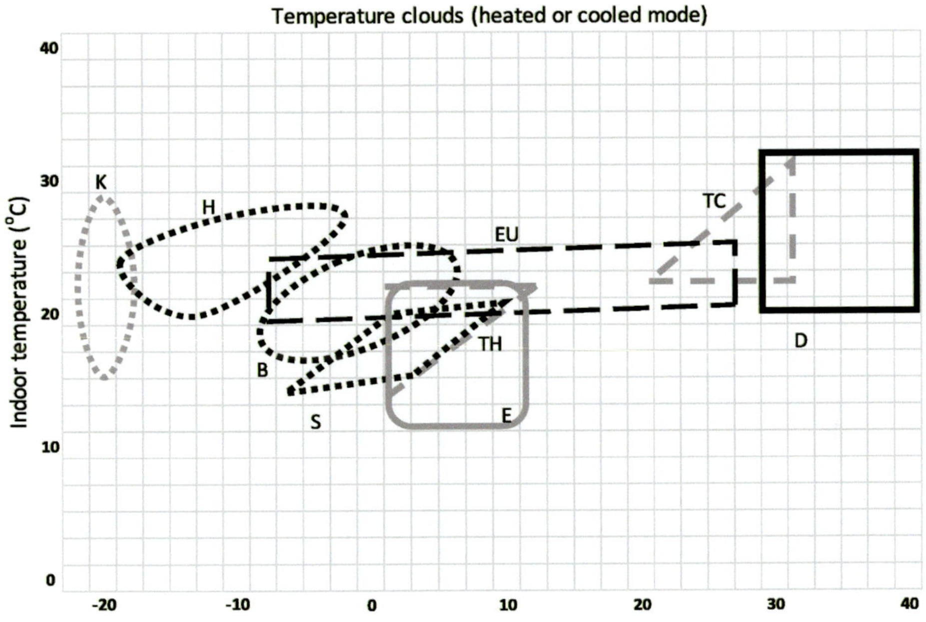

FIGURE 6.4 Temperature cloud outlines for a number of buildings in comfort surveys and which were conditioned (heated or cooled) at the time of the surveys throughout the world. **Key:** Black-dotted: China (Harbin (H), Beijing (B), Shanghai (S) in winter). Dotted grey: Russia (Khabarovsk winter (K)). Grey line: England (E). Dashed grey: Japan (Tokyo; (TH) heated, (TC) cooled). Black line: Saudi Arabia (Dammam (D)). Black dash: European offices.

Source: Nicol et al. (2020).

come in a wide variety of shapes, depending on the physical characteristics of the buildings, the method of heating or cooling, and the range of outdoor temperatures. Thus, the European buildings which are air-conditioned year-round form a narrow slab (c.f. Figure 6.2), indicating a more or less constant range of indoor temperature. Such data clouds from most climates overlap sometimes in the low 20°C indoor temperature range.

Indoor Temperatures in Unconditioned Buildings in Pakistan

Pakistan stretches from the Indian Ocean in the south to the Himalayas in the north (Figure 6.5). In winter, mean monthly outdoor air temperatures range from 5 to 18°C, and in summer, they range from 27 to 35°C and much higher in recent years. Comfort surveys in five cities in the different climate zones of the country were undertaken between 1993 and 1995.[8,9]

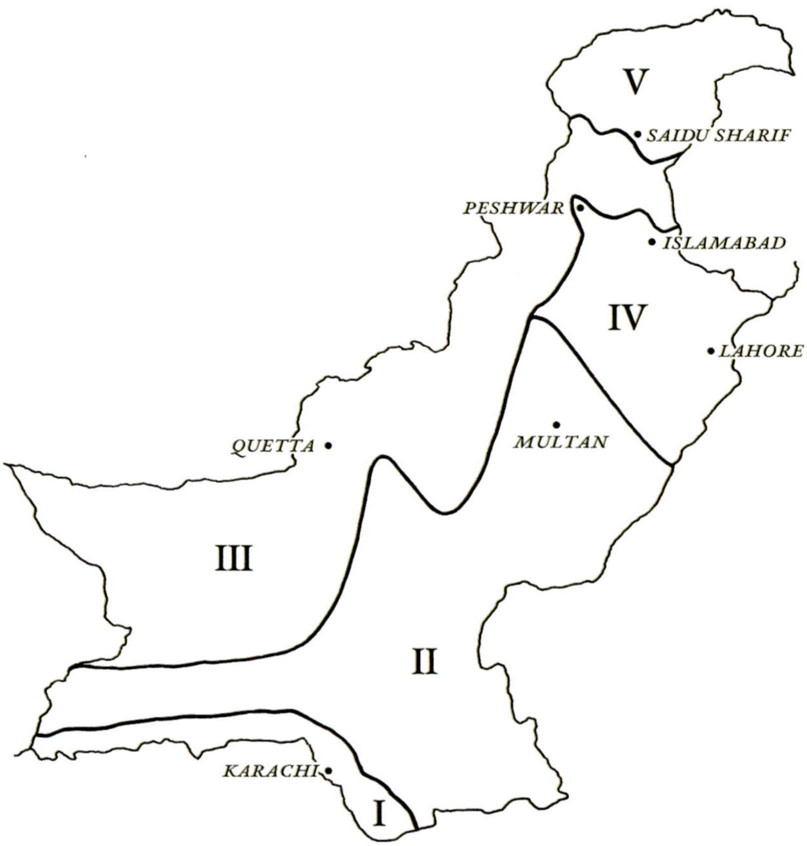

FIGURE 6.5 (a) Map of Pakistan indicating the positions and climate types of the cities included in the surveys. **KEY:** HOT, 32°C or more; WARM, 21–31°C; MILD, 10–20°C; COOL, 0–9°C; COLD, below 0°C. **CLIMATE TYPES:** (I) warm summer + mild winter; (II) hot summer + mild winter; (III) warm summer + cool winter; (IV) hot summer + mild winter; (V) mild/warm summer + cool/cold winter.

Source: Susan Roaf.

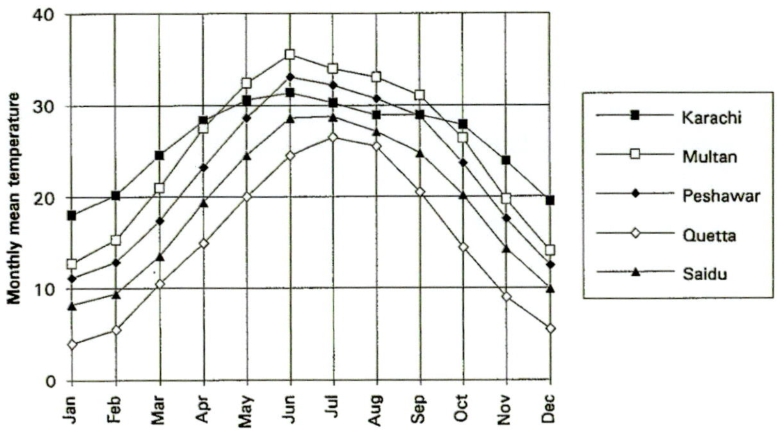

FIGURE 6.6 The historic temperatures of five cities in different regions of Pakistan by month.

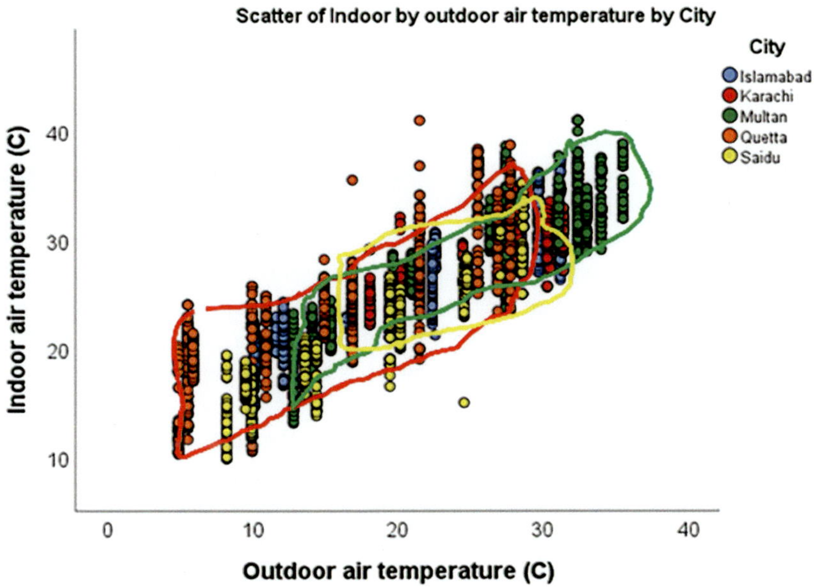

FIGURE 6.7 Coloured outlines indicate the approximate comfort clouds for Multan (hot desert), green; Quetta (cold desert), orange; and Karachi (Coastal area), yellow. Note that the clouds are almost parallel but have limits, reflecting the climate of the cities.

Source: Fergus Nicol.

Figure 6.6 shows the daily mean indoor air temperatures recorded in offices in five cities. Instantaneous outdoor temperature was not recorded by the local met stations, so the daily mean outdoor temperature was collected from a local weather station. This accounts for the vertical data columns in Figure 6.6. as they were all recorded with the same outdoor temperature.

This figure gives an indication of the extreme range of indoor temperatures occupied in offices in different climates in one country over one year. There are urgent calls for more 'resilient' buildings to be required and tested at planning and building permission stages by modelling them as 'HVAC-Off' functionality, to see if a building can still be occupied when the grid energy supplies fail.[10] Even in the extremes of Pakistan's climates, buildings can provide adequate thermal protection in typical weather conditions. This graph shows clearly the measured buildings that work best in both the heat and the cold in a particular city.[11]

The Geography and Economics of Comfort in Pakistan

Climate proved to be the major influence on comfort in the Pakistan study, so as geography influences climate, comfort is directly related to the morphologies of Pakistani regions. The hottest city, Multan (122 m altitude), in the centre of the country, has a continental desert climate that heats up very quickly in spring (32°C), gets very hot in summer (44°C), and cools down to low temperatures in winter (13°C). The coldest city, on a high plateau in the west, is Quetta (1,679 m altitude), with the coldest mean outdoor temperatures all year. At two-thirds Quetta's elevation, Saidu Sharif is not far behind at 970 m in the foothills, leading up to the great Hindu Kush and the Himalayas (Figure 6.6).

Buildings in Islamabad (620 m elevation), the modern capital of the country, appear to be better built than elsewhere, judging from the narrower range of temperatures recorded at any time of year in them. Contrast that with the poorer cities of Multan and Quetta, having a much larger range of indoor temperatures. Karachi sits at an altitude of only 10 m above sea level and has seasonal mean temperatures in winter of 18°C, in spring of 31°C, with the summer mean a little different at 32°C. Its proximity to the moderating influence of the Indian Ocean is reflected in the annual temperature range of only 14°C.

The comfort clouds of these cities highlight the impact of the geographical location of each on the ranges of the mean outdoor and indoor temperatures experienced in and around them. Approximate clouds from surveys in Quetta, Karachi, and Multan are shown in Figure 6.7. Note that although there are variations in the depth (indoor air temperatures) of the clouds, their slope is remarkably constant, excepting Karachi, which has a very different climate.

This characteristic slope of the data of buildings which are not heated or cooled is found throughout the world, as is shown in Figure 6.8 from Nicol et al.'s 2020 paper, which compares the indoor temperature clouds from seven very different climates and cultures. Note the similarity of the slopes of the temperature clouds in a number of countries where, for at least part of the year, they resemble the slopes found in summer data from the French offices in Figure 6.2.

Any relationship between indoor and outdoor temperatures in buildings without conditioning is caused by similarities in the heat flows through the structure of the building, arising from the materials and geometry, for instance, the size and orientation of the windows, and from the use of passive controls by building occupants.

Costs of Comfort and Discomfort in Conditioned Buildings

Figure 6.3 shows a temperature cloud from five Swedish buildings which are air-conditioned throughout the year. Here, the indoor temperature is kept within a limited temperature range. The contrast with other buildings shown in Figure 6.4 is remarkable. The mean indoor temperature in Figure 6.3 is 23.3°C (standard deviation of ±1.1). Despite the narrow indoor temperature range, the overall frequency of discomfort votes is 6.5% (2.8% from cold and 3.7% from heat), indicating that whilst the free-running Pakistani office workers are subjected to many times the

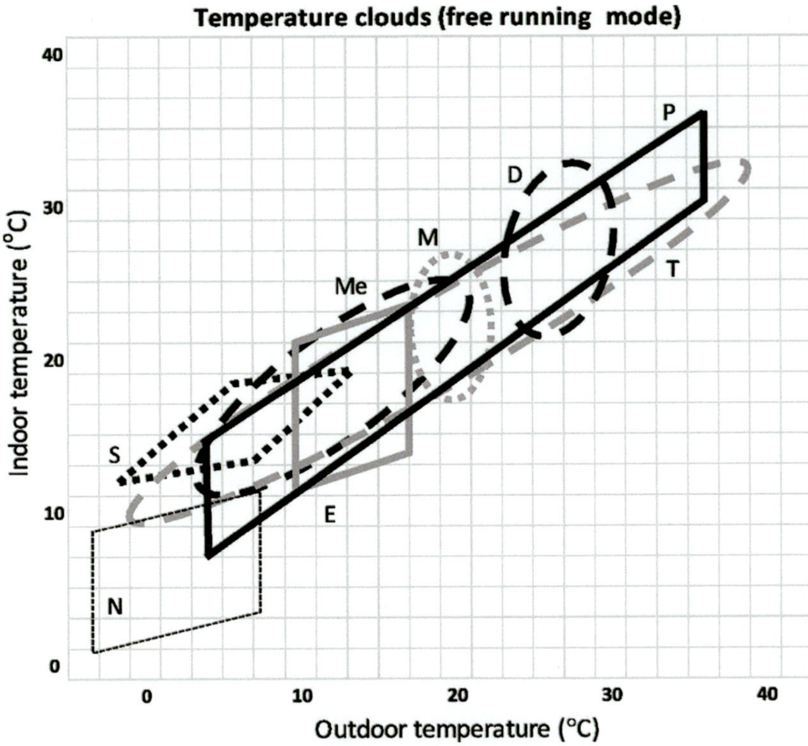

FIGURE 6.8 Temperature clouds from naturally ventilated buildings without heating or cooling at the time of the survey from comfort surveys throughout the world. **Key:** Dotted black, China (Shanghai (s)); dotted grey, Russia (Moscow summer (M)); grey line (England summer (E)); dashed grey, Japan (Tokyo (T)); dashed black, Australia (Melbourne (M), Darwin (D)); black, Pakistan (P); narrow black dashed, Nepal (N).

Source: Hom Rijal.

indoor temperature range of the Swedish, the frequency of discomfort in the Pakistani data, at about 20%, is just about the level considered acceptable by ASHRAE for NV buildings. Additionally, of course, the Swedish buildings require considerable amounts of energy to keep the temperature more constant throughout the year, whereas the Pakistani buildings use relatively little.

The potential to control temperatures of the heating, cooling, and probably ventilation systems of conditioned buildings is quite clearly far greater than in passively heated, cooled, or ventilated buildings. Thermostatically controlled systems enable temperatures indoors to be kept within levels deemed acceptable by the system operators rather than the building occupants themselves. Narrow indoor temperatures could be theoretically maintained even when the range of outdoor temperatures around the buildings was much wider. In Figure 6.4, outdoor temperatures in conditioned buildings range from −25°C to +40°C, compared to about −5°C to +35°C in the unconditioned buildings in Figure 6.8. The range of accepted indoor temperatures is almost the same in both circumstances, at about 10°C to 35°C. These figures highlight the difficult truth that when designers rely on mechanical systems to maintain adequately comfortable temperatures indoors, the need to design to minimise temperature swings may seem less urgent, so energy profligate buildings can seem to be acceptable. But this is at a cost which increases as climate change increases the frequency of extreme weather.

Other Cloud Formations

In most comfort surveys, the data set will contain information which is far more diverse than just the indoor and outdoor temperatures. In effect, the points on the temperature cloud will be multidimensional, including measurements of humidity, air movement, and so on, as well as measurements of the sensual responses of the subjects and their use of various passive and mechanical thermal controls, as shown in Figure 6.9. These dimensions can often be used to shed light on the meaning of relatively simple temperature measures.

Humidity and Air Movement Clouds at Higher Temperatures

In warmer climates, apart from temperature, two key climatic factors have a major impact in comfort: humidity and air movement. By exploring data from thermal comfort surveys in two-dimensional clouds, some answers can be provided about how real people respond to the indoor environments they occupy in different climates, as opposed to results from the artificial conditions set up in laboratory experiments.

For example, in Figure 6.9, the perceived levels of skin moisture or sweating recorded in the Pakistani surveys in the 1990s are indicated by the darkness of the markers, with the darker the marker, the more extreme the skin moisture. Sweating is not only a physiological response to the environment but also triggers a person's awareness of their own response to local conditions, in turn informing their feelings of comfort.

Looking at the results from such a field survey can help pinpoint a set of indoor conditions which locals will find comfortable. The clouds in Figure 6.9 can help answer questions like:

- How does the subjects' clothing vary with temperature or humidity?
- What part does humidity play in how people feel in hot climates?
- What is the interaction between temperature and air movement?
- How does the subjects' skin moisture vary with temperature and humidity?

Figure 6.9a shows that people tended to wear less clothing as humidity rose, and skin moisture was highest when they were wearing fewer clothes. In the Pakistan surveys, water vapour pressure was typically higher at higher temperatures. Figures 6.9c and d show clearly that moderate or profuse sweating rarely occurred at temperatures below 30°C and that temperature clearly related to both how much clothing people wore and the humidity water vapour pressure. Figure 6.9d shows the importance of high levels of air movement at temperatures above 30°C, emphasising the importance of fans to help counteract moderate or profuse sweating.

Heating in Cooler Climates

Figure 6.10 shows data from UK offices indicating the extent of heating used.[12] A wide range of indoor temperatures are recorded in heated offices, as demonstrated in this survey of supposedly temperature-controlled offices done in 1996–1997. Heating systems were on in almost all rooms when outdoor temperatures fell below 15°C, and those indoor temperatures were largely

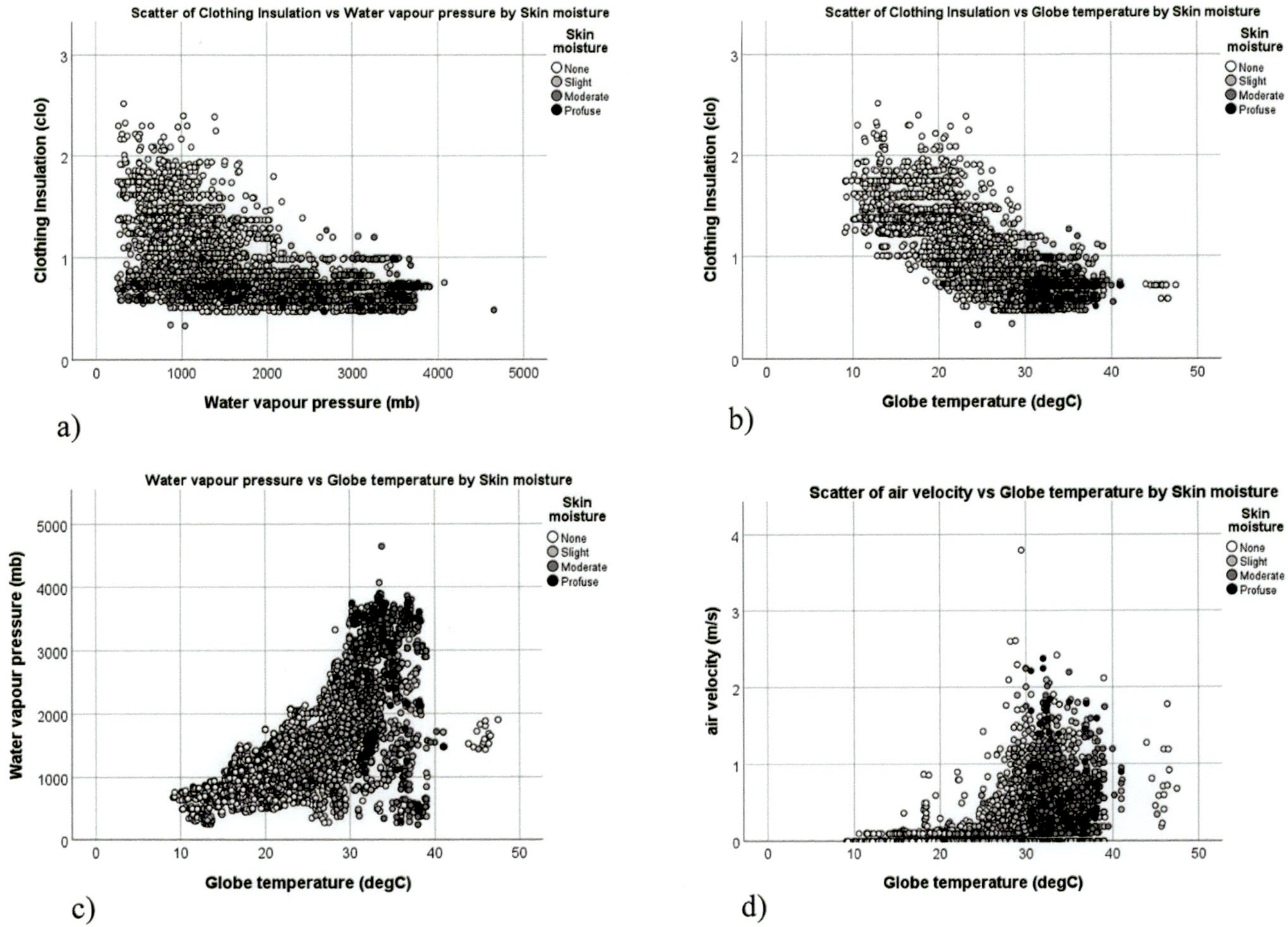

FIGURE 6.9 Two-dimensional clouds between water vapour pressure or globe temperature and clothing insulation and air velocity in the Pakistan surveys. All shown as function of skin moisture (sweating) reported by the subjects on the scale of 'none', 'slight', 'moderate', and 'profuse'.

Source: Fergus Nicol.

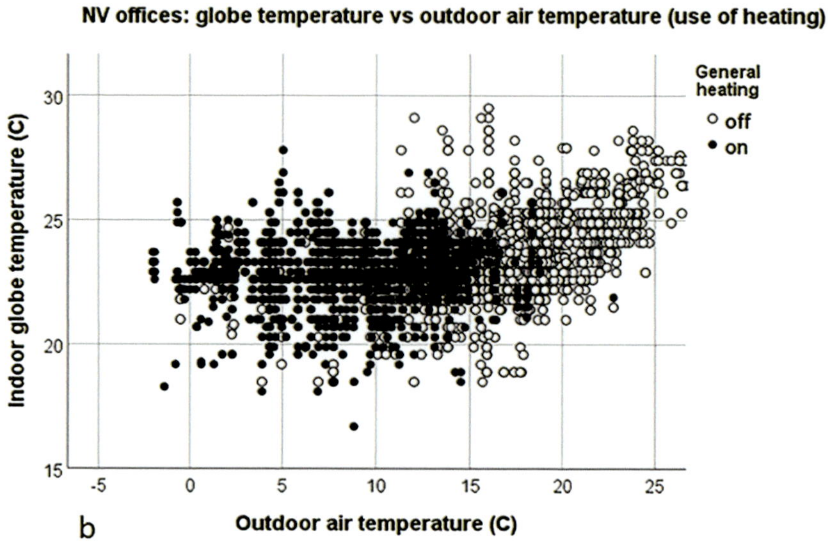

FIGURE 6.10 Data from surveys in offices in the UK showing clearly that heating is used (filled circles) over a wide range of outdoor temperature between −2°C and 17°C and indoor temperatures in heated buildings ranging from around 17°C to 28°C.

Source: Fergus Nicol.

between 20°C and 25°C, but some offices kept temperatures as low as 17°C, while others rose as high as 27°C and some offices were unheated when the indoor temperature was around 18°C.[13]

This again is now historic data, but it should also be seen in light of a British culture that has developed many different strategies for staying warm in cooler buildings, breeding skills that can be transferred between the home and the office.[14] The push to increasingly mechanical comfort in buildings has inevitably resulted in the de-skilling of the population in how to remain comfortable at lower temperatures.

Using Clouds to Explore Resilience

An important measure of resilience is the extent to which a building can protect its occupants in thermally dangerous situations caused by a breakdown in the normal functioning of its systems and energy supplies. When 'the machine stops',[15] the poor climatic design of buildings can result in catastrophic situations. On 24 August 2023, a near countrywide blackout affected Sharjah, one of the richest countries in the Gulf. Residents soon became extremely distressed:[16]

> The entire building is completely dark, and we can't bear the heat since all air conditioning in the building is off.

With temperatures outdoors over 45°C during the day and 33°C at night, residents' health was affected. Some took to air-conditioned cars, others to shopping malls. Companies sent employees home, and those in cars cooked in traffic jams because traffic lights failed. Some disabled residents were stuck on higher floors because lifts stopped, so they could not leave their buildings at all. Many flooded out onto the baking streets from buildings that had become death traps once the grid failed.

The extent to which the design and construction of a building can modify the thermal amplitude of the indoor climates of buildings can become apparent in a suitably designed temperature cloud. Clouds can also be used to map how close to thermal breaking points populations are, in effect their vulnerability to heat-related stress, health, and mortality. These types of insights are not available from comfort analyses that use regression lines derived from complete data sets. Clouds clearly identify local normal temperatures and outliers of less resilient buildings.

Clouds as Vulnerability and Resilience Indicators

Further work is needed to help develop comfort clouds into new tools to enhance the use of field data in the design and categorisation of comfortable and resilient buildings in different climates. Clouds can be formed of data from one or more comfort surveys they have been shown to provide insights into the geography of climate, and social, political, and economic insights into the nature of comfort in different cultures. The clouds demonstrated, for instance, that some cities in Pakistan had poorer buildings than others, and that in some regions, populations were far more likely to experience thermal stress than in others.

In our rapidly heating world, it will become ever more important that decision- and policymakers have a clear understanding of the vulnerabilities of populations in relation to more extreme weather. Temperature and other comfort clouds can provide clear vulnerability indicators of which people in a region are more exposed to the worst effects of high or low temperatures in poor buildings and when and where critical thresholds for health and mortality are approached and might be crossed.[17]

Thermal vulnerability comfort surveys will have to be done not once but periodically, as populations become increasingly exposed to killing climate events and trends. India is a subcontinent that is already suffering badly from rising temperatures, and experts there have already begun the systematic collection of the necessary data, 2020–2022, from eight cities in the five different climatic regions.

Rawal et al. found that more than 80% of the Indian residential occupants then surveyed experienced a neutral thermal sensation in the indoor operative range of 16.3–35°C,[18] which concurs well with our earlier Pakistani data. PMV, in his study, severely underestimated the extent of adaptation at play in the measured Indian homes. The data collected already exceeded the temperatures prescribed by the recent ASHRAE-55 and EN 16798-1 models by 2°C and 0.3°C, respectively. We can expect that temperature means and averages will continue to be exceeded as the climate heats up.

The IMAC-R project is being continued as a running assessment of the thermal comfort needs of the Indian population, providing vital information for the modelling of strategies for long-term energy savings and climate action on the sub-continent. Temperature and other related clouds can provide a new methodology for using the data collected to:

1. Map vulnerabilities
2. Evolve ongoing indicators based on changing data
3. Inform bottom-up action planning and implementation programmes for the protection of population in a hotter future

Energy and carbon issues have driven related policy thinking to date, but as extreme weather deaths and health-related economic impacts increase, a higher priority must be given to making sure programmes optimise the comfort, health, and mortality prospects of populations as well. Cloud analysis can help the precision targeting of policies to where help is most needed.

The Shifting Shapes of Clouds

Comfort clouds clearly demonstrate that the thermal comfort experiences of citizens of one city in the same country, or continent, can be radically different from those of other cities with the same nationality, or a culture in different geographic or climatic zone, even when they are as close together as are high desert Quetta and hot desert Multan.

International comfort standards with their narrow comfort zones are patently wrong. Why would one specify the same indoor optimal operating temperature in winter or summer, in Sweden or Portugal, or in Islamabad or Multan? Using international thermal comfort standards for such disparate climates makes no sense at all, unless you want to sell air conditioning machines with very simple and uniform operating manuals. Much better to identify the seasonal thermal preferences and acceptances for each major population grouping and then use them to inform planet-friendly design and servicing solutions. Exploration of the occupied temperature clouds of different cities at different times of the year provides a tool to inform optimal design solutions for safe and comfortable buildings in any climate. Perhaps best of all is to put the control of indoor temperatures into the hands of the people who actually inhabit their own very different buildings in different locations and climates. Let them strike their own balance between comfort and costs.

Figure 6.11 indicates the narrowness of the temperature range recommended in official guidance (red lines) compared to the range found in actual dwellings (yellow lines). It is clear that

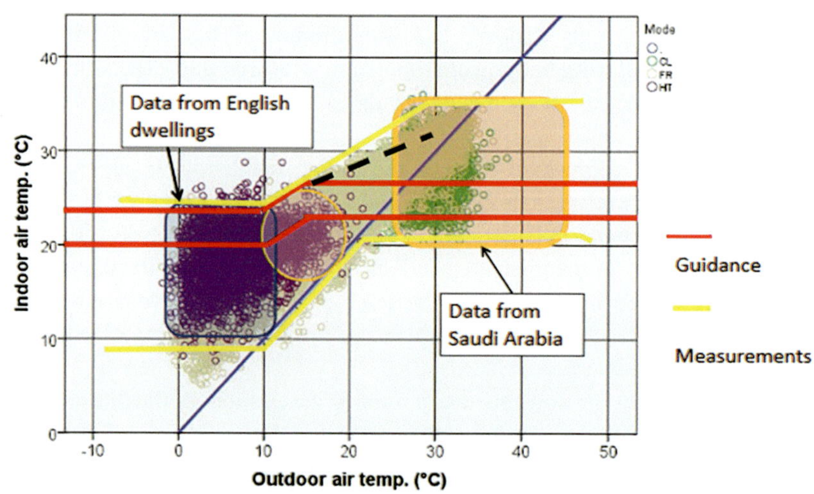

FIGURE 6.11 Clouds of points from surveys in England (purple), Japan (brown), and Saudi Arabia (yellow) show that the indoor temperatures from a range of climates can give rise to a liveable indoor climate ranging from 10 to 35°C, outlined in yellow. The temperature ranges recommended in guidance from climate scientists and engineers are outlined in red.

Source: Fergus Nicol.

there is surprisingly as much or more variation in the indoor temperature when a building is using heating or cooling than when it is using no conditioning.

The only way of truly understanding the temperatures people are actually comfortable at is to ask them how they feel in buildings they live and work in where the temperatures are measured. Comfort clouds show that people occupy indoor temperatures from 10°C (50°F) to 35°C (95°F). The next chapter explores ways in which comfort is achieved in different climates around the world through the co-evolution of the buildings along with the local comfort cultures of a region.

Notes

1 Fanger, P.O. (1970). *Thermal Comfort*, Copenhagen: Danish Technical Press.

2 Rijal, H.B., M.A. Humphreys and J.F. Nicol (2017). Towards an adaptive model for thermal comfort in Japanese offices, *Building Research & Information*, 45(7), pp. 717–729. http://dx.doi.org/10.1080/09613218.2017.1288450

3 Nicol, F. (2022). The shapes of thermal comfort and resilience, in *The Routledge Handbook of Resilient Thermal Comfort*, Edited by F. Nicol, H. Bahadur Rijal and S. Roaf, London: Routledge, ISBN: 978-1-032-15597-5

4 Nicol, F., and K. McCartney (2001). Final report (public) Smart Controls and Thermal Comfort (SCATs) (also subsidiary reports to Tasks 2 and 3 and contributions to reports to Tasks 1, 6 and 7), *Report to the European Commission of the Smart Controls and Thermal Comfort project (Contract JOE3-CT97-0066)*, Oxford: Oxford Brookes University.

5 Nicol, F. (2022). The shapes of thermal comfort and resilience, in *The Routledge Handbook of Resilient Thermal Comfort*, Edited by F. Nicol, H. Bahadur Rijal and S. Roaf, London: Routledge, ISBN: 978-1-032-15597-5

6 Nicol, J.F. (2017). Temperature and adaptive comfort in heated, cooled and free-running dwellings, *Building Research and Information*, 45(7). https://doi.org/10.1080/09613218.2017.1283922

7 Nicol, F., H.B. Rijal, H. Imagawa and R. Thapa (2020). The range and shape of thermal comfort and resilience, *Energy & Buildings*, 224(October), p. 110277. https://doi.org/10.1016/j.enbuild.2020.110277

8 Humphreys, M., F. Nicol and S. Roaf (2015). Fieldwork in Pakistan, Chapter 11, in *Adaptive Thermal Comfort: Foundations and Analysis*, London: Routledge, pp. 73–84.

9 Nicol, F. (2019). The limits to accepted indoor temperatures, *Proceedings of Comfort at the Extremes Conference 2019*, pp. 22–30.

10 Overbey, D. (2023). Towards passive survivability, *Proceedings of the 2023 ICARB Conference: Measuring Net Zero – Carbon Accounting for Buildings and Communities*, Edinburgh, pp. 376–398, Accessed on 08/01/2024 at: https://icarb.org/2023-conference/papers/

11 Nicol, J.F., I.A. Raja, A. Allaudin and G.N. Jamy (1999). Climatic variations in comfortable temperatures: The Pakistan projects, *Energy and Buildings*, 30(3), pp. 261–279.

12 Raja, I.A., J.F. Nicol and K.J. McCartney (1998). Natural ventilated buildings: Use of controls for changing indoor climate, *Renewable Energy*, 15, pp. 391–394.

13 McCartney, K.J., and J.F. Nicol (2002). Developing an adaptive control algorithm for Europe, *Energy and Buildings*, 34(6), pp. 623–635.

14 Humphreys, M., F. Nicol and S. Roaf (2011). Keeping warm in a cool house, *Historic Scotland Technical Paper 14*, Edinburgh, Accessed on 08/01/2024 at: http://www.historic-scotland.gov.uk/technical-paper14.pdf

15 Forster, E.M. (2011). The machine stops, *Penguin Classics*, ISBN: 0141195983. This short story was first published in *The Oxford and Cambridge Review in 1909*.

16 Accessed on 08/01/2024 at: https://www.utilities-me.com/news/breaking-news-widespread-power-outages-reported-in-sharjah

17 Roaf, S., and K. Beckmann (2017). Monitoring climate change adaptation: Lessons from Scotland, in *Sustainable Buildings and Built Environments to Mitigate Climate Change in the Tropics – Conceptual and Practical Approaches*, Edited by T. Harso Karyono, R. Vale and B. Vale, London: Springer.

18 Rawal, R., Y. Shukla, V. Vardhan, et al. (2022). Adaptive thermal comfort model based on field studies in five climate zones across India, *Building and Environment*, 219, p. 109187. https://doi.org/10.1016/j.buildenv.2022.109187

Comfort and Cultures

7

COMFORT, CULTURES, AND CUSTOMS

Introduction

People around the world adapt to those thermal conditions they usually occupy using involuntary physiological as well as conscious behavioural actions to do so. The comfort clouds in Chapter 6 demonstrated that climate is a key influence on comfort. Pakistanis share a single national cultural identity but occupy, and adapt to, very different temperatures in homes and offices at different times of year in their maritime, continental, or mountain climates. Even in one city at one particular time of year, not only can there be a large range of indoor temperatures in occupied buildings, but individuals also move between them on their own well-understood thermal pathways through a day: constantly moving between micro-climates within the buildings where they live and work and otherwise experience, from the shops, restaurants, transport, cinemas, to the places of worship they visit.

People acclimatise physiologically and adapt behaviourally by tailoring the customary thermal pathways they follow through a day to suit their own preferences and circumstances. It is that tailoring which matters so much, the fitting of the physiological requirements of an individual's body to their own means and opportunities, as well as the physics of the environments and the climates they occupy. They evolve habits that shape their own preferred thermal lifestyles to the world around them.

Comfort and the Habits of Hobbits

Michael Humphreys pointed out that if you wanted to design a comfortable home for a Hobbit, it is not necessary to have a PhD in physiology or physics.[1] You would simply have to follow Hobbits around for a while and perhaps ask them a few questions on the conditions that make them comfortable. Then by simply replicating those conditions, the result would be happy Hobbits in a comfortable, earth-sheltered home, perhaps like the modern Hobbit house in Figure 7.1.

The variations in the thermal habits of people living in very different or similar climates may amaze us, but they also reflect the richness of the human condition. Try explaining to an Inuit from the Arctic how the desert dwellers of the Sahara spend their days, or vice versa, and each would find the other's lifestyle difficult to comprehend. The same might be said of different

DOI: 10.1201/9781315645070-11

FIGURE 7.1 An earth-sheltered Hobbit home designed and built by Simon Dale in West Wales in 2010 by building on the appearance of Hobbit homes in the films of The Lord of the Rings Trilogy.

Source: www.Simondale.net.

approaches to comfort practiced by neighbours or grandparents and grandchildren in the same location. We are all the same, but we are also all different.

The evolution of thermal cultures over decades and millennia is simply a part of the evolution of the human species, who have learnt how to survive, even in the most extreme climates on the planet.[2] Customary lifestyles evolved not only to help people survive but also to thrive around the planet. The quality of a life in different climates might be immeasurably enhanced by sitting in the path of a cool breeze on a hot day, or in front of a warm fire on a cold one. The reward of the alliesthesia response is to nudge humans into acting in the best interests of the body's thermal condition by tempting episodes of thermal delight, as well as directing it through the punishing experience of thermal distress. The following tales show some unique thermal lifestyles of people in different climates to illustrate how the thermal balance in a lifestyle is maintained using the carrots and sticks of thermal delight, comfort, and distress.

Nomadic Life in Siberia

An insight into traditional lives in an extremely cold climate is afforded a study of a Siberian herdsman in his late 40s who lives in Tuva in South Siberia.[3] The yurt is the traditional tent type his forebears have occupied for generations (Figure 7.2). The family also have a base in the local

FIGURE 7.2 A Tuvan yurt in north-eastern Russia, shaped rather like spaces in a Hobbit home in internal dimensions, but more mobile.

Source: Dolaana Khovalyg.

town, where their children go to school. Husband and wife occupy the yurt while he tends to their extensive herds of sheep and cattle. Their fellow citizen, a professor of civil engineering in Lausanne and two physiologists, monitored indoor and outdoor environmental variations in the tent over two weeks in January 2020, along with the heat flows through the structure, making a record of his activities over those days.

Yurts are designed for mobile living as pastoralists migrate between seasonal grazing pastures two to six times per year. A *yurt* is a temporary structure, light and easy to assemble and disassemble. This yurt has flexible lattice walls in six sections, with long poles supporting a timber crown, from which the tent cloth is suspended. The lattice wall is set low to contain heat where people are seated or sleeping in winter, while in the hot summer, the tent diameter decreases and the lattice walls are raised to help the stack effect remove heat from the crown and help with cooling indoors. There are no windows in the yurt, while a skylight in the crown provides sufficient daylighting indoors. The tent envelope consists of different layers of materials; the outermost material is a waterproof tarpaulin, then layers of felt over the structure, while the innermost material suspended from the structural skeleton is lighter and typically decorative. The floor of the yurt is made up of multiple layers of thin materials, predominantly carpet, felt, and linoleum. In the seating area, the earth beneath them is warmed over time when the central stove is lit, activating the mass of the floor as an important thermal store over winter. He and his wife sleep in a bed to the right of the door, while guests, when they stay, occupy the opposite area across from the stove.

At night in January, the outside air temperature dropped during the study to −32°C, rising to −15°C during the day, with the indoor air temperature swinging between −10°C at night and up to 40°C whenever the stove was fired up when the yurt was occupied during the day. On a typical winter day, they wake up around 6–8 o'clock, and after she has lit the stove and made breakfast, he goes to shovel fresh manure around the animal shed and herd his livestock for up to six hours (10:00 to 16:00 hours). The wife, who remains working on and around the tent, occasionally tops up the firewood or reignites the stove as required. Sometimes, he goes back himself to the yurt a couple of times during the day to do the same. At its hottest in the morning, the temperature in the stove reaches 750°C, dropping down to −10°C at night. The indoor air temperature typically ranges between 0 and 50°C between the firing events during the day.

In this extreme environment with a scarcity of fuel for the stove, the yurt is only warmed when people are in it. When occupants are in bed, it is allowed to go out, and thus, valuable heat is supplied only when humans, or vulnerable cattle, need it. This is obviously a useful lesson from this story. It is important to note that the cast-iron stove almost instantaneously heats the 50 m^3 indoor volume of the tent, with a low ceiling to keep the heat near the seated or standing people and small floor area of 25 m^2 and a minimum wall-to-volume ratio. The very fast heating response time of the stove is critical here, as people entering the yurt may have lost a lot of the heat out of their bodies while outside in the extreme cold.

For the human body in yurts, there are three key factors in the 'keeping warm' equation:

1. *Calorific (energy).* Input needed to generate heat when needed, from food and drink.
2. *Energy in the body battery.* Heat loss and gain from the body are managed with clothing, activity, and behaviours.
3. *Ambient temperatures.* Affected by moving from outside to inside and by heat from the stove.

The calories taken in as food and drink provide the energy to keep the metabolism going. Intermittently imbibed chemical energy from those calories is stored for use when needed in the energy battery of the body. The amount of energy that can be stored around a body depends on the amount of body fat. The heat balance depends on the rate that energy is generated by the resting metabolic rate and through activity, the metabolising of food, cold-induced thermogenesis, and the rate at which heat is lost from the clothed body. Heat can also be gained from the sun or fire through the expert use of weatherproof and insulating clothing.

Acclimatisation to different climates occurs largely through physiological changes in the body over time, but adaptation to different thermal conditions also involves changing behaviours, including locations, eating, and clothing regimes. In established thermal lifestyles, these will be based on deep-rooted and innate learning that may have evolved as they were handed down through generations. For instance, populations indigenous to cold areas, like the Inuit, have been shown to adjust their activity level to avoid generating too much sweat inside the clothing, as the latter negatively affects its insulation properties and can lead to unwanted body cooling as the sweat evaporated after activity causes short-term discomfort or long-term health effects. Knowing when and how to adjust clothing for lower sweat accumulation before activity is increased can be an evolved, largely subliminal behavioural response and skill.

Adapted individuals have a profound understanding of the balance of voluntary and involuntary factors needed to maintain homeostasis over time. Back in the yurt, a herdsman knows when their energy battery is running low and returns to the yurt to have it rapidly topped up with food and drink intake and with heat from the stove. In other extremely cold regions, enforced warm-up breaks are mandated for workers, as the example from Canada in Table 7.1. With the growing incidence of more extreme hot and cold spells, it is important to develop locally appropriate temperature exposure limits, not only for workers, but also for any group outdoors during extremely hot or cold spells. This practice will be increasingly important in minimising health impacts and death rates from people in extreme climates.

TABLE 7.1 The Saskatchewan Occupational Health and Safety Division work warm-up schedule mandates required breaks for different temperatures and wind velocities at the worksite

Sunny sky air temperature		No noticeable wind		Wind 8 km/h (5 mph)		Wind 16 km/h (10 mph)		Wind 24 km/h (15 mph)		Wind 32 km/h (20 mph)	
°C below zero*	°F below zero*	Max. work period	Number of breaks**	Max. work period	Number of breaks**	Max. work period	Number of breaks**	Max. work period	Number of breaks**	Max. work period	Number of breaks**
26 to 28	15 to 19	120 minutes	1	120 minutes	1	75 minutes	2	55 minutes	3	40 minutes	4
29 to 31	20 to 24	120 minutes	1	75 minutes	2	55 minutes	3	40 minutes	4	30 minutes	5
32 to 34	25 to 29	75 minutes	2	55 minutes	3	40 minutes	4	30 minutes	5	Non-emergency work should stop	
35 to 37	30 to 34	55 minutes	3	40 minutes	4	30 minutes	5	Non-emergency work should stop			
38 to 39	35 to 39	40 minutes	4	30 minutes	5	Non-emergency work should stop					
40 to 42	40 to 44	30 minutes	5	Non-emergency work should stop							
43 and below	45 and below	Non-emergency work should stop									

Note: **Includes a normal break after two hours and the number of additional warm-up breaks needed. The green zone is 'work as usual', 'apply normal break schedule', 'dress appropriately for the weather and work duration'. Orange zone, 'work with precautions', and red zone, c'ease all but essential work'.[4]

Lightweight or Heavy Buildings at the Extremes?

The Tuvan yurt has not gained extra insulation layers in the envelope over the millennia, perhaps due to the fact that its highly efficient structure has been tailored in form and weight to the carrying capacity limits for the migrating animals, that carry it between campsites, and the lack of affordable, higher-performance alternatives for the traditional tent materials. The temperature inside drops gradually at night, when the occupant sleeps and the fire in the stove dies out. However, the indoor environment does not become as cold as outdoors due to the heat released back into the tent by the occupant(s) and from the ground when indoor temperature drops below +8°C at night. Nomads enable the heat storage capacity of the ground to be utilised by not obstructing the floor with heavy insulation materials and letting the daily heat from the stove be lost to the ground and re-released back into the tent at night, when there are no active heat sources.

A good lesson here is that, in extreme conditions, every single unit of energy must be rationed wisely to ensure that not only are temperatures pleasant when sat before the fire after a hard day's work outdoors but there is also enough heat in the system, and the thermal properties of the sleeping arrangements are adequate to ensure that safe temperatures indoors can be maintained when the fire goes out at night.

The thermal performance of the yurt is highly dependent on the relationship between it and the morphology of the landscape around it, not least in terms of prevailing winds and wind sheltering features in the landscape. Modern designs are focussed only on the U-values of walls, but in a yurt, more important to its success as a structure than the U-value of its envelope may be simply the direction in which the yurt and its door face in, to make best use of local winds, precipitation patterns, rainfall and snow build-up, while avoiding their worst impacts. Simple U-values are not the most critical factors in buildings that are also adapted for extreme weather, as they are often presented in modern regulations.

In a lightweight, cold envelope tent, lots of energy is needed to keep it thermally habitable when occupied by people who (1) are not sleeping, (2) are not very active, and (c) need to recharge their heat battery quickly, requiring a high-temperature heat source as provided by the hot stove. That is rather different from Nepalese families who live in high-mass homes with thick envelopes that themselves are constant, low-temperature radiators continually heated by a slow-burning stove that is on all day. Having either a low-mass cold envelope or a high-mass warm envelope dictates the types of heating regimes required to service thermally safe and often comfortable lifestyles inside the structures. The very different systems work through that symbiosis between building, climate, and culture.

Inside–Outside Lifestyles in Nepal

In a slightly less cold environment in upland Nepal, the buildings are constructed with thick heavy walls. People in the Mustang district in the Himalayas live in homes where the mean air temperature in the living rooms was measured at 10.7°C.[5] Men, and some women, spend much of their day out of doors, involved in agriculture and herding in temperatures that can fall to −10°C to −20°C in winter, but they know that when they do return home for lunch, they can sit around a hot stove and get warm and store up enough heat in their bodies to go back out into the cold for the afternoon and eat enough to provide the energy to top that heat up when necessary.[6]

The difference between the living environments of these Nepali villagers and a yurt dweller is that the Nepali homes have thick mud brick walls that provide a longer-lasting heat store when compared to the relatively thin material of a yurt tent cloth. In the more exposed tent environment, much more pressure is put on the delicately balanced management of the different elements of the thermal sphere, whereas in the warmer Nepali climate, with thicker walls and roofs in homes, a more relaxed attitude to the opening and closing of doors and the mixing of indoor and outdoor climates can be absorbed by the thermal inertia provided by the mass of the dwelling. Interestingly, here the clothing, too, can be much bulkier and heavier for the women, working predominantly indoors than for the men, who spend their days in more active farming pursuits outdoors, as shown in Figure 7.3.

What is clear for these villagers is that the high levels of effort put into travelling long distances to gather wood or dry dung for the fire mean that maximum benefit must be derived from the available heat, so what is there must be shared by the whole family in the one living room and eked out over the day to keep each generation, and gender, acceptably comfortable in the home.

Cold Comfort Farm in 1950s England

Comfort temperatures have changed radically over the last century for many around the world, with the introduction of piped and wired gas and electricity into homes replacing solid wood or coal fires in many areas, not least in Britain.

It is still common to hear older people say in the UK, 'I used to have ice on the inside of my bedroom windows in winter where I grew up'. Here is a short thermal history of Catherine Milner, who spent her childhood at East Farm near Sudbury in Suffolk in the early 1950s. She remembers it being 'pretty grim in retrospect' in a house without any modern appliances, although there were improvements over time. On marriage, her mother moved into the

FIGURE 7.3 Inhabitants of the Mustang region of Nepal at home, where the average indoor temperatures of living rooms is 10.7°C. Note the number of people in the room, use of hats, thick clothing, blankets for seated people, small windows, thick walls, cushions between sitters and the external wall, a central stove, and the researchers' temperature monitoring equipment.

Source: Photo by Hom Rijal.

farmhouse in 1947 without electricity, running water, bathroom, or flushing lavatory. Catherine says of it:

> *I did not remember the lack of bathroom, or running water, as this was installed when I was very young, but the arrival of electricity when I was 7 in 1955 was a highlight and a great excitement. Before this, Mother lit the lamps in the farmhouse on winter evenings as I cycled home from school. Heat in the house was open fires, paraffin heaters, and a boiler in the kitchen which heated the hot water. The boiler used solid fuel and was allowed to go out by the end of the afternoon to save fuel. I remember always washing in tepid or cold water in the mornings. The boiler heated the kitchen, although due to draughts, it was usually cool by the evening in cold weather. A fire was lit in the main living room which burnt mostly wood harvested from the farm, and some coal was used from time to time. We used to sit really close to the fire as the room was draughty. I always had a hot bath before going to bed, and a paraffin heater was used to warm the bathroom, which was cosy. But we were not allowed much water, as this was pumped from a bore and cost petrol. I was provided with a hot water bottle to put into my bed in the unheated bedroom. My mother believed in fresh air, and it was only in extremely cold temperatures that my bedroom window was closed. I had a chamber pot, the contents of which froze in winter. One winter, I remember my snowstorm toy freezing and breaking. I do not remember being cold in bed once I had warmed it up. Getting dressed in the morning was a speedy job in the winter. But the kitchen was usually warm for breakfast,*

as a paraffin heater would be lit while the boiler got going. I always had fresh cream from the house cow on cereal, followed by a cooked breakfast bacon and egg or scrambled eggs, fish on Fridays and Saturdays, and sausages on Sundays. I do not remember being ill apart from the childhood illnesses. I never believed I was hard done by as a child, although East Farm did have the wider family reputation of being a cold house.

Few people are still living in such conditions in Britain today, as times have changed radically, but here we see the use of effective thermal practices making the rather-extreme conditions work effectively over a day: a good hot breakfast and the injection of energy calories at the start to the day and a hot bath and hot water bottle in bed as a good thermal end to it. Many different cultures use the benefits of having a hot bath before going to bed as a way of setting the night up for a peaceful sleep in unheated quarters. Most families then spent much of the day in a single room, generally the kitchen. Other rooms would be heated only if there was a special reason for it, perhaps illness or visitors.

As energy prices soar, many people will be re-learning such tricks for making cold buildings bearable over the day and in different seasons, without wasting energy on building conditioning. The day can be successfully run charging the human thermal battery within a cold climate by injecting it with intermittent bursts of ambient heat to warm the body from the outside by heaters or hot water and internally using calories or hot food or liquid to feed the internal heat engine of the metabolism. The charging and discharging of the system with energy is carefully balanced across the 24 hours of the day to avoid getting uncomfortably cold and the catastrophic effects of hypothermia.

In East Farm, where people only slept in the uninsulated upstairs, the exact temperatures in the different thermal micro-climates downstairs were important, where rooms were warmed as necessary downstairs during the day. Different heat sources were used for different purposes: instant heat gained from the paraffin stoves and the lighting of a fire in a room and longer-burning appliances for water heating, cooking, and the charging of the thermal stores around the home, like the solid mass of chimney breasts and room surfaces, and the hot water in the tank. This case study of the thermal lifestyle of a family in the 1950s and the 1960s sheds some light on the lived experiences of comfort in other English homes.

Learning to Live With Heat and Cold

At the Central Building Research Institute (CBRI) in Roorkee, Dr M. R. Sharma noted in 1974 of his day in the office in summertime:

The room is full of warm air in the mornings. The windows are opened and the fans run at full speed to churn cool air into the room. Within half or three-quarters of an hour the air is cool enough for work to begin. Conditions remain comfortable with the fans running throughout the forenoon.[7]

The user of Dr Sharma's office had learnt how to overcome the fact that in the hot season, the windows and doors of the office had been shut overnight, so the heat had to be purged from it before work could begin. Here, the adaptive opportunities to hand included fans and opening windows. No one had thought to redesign a secure window type that could be left open overnight, perhaps with a metal grill to keep out thieves and insects and adding an external shade awning or shutter to enable and improve comfort indoors. Perhaps a valuable new profession might be started to change the furniture of buildings around the passive provision of comfort and training occupants to use new features rather than simply solving problems of the poor thermal performance by putting in more machines into building. People can appear incapable of making decisions to change the thermal infrastructures of their lives, often with uncomfortable or tragic outcomes, as heatwaves and cold storms become more widespread in our changing climates.

Comfort Cultures: Keys to Long-Term Climate Adaptation

Recognising and recording changes in the comfort cultures of a region over time is not easy. Conversely, it is easy to see when a climate-dependant technology ceases to be viable because of climate change. For example, during the 'Little Ice Age' that occurred across Northern Europe between the 1500s and the late 19th century, all seasons were generally cooler, resulting in wetter summers and longer, more severe winters. During this period, *icehouses* flourished. Plentiful ice was collected from ponds, lakes and rivers in winter and stored in typically underground icehouses designed to provide the frozen means to cool food, drinks, spaces, and people in the salad days of summer.[8] By the end of the 'Little Ice Age' in the early decades of the 20th century, winters were no longer cold enough in many places to produce enough ice regularly to stock the icehouses. The young men who had harvested the ice before the First World War often never returned from it, and many of those who did moved to cities. Icehouses were virtually obsolete by 1918, their existence almost forgotten as their function was superseded by refrigeration technologies. When icehouses disappeared, the skills and equipment used, and the buildings and communities involved in ice collection, transport, and storage also passed away.

It is more difficult to trace the impacts of climate change on less solid entities, like comfort cultures. Buildings, however, can reflect their evolving thermal histories quite well. One study of a house located in Visby on the Swedish island of Gotland in the Baltic Sea provides evidence of the huge changes that occurred in the region over 350 years. Built in the 1660s, the Burmeister House shown in Figure 7.4 was adapted over the course of 450 years to satisfy ever-changing expectations of thermal comfort. The medieval great hall was gradually reduced in size to a series of smaller, more easily heated and selectively used rooms. The relationship between owners and servants clearly changed, until by the 20th century, even the staff quarters were heated as well. Fire and stove types changed regularly, and the interior decor and furnishings evolved with the changing lifestyles and expectations of inhabitants.[9]

The house changed slowly with the steadily altering climate. Slow change could be coped with, whether good or bad. Generally, beneficial climatic effects tend to improve what can be achieved by human societies, whereas more extreme climatic shocks tend to restrict their options and outputs.[10] Populations are not culturally prepared for, or able to deal with, never-before-experienced extreme weather events, like unprecedented ice-storms or heat waves. The amount of damage done by such climate shocks will depend locally on both how exposed and vulnerable to them local populations are, often influenced by how rich, poor, healthy, old, or young they are.[11] We now know we are approaching what appears will be catastrophic climate change,

> *'Yet, as collective humanity we are only just beginning to take the steps needed to respond to the threats that we and the succeeding generations are confronting.'*[12]

FIGURE 7.4 The Burmeister House on the Swedish island of Visby.

Source: Arkland at Wikimedia Commons.

In a world regulated by the interests of technocrats, corporations, and their friends in politics, policies tend to focus on putting hopes for emissions and energy reductions and transitions into grid, community, and building-scale mechanical solutions that yield the highest profits for the most effective lobbyists.[13] Who can make a profit from simply changing a culture? Or a building that is modestly modified over hundreds of years, like the Burmeister House?

The challenge of protecting people in buildings against climate shocks now requires the development of not only a new vocabulary of adaptive opportunities but also new understandings of the need to store heat and cold passively and step changes in the way we regulate building performance. Populations also need education to help them survive rapid changes in climates and weather. This means resetting social expectations of what constitutes comfort and how to achieve it. Powerful programmes are needed to teach children in schools, colleges, and universities; managers in offices; and families in local community centres or libraries how to access and effectively use the many different adaptive opportunities at hand to remain comfortable over time. This is an urgently needed first step to lowering impacts of societal vulnerability from climate variability. Invest in changing behaviours and upgrading buildings and resilience education in the first place.

The machines will sell themselves. Ultimately, it is buildings themselves and the ways we use them that are our first line of defence against evermore extreme weather events, not machines, or the grid.

Notes

1 Humphreys, M. (1995). Thermal comfort and the habits of hobbits, in *Standards for Thermal Comfort Indoor Air Temperature Standards for the 21st Century*, Edited by F. Nicol, M. Humphreys, S. Roaf and O. Sykes, London: Taylor and Francis.
2 Vellinga, M. (Ed.) (2022). *Encyclopaedia of the Vernacular Architecture of the World*, 6 volumes, London: Bloomsbury Publishing.
3 Khovalyg, D. (2022). Resiliency lessons of traditional living in nomadic yurts, in *The Routledge Handbook of Resilient Thermal Comfort*, Edited by F. Nicol, H. Bahadur Rijal and S. Roaf, London: Routledge.
4 https://www.worksafesask.ca/wp-content/uploads/2019/03/Working-in-Cold-Conditions-Fact-Sheet.p
5 Rijal, H.B. (2021). Thermal adaptation of buildings and people for energy saving in extreme cold climate of Nepal, *Energy & Buildings*, 230, p. 10551. https://doi.org/10.1016/j.enbuild.2020.110551
6 van Marken Lichtenbelt, W., J. Vanhommerig, N. Smulders, J. Drossaerts, G. Kemerink, N. Bouvy, P. Schrauwen and G. Teule (2009). Cold-activated brown adipose tissue in healthy men, *New England Journal of Medicine*, 360(15), pp. 1500–1508. https://doi.org/10.1056/NEJMoa0808718
7 Nicol, J.F. (1974). An analysis of some observations of thermal comfort in Roorkee, India and Baghdad, Iraq, *Annals of Human Biology*, 1(4), pp. 411–426.
8 Beamon, S., and S. Roaf (1990). *The Ice-Houses of Britain*, Routledge.
9 Eriksdotter, G., and M. Legnér (2015). Indoor climate and thermal comfort from a long-term perspective, *Home Cultures*, 12(1), pp. 29–53. https://doi.org/10.2752/175174215X14171914084692
10 Pfister, C. (2010). The vulnerability of past societies to climatic variation: A new focus for historical climatology in the twenty-first century, *Climatic Change*, 100, pp. 25–31.
11 Roaf, S., D. Crichton and F. Nicol (2009). *Adapting Buildings and Cities for Climate Change*, Routledge.
12 Giddens, A. (2009). *The Politics of Climate Change*, Cambridge: Polity, p. 260.
13 Roaf, S., and F. Nicol (2022). Resilient comfort standards, in *The Routledge Handbook of Resilient Thermal Comfort*, Edited by F. Nicol, H. Bahadur Rijal and S. Roaf, London: Routledge.

8

COMFORT COLONIALISM

Introduction

As Europeans sailed through the centuries to conquer new worlds, enriching themselves and their empires, they took with them their customary lifestyles and comfort requirements. At first, they developed new building archetypes to help meet their own social and physical needs in alien climates. Then, as Winston Churchill later noted, the buildings gradually shaped them. Lifestyles and customs had to co-evolve with their buildings in local climates to both enable the incomers to avoid thermal stress, ill-health, and possibly death, and also to increase the thermal comfort and delight experienced in them. Bremner et al. noted in their 2016 book on architecture and urbanism in the British Empire how well-designed many colonial buildings were for their climates.[1]

This chapter is about the next phase of comfort colonialism that occurred across the 20th century, powered by the miracle of electricity. Great lengths were gone to in order to persuade the upper echelons of 'natives' in less-developed, hotter climates that what they really wanted, and needed, to be part of the lucrative international mainstream economy was the magic Western technology of air conditioning. Along with machines came the codes and standards that dictated the required air temperature settings needed to keep people comfortable. It was assumed that, globally, people preferred the same temperatures as the middle and upper classes in America, northern Europe, and Scandinavia. Unfortunately, all this came at a not-insubstantial-cost to the masses in the cultures involved, and to the global climate itself.

It is important to understand that 20th-century notions of what constitutes comfort were developed when 'White men' ruled the world. This chapter outlines some of the strategies used by a small group of professions and industries of European heritage to colonise global ideas of what constitute comfort and how to achieve it. The term *techno-colonisation* has been applied to processes like this where an original operating model, like that of comfort provided in climatically appropriate buildings and lifestyles, is then hijacked by an increasingly *industry-normalised* alternative reality, here of comfort provided exclusively by machines. Over the decades of the 20th century, this new reality was drip-fed into the *functioning of states* by the air conditioning

DOI: 10.1201/9781315645070-12

industry, until building standards, codes, and regulations around the world were inserted into building legislation that increasingly forced designers to include mechanical systems in buildings, whether they are needed or not.

The obvious impacts of this *comfort colonisation* are excessive and often unnecessary energy use and related GHG emissions. Also at stake are millennia of accumulated climatic design wisdom in the extraordinary range of vernacular buildings that are lost to the world every year.[2] Once lost, that wisdom is often gone forever, despite the best efforts of those trying to record, understand, and reclaim it.[3] Those who try to preserve and refurbish historic buildings are typically stymied by the legal straitjackets imposed by the industry responsible for, and profiting from, the cut-throat marketisation of comfort globally, and their friends in government. Here is our understanding of how all this happened.

Imperial Colonialism

Michael Humphreys recalled that while Charles Webb was working with the Building Research Establishment in the 1950s, the city state of Singapore introduced its first building regulations. They were, in fact, the British building regulations, almost word for word. A major problem was that in the hot, humid climate of Singapore, comfort is dependent on air movement for cooling, while the required air flow rates in the imported building regulations were very low. In cooler Britain, air movement was deemed to be unwanted 'cold draughts'. In Singapore, with its tropical climate, European, Malays, and Chinese were comfortable at temperatures rising to around the centre of the thermoneutral zone (TNZ – say, 28–32°C here, for argument's sake, with minimal clothing) at around 30°C, as shown for 1958 in Table 8.1. At temperatures around and above the TNZ, more air movement is essential for comfort. In many areas of Southern Asia with similar climates, people relied on desk or ceiling fans and natural breezes drawn through mosquito-screened doors and windows to enhance comfort.

The imposition of UK regulations in Singapore was a clear example of 'comfort colonialism', a policy designed to extend a country's power and influence through colonisation, use of military force, or other means. Building regulations are the 'other means' used to promote the ideas and interests of the ruling colonial groups over the Indigenous populations'. The thinking and publications of the 20th-century *climate determinists* like Ellsworth Huntington in the United States and S. F. Markham in the UK made clear what the benefits would be to 'White men'.[4]

TABLE 8.1 Summer comfort of clothed people at rest in still air for different nationalities, given as dry bulb temperatures in the 1950s

PLACE	*Sydney*	*Sydney*	*England*	*Singapore*	*Malaya*	*USA*	*USA*
RACE		*European*	*European*	*European*	*Malaya + Chinese*	*European*	*European*
PLACE	Outdoor	Indoor				South	North
DBT°F	70–87	72–82	63–68	76–86	80–86	72–85	70–80
DBT°C	21.1–30.5	22.2–27.7	17.2–20	24.4–29–30	26.6–30	22.2–29.4	21.6–26.6

Source: Macfarlane (1958).[5]

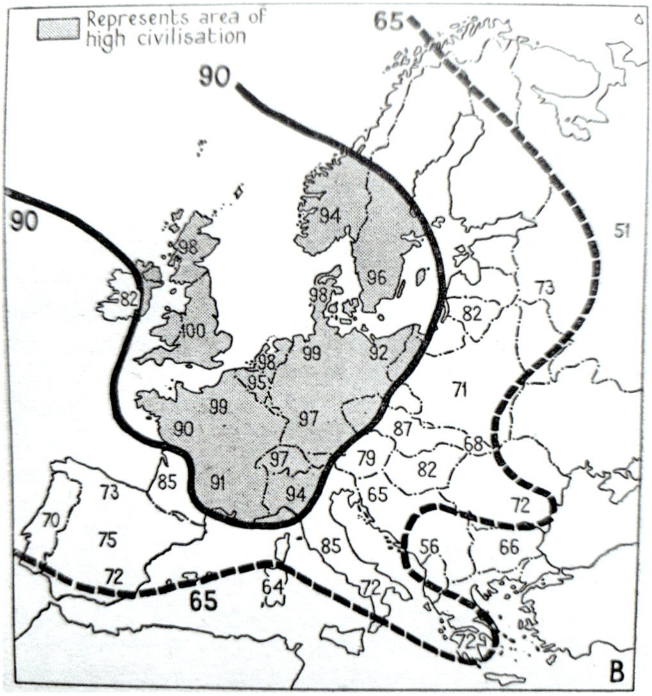

FIGURE 8.1 Map of areas of 'high civilisation' from Markham 1942, after Huntington. Note that 100% civilisation is roughly over Oxford, where S. F. Markham went to university at Wadham College.

Source: Susan Roaf.

Such authors were still being widely read in the 1950s, and the toxicity of the sentiments they state so firmly no doubt still reverberates around the inner thinking of some today. Huntington wrote in 1914:[6]

> *The tropical portions of America and Africa, as everyone knows, are the richest unexploited regions in the world. If ever they are to be developed the work must apparently be done by people of European origin, for the native races seem incapable of doing it alone and Europe and America are scarcely willing to leave the task to Asiatics. . . .*
>
> *Possibly we shall devise a plan whereby some means of creating the stimulus which now comes from the optimum temperature and from frequent changes shall be as much a part of a tropical house as a stove or furnace is a part of a house in regions with cold winters. . . . We have conquered low temperature in large measure. Our next great task is to conquer the uniform heat of the lands within the tropics.*

The magic stimulus to enable such cool temperatures in tropical buildings was, of course, air conditioning systems, and a lot of money was made selling them into the tropics by corporations from cooler, richer countries. One of the marketing ploys used to persuade building designers was that for 'people' to be comfortable and productive, they had to continually occupy temperatures within the Western comfort zones. That meant they now needed machines to achieve those temperatures to be part of the international movement in buildings and lifestyles, to be considered as 'developed', and also to benefit from the miracle of capitalism, a system in which people at the top are paid obscenely high wages, irrespective of the working conditions and pay of those at the bottom of the wealth pyramid. What is not to be liked about that idea for those in a position to benefit from it? The alternative to an air-conditioned future was presented by Markham, an influential British Nazi sympathiser, in his 1942 chapter on the Jews and the 'poor Whites':[7]

> *When white men settle in warm climates the first generation appears to maintain its energy practically unimpaired, but those who follow show a gradual deterioration, a social and economic retrogression, until . . . the white man becomes not only lazy, but also 'something considerably lower than a decent native.'*

Techno-Colonialism

As empires shrank, the shift in emphasis moved from the political supremacy that enabled the plundering of resources from vassal states to the ambition of technological supremacy and the domination of the lucrative forces that lubricated international markets. *Techno-colonialism* was a term coined by Randy Bush back in the 1990s to describe the exploitation of poorer cultures by richer ones through technology.[8] He was talking largely about developments in digital markets, but he pointed out that like ice cream, it comes in many different flavours. Of his seven different types of exploiter, perhaps the best fit for the air conditioning industry was originally as the *exploiting hero*, who brings often inappropriate technology in order to gain leverage for personal power or money. 'They care little for the locals or their needs. They have the righteous solution and will break anything to insert it, take whatever they came for, and leave'.

The history of techno-colonialism in relation to air conditioning has deep roots, and a range of strong sticks and carrots developed over the 20th century to push products for profit, and also to enable those from cooler developed countries to operate more effectively in warmer climates.

By the mid-20th century, S. F. Markham, by then a well-respected and influential Conservative member of the UK parliament from 1951 to 1964, peddled the narrative that the spread of the technology was an indicator of 'civilisation', along with later framings of 'high-tech' buildings being essential for economic growth in emerging markets. A new narrative now has more impetus as it is inevitably that in our heating world buildings everywhere will, at some times of the year, increasingly need efficient and affordable cooling and heating. The problem is that national and international comfort standards over-ambitiously sought to completely cut off indoor from outdoor climates around buildings to maximise sales of their own products.

This is visible in the selling of green rating schemes like BREEAM[9] or Passivhaus[10] around the world to clients in very different climates. Even when local versions of such schemes are developed to reflect local market priorities, they continue to espouse the tenets of their techno-colonial blueprints. In America, the US air conditioning industry rather judiciously hijacked the US Green Building Council agenda (USGBC – formed in 1993) and, with it, launched the LEED rating system in 2004 under the directorship of one high-end commercial property developer and the then director of environmental marketing and communications at United Technologies in the Carrier air conditioning corporation. To be even eligible for *LEED Platinum* status originally required full central air conditioning systems of the type sold by Carrier and actually precluded the gaining of points for the natural ventilation of buildings by simply opening windows.[11] LEED has been adopted in many countries by clients who think they are buying 'green' buildings with its accreditation that helps sell buildings and spaces to corporate clients in high-end property markets. Leading users of the LEED product are China, Canada, India, and of course, the USA.[12]

Some see the worst impacts of the global techno-colonisation of building markets as the soaring levels of global greenhouse gas emissions from the buildings they mandate. In the post-COVID global downturn in building markets, the rising energy costs of constructing and operation became very problematic. Many are also turning away from the often-poor environmental qualities in and around high-tech neighbourhoods, as noted in what follows for London's Canary Wharf. By 2023, the falls in the value of high-end commercial properties were unprecedented, even before the extreme heatwave summer of 2023 that significantly increased the air conditioning costs of large glass buildings. As interest rates and energy costs rise and demand for office space falls, building investors, owners, and users stand to possibly lose huge amounts of money in proportion to the size of the buildings as a consequence, especially in the hottest regions of the world.[13] Large corporate investors may well follow data server centres into the cooler regions of the world, where the air conditioning costs of running large organisations are significantly lower.

Are They Really Comfortable?

Experts, professionals, and industrial organisations are very important in lobbying for, and steering, policy directions for non-expert politicians. Many still actively push often physiologically and culturally inappropriate narrow comfort zones that require air conditioning systems to maintain them. Comfort experts have historically pooh-poohed the fact that adapted people can be comfortable at temperatures outside the narrow range of the heating, ventilating, and air conditioning (HVAC) industry comfort temperatures of around 20–26°C: 'Oh, but are they *really* comfortable?' The culturally patronising implication is that the expert, possibly from a completely different culture and climate, knows more than that person or population about how they feel.

In 1977, Ole Fanger, the godfather of narrow comfort zones, said, when asked whether people's comfort temperatures vary in different parts of the world, that he had done identical experiments in the United States and Denmark that showed decisively no difference more than about 1°C in the comfort preferences of Europeans or Americans.[14] He then extrapolated from those findings and others to cover the comfort temperature preferences for populations 'around the world', stating that:

> *People from Tropical countries were tested in Copenhagen shortly after their arrival by air, and they were found to prefer a temperature only about 1°C higher than that preferred by Europeans. . . . [D]wellers in hot climates may prefer a lower temperature than the one they describe as comfortable.*[15]

Fanger appeared to have no compunction in ignoring the findings of scientists like W. V. Macfarlane of the University of Queensland, who, in 1958, had shown clearly that people in different places were comfortable at very different temperatures, even across America. In today's world, we might claim to see such blinkered thinking as exhibiting an 'affiliation bias', but such manipulation of the facts still goes on in most committees dealing with temperature standards, and associated research includes numbers of 'experts' whose role it is to keep tabs on the proceedings of related committees, as we have experienced when working on Standards TM59 and even CIBSE Guide A in the UK.[16] There are some strong forces at work pushing techno-colonialism.

WEIRD Science

WEIRD here stands for 'Western, educated, industrialised, and democratic' societies. In relation to 'comfort', the fact that the real, everyday thermal experiences and preferences of the global masses have been sidelined for so long by the few in power in the 'thermal comfort industry' may also have something to do with the phenomenon of 'WEIRD science'. This term is used by sociologists to denote an objectivity bias that results from the fact that most of the research that underpins such standards has been predominately conducted in just such highly industrialised, educated, high-income democratic Western countries.[17] In the case of much comfort research, it was also largely undertaken by White men using small cohorts of White male students in laboratory experiments of limited scope and reach. The forcing of Western comfort standards onto the world was not only influenced by the thinking of politicians in thrall to the climate determinists and the air conditioning industry, but also the cultural determinists and the elitist White minorities who have used effective carrots and sticks to mould developing nations into market-ready entities and opportunities, justified by some as being 'for the good of humanity' to turn them into areas of 'high civilisation', as shown in Figure 8.1.

Since the mid-1990s, the development and widespread adoption of the adaptive model of thermal comfort has shifted this dial away from the idea that all people experience events and sensations in the same way. There is now a wealth of evidence that human brains are shaped to some extent by features of their environments. The world we all experience is, according to Anil Seth, always an active construction, a kind of 'controlled hallucination in which the brain uses sensory stimulations to update and calibrate its best interpretation of what is going on'.[18] He maintains we live in a perceptual echo chamber, just as we do in social media nowadays. The idea that the perceptual echo chamber of a 19th-century inhabitant of a house on the shores of the very hot, humid Gulf might bear any resemblance to that of the college students in Middle America in the 1950s laboratory comfort experiments is risible. Yet mesmerised researchers still compare the comfort experiences of people in those two situations by the same yardsticks.

The PMV method developed by Ole Fanger has already been robustly scientifically discredited in recent decades by comfort researchers across the board.[19] It is now widely agreed that

> *[e]mpirical research has led to mixed results and discussion of interpretation. Laboratory studies have often supported the validity of Standard 7730 (PMV method) whereas field studies have not.*[20]

PMV has not actually been validated in real buildings, as clarified earlier by Bjarne Olesen and Ken Parsons. Encouragingly, the understanding of the comfort perceptions of real people in real buildings using the adaptive comfort research has resulted in a strong and steady increase in publications from a diverse range of countries, climates, and more women, indicating possibly a positive trend from the impacts of globalisation in this socio-scientific research field. Early comfort research concentrated on the experience of comfort by White men in laboratories, in offices, on parade grounds, or in industrial situations. That is no longer true, with many women and men now championing diverse approaches to comfort research on gender, age, and culture experienced in the real lives of people in different climates. The first step to escaping from an echo chamber is to realise that you are in one.

Affiliation Bias in Comfort Research

Perhaps people are incentivised not to take notice of good or bad research findings unless there is a clear benefit to themselves? For instance, apparently credible studies were produced by researchers in the 1980s and the 1990s to persuade building owners and occupants it was actually healthy to be in spaces with the windows closed when others were smoking in those spaces. Research was produced in the US courts in the 1990s 'proving' that adequate air quality could be maintained while people smoked indoors if mechanical equipment was installed and air flow rates increased throughout the spaces. Does this sound familiar? This is what was claimed also during the COVID pandemic. An exhaustive study published in 2002 by Bero[21] on the subject of the public commentary published on the California environmental tobacco smoke risk assessment, and also the Washington and Maryland indoor air regulations in the 1990s, found evidence of *systemic biases* in the use of the risk assessment process by supporters of regulations to allow passive smoking indoors. He found that various interests *socially constructed evidence* to support their predefined positions. At one point, the tobacco industry reinforced the belief that mechanical ventilation was actually a very good control measure for environmental tobacco smoke, referencing the American Society of Heating and Air-Conditioning Engineers (ASHRAE) Standard 62–1989.

Bero's study confirmed the findings of previous research that the attitudes of experts were influenced by *affiliation biases*. He recommended that their data and the sources of research controversies they promoted should be independently investigated. A key ploy was the use of different criteria by different interest groups to increase confusion. It was proposed that all researchers should fully disclose their contacts with special interest groups, ranging from corporate, public health, and environmental groups, to eliminate any *affiliation bias* in their conclusions. A highlighted problem then was that ordinary people were deliberately being blinded by the highly technical pseudoscience being peddled by some representatives of the HVAC research industry paid for by the tobacco industry in order to keep people smoking in buildings despite the blatant health risks of doing so. Two marketing maxims may apply here: Say something often enough and people will believe it, and make things confusingly complex enough and no one will be able to contradict it.

Unreliable Research

In a world where science often appears to be for sale, much research is paid for by industry or governments that are keen to ensure the funded work produces desired outcomes. Academic publishing has become highly marketed, with scholars largely having to pay to have papers in journals, as scholarly outputs are traded as commodities.[22] For universities increasingly managed as businesses, many courses and research fields are cut to make way for more highly funded projects and programmes. Too often, public-interest science that has no product-reinforcing outputs is ousted while poor science product–focussed studies are promoted, as the current authors have found themselves.[23] At worst, apparently credible science is expensively couched in apparently correct but difficult-to-access algorithms or simulation pathways to market, for instance, the benefits of fully HVAC-serviced buildings. An example of this is some research published to prove that the *productivity* of workers is higher in air-conditioned environments rather than in naturally ventilated spaces. However, independent researchers at the University of California–Berkeley recently tested 35 different published methodologies for evaluating employee productivity and found no clear relationship was proven between indoor temperature and office work performance either for the range of temperatures measured in most of the office buildings (20–30°C) or a wider range (18–34°C). They plausibly suggest that this is because temperature alone may not fully describe how the thermal environment affects building occupants. They recommended that, while there must be temperature effects that do affect performance in office work, none of the current methods of proving so were scientifically correct.[24] So a possible myth peddled through apparently credible and expensively funded and published research may well not be worth the paper it is printed on. In a world where it is often difficult to know whom to believe, unreliable research is a huge problem. It cannot be solved by simply installing benign players, like not-for-profits or publicly owned organisations, in these academic marketplaces in place of hard-nosed commercial businesses and professional bodies and their friends in government. Even philosophers are now asking, '[i]f parts of science are widely understood to suffer from bad methodology, profiteering interests, and status-seeking, why trust science?'[25]

The Turning-a-Blind-Eye Trick: Sick Building Syndrome

Sick building syndrome (SBS), otherwise known as tight building syndrome, has almost disappeared from research funding programmes despite the fact that some indoor environments can be extremely hazardous to building occupants. People have begun to again ask, 'Are we doing enough about it?'[26] The obvious answer is *no*; it is an inconvenient truth about some high-end building types that some consider best buried.

Some signs and symptoms of the SBS are as follows: headaches; dizziness; nausea; eye, nose, or throat irritation; dry cough; dry or itching skin; difficulty in concentration; fatigue; sensitivity to odours; hoarseness of voice; allergies, cold, and flu-like symptoms; increased incidence of asthma attacks; and personality changes. Signs and symptoms of associated building-related disease are as follows: coughs, chest pain, shortness of breath on mild exertion, oedema, palpitations, nosebleeds, cancers, pregnancy problems, and miscarriages. Extrinsic allergic alveolitis, Legionnaire's disease, humidifier fever, pneumonia, and occupational asthma are also known to occur. Often, the causes of such symptoms are obscured by the complexity of factors in the indoor environment, but generally, they reduce work efficiency and increase absenteeism. Many of the complainants report relief soon after leaving the building, although lingering effects of neurotoxins can occur.[27]

By the 1990s, it was well established that people who work in fully air-conditioned environments suffered much higher rates of SBS ailments[28,29] than those in naturally ventilated offices.[30,31] Back then, leading experts made cogent recommendations as to what steps are needed to be taken to ensure people could occupy healthier, more comfortable, and lower-impact buildings.

A seminal study by Mark Mendell et al. in 2002 proposed addressing the issue with a list of actions that would have benefits for business, government, and other organisations, and also for nonworking building occupants, such as customers, patients, and students. Then they listed the potential public health benefits of acting in a timely fashion in the USA, including a reduction in the 6–11 million mild to severe cases of SBS annually, the prevention of a number of deaths associated with communicable diseases and asthma, the avoidance of frequently experienced acute symptoms among 8–30 million US workers, and an improved quality of life in the work environments where most adults spend one-fourth of their lives. That study was published nearly two decades before the COVID outbreak in 2020, when changes, had they been effected then, might have fundamentally improved health and mortality outcomes from that pandemic.[32] Their recommendations included the need to ensure that committees deliberating on indoor air quality (IAQ) issues be peopled by a wider range of experts from health and comfort fields and not just from the HVAC industry that has a vested interest in promoting highly mechanised buildings.

Bill Bordass and Adrian Leaman concluded in 2017, after some 30 years of research on building performance reinforced their earlier findings that occupants prefer to occupy shallower building plan forms that allow for natural ventilation, like to be able to control their own comfort conditions, and are not sold on working in large open-plan offices. Buildings need to be properly cleaned, maintained, and managed, and if people have concerns about their own comfort, they need to be dealt with quickly. Such recommendations do not involve rocket science, and they emphasised that the energy and comfort performance of many buildings are not taken seriously by managements unwilling to learn from building performance experiences in practice.[33] Few such recommendations from trustworthy sources, like Mandell et al. or Bordass and Leaman, have resulted in major changes in design approaches of 'build and sell' developers who opt for the 'fit and forget' certainty of standards-driven construction solutions.

The way in which UK IAQ and comfort levels produced by the Chartered Institution of Building Services Engineers (CIBSE) in 1986 in Table 8.2 were written completely ignored the health and financial benefits and opportunities for natural ventilation. They obviously pushed designers into adopting fully air-conditioned buildings to meet 'recommended levels'. CIBSE currently recommends air flow rates based on an expanded table that includes descriptions for different occupations in both summer and winter in which no distinction is made between fresh and reused air. The requirement to guarantee a certain flow rate technically eliminates the use of unpredictable natural ventilation rates to meet the standards.

TABLE 8.2 Comfort levels as recommended by the Chartered Institution of Building Services Engineers in 1986.

Parameter	*Recommended level*
Temperature (dry bulb)	19–23°C
Relative humidity (RH)	40–70%
	More than 55% RH is needed in carpeted buildings with under-floor heating to avoid electrostatic shocks.
Ventilation	
Delivery of fresh air	8 l/s/person (minimum); 16 l/s/person where some smoking; 25 l/s/person where heavy smoking
Total air supply	4–6 air changes/h
Airspeed	0.1–0.3 m/s: less than 0.1 m/s causes stuffiness; more than 0.3 m/s causes draughts; for air speeds higher than 0.1 m/s, CIBSE recommends increase in air temperature to take account of air movement
Sound	46 dBA upper limit for general office work
Lighting	500 lux for general office work; 750 lux for deep-plan offices and work at drawing boards, proofreading, etc.

Source: CIBSE (1986).[34]

A major tipping point in the acceptance of poor indoor and outdoor working conditions and climates came during the COVID pandemic in high-end buildings and neighbourhoods. People often did not feel physically safe in many office settings, fearing they would catch COVID from the pathogen-containing aerosols circulating in the blown air.[35] Tragically, this proved to be the case, especially in some hospitals. In the first wave of COVID in the UK, around a fifth of all those who died from SARS-Cov-19 had actually caught the virus in hospital.[36] Expert advice was to increase air speeds in ducts and to use filters, rather similar to the passive smoking advice. Trust weakened between employer and worker, with many staying at home for as long as possible, and even leaving jobs in what they felt were unsafe workplaces.

Research showed that simple regimes like opening ward windows in hospitals for 15 minutes three times a day radically reduced virus transmission rates and mortality levels in wards.[37] Many modern hospitals do not even have openable windows through which the pathogens can be effectively purged,[38] despite the fact that opening windows can save lives during power failures, which are increasingly occurring in hospitals for a range of reasons. Even some 'green' building standards do not require opening windows, pushing designers into using 'efficient' ducted ventilation systems, even on greenfield sites. The effectiveness of even short bursts of window opening are clear from a number of studies,[39] but what is too often sacrificed in buildings with non-opening windows is not only physical health but also the mental health of people locked indoors, as many of us will know from personal experience, having felt imprisoned in fixed-window rooms in hospitals, yearning for the sunlight and fresh air that Florence Nightingale believed were so essential for the health and recovery of patients.

It is understandable that employers wanted to move away from terms like SBS and to reframe related issues against indicator sets of 'well-being' that can be made to fit specifically to conditions in fixed-envelope buildings. COVID made it very clear that some building types are less healthy to occupy than others, not just in how they are designed, but also in how they are managed and operated.

Eliminating Competition: De-Valuing Thermal Mass in Design

Between 1930 and 1958, those involved in the General Motors streetcar conspiracy were convicted of creating the National City Lines bus company that bought, and closed down, streetcar companies in 25 American cities to remove competition to their own trade. GM was fined $5,000 for violating the Sherman Antitrust Act for this activity. The attempt to monopolise surface transportation was largely successful, and in only seven of the targeted cities did legacy city tram systems survive.[40] The conspiracy provided a plotline in the popular film *Who Framed Roger Rabbit?*

In a bid to promote air-blown air conditioning systems, the need to keep sizing and performance calculations as simple as possible and the ambition to eliminate natural ventilation systems resulted in a downplaying of both natural ventilation and the role of thermal mass in the thermal design of buildings. Traditionally, in more extreme climates, storing heat/coolth energy in the thermal mass of the structure and materials of buildings was used to store energy to both reduce indoor temperature swings and also make heat or coolth available later, when and where required. Well-designed thermal mass can reduce the amount of energy required by both bringing temperatures into an acceptable thermal range (temperature shaving), so reducing the need for heating and cooling, and making it available when needed (temperature shifting). However, the charging and discharging of the mass with solar energy during the day for passive heating at night in winter, or with cold night air in summer for passive cooling, often requires clever management of natural ventilation pathways through the structure. If the aim is to have buildings that are fully air-conditioned 24/7, 365, then the crucial role of mass interferes with this aim.

In the mid-19th century, the radiant heating and cooling of buildings predominated in Europe and also still the USA. Some interesting research was undertaken on radiant systems, for instance, by Clarence Mills, who, in 1946, detailed the health risks of air-conditioned buildings and listed many of the modern SBS symptoms but noted:[41]

> *Fortunately there exists a surprisingly superior method which does not require the warming of air in winter or the cooling of it in summer. I refer to reflective radiant conditioning, as demonstrated a few years ago in my laboratory, in which all the warming and cooling of the room occupants is done entirely through radiant channels.*

Innovative radiant heating research was almost entirely phased out by the 1960s, and the importance of thermal mass has been largely sidelined since then in building standards and regulations and even actively discouraged. For instance, in the ASHRAE thermal comfort database, measurements are embedded that systematically diminish the usefulness of designing with thermal mass. Eliminating definitions of comfort achieved by having room surfaces that are warmer than room air in winter and cooler in summer, that is, using a higher potential separation of mean radiant temperatures (MRT) from air temperatures, means that the radiant heating and cooling capacity benefits of buildings are lost from the designer's palette of options enforced by regulations. This promotes the benefits of air-blown systems and prevents people from understanding and reusing the strengths of vernacular high-mass buildings.[42] Such skewing of standards might be seen by some as a way of eliminating competition in this market.

Comfort in the Inferno Climates of the Future

The inferno summer of 2023 in the Northern Hemisphere foretold of a very different future. Yes, the need to radically reduce emissions to try to damp down the soaring increases in global temperatures will mean the move to the use of local energy in naturally ventilated buildings for as much of the year as possible, but it will also mean that for some periods of the year, extreme heat will have to be combatted with more fan use and air conditioning in cool refuge spaces in buildings. That can be countered in many cases by rooftop solar to provide the energy for fans and machines, but it also requires the throwing out of 20th-century standards and regulations that currently hamper adaptation of buildings. It will mean more shade and more mass used in buildings to protect occupants from the heat and enable the use of night-time cooling of structures. Thermal mass has been shunted into the sidings by HVAC organisation–led regulations, not least in the ASHRAE standards. The world is beginning to understand the vital role to be played in future buildings by thermal mass, so time to reshape new standards in relation to its use that are *fit for purpose* in what is shaping up to be an inferno-riddled future.

Confounding Variables: Carbon Dioxide

Carbon dioxide levels are currently used by many, and in some building regulations, as a proxy for IAQ. However, CO_2 is not a mild pollutant or even a dangerous gas like carbon monoxide (CO) that regularly kills people indoors. Andrew Persily of the US National Institute of Standards and Technology (NIST) pointed out in 2022 that the relationships between indoor CO_2, ventilation, and IAQ have been studied for decades, but misinterpretations and misapplications of CO_2 as an indicator of IAQ and ventilation are still too common.[43]

Unsubstantiated 'safe levels' of CO_2 indoors are regularly promoted in regulations despite the fact that the 2022 ASHRAE Position Document on Indoor Carbon Dioxide recommends that indoor CO_2 concentrations, although useful, do not provide an overall indication of IAQ or have proven impacts on health, comfort, or performance.[44] CO_2 does indicate how many people are breathing in a space, how stuffy it feels to both individuals and whole cultures. Nicol found during the European SCATS project that CO_2 levels in Portuguese offices regularly rose to over 3,000 ppm without occupant complaints.[45] Submarine crews regularly work in 2,000–5,000 ppm CO_2. Levels in crowded classrooms often rise to 4,000 ppm. The average CO_2 concentration in an aircraft cabin is claimed to be around 1,350 ppm but can reach 3,000–4,000 ppm at take-off and landing, when the air conditioning is off, and up to 5,000 ppm in cabins during emergencies in flight.[46]

The HVAC industry typically deems 1,000 ppm CO_2 to be a trigger for the need for increased ventilation, a figure proposed as a safe level for the gas in rooms in 1858 by Max von Pettenkofer and used ever since.[47] However, using indoor levels of CO_2 to assess airborne disease transmission proved fatal during COVID in one hospital ward where the CO_2 trigger points to recirculate air were set to below 1,000 ppm. Very ill elderly patients were breathing so lightly that the CO_2 set points for turning the ventilation system on were never reached, so the viral load in the space soared, leading to patient deaths.[48]

Some in the industry then recommended lowering advised CO_2 levels to the ridiculously low level of 600–800 ppm. Such low levels can be difficult to achieve in multi-occupancy spaces, and also, as identified by Satish et al. in 2012, 'if you want more productive and healthier people indoors using HVAC this requires more machines and energy to run them so considerably raising emissions from buildings'.[49] It is those emissions that are now the dangerous drivers of climate change, hence the slow move back to buildings with opening windows to reduce stuffiness and purge pollutants and pathogens when and where needed. The naturally ventilated Ng Teng Fong Hospital in Singapore, shown in Figure 8.2, provides an inspirational model for leading-edge naturally ventilated hospitals around the world.[50]

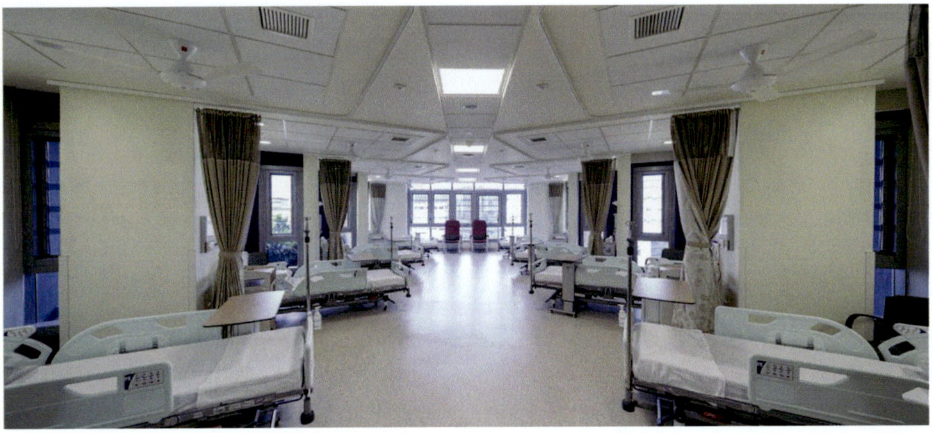

FIGURE 8.2 The 2020 Ng Teng Fong Hospital in Singapore, with naturally ventilated wards with ceiling fans, designed as a healing environment by a team led by architect Lim Lip Chuan of CPG Healthcare Ltd. in Singapore by the San Francisco office of HOK Architects.

Source: Lim Lip Chuan.

Confusing Complexity: Unworkable Standards

Measuring and managing the IAQ of buildings is confusingly complex and almost impossible to police.[51] Not only is it difficult to accurately estimate *dosage indicators* of individual substances like pathogens, formaldehyde, or VOCs, but they can also be expensive to measure in real buildings. CO_2 is cheap and easy to record but is of little use in measuring viral load or toxicity of air in a space on the whole. IAQ, like temperatures and comfort indoors, results from conditions specific to a particular time, person, place, and weather conditions. Philomena Bluyssen makes this clear:[52]

> *IEQ is still described with quantitative dose-related indicators, expressed in number and/or ranges of numbers for each of the factors (indoor air, lighting, acoustics and thermal aspects). Building and occupant-related indicators are overlooked. Interactions of stressors and effects at and between human and environment level are ignored. Individual differences in needs and preferences of occupants (over time) are not accounted for.*

The building–user–climate interface is complex and not necessarily predictable. Standards are useless unless they are applied within a clear review process that can result in robust actions to improve building performance and occupant perceptions over time. As with people visiting a doctor, it is best to treat the whole person. As Bluyssen states, the issues involved are not just about thermal comfort; they include noise, light, and air pollution as well. For instance, chronic exposure to moderate levels of sound from mechanical systems and other sources can induce chronic stress, which, in turn, can lead to a range of diseases. Light pollution at night disturbs people's circadian rhythm, and a view of blue sky is important to one's daily healthy light exposure. The health impacts of poor design are often interconnected and require careful and intelligent attention. Indoor environmental quality is more than the sum of its parts, and to ensure good IEQ can be achieved, other, different indicators to prevent (integrated) long-term health effects are needed.

There is growing interest in Europe in the benefits of annual health checks for multi-occupant buildings where the full range of environmental stressors affecting a building's population can be measured, evaluated, and acted upon to regularly monitor and gradually fine-tune the building systems to their populations and also adapt to the changing circumstances around the building.[53] Annual health checks could improve occupants' experiences and satisfaction with the indoor conditions and also minimise energy use in and emissions from buildings, sifting out obsolete or inappropriate systems, controls, and behaviours. Rather than rigid proscriptive standards, a flexible and regular review process would enable a rapid evolution of improvements in line with rapidly changing circumstances in and around the building.

The Hyper-Precision and Pseudo-Precision Problem

One of the most effective strategies for industries trying to persuade populations to agree to accepting their version of reality is the use of selective science studies combined with mind-numbing levels of *hyper-precision* and *pseudo-precision* to blind non-experts with a preferred version of science. The authors have been amused over the years at the degree of precision in the temperatures required by the HVAC industry guidelines.[54] In first-class buildings, standards are presented to one or two decimal places when everyone knows that such accuracy in thermal environments indoors, or out, is impossible to achieve in real spaces with real people.

TABLE 8.3 Allowable PMV and PPD

Category			*Allowable predicted mean vote*	*Allowable predicted percentage dissatisfied*
ISO 7730–2005	*EN 15251–2007*	*Description*		
A	I	Recommended for spaces occupied by very sensitive and fragile persons with special requirements (very young children, elderly, ill)	−0.2 < PMV < 0.2	<6%
B	II	Suitable for most new buildings and renovations	−0.5 < PMV < 0.5	<10%
C	III	Suitable for existing buildings	−0.7 < PMV < 0.7	<15%
	IV	Values other than the preceding; acceptable for only part of year	PMV < −0.7 or PMV > 0.7	>15%

Note: In this table, *PMV* is the predicted mean vote (index of thermal comfort), and *PPD* is the predicted percentage (of people) dissatisfied, both adapted and modified from the international standard ISO 7730:2005 and the European standard EN 15251:2007.*Source:* Susan Roaf. Modified from the US Green Building Council website.

Table 8.3 is a good example of the numerical blinding attempts of the HVAC industry to persuade populations of their credibility. Abstracted PMV values for comfort listed by regulation drafters to one decimal point literally means nothing to most ordinary engineers, architects, or heaven forbid, real people, or even regulation drafters. The comfort standards are riddled with such pseudo-scientific numbers with *hyper-* and *pseudo-precision.*

The distinction between what is 'precise' and what is 'useful' is clear in the example of the admittance method for predicting indoor temperatures, which uses the interactions between radiant and convective heat flow to predict indoor temperatures and therefore includes consideration of thermal mass in the calculation. The analogy used between electrical circuits and heat flow used in this method to calculate indoor temperatures is less 'precise' than the results from digital calculations, but often more realistic, and, importantly, enables designers to get a feel for the scale of the contribution to indoor comfort that is made by mass in a design (see Appendix 1). Many popular design simulation tools sideline the use of both thermal mass and natural ventilation in their inputs and outputs because of the complexity of their interactions and impacts on indoor temperature that are impossible to put a 'precise' number on. The result is too often precisely calculated: lightweight, fixed windows, and mechanically ventilated buildings that can overheat and are unnecessarily expensive to run.

Psycho-Technics

In 1942, political economist Joseph Schumpeter warned that rational argument is mostly eclipsed in modern life by a barrage of advertising, creating a political reality whereby 'direct attacks upon the sub-conscious' are at work.[55] In the field of comfort research, ordinary people want to know they and their loved ones and employees will be comfortable and healthy indoors and are happy to accept the advice of experts who tell them that this level of temperature or CO_2 will keep them safe. Schumpeter thought that successful advertising and decision-making policy in politics are shaped by unconscious wishes and fears. This was clear during COVID, as governments clutched at any apparently credible strategy to reduce deaths. So conflicting policies promoting the lockdown (China) or opening-up of society (Sweden) to manage infection transmissions were developed using similar scientific models to map the spread of the coronavirus SARS-CoV-2, estimate the course of the outbreak, and calculate the impacts of measures on managing it. Schumpeter presciently warned:[56]

> *The psycho-technics of party management and party advertising, slogans and marching tunes, are not accessories. They are of the essence of politics.*

Air conditioning is evermore essential in a heating world and has liberated populations to be, and remain, comfortable in many warm and hot regions. It has helped designers create ever-larger buildings and underpinned the expansion of economies. The air conditioning industry has delivered great benefits to humanity and will increasingly do so in a hotter future. It has worked assiduously to lower GHG and ozone-depleting emissions from its mechanical systems and make them more user-friendly and healthier over time. Unfortunately, national and international comfort and IAQ standards have also been used as marketing tools, moulded with a range of hooks and levers, to arrive gradually and undemocratically to a point where regulatory strait-jackets exist for buildings in order to force into them a narrow range of acceptable conditioning indoors that require HVAC equipment to achieve.

Comfort Hooks Like Thermal Shocks

In 1946, Clarence Mills, in his book titled *Climate Makes the Man*, wrote:[57]

> *Great strides have been made in the mechanics of air conditioning but the over-enthusiastic hopes people once had for it have faded considerably as experience has emphasized its handicaps. Chief disadvantage is the sharp contrast between air conditions indoors and out which must result either from adequate winter heating or effective summer cooling. Everyone has experienced the disagreeable shock of going from a cooled building out into the choking air of a severely hot day. People have collapsed from the sudden change.*

By 1955, the ASHRAE Guide was keen to stress that the 'sudden chilling (shock) of persons entering a cooled and air-conditioned buildings during the summer months may at times be important'. They pointed out that this 'shock is not usually harmful, but under some conditions it may be unpleasant or even harmful'. Some confusion here as the discomfort caused by the transition between outer and inner climates is talked down as a problem. They agree that 'an exit shock upon-re-entering a warm environment is equally plausible'. *Plausible* is an interesting word that implies some people professing to this sensation may not be believed. They so reassure people that a healthy individual will not suffer demonstrable harm.[58]

When air conditioning was not a common feature of buildings, people who spent most of their lives without it would feel that it was the air conditioning that caused discomfort. Nowadays, people who spend most of their lives in air-conditioned buildings or transport consider that it is the outdoor environment that is uncomfortable. The scale of thermal shock felt may well drive an enhanced preference to be in air-conditioned spaces, so thermal shock becomes a behavioural driver to stay indoors. The adaptive principle says that people adapt to those temperatures they normally occupy, and the more often they are in air-cooled environments in hot climates and bring clothing items such as cardigans to avoid discomfort, the more they think those are what they prefer. People actually get addicted to being in air-conditioned buildings, for various reasons, sometimes simply because they are associated with higher-income occupations, lending a higher status to their occupants.[59]

Standards: From Ratchet to Straitjacket

A key reason that globally GHG emissions from buildings still grow annually, despite billions being spent on programmes to reduce emissions from them, is that often well-meaning building standards implemented to reduce them can have unintended consequences that result in escalating current energy-use requirements to be met.[60] Standards are obviously indispensable as guides for building designers and actors. They represent the current thinking and policy ambition on what is needed as different technologies become normal and the most effective strategies to transform markets in a desired direction. Standards are based on ideas of 'normal' provision, and far from being neutral, standards are forged in the heat of competing interests and ambitions in the marketplace. Cass and Shove, in 2018, pointed out the 'dark sides' of the locking-in and ratcheting-up of requirements, not least in the office sector. Through an examination of ten case studies of speculative 'grade A', 'high' specification, corporate office developments in London, they were able to show a glaring mismatch between the regulatory image of what is normal for a very good office and the views of the ordinary occupants who worked in them every day.[61]

They found that guidelines designed to limit consumption routinely resulted in overspecification, and higher energy demands, for four reasons:

1. They locked in dominant developer-driven solutions, such as high levels of lighting, suspended ceilings, and the use of air conditioning, etc.
2. They ruled out alternatives like passive and non-mechanical options for design and servicing because they reproduce a market ideal which equates quality with air-conditioned offices.
3. They did not keep pace with changing practices, such as innovations in digital office technology, which may have meant that passive cooling systems are sufficient, and additional air conditioning is not 'needed'.
4. They act as carriers of conventions, such as the largely discredited air conditioning industry's narrow limits on comfort temperatures. The standards 'normalise' industry-promoting design features that often make anything else, for instance, what ordinary people like and want, seem odd or deviant.[62]

Such guidelines, over time, can tighten up the straps of the straitjacket and ratchet up the disconnect between what industry and, conversely, ordinary building occupants want. What occupants want is, in part, a function of what they are given because they have adapted to them, but comfort studies show that although people may report being comfortable indoors, they often simultaneously report they would personally prefer warmer or cooler conditions.

This disconnect between the industry ambitions and ordinary people's preferences began to prove costly for those owning high-end offices in business districts. In the London business hub at Canary Wharf, a range of 'events' pushed some corporate businesses to abandon premises in tall glass tower buildings for *smaller, greener, more flexible, and convenient buildings.* HSBC has cancelled their tenancy of one whole building when their lease expires in 2027 in what

Corporation of London called a 'flight to quality' back into the old city in June 2023.[63] They might also have said a flight to healthier, and thus happier and more productive, workplace, perhaps?

The Widening Credibility Gap

Standards only work if they are believed, credible, and trusted. The waning of confidence in the healthiness of high-end open-plan office during COVID was then hit by a further blow. On 22 February 2022, Russia invaded Ukraine. Countries across Europe, particularly those heavily dependent on Russian gas for heating, grew alarmed, and several took an early and effective step: They abandoned their own national comfort standards. On 2 April, the Dutch government launched a new campaign, called *Zet ook de knop om* (Flip the Switch). It mandated that the heating thermostat in all central government buildings be set permanently to only come on at temperatures of 19°C and below. They calculated that this simple move would result in savings of 15% on heating energy consumption. A number of other organisations followed suit to save the soaring energy costs, rather than carbon emission reductions. It was admitted that staff and students at the University of Amsterdam, who were used to 21°C, would unfortunately be less comfortable.[64] No doubt they adapted rapidly enough with a vest or extra jumper.

A similar seismic shift had already occurred in Japan on 11 March 2011 when the Great East Japan earthquake caused enormous damage, leading to a 15% peak-power reduction imposed to address power shortages. Tanabe et al. investigated occupants' comfort and productivity in five office buildings in Tokyo during the subsequent summer season under mandatory electricity

FIGURE 8.3 Typical home construction in much of America is of lightweight timber frame with chipboard walls nailed inside and out. Minimal insulation means that when the temperature outdoors reaches 45°C (115°F), as it does in Phoenix, Arizona, the only way to become and stay comfortable is to run the air conditioning all day.

Source: Susan Roaf.

savings implemented after the earthquake. Most offices followed the Super Cool Biz campaign's recommendation for employees to wear light summer clothing at work and only turn on air conditioners when temperatures reached 28°C. They found that in the circumstances occupants were happy to experience lower illumination levels and less happy with increased temperature, but 90% of people in the study accepted the 'poorer' indoor environment but did feel they were slightly less productive in them.[65]

The Super-Wicked Problem of Vested Interests

Writing on the building industry in the *Harvard Law Review*, Richard Lazarus pointed out that the challenge of building 'resilience' into our world is a *'super wicked'* problem, riddled with barriers to the evolution of the system imposed by those with vested interests in maintaining the status quo in a non-evolving system. He proposed that[66]

> *to be successful over the long term, climate change legislation will need to include institutional design features that significantly insulate programmatic implementation from the undue influence of powerful political and economic interests propelled by short-term concerns.*

The housing boom to bust in Arizona in 2007–2010 was largely driven by the simple fact that the main buildings players in the state were, if not in the pockets of, then certainly in hock to, its main housing developers. Politicians relied on their donations. Universities benefitted from new buildings and subventions, even to the extent that they discouraged their own staff from pursuing research that may interfere with new developments, like questioning if developers actually had the water rights when they lured purchasers to buy homes from across the desert state. The incomes of city planning authorities were boosted by each new development, and the mayor's election campaigns profited from their generous donations.[67] This is still happening today, and developers roll their carpets of new developments of unsuitable housing out across the rapidly heating desert state like those in Figures 8.3 and 8.4. Unless the citizens of the state are significantly insulated from the undue influence of powerful political and economic interests, housing crash after crash will continue to happen there, in a region where soaring energy costs are exacerbated by ever hotter weather.[68] People are reluctant to pursue novel, adventurous, and untried avenues when 'business as usual' approaches work well enough for themselves and their sectors, as is the case with the Arizona developers. But:

> *Placing a system in a straitjacket of constancy can cause fragility to evolve*[69]

State Capture of Comfort

This term describes the process by which the decision-making functions of regional, national, and international governments are taken over by vested interests. The Arizonan example is an excellent example of state capture. Diesendorf and Taylor wrote in 2023 that *state capture* describes actions that go far beyond one person paying another to influence political decisions:[70]

> *State capture is systematic and well-organised by people who have an established relationship with one another. It involves repeated transactions, often on an increasing scale, and can involve corporate management of all major political parties, public officials, mass media, social media,*

> *police and institutions originally designed to regulate the industry concerned. It involves political donations, lobbying, 'revolving door' jobs between governments ministers and their political advisers on one hand and corporations on the other, re-purposing institutions, astroturfing, research by self-styled 'think-tanks', policymaking and public influence campaigns.*

The complete subversion of national building regulations and international standards around comfort solely into the language of mechanical solutions has been the result of a very successful campaign by the HVAC industry over more than a century. In 1894, the American Society of Heating and Ventilating Engineers held its first meeting in New York with 75 chartered members. They agreed on bye-laws to effectively develop a monopoly around the range of services that those engineers offered,[71] by not only fixing prices to be paid for their services, but also aggressively pursuing, amongst others, the following aims:[72]

- To use legislation to compel the ventilation of all public buildings in accordance with the standards of this society
- To push legislation favourable to improvement in the arts of mechanical heating and ventilation and to oppose legislation that promotes other methods of conditioning buildings

We see the results of their success all around us in the thin, tight-skinned buildings with no opening windows that are unable to function without electricity.

We see their success in the deliberate refusal of the World Health Organisation to admit that SARS-CoV-2 was an airborne pathogen until five months after the pandemic became a public threat in February 2020. On 7 February, the ostensibly COVID-sceptic US president Donald Trump said of COVID:[73]

> *You just breathe the air and that's how it's passed and so that's a very tricky one. That's a very delicate one.*

And if one owns numerous large thin, tight-skinned buildings like Trump Towers, that is a very tricky one, because the last thing you need is for it to be proven, as it subsequently was, that COVID was transmitted through the HVAC ducts around such buildings.[74] Conversely, actual research in hospitals showed clearly that the dramatic reductions in viral load indoors resulted from simply opening windows on wards.[75] Hospitals are internationally being built now with no opening windows, persuaded by often flawed research. This is happening even as grid failures during extreme weather are increasingly taking down hospital-wide energy and server systems during extreme heat and cold events.[76] It is obviously sensible for professional bodies to aim to ensure and improve standards of performance and service for their own product ranges. It might be deemed to demonstrate the state capture of comfort when the HVAC profession succeeds in ensuring that the only types of buildings allowed to be built under government regulations and standards are those that require ever more of the mechanical equipment, and those in which opening windows is prohibited.

Is Techno-Colonialism Killing the Vernacular?

The fragility, or brittleness, in the fabric of our built environment is made worse in a world of more extreme climates. Many may say, 'If we can keep our heads down, we can squeak through

FIGURE 8.4 The Tartesso planned community is designed to reach 49,000 people of lightweight timber-frame homes. Occupied first in 2008, it is located 40 mi from Phoenix, Arizona, its main source of employment.

Source: © Thomas Hawthorne and Michael Chow. USA TODAY NETWORK.

OK, at least for my working life/lifetime'. In the face of the growing strength of weather abnormalities, other voices are growing, saying, 'Better to be safe than sorry – let us try change'. Is the tide turning on the American dream? Is the appetite for modern buildings and the international style of architecture diminishing? Have the skills and knowledge necessary to design for local climates been lost? Certainly, the vernacular architecture in many parts of the world has been quietly killed off and replaced by more fashionable archetypes and construction materials and methods. Gone from many modern buildings are the passive technologies and approaches that enabled buildings to be run on free local energy across the year and to be simply decoupled from the need to constantly draw on imported grid energy. The need to halt the destructive impacts of comfort and techno-colonialism on regional building designs and lifestyles is clear. In a heating world, mechanical cooling will be increasingly needed at the hottest times of the year, but the processes at play that eliminate alternatives and coerce populations into needing only certain technologies, all the time, must be reviewed and reigned-in. Robust alternatives for cooling without machines for as much of the year as possible need expanding, not least for reasons of affordability and also power insecurity.

Adieu to the Siesta: Killing Off Comfort Cultures

In pre-COVID Spain, corporate management of large companies lobbied the government hard to ban the afternoon siesta for workers. They argued the need for more productivity from the workers of southern Spain to fuel economic growth. Things came to a head in 2016 with the declaration of the president of Spain that the siesta was cancelled. 'People want to work, but they also want a life', said the president of the National Commission for the Rationalization of

Spanish Schedules, who proposed that employers who introduced flexitime and homeworking should be rewarded with tax breaks, while also promoting the responsibility of the government to provide more nursery places and the opening of schools earlier in the morning to enable workers to spend more time at the office.[77]

It was even argued that the low Spanish birth rate of 1.32 children per woman, compared to 1.58 for the whole of the European Union, could be put down to the state of the economy and 'because people are tired even when it comes to human relationships'. Comfort has a lot to answer for! The downside of people incorporating a siesta into their working day may be seen as resulting in the enforced extension of working hours into the evening, eating into family time. The siesta debate, mixed with new trends of working from home (WFH) and against the backdrop of soaring energy costs, rising global temperatures, and the urgent need to significantly reduce GHG emissions, really opens up a wide landscape of opportunities to develop new thermal pathways through daily lives and cultures for working days in warming climates. The siesta is still alive and snoozing in many hotter countries in non-prestige buildings and may yet provide a vital lifeline for populations in heating climates.

Quilty et al. published a study in 2023[78] showing that non-Indigenous people in the Northern Territory of Australia were 18% more likely to die in hot weather than Indigenous people despite the latter having severe health disadvantages. The former resorted to ever-higher use of air conditioning (52% in 1980 and 92% in 2010), while the latter used lowered activity levels during siestas to ride through the high temperatures. They admit that technological and infrastructural adaptations are crucial in preparing for hotter climates but stress that social and cultural adaptations to increasingly hot weather are potentially powerful mechanisms for protecting human health. Cool shelters during extreme heat were vital, and they posited that excessive exposure to air-conditioned spaces might well impair physiological acclimatisation to ambient thermal conditions. They cited the constant political pressure to remove the siesta in Spain in order to improve economic productivity as a clear example of a society in which the health of an economy is considered more important than the health of the planet and the population. They concluded that perhaps the most important tool at hand today to combat climate change is culture change, a reality that needs to be recognised by global society as a whole.

Cultural change on a global scale is exactly what happened during the COVID pandemic, when people were first locked down in their homes then encouraged to work at home to avoid spreading or catching the virus. Many groups enjoyed not having to go into the office but then had to adjust their lifestyles to remain affordably comfortable at home against a backdrop of rising energy prices. New behaviours were developed, exploiting their knowledge of the thermal landscapes of their homes and developing new accoutrements for comfort, small local heaters or fans, foot warmers, and comfort clothing and eating. The WFH (working from home) movement was born, and while some actually feared going back to work for health and commuting cost reasons,[79] others, especially higher-paid workers, really appreciated the opportunity to retreat from noisy, overheated or overcooled, open-plan offices. Benefits included being happier, more comfortable, and able to concentrate better in their own homes. Many young workers living alone and away from family homes did suffer, as did those with unhappy home environments. The expansion of electronic working from home can be isolating, and becoming familiar with workmates in just the two dimensions of the computer screen became alienating for some. We are now 'stuck with' the WFH culture, but research shows that actually WFH increased productivity by around 5% according to one study. Inner-city environments benefitted from having fewer people commuting into them but suffered economically as well.[80]

Designers now wanting to move beyond creating 'workplace zoos'[81] have to think deeply about how to re-humanise work–life balances and environments. Ideally, all buildings will inevitably have to be designed to run for as much of the year as possible on local natural energy, and only when there is a perceived need will they be heated or cooled. In a heating world, the length of time a building runs in conditioned or natural energy modes of operation will change,[82] requiring the rapid evolution of new climatic design solutions to minimise GHG emissions from them while providing thermally safe places for people to occupy indoors. Perhaps a useful mantra for the next design era will from now on be 'Building First'. The techno-colonialists must step aside now to let this happen to enable us all to survive.

Quilty et al. spoke well when quoting Australian First Nations scholars, who emphasise the need to reject colonial and capitalist ideologies and return to Indigenous epistemologies that emphasise the importance of respectful relationships with country and the natural environment.

Comfort Justice

The flip side of *comfort colonialism*, where the standards were developed and aggressively marketed to promote the interests of the few, is the growing need for more *comfort justice*, for policies and regulations that are designed to improve the lives of the many. The process of designing a comfortable building requires designers to relate the task to a site in a region, for the time and a particular population, culture, and economy. The 20th-century regulatory straitjackets and design prototypes created for the 'international movement' and ignorantly or cynically imposed by the techno-colonialists on global markets urgently need to be sidelined for a generation of new vernacular, locally and climatically appropriate buildings. Far-sighted political willpower is needed to release the ratchets and hooks of the techno-colonialist, reject their affiliation biases and pseudo-precision and pseudoscience to fairly enable future generations to build structures that will keep them safe in a different future. As Diesendorf and Taylor say, 'a sustainable civilisation needs an economic system that fosters ecological sustainability and social justice'.

In a world where even the middle classes are falling into fuel poverty, the growing calls for *climate and comfort justice* must focus the minds of designers and decision-makers on the real-world purposes of codes, regulations, and standards, not least because it is well understood that thermally stressed populations can become violent very quickly.[83]

Notes

1 Bremner, G.A. (2016). *Architecture and Urbanism in the British Empire*, Oxford University Press.
2 Vellinga, M. (Ed) (2026). *Encyclopedia of Vernacular Architecture of the World*, London: Bloomsbury Publishers.
3 Sengupta, T., and S. King (2024). *Reclaiming Colonial Architecture*, London: RIBA Publishing, ISBN: 978-1915722362
4 Macfarlane, W.V. (1958). Thermal comfort zones, *Architectural Science Review*, 1(1), pp. 1–14.
5 Markham, S.F. (1942). *Climate and the Energy of Nations*, Oxford: Oxford University Press, p. 98.
6 Huntington, E. (1914). The adaptability of the white man to tropical America, buildings and restaurants, *Sustainable Cities and Society*, 27, pp. 32–41.
7 Markham, S.F. (1942). *Climate and the Energy of Nations*, Oxford University Press, p. 122.
8 Bush, R. (2015). *On Techno-Colonialism*, Accessed on 06/07/2023 at: https://psg.com/on-technocolonialism.html
9 Accessed on 19/07/2023 at: https://bregroup.com/products/breeam/
10 Accessed on 19/07/2023 at: https://bregroup.com/a-z/the-passivhaus-standard/

11 Shaviv, E. (2008). Passive and Low Energy Architecture (PLEA) vs green architecture (LEED), *Paper 371 in Proceedings of the PLEA 2008 Conference*, Dublin.
12 Accessed on 19/06/2023 at: https://www.usgbc.org/articles/usgbc-announces-top-10-countries-and-regions-leed-2021
13 Accessed on 19/07/2023 at: https://www.thebanker.com/Commercial-property-market-under-stress-1686651435
14 Fanger, O. (1970). *Thermal Comfort*, New York: McGraw-Hill Book Company, pp. 81–84.
15 National Bureau of Standards (1997). Thermal analysis, human comfort, indoor environments, *NBS Special Publication*, 491, Washington DC: US Department of Commerce.
16 Roaf, S., and F. Nicol (2022). Resilient comfort standards, in *The Routledge Handbook of Resilient Thermal Comfort*, Edited by F. Nicol, H. Bahadur Rijal and S. Roaf, London: Routledge.
17 Hendriks, T., M.A. Warren, M. Schotanus-Dijkstra, A. Hassankhan, T. Graafsma, E. Bohlmeijer and J. de Jong (2019). How WEIRD are positive psychology interventions? A bibliometric analysis of randomized controlled trials on the science of well-being, *The Journal of Positive Psychology*, 14(4), pp. 489–501. https://doi.org/10.1080/17439760.2018.1484941
18 Seth, A.K. (2021). *Being You: A New Science of Consciousness*, London: Dutton, an Imprint of Penguin Random House LLC.
19 Humphreys, M., F. Nicol and S. Roaf (2016). *Adaptive Thermal Comfort: Foundations and Analysis*, London: Earthscan/Routledge, March 2012, ISBN: 978-0-415-69161-1
20 Olesen, B.W., and K.C. Parsons (2002). Introduction to thermal comfort standards and to the proposed new version of EN ISO 7730, *Energy and Buildings*, 34(6), pp. 537–548.
21 Bero, L.A. (2002). Case studies of the role of research in the development of indoor air policies, *9th International Conference on Indoor Air Quality and Climate*, Monterey, California, June 30–July 5, 2. pp. 1–11.
22 Bauwens, T., D. Reike and M. Calisto-Friant (2023). Science for sale? Why academic marketization is a problem and what sustainability research can do about it, *Environmental Innovation and Societal Transitions*, 48, p. 100749, ISSN: 2210-4224
23 Roaf, S., and F. Nicol (2022). Resilient comfort standards, in *The Routledge Handbook of Resilient Thermal Comfort*, Edited by F. Nicol, H. Bahadur Rijal and S. Roaf, London: Routledge, ISBN: 978-1-032-15597-5
24 References: Porras-Salazar, J.A., S. Schiavon, P. Wargocki, T. Cheung and K.W. Tham (2021). Meta-analysis of 35 studies examining the effect of indoor temperature on office work performance, *Building and Environment*, 203, pf. 108037. https://doi.org/10.1016/j.buildenv.2021.108037
25 Morgan, A. (2024). Science, anti-science, pseudoscience, truth, in *The Philospher*, Published by the Philosophical Society of England.
26 Ghaffarianhoseini, A., H. AlWaer, H. Omrany, et al. (2018). Sick building syndrome: Are we doing enough?, *Architectural Science Review*, 61(3), pp. 99–121. DOI: 10.1080/00038628.2018.1461060
27 Joshi, S.M. (2008). The sick building syndrome. *Indian Journal of Occupational Environmental Medicine*, 12(2)(Augus), pp. 61–64. https://doi.org/10.4103/0019-5278.43262. PMID: 20040980; PMCID: PMC2796751.
28 Fisk, W.J., A.G. Mirer and M.J. Mendell (2009). Quantitative relationship of sick building syndrome symptoms with ventilation rates, *Indoor Air*, 19(2), pp. 159–165, Accessed on 11/12/2019 at: https://www.ncbi.nlm.nih.gov/pubmed/19207289
29 Mendell, M.J., G.M. Naco, T.G. Wilcox and W.K. Sieber (2003). Environmental risk factors and work-related lower respiratory symptoms in 80 office buildings: An exploratory analysis of NIOSH data, *American Journal of Industrial Medicine*, 43(6), pp. 630–641.
30 Ghaffarianhoseini, A., H. AlWaer, H. Omrany, A. Ghaffarianhoseini, C. Alalouch, D. Clements-Croome and J. Tookey (2018). Sick building syndrome: Are we doing enough?, *Architectural Science Review*, 61(3), pp. 99–121.
31 Seppanen, O., and W.J. Fisk (2002). Association of ventilation system type with SBS symptoms in office workers, *Indoor Air*, 12(2), pp. 98–112.
32 Mendell, M.J., W.J. Fisk, K. Kreiss, H. Levin, D. Alexander, W.S. Cain, J.R. Girman, C.J. Hines, P.A. Jensen, D.K. Milton, L.P. Rexroat and K.M. Wallingford (2002). Improving the health of workers in indoor environments: Priority research needs for a national occupational research agenda, *American Journal of Public Health*, 92(9), pp. 1430–1440.
33 Leaman, A., and B. Bordass (2017). Productivity in buildings – 20 years on, in *Creating the Productive Workplace*, Edited by D. Clements-Croome, 3rd Edition, Ch. 15, London: Taylor & Francis.

34 CIBSE 1986 Guidelines published in Joshi, S.M. (2008). The sick building syndrome, *Indian Journal of Occupational Environmental Medicine*, 12(2), pp. 61–64. https://doi: 10.4103/0019-5278.43262
35 Roaf, S. (2022). COVID-19: Trust, windows and the psychology of resilience, in *The Routledge Handbook of Resilient Thermal Comfort*, Edited by F. Nicol, H. Bahadur Rijal and S. Roaf, London: Routledge.
36 Knight, G.M., T.M. Pham, J. Stimson, et al. (2022). The contribution of hospital-acquired infections to the COVID-19 epidemic in England in the first half of 2020. *BMC Infectious Diseases* 22, pf. 556. https://doi.org/10.1186/s12879-022-07490-4
37 Dancer, S.J., K. Cormack, M. Loh, et al. (2022). Healthcare-acquired clusters of COVID-19 across multiple wards in a Scottish health board, *Journal of Hospital Infection*, 120, pp. 23–30.
38 Su, W., Z. Ai, J. Liu, B. Yang and F. Wang (2023). Maintaining an acceptable indoor air quality of spaces by intentional natural ventilation or intermittent mechanical ventilation with minimum energy use, *Applied Energy*, p. 121504. DOI: 10.1016/j.apenergy.2023.121504
39 Su, W. Z. Ai, J. Liu, B. Yang and F. Wang (2023). Maintaining an acceptable indoor air quality of spaces by intentional natural ventilation or intermittent mechanical ventilation with minimum energy use, *Applied Energy*, 348, p. 121504. DOI: 10.1016/j.apenergy.2023.121504
40 Kwitny, J. (1991). The great transportation conspiracy, Chapter 19, in *Controlling Technology: Contemporary Issues*, Edited by W.B. Thompson, Buffalo, New York: Prometheus Books.
41 Mills, C. (1946). *Climate Makes the Man*, London: Victor Gollancz Ltd., p. 117.
42 Teitelbaum, E., and F. Meggers (2022). Rethinking radiant comfort, in *The Routledge Handbook of Resilient Thermal Comfort*, Edited by F. Nicol, H. Bahadur Rijal and S. Roaf, London: Routledge.
43 Persily, A. (2022). ASHRAE indoor carbon dioxide position document: What's next?, *Proceedings of the 42nd AIVC Conference Rotterdam*, ISBN: 2-930471-63-1, pp. 33–35.
44 ASHRAE (2022). Position document on indoor carbon dioxide. *American Society of Heating, Refrigerating and Air-Conditioning Engineers*, Accessed on 01/08/2023 at: https://www.ashrae.org/file%20library/about/position%20documents/pd_indoorcarbondioxide_2022.pdf
45 McCartney, K.J., and J.F. Nicol (2002). Developing an adaptive control algorithm for Europe: Results of the SCATs project, *Energy and Buildings*, 34(6), pp. 623–635, ISSN: 0375 7788
46 Accessed on 19/06/2023 at: https://airpophealth.com/blogs/air-resource-centre-arc/carbon-dioxide-on-planes
47 Porteous, C. (2011). Sensing a historic low-CO_2 future. *Chemistry, Emission Control, Radioactive Pollution and Indoor Air Quality*. https://doi.org/10.5772/16918
48 de Man, P., S. Paltansing, D.S.Y. Ong, et al. (2020). Outbreak of coronavirus disease 2019 (COVID-19) in a nursing home associated with aerosol transmission as a result of inadequate ventilation, *Clinical Infectious Diseases*, 73, pp. 170–171. https://doi.org/10.1093/cid/ciaa1270
49 Satish, U., M.J. Mendell and K. Shekhar, et al. (2012). Is CO_2 an indoor pollutant? Direct effects of low-to-moderate CO_2 concentrations on human decision-making performance, *Environmental Health Perspectives*, 120(12), pp. 1671–1677.
50 https://www.hok.com/news/2017-07/architectural-record-explores-natural-ventilation-at-singapores-ng-teng-fong-general-hospital/
51 Bluyssen, P.M. (2009). *The Indoor Environment Handbook: How to Make Buildings Healthy and Comfortable*, London, UK: Earthscan, ISBN: 978-1-84-407787-8
52 Bluyssen, P.M. (2025). The need to go beyond the comfort-based dose-related indicators in our IEQ-guidelines, *Building and Environment*, p. 113349, ISSN: 0360-1323. https://doi.org/10.1016/j.buildenv.2025.113349
53 Kabele, K., Z. Veverková and M. Urban (2022). Methodology of IEQ assessment in energy efficient buildings, in *The Routledge Handbook of Resilient Thermal Comfort*, Edited by F. Nicol, H. Bahadur Rijal and S. Roaf, London: Routledge, ISBN: 978-1-032-15597-5
54 Accessed on 17/06/2023 at: https://www.usgbc.org/resources/table-1-comparison-comfort-acceptability-ranges-iso-7730%E2%80%932005-and-en-15251%E2%80%932007
55 Pick, D. (2022). *Brainwashed: A New History of Thought Control*, London, Profile Books Ltd., p. 195.
56 Schumpeter, J. (1942). *Capitalism, Socialism and Democracy*, London, 2003, pp. 258 and 283.
57 Mills, C. (1946). *Climate Makes the Man*, London: Victor Gollanc Ltd., p. 116.
58 ASHAE (1995). *Heating Ventilating Air Conditioning Guide*, Vol. 33, New York: Published by the American Society of Heating and Air-Conditioning Engineers, Inc., pp. 127–128.
59 Cândido, C., R. de Dear, R. Lamberts and L. Bittencourt (2010). Cooling exposure in hot humid climates: are occupants 'addicted'?, *Architectural Science Review*, 3(1), pp. 59–64. DOI: 10.3763/asre.2009.0100

60 Shove, E. (2018). What is wrong with energy efficiency?, *Building Research & Information*, 46(7), pp. 779–789. https://doi.org/10.1080/09613218.2017.1361746
61 Cass, N., and E. Shove (2018). Standards? Whose standards?, *Architectural Science Review*, 61(5), pp. 272–279. https://doi.org/10.1080/00038628.2018.1502158
62 Shove, E., G. Walker and S. Brown (2014). Transnational transitions: The diffusion and integration of mechanical cooling, *Urban Studies*, 51(7), pp. 1506–1519.
63 Moore, R. (2023). As HSBC quits and hybrid working stays, is this the end for HSBC Wharf?, *The Observer*, 02/07/2023, Accessed on 02/07/2023 at: https://www.theguardian.com/artanddesign/2023/jul/01/hsbc-quits-canary-wharf-hybrid-working-london-docklands
64 For example: https://www.uva.nl/en/content/news/news/2022/04/temperature-in-uva-buildings-to-be-gradually-lowered.html?cb
65 Tanabe, S.-I., Y. Iwahashi, S. Tsushima and N. Nishihara (2013). Thermal comfort and productivity in offices under mandatory electricity savings after the Great East Japan earthquake, *Architectural Science Review*, 56(1), pp. 4–13. https://doi.org/10.1080/00038628.2012.744296
66 Lazarus, R.J. (2009). Super wicked problems and climate change: Restraining the present to liberate the future, *Cornell Law Review*, 94(5), pp. 1153–1234, Accessed on 04/07/2023 at: Super Wicked Problems and Climate Change: Restraining the Present to Liberate the Future (cornell.edu)
67 Roaf, S. (2014). Transitioning to eco-cities: Reducing carbon emissions while improving urban welfare, Chapter 7, in *Secure & Green Energy Economies*, Edited by Y.-D. Wang and J. Byrne, Washington: Transaction Publishers, ISBN: 978-1-41-285375-0
68 Roaf, S., F. Nicol and R. de Dear (2013). The wicked problem of designing for comfort in a rapidly changing world, *Architectural Science Review*, 56(1), pp. 1–3. http://dx.doi.org/10.1080/00038628.2012.753783
69 Holling, C.S. (Ed) (1978). *Adaptive Environmental Assessment and Management*, Chichester: John Wiley and Sons Inc, p. 105.
70 Diesendorf, M., and R. Taylor (2023). *The Path to a Sustainable Civilisation: Technological, Socioeconomic and Political Change*, Singapore: Palgrave Macmillan, p. 130, ISBN: 978-9-81-990662-8
71 ASHRAE (2020). *Proclaiming the Truth*, 2nd Edition, Princeton Selling Group.
72 Comstock, W.S., and B.J. Spanos (1995). *Proclaiming the Truth: An Illustrated History of the American Society of Heating, Refrigerating and Air-conditioning Engineers, Inc.*, Atlanta, GA: ASHRAE.
73 Woodward, B. (2020). *Rage*. New York: Simon & Schuster.
74 Nissen, K., J. Krambrich, D. Akaberi, et al. (2020). Long-distance airborne dispersal of SARS-CoV-2 in COVID-19 wards, *Scientific Reports*, 10, p. 19589. https://doi.org/10.1038/s41598-020-76442-2
75 Dancer, S.J., K. Cormack, M. Loh, et al. (2022). Healthcare-acquired clusters of COVID-19 across multiple wards in a Scottish health board, *Journal of Hospital Infection*, 120, pp. 23–30. https://doi.org/10.1016/j.jhin.2021.11.019
76 Guardian, The (2022). https://www.theguardian.com/society/2022/jul/21/london-nhs-trust-cancels-operations-as-it-system-fails-in-heatwave
77 See: Accessed on 06/04/2023 at: https://www.bbc.co.uk/news/magazine-35995972
78 Quilty, S., N.F. Jupurrurla, L. Aparna, et al. (2023). The relative value of sociocultural and infrastructural adaptations to heat in a very hot climate in northern Australia: a case time series of associated mortality, *The Lancet, Planetary Health*, 7(8), pp. e694–e693.
79 Roaf, S. (2022). COVID-19: Trust, windows and the psychology of resilience, in *The Routledge Handbook of Resilient Thermal Comfort*, Edited by F. Nicol, H. Bahadur Rijal and S. Roaf, London: Routledge.
80 Barrero, J.M., N. Bloom, S.J. Davis (2021). Why working from home will stick, *Centre for Economic Performance*, Discussion Paper No. 1790.
81 Oseland, N. (2022). *Beyond the Workplace Zoo: Humanising the Office*, London: Routledge.
82 Roaf, S. (Ed) (2012). Innovative approaches to the natural ventilation of buildings, *Architectural Science Review*, ASR, 55(1), pp. 1–3. http://dx.doi.org/10.1080/00038628.2012.642190
83 Anderson, C.A. (2001). Heat and violence, *Current Directions in Psychological Science*, 10, pp. 33–38.

Thermal Heartbeats of Buildings

9

THE THERMAL HEARTBEATS OF BUILDINGS

Introduction: Thermally Well-Behaved Buildings

Passive survivability in design is defined as the ability of a building to maintain habitable conditions during an extended power outage or loss of heating fuel. The term was defined first by Alex Wilson of the US Passive Design Institute,[1] who advocated for much more resilient buildings after Hurricane Katrina struck New Orleans, Louisiana in 2005. The idea is currently gaining traction, with many people who occupy buildings that fail during extreme weather events, be it a heat dome over Vancouver (06/2022 to 01/07/2022), an ice storm in Texas (31/01/2022 to 02/02/2021), or a city-wide power outage in Sharjah in the United Arab Emirates (24/08/2023) when the temperature is 48°C (119°F) in the street outside, the lift does not work, and there are no opening windows in the apartment. Add into the equation the soaring costs of energy and the modern trend of building lightweight, over-glazed buildings with poor or no natural ventilation options and buildings really can become lethal for occupants. Alex Wilson and Daniel Overbey[2] are now calling for people to design climate-safe buildings and propose that to do so three things are needed:

1. People need to know what future climates will probably be like and have free and easily accessible data on them to test future buildings' performance.
2. New buildings must be simulated with the 'HVAC/off' function to prove that people are able to survive in them without mechanical heating or cooling when grids fail.
3. All building simulations to have easy-to-use passive solar heating and natural cooling simulation options from day 1 of design.

We saw in the Mongolian yurt example that some lightweight structures can be occupied in the extreme cold, but only with a good supply of fuel to heat people when occupied. In very hot climates, a tent can simply be used as a shade awning to keep people comfortable beneath it when air temperatures are within customary safe ranges for the adapted occupants. But for a good *passive survival* building that will keep people healthy and even alive during an extreme temperature event when the power does fail, this requires a steady thermal environment provided by a 'well-behaved' indoor climate with a safe thermal heartbeat in at least spaces in a building.

DOI: 10.1201/9781315645070-14

Thermal Heartbeats of Buildings

The amplitude in the fluctuation of room temperatures and the speed at which temperatures change indoors can be thought of as akin to the thermal pulse, or the heartbeat, of a building. This is affected by the climate and weather outside, of course. A traditional building with a thick overhanging roof in a tropical climate where there is little variation between day and night may have a slow thermal pulse with a small amplitude over most of the year. Of course, put a glass box building in that climate and it will have huge diurnal swings as the intense overhead sun will send temperatures indoors soaring to feverish levels when not impeded by clouds. How 'well-behaved' the core thermal character of a building is can be reflected in the amplitude of its temperature pulse over a day or more. This is related to both the outdoor temperature and the extent to which the building itself modifies that temperature over the changing seasons that is dependent on its form, the quality of its construction, and the extent to which it can absorb and store heat in its structure to slow down that thermal pulse (Figure 9.1).

The internal pulses of different spaces around a building are dictated in part by the diurnal and seasonal heat or cold inputs from the local climate and the amount of accessible internal mass there is to absorb, store, and temper excess energy inputs. In the Mongolian yurt in Chapter 7, temperatures inside the lightweight tent see-saw rapidly, depending on whether the stove was alight or not. In the thick-walled Nepalese house where the stove is always alight, the thermal mass of the house slows the thermal pulse indoors as excess heat is stored in the walls, roof and floors and radiated back into the room steadily over the day. Like the herder in his tent, they will use just as much fuel as needed to keep occupants thermally safe and to fuel domestic chores like cooking and heating water.

The third heat source involved in dictating comfort temperatures is that generated by the people themselves in the spaces. If they have come indoors after strenuous activity, they will not need such high room temperatures to maintain homeostasis so will be satisfied with lower

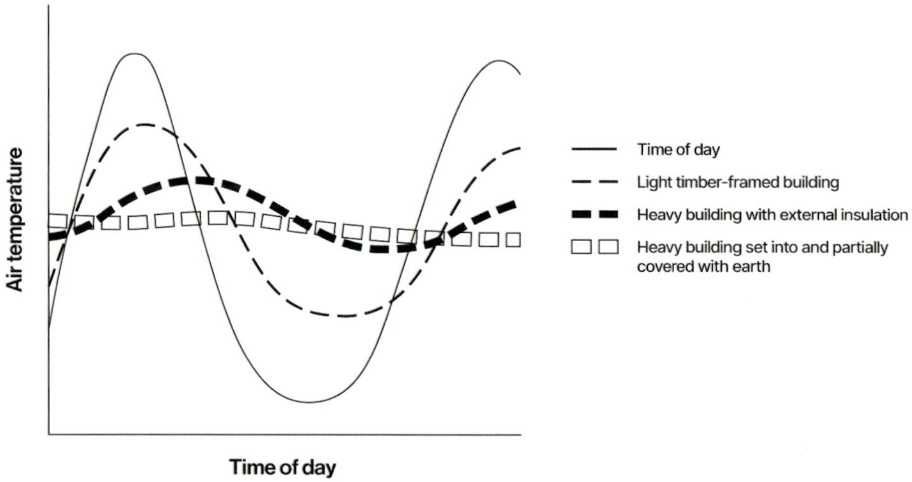

FIGURE 9.1 The thermal pulse of a building is affected by the ability of its materials to absorb and store heat over time and the extent to which it is coupled thermally to the steady temperatures of the ground beneath it, rather than the more rapidly fluctuating temperatures of the air around the site.

Source: Susan Roaf.

temperature and so may not feel the need to stoke up the fire or have a hot drink or food until they have cooled down a little. The level of innate skills required to time-manage the combination of weather, building design and materials, clothing, activities, food and drink, internal heat sources, etc. to maintain an agreeable thermal pulse indoors and in the metabolism of the occupant in a traditional building is impressive. It is aided by occupants having been programmed through adopted habits and customs to know exactly what to do to remain comfortable.

Chapter 7 showed alternative ways of staying thermally safe, even in similar extreme conditions. The same applies for animal species in extreme conditions. Cattle in northern Russia use a range of highly evolved adaptive strategies to survive outdoors in the extreme cold, such as developing thick fur coats and more body fat in which to store and conserve energy (heat).[3] Black bears survive thorough winter by hibernating to avoid the worst months of the Arctic winter. There is not only one way of adapting to a more extreme climate.

Many of us are programmed to be innately aware of the thermal heartbeats of buildings and how they will feel in the climates they occupy. Look at the two buildings in Figure 9.2. It is almost as if one immediately senses how they will feel thermally inside, without the aid of mechanical conditioning systems. The deep, steady coolth offer by the total protection from the sun by the roof and the massive walls of the (1) Sri Lankan bungalow will produce a steady, even cool feeling indoors, while in the (2) Phillip Johnson Glass House, the internal climate inevitably and unrelentingly tracks the outdoor temperatures, separated only by its glass walls and thin roof. It must arise from a very basic survival urge that enables us to make such innate thermal judgements, a need to know what we are letting ourselves in for.

FIGURE 9.2 (a) A European bungalow, the Kandy House Hotel, outside Kandy, Sri Lanka. (b) The Phillip Johnson Glass House in New Canaan, Connecticut, USA.

Source: Susan Roaf and the same as Figure 2.7a.

The difference between the two homes is that Glass House was designed to be kept comfortable using various levels of mechanic conditioning, while the bungalow was designed to stay comfortable without them. Each building has an inherent thermal rhythm. Such rhythms were part of the traditional builder's pallet of design strategies to be played in the thermal design of buildings to make the most of the available heat, coolth, light, and air to enhance thermal comfort and delight indoors. Thermal rhythms obviously do still exist in modern buildings but are overridden and disconnected from the weather outside by the mechanical systems that can rely on imported energy all day and year to remain habitable.

Thermal Pulses in Cool English Homes

People, when they can, control indoor temperatures in their own homes to suit themselves. Many people in English homes, for example, occupy what might be considered very low temperatures compared with international comfort standards.[4] While building stock and design models almost universally assume that all households turn heating on and off at the similar times of day in winter; in reality, people use heating in their homes in many varied ways.[5] One study, published in 2013, used hourly mean temperature data from 275 living rooms in English homes over three winter months, as shown in Figure 9.3.[6] People actually lived in average wintertime temperatures ranging from 11°C to nearly 27°C, with spikes in one home in the evenings up to 29°C. Temperatures experienced when the heating is on will very much depend on the type of heating used. Storage radiators using off-peak electricity are still common in UK homes using concrete bricks inside the radiators to store the heat from the overnight cheap rates to provide gentle background heat through a space across the day, providing a steady background temperature.

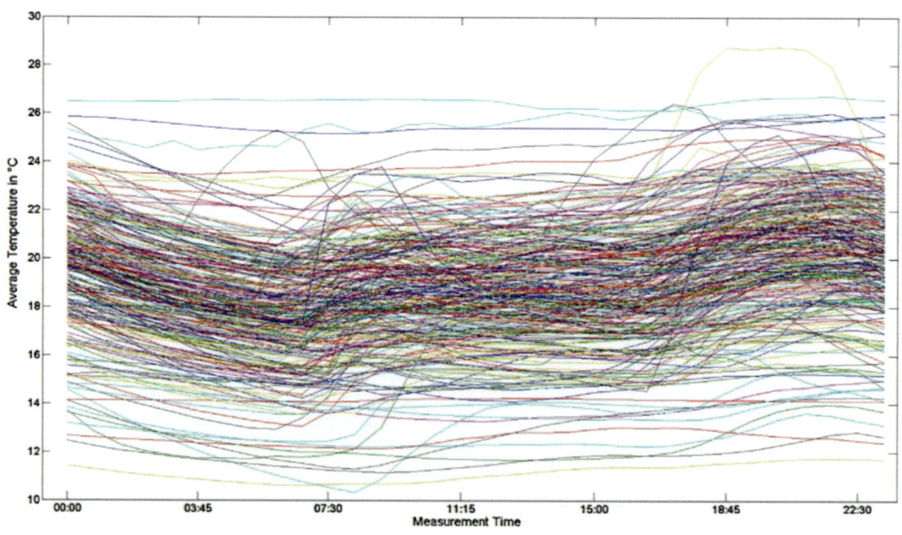

FIGURE 9.3 Mean hourly averaged temperatures in 248 English homes recorded between mid-July 2007 to early February 2008. The individual lines correspond to average weekday winter temperature data.

Source: Gesher Heubner.

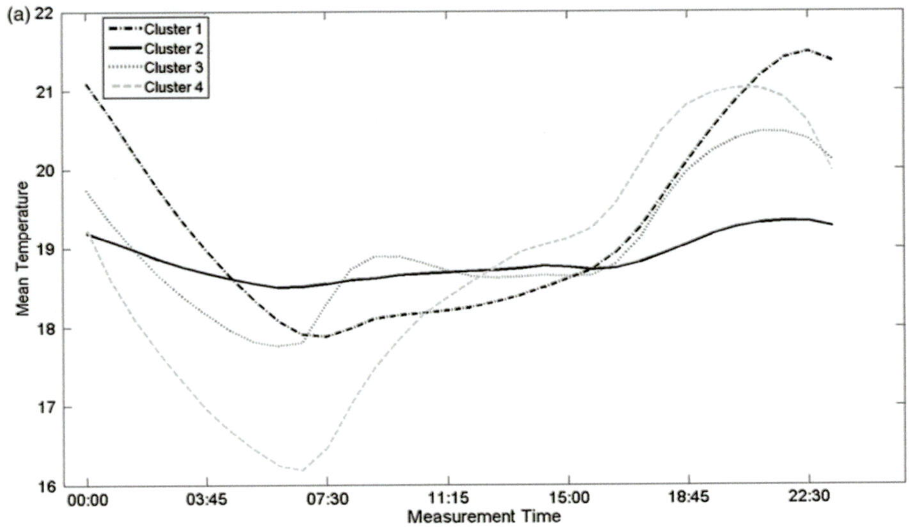

FIGURE 9.4 Average temperature profiles of the four cluster patterns, identified in the 2007–2008 English homes, and survey data for weekdays.

Source: Gesche Heubner.

Some people still use gas fires to provide radiant heat in living rooms, and this provides a 'surge' heat that warms the room quickly and continuously, so the temperature ramps up in it over time. People entering the room from the cold will positively respond to the visual and physical warmth provided by a gas fire, and as temperatures in the room rise, they will only turn the fire off when they get overly hot, often at temperatures in the high 20s centigrade.

Central heating systems, typically still use gas boilers in the UK and are often set to come on in the morning as people get out of bed, shutting off as they go to school or work. The boiler fires up again as people come home from work and settle down for the evening.

Heubner et al. then reduced that data into four different clusters of temperature profiles that were significantly different in their shape and varied in their minimum and maximum temperatures occupied.[7] Some 60% of the 'thermal pulses' of the homes showed atypical profiles that are not used in standard stock or design models. The study demonstrated that one size does not fit all in terms of the home heating habits of the English, which may explain why the UK comfort data clouds shown in Chapter 6 are so idiosyncratic, very different from the more uniform ranges of temperatures measured in the heat pump dominated, temperature clouds of Sweden? This reality reinforces the fact that it might be a little difficult for a Swedish researcher, or regulator, to understand, let alone relate to, the comfort customs and practices in the homes or offices of the UK, France, or Portugal, as we saw in Chapter 6.

Thermal Pulses in Hot Saudi Homes

Contrast the heartbeats of cool English homes with a mean temperature of under 19°C with those in Dammam with means of around 28°C, as shown in Figure 9.5. Living rooms in 17 middle-class homes in the Saudi Arabian Gulf port city of Dammam were measured over a year.

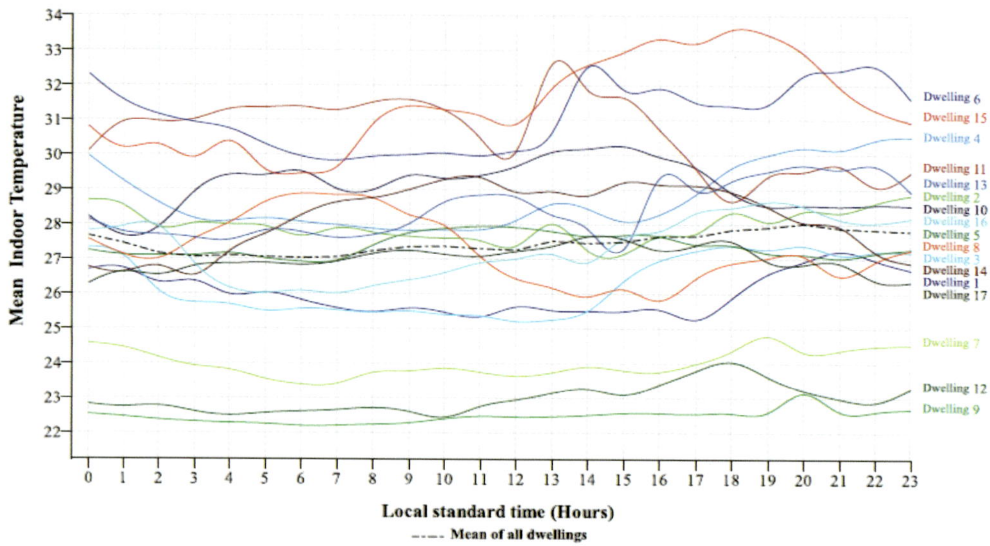

FIGURE 9.5 Temperature traces of mean indoor temperature in Dammam homes during the summer season.

Source: Abdulrahman Alshaikh.

In the summer, the mean indoor temperature was unsurprisingly around what might be expected for lightly clad people as at the lower end of the thermo-neutral zone.[8] Several homes appear to have been cooled most of the day, while at the upper end of the scale, perfectly comfortable temperatures at the top end of the thermoneutral zone may have been made even more so with the use of local fans.

Ramp and Jump Comfort

These profiles clearly show the ramp and jump of temperatures experienced in both UK and Saudi Arabian housing. The *ramp* – the rate of decay or increase of the temperature in a house once the heating or cooling is turned off or on – is an indicator of the quality of the building. If temperatures slide down or up slowly, it indicates that heat is lost or gained slowly either because the space is surrounded by well-insulated surfaces (and not too much glass), there are few draughts sapping heat on the air, and there is enough thermal mass to store heat in to slow the rate of total heat loss from the system. Buildings that loose or gain heat quickly may be draughty, poorly insulated, and possibly with little effective thermal mass.

Jumps in temperatures indoors are mostly created by too close a connection between changing outdoor weather. Over-glazing in many modern buildings causes temperatures to soar when the sun hits the glass, with overheating exacerbated by too few opening windows available to dump heat out of when that happens. Another problem is in poorly draught-proofed buildings, when a strong cold wind arises and permeates the structure, dropping temperatures rapidly.

The quality of the building itself will influence the jumps and ramps in the temperatures indoors, shown in Figures 9.3, 9.4, and 9.5. The heating and cooling preferences and habits of occupants displayed may also provide an indication of the wealth of the household, with people not even turning the heat on in the morning possibly being too poor, or parsimonious, to do so, and those who occupy higher temperatures when at home may be presumed to be richer, or more profligate or generous, to pay for the extra energy. Occupants might also be ill or disabled and have no alternative but to keep heating high or damage their health and well-being. Some may even be eco-warriors trying not to waste energy when they can simply put on another jumper.

Hot Desert Lifestyles

In cold climates, the trick is to keep the body topped up with as much heat as it needs to maintain comfort. In very hot climates, the opposite is true. In cold climates, people can put on more clothes to prevent heat loss from the body, but as temperatures rise, there comes a point, around about the thermoneutral zone limits of a body, say, c. 28–33°C, where there is little more clothing that can be shed, for comfort or cultural reasons. Other processes then take over to reduce heat strain on the body. Moving more air over the skin is effective, as discussed in Chapters 2, 3, and 5, but at very high temperatures, the ambient conditions provided by the building itself become an essential last resort to keeping people thermally safe indoors.

The extraordinary architecture of the deserts of the Middle East and Asia, from Iraq to India, enables people to live in real comfort over a summer's day, even when local temperatures climb to 45–50°C outdoors. The secret is to get as far away from solar gain as possible during the hottest times of the day. Traditionally, this is achieved in basements that provide thermal refuges against the scorching afternoon sun. Simply by building down into the earth, a drop of 20°C or more can be achieved; from the roof of a house in Baghdad, for instance, at 50°C, the basement three floors below can be at 30°C.[9] This temperature is approximately in the middle of the thermoneutral zone and can be perfectly comfortable for a resting body with some air movement.

Figure 9.6 shows the thermal pathways over 24 hours of three different housewives in different houses in the desert city of Yazd in Iran in 1976.[10] Summer temperatures climbed regularly to over 40°C on roofs, and to avoid the scorching afternoon heat, the whole family retired to the 3–5 m deep basements at around noon to eat a full lunch and then sleep in the coolth until around 5:00 p.m. When the sun left the courtyard, they emerged into early evening breezes and often sprinkled water over the courtyard and raised the summer terrace (*aivan*) to cool its surfaces. While the morning and evening were spent in and around the courtyard, and the afternoon in the basement, at night the family climbed the stairs to the roof and slept under eiderdowns beneath the cold, star-studded sky. The sun hitting the roof again around 06:00 a.m. was the sign for all to roll up their bedding and descend to breakfast on the shaded courtyard floor and start the housework or head off to school or the office or bazaar for work. Movement between the four thermal transitions of the day, two steps and two ramps, is associated with a change triggered by some discomfort. As the sun struck the roof, the sleepers on it were awoken by heat on their faces. As the court got too hot, the descent to the basement was matched by the delight of the delicious cool below. By 5:00 p.m., the air below felt a bit dank, prompting a rise to the fresh, sun-free court with its fresh early evening breezes, and by midnight the family were happy to ascend to sleep beneath their warm quilts under a heaven full of stars.

Such thermal pathways enable people to live comfortable and civilised lives in hot desert climates through the timely use of the thermal landscapes of the spaces around the house, the use of planting and of water in a central pond in the court to lower the temperatures in those spaces. The morning and evening breezes were also important in the courtyard, and those that refreshed people in and around the windcatchers in the summer living rooms provided a source of thermal delight. In hot climates, the radiant temperatures of surfaces become critical. With the

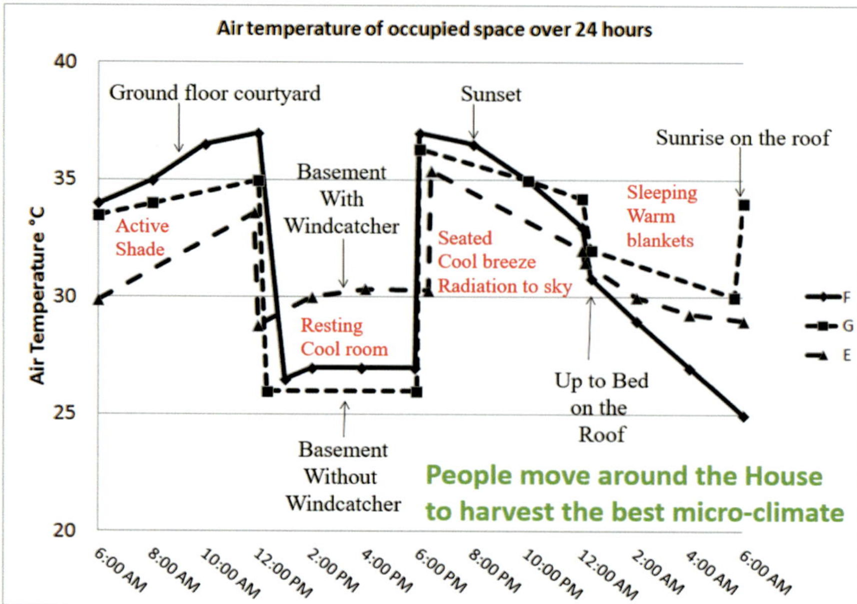

FIGURE 9.6 In the central Iranian desert, the three lines show the temperatures occupied by three women in different households going about their daily routines: getting up at 6:00 a.m. with the sun, moving around the ground floor courtyards till noon, descending to the basement for lunch and rest, then emerging back into the courtyard once the sun begins to set for the evening, until they go up to the roof to sleep at around midnight.

Source: Roaf (1989).

heavyweight mud-walled buildings in Yazd retaining their heat, the roof was a fresh air haven at night. The coolest rooms in the buildings were suitable for sleeping too if they could also have fresh, cool night air passing through them to compensate for the higher radiant temperatures from the walls. In Yazd, many old houses had blank 'Neemiyeh' walls without rooms behind them on the east of the courtyards, facing west towards the extreme heat of the low afternoon sun that is so difficult to shade against. Note that the main meal and the largest calorie input of the day is at mid-day, when people are in the cool basements, relaxing, and followed by rest during, which the extra heat of digestion can be dissipated while sleeping. This is similar to the siestas taken after lunch in many other hot parts of the world. So the thermal pulse experiences by our housewives in Yazd were not those of a single room but those they encountered over the course of their ordinary day, their individual thermal pathways that will be modified with the changing of the seasons as they modify their lifestyles to harvest comfort where and when they can find it around the house.

Are Temperature Jumps and Ramps Healthy?

The HVAC industry's notion of a comfortable building is one with a steady temperature all day and year, provided by mechanical equipment. However, such conditions can actually be physiologically unhealthy too. Sitting all day in a single thermal comfort soup of an indoor climate, as many older people in care homes do, for instance, is now thought not to be such a good idea. Recent research shows that well-understood, customary excursions outside the comfort zone can significantly improve the health of occupants.[11] Changes in the ambient thermal environment can increase metabolic activity, for instance, through shivering or sweating, so stimulating the different processes in our thermoregulatory system, such as the vasoregulatory expansion and contraction of the blood vessels. Studies have shown that key health risks like obesity and insulin sensitivity in patients, as well as the cardiovascular parameters that can be positively affected by regular exposure to heat and cold, can be positively impacted by occupying more dynamic thermal environments. These result in acclimatisation to thermal changes that enable people also to be comfortable in a far wider range of temperatures, leading to greater enthusiasm for window opening and exposure to a broader range of temperatures, perhaps not the full 'Wim Hof'[12] treatment but a return to the joys of fresh air enjoyment, when undertaken sensibly.[13]

The importance of fresh air and sunlight to the health of patients lay at the heart of traditional approaches to the design of hospitals that exploited them to improve the health of patients,[14] a lesson whose importance has been made clear during the COVID-19 pandemic, where hospitals without opening windows have been found to have far higher incidences of viral cross-infection.[15] At the peak of the pandemic in early 2021, a study in a Lanarkshire hospital in Scotland reported that fatalities ceased completely in a ward with COVID patients where a window opening regime was instigated for a mere 15 minutes morning, noon, and evening. Not a single person reported any sense of discomfort despite being in mid-winter as the short sharp ingress of cold air that purged the space of pathogens was thought to be a very good thing by all involved.[16] As always with comfort, 'it all depends'.

There is a growing movement in the passive conditioning of buildings that encourages well-understood and dynamic temperature drifts and jumps, exploiting the thermal landscapes of buildings. Using a combination of behaviours and passive opportunities, like the use of local, personal comfort systems, such as local heaters, windows, or fans, and the use of clothing and food and drink, is taken to radically reduce energy consumption in buildings[17] and improve the comfort of occupants within them.[18]

However, if each of us has our own individual comfort requirements and understandings of how we achieve comfort, how can architects and engineers design to please all when multiple occupants around buildings are involved? It helps that, in reality, there are commonly shared psychological responses to temperature changes within buildings.

Avoiding Discomfort in Shared Spaces

Marcel Schweiker pointed out in 2022 that in order to increase human–building resilience, people who design, operate, and occupy buildings will have to work hard to keep abreast of the rapidly changing circumstances around buildings and continually evolve their designs to enable people to remain safe and acceptably comfortable in them.[19] People in shared spaces cooperate to tolerate wider ranges of temperatures. When alone in spaces, they can access and operate many more features, like opening windows and doors and operating blinds and shades, than when constrained by others. In offices, people tend to use more personal comfort technologies like desk fans or heaters. They make allowance for different comfort requirements to coexist more easily with their neighbours in the space.[20]

In households, similar negotiations and compromises on comfort conditions are made, with a far greater diversity of actions and behaviours on offer. In rapidly rising global and local temperatures, it is important to understand what works today and why and how adaptive opportunities can be extended in the future to keep people thermally safe indoors over time, not least to meet the complex needs of different households and other building occupants.[21] In any public setting, like schools, conditions indoors will result from the actions that occupants are able to take to maximise overall benefit to themselves as far as they can. Of course, comfort is complex, and their desire and ability to act may be tempered by other factors, like humidity, sound, light, and social conditions, in the context of financial costs and personal or group agency to act. Rarely is perfection achieved across the board, so there has to be compromise.[22]

How much thermal change can people accommodate? It is clear that how hot or cold it gets is important, the amplitude of the thermal pulse in spaces, but so too is how long it is too hot or cold for. People can cope with big step changes in temperatures they expect to encounter over a day, and they know how to cope with smaller thermal ramps on their thermal pathways through a day with small adaptive actions. Sedentary people, however, often find thermal variation in a space uncomfortable if it requires them to change their clothing or behaviours too often. It is important to understand comfort time frames.

Comfort Time Frames

Normally, in Western lifestyles, there are five different thermal time frames that people operate in and adapt to in order to avoid discomfort:[23]

1. *Instantaneous changes* in indoor temperatures, for example, when the sunny sky turns cloudy, lead to people changing clothing and their local conditions by harnessing passive comfort fittings, like windows, blinds, and shades. Such swift changes will also occur when people move from one space to another, for instance, between indoors and outdoors or from room to room. These changes will often be accompanied by a change of clothing (e.g. by putting on or removing an overcoat). In a familiar setting, they can often be dealt with and not give rise to discomfort.
2. *Within day* responses to changing temperatures in a space over several hours with changing external weather or internal thermal conditions are particularly important for sedentary occupants in a classroom or office. These may be responded to by altering clothing, fittings, locations, personal comfort systems (PCS), as well as using heating or cooling equipment. Again, there will be changes: for instance, on a journey from work, which will generally be understood in a familiar setting and can be compensated for by clothing.
3. *Short-term, day-to-day* reactions to weather conditions, like changing cold or warm fronts in high- or low-pressure systems. People will choose clothing ensembles for the day based on experiences from the day before. People learn to anticipate and cope with usual temperature shifts with mechanisms chosen from a well-understood pallet of possible strategies.
4. *Medium-term, month-to-month.* Inter-seasonal shifts in temperature are associated in all climates (except possibly in the temperate tropics) with major changes in clothing, behaviours, lifestyles, and the use of passive adaptive fittings and heating or cooling devices of systems.
5. *Long-term, year-to-year* evolution of behavioural patterns of individuals and populations as the consequence of migration, the heating climate, more extreme weather, technological or economic advances in a region, or changing social and economic trends and opportunities.

Thermal Shocks – Instantaneous Temperature Changes

In many offices in the Gulf states, employees commonly use words like *freezing* to describe the indoor temperatures they work in that are as low as 18–22°C when it is 45°C (113°F) or more outdoors. Similar thermal shocks are reported in overcooled offices in South-East Asia.[24] In mid-20th-century America, thermal shocks on entering air-conditioned buildings were seen as a downside of the technology. In 1946, Clarence Mills wrote that the great hopes for air conditioning 'had faded considerably as experience has emphasized its handicap's chief disadvantage is the sharp contrast between air conditions indoors and out'. It was the contrast between one temperature and another that was thought so harmful.[25]

Americans traditionally associated annual temperature shifts as contributing to *seasonal vitality*, as colder winter and spring conditions made people work harder to keep warm. In the 1955, ASHAE Guide (33rd edition) engineers were instructed to use their own judgement and experience to design for appropriate indoor temperatures in a given locality and climate. Their decisions should be influenced by the heat capacity of the building, the degree of insulation, the amount of protection from the wind, the length of time minimum night-time temperatures persist, whether the building is used day and night, and other factors, including what temperatures people preferred indoors in that climate. The ASHRAE Guide provided tables of the climatic conditions for the main cities of all US states to help inform those judgements and to reduce the incidence of thermal shock between the inside and the outside of buildings.[26]

Most people can take such instantaneous indoor/outdoor thermal shocks in their stride, but they can be particularly dangerous for two groups: those with a range of health conditions that may be impacted by heat or cold and those who are adapted to colder conditions indoors and then have to spend periods in high outdoor temperatures, and vice versa, leading to acute thermal stress, and in some cases death, as in strenuous outdoor exercise, forced marches, or pilgrimages.

Diurnal Temperature Drifts: +/–2°C (+/–3.6°F) Okay in the UK

What sedentary occupants are most thermally sensitive to is short-term temperature variations in a space, particularly where temperatures over which they have no control see-saw up and down. Such conditions are common in highly glazed spaces when the sun and clouds alternate outdoors. One minute, occupants are hot, so jackets are removed, and then as clouds move across the sky, the jacket goes back on again. These actions result from people registering that they feel uncomfortable and move to act upon that fact to return to comfort. In a hot room, a window might be opened, leading to an immediate drop back down to cool enough temperatures to improve the comfort condition of the person beside it, or some clothing taken off to increase heat loss from the body. In a fully air-conditioned building, the indoor climate is detached from that outside, so no action is theoretically needed, just more energy.

Over a longer period of several days, people plan suitable clothing to facilitate comfort in the anticipated forthcoming conditions so barely notice that temperatures are drifting up or down over days. However, if there is a sudden change in outdoor temperature, for instance with a passing weather front, effort will be put into planning different clothing for the following day.

Nicol and Humphreys[27] found that for 'within day' surveys, people's voting in English schools and offices showed that every degree centigrade of temperature change corresponded to half of one comfort vote. So to go from slightly or comfortably warm to slightly or comfortably cool, a range of around +/–2°C (+/–3.6°F) was generally accommodated without undue complaint across a working day. But if the shift occurs from slightly cool or warm to the extremes of hot or cold, then people often reported feeling uncomfortable.

Children recorded by Humphreys[28] rarely changed their clothing during a school day in response to the often quite large swings of classroom temperature. Their response was only about a tenth of that needed to fully adapt to the temperature changes they experienced. They got rather warm instead of taking off their jumpers. The changes from one day to the next were rather greater, so the next day, when it was warm, they left the jumper in their school bag. From one week to the next, changes were greater still when they might have left the jumper at home. The clothing adaptations tend to lag behind the changes of indoor temperature. Office workers will also put up with a bit of discomfort on any one day but will change their clothes on the following day to ensure they are more comfortable in the changing weather.

People do not like occupying a continually fluctuating thermal environment that they cannot control. Situations in which people are forced to register that they are uncomfortable and required to take on and off clothes, change thermostat settings, or turn on and off personal comfort technologies like local fans or heaters become annoying. They then register they are uncomfortable and act upon it. If they successfully become more comfortable, the body may reward itself with feelings of pleasure for moving back into a safer thermal range. Such background see-sawing of thermal conditions can be avoided or modulated through the form and orientation of the building itself, the inclusion of well-designed thermal mass, or the use of passive or mechanical controls and systems, that is, if occupants are able to access and use them.

Seasonal Thermal Drifts: +/−4°C (7.2°F) Okay in the UK

Field surveys of sedentary office workers and of schoolchildren by Humphreys showed that populations seem to cope well with seasonal changes of indoor temperature in the UK of around ±3 or 4° from their annual average indoor temperature, provided there are no fixed dress codes.[29] So a modest day-to-day drift, up a few degrees indoors in summer and down in winter below that annual median temperature, is fine if that median temperature is reasonable for the climate. To cope with an annual range of mean indoor temperatures of up to 8°C, *customary behaviours* are extremely important in supporting the body's thermoregulatory system with habitually engrained adaptive strategies that can evolve over generations to maintain safe body temperatures. In the UK, indoor temperatures ranging from around 19°C to 27°C are not uncommon in schoolrooms and homes, and the drift between may be barely noticeable over the year. But there are no hard and fast rules on this. In continental climates, where outdoor temperatures like those in the Siberian case study may range from −30°C to +40°C, people will adapt to a wider range over the whole year as they do in Pakistan (see Chapter 5) with an outdoor and indoor temperature range of 20–30°C or more. In equatorial regions, where temperatures vary only slightly over a year, people feel comfortable in much narrower thermal drifts.

People-Centred Buildings

Studies done by the Usable Buildings Trust in the UK highlighted that successful buildings were able to respond quickly to changes in perceptions of comfort levels. The more quickly actions are taken, the more comfortable occupants feel.[30] So whether temperature control involves passive actions, like opening a window or putting down a shade or blind, or low-cost actions, like using a fan or hot water bottle, to fully mechanical solutions, like turning on a heating or cooling systems, what matters is the change can be effected promptly to improve the comfort of occupants indoors.

People-centred buildings are either those that have well-behaved thermal heartbeats indoors to minimise occupant discomfort or a co-evolved thermal lifestyle in which the daily routines of their occupants are able to be orchestrated to tune in with the thermal life of the building to minimise discomfort.

Notes

1 Overbey, D., and A.Wilson (2023). Toward passive survivability: Barriers and opportunities to optimize the built environment for greater thermal safety, in *Proceedings of ICARB 2023: Measuring Net Zero*, Edited by S. Roaf and W. Finlayson, pp. 376–397, Accessed on 22/01/2024 at: www.icarb.org/2023-conference/papers/
2 Overbey, D., and A. Wilson (2023). Toward passive survivability: Barriers and opportunities to optimize the built environment for greater thermal safety, *Proceedings of ICARB 2023: Measuring Net Zero*, Edited by S. Roaf and W. Finlayson, pp. 376–397, Accessed on 22/01/2024 at: www.icarb.org/2023-conference/papers/
3 Buggiotti, L, A.A. Yurchenko, N.S. Yudin, et al. (2021). Demographic history, adaptation, and NRAP convergent evolution at amino acid residue 100 in the world northernmost cattle from Siberia, *Molecular Biology and Evolution* (March 30). https://doi.org/10.1093/molbev/msab078
4 Huebner, G., M. Mcmichael, D. Shipworth, M. Shipworth, M. Durand-Daubin, A.J. Summerfield (2014). The shape of warmth: Temperature profiles in living rooms, *Building Research & Information*, 43, pp. 1–12. https://doi.org/10.1080/09613218.2014.922339
5 Kelly, S., M. Shipworth, D. Shipworth, M. Gentry, A. Wright, M. Pollitt, D. Crawford-Brown and K. Lomas (2013). Predicting the diversity of internal temperatures from the English residential sector using panel methods, *Applied Energy*, 102, pp. 601–621. https://doi.org/10.1016/j.apenergy.2012.08.015
6 Huebner, G.N., M. McMichael, D. Shipworth, et al. (2013). The reality of English living rooms – A comparison of internal temperatures against common model assumptions, *Energy and Buildings*, 66, pp. 688–696, ISSN: 0378-7788. https://doi.org/10.1016/j.enbuild.2013.07.025
7 Huebner, G., M. McMichael, D. Shipworth, M. Shipworth, M. Durand-Daubin and A. Summerfield (2015). The shape of warmth: Temperature profiles in living rooms, *Building Research & Information*, 43(2), pp. 185–196. https://doi.org/10.1080/09613218.2014.922339
8 Al Shaikh, A.R. (2016). *Design Principles for Thermally Comfortable and Low Energy Homes in the Extreme Hot-Humid Climatic Gulf Region, with Reference to Dammam, Saudi Arabia*, Doctoral thesis at Heriot Watt University, ISNI: 0000 0004 6347 5564, Accessed on 19/03/2023 at: https://ethos.bl.uk/OrderDetails.do?uin=uk.bl.ethos.712727
9 Nicol, F., M.A. Humphrey and S. Roaf (2012). *Adaptive Thermal Comfort: Principles and Practice*, London: Routledge, p. 70.
10 Roaf, S. (1989). *The Windcatchers of Yazd*, PhD thesis at Oxford Brookes University.
11 van Marken Lichtenbelt, W., M. Hanssen, H. Pallubinsky, et al. (2017). Healthy excursions outside the thermal comfort zone, *Building Research & Information*, 45, pp. 819–827. https://doi.org/10.1080/09613218.2017.1307647
12 Hof, W. (2020). *The Wim Hof Method*, London: Penguin Random House.
13 van Marken Lichtenbelt, W. (2017). Who is the Iceman?, *Temperature*, 4(3), pp. 202–205. https://doi.org/10.1080/23328940.2017.1329001
14 Hobday, R.A., and S.J. Dancer (2013). Roles of sunlight and natural ventilation for controlling infection: Historical and current perspectives, *Journal of Hospital Infection*, 84, pp. 271–282. https://doi.org/10.1016/j.jhin.2013.04.011
15 Roaf, S. (2022). COVID-19: Trust, windows and the psychology of resilience, in *The Routledge Handbook of Resilient Thermal Comfort*, Edited by F. Nicol, H. Bahadur Rijal and S. Roaf, London: Routledge.
16 Dancer, S.J., K. Cormack, M. Loh, et al. (2022). Healthcare-acquired clusters of COVID-19 across multiple wards in a Scottish health board, *Journal of Hospital Infection*, 120, pp. 23–30.
17 Brager, G., G. Paliaga and R. de Dear (2004). Operable windows, personal control, and occupant comfort, *ASHRAE Transactions*, 110, pp. 17–35.

18 Zhang, H., E. Arens and Y. Zhai (2015). A review of the corrective power of personal comfort systems in non-neutral ambient environments, *Building and Environment*, 91, pp. 15–41. https://doi.org/10.1016/j.buildenv.2015.03.013

19 Schweiker, M. (2022). Rethinking resilient thermal comfort within the context of human-building resilience, in *The Routledge Handbook of Resilient Thermal Comfort*, Edited by F. Nicol, H. Bahadur Rijal and S. Roaf, London: Routledge, ISBN: 978-1-032-15597-5

20 Marín-Restrepo, L., M. Trebilcock and J.A. Porras-Salazar (2020). Adaptation by coexistence: Contrasting thermal comfort perception among individual and shared office spaces. *Architectural Science Review*, 63(3–4), pp. 235–247. https://doi.org/10.1080/00038628.2019.1708257

21 Raw, G. (2018). What do households do to keep cool? *Proceedings of the 10th Windsor Conference on Rethinking Comfort*, pp. 240–253. ISBN: 978-1-9161876-4-1 Accessed on 07/08/2024 at: https://windsorconference.com/proceedings/

22 Raw, G. (2022). Varying comfort: A different kind of challenge, *Proceedings of the Comfort at the Extremes Conference: Comfort, Covid and Ventilation*, Edinburgh, pp. 348–356. ISBN: 978-1-9161876-4-1 Accessed on 07/08/2024 at: https://mosser.scot/CATE2022/CATE2022%20Proceedings%20(web_220930).pdf

23 Nicol, F. (2000). Time and thermal comfort: Evidence from the field, in *Proceedings of the World Renewable Energy Congress VI (WREC2000)*, Brighton: Elsevier Science.

24 Alnuaimi, A., S. Natarajan and T. Kershaw (2022). The comfort and energy impact of overcooled buildings in warm climates, *Energy and Buildings*, 260, pf. 111938, ISSN: 0378-7788

25 Mills, C.A. (1946). *Climate Makes the Man*, London: Victor Gollancz Ltd., p. 116.

26 ASHAE (1955). *Heating and Air Conditioning Guide*, Vol. 33, New York: Published by the American Society of Heating and Airconditioning Engineers, p. 244.

27 Nicol, F., and M. Humphreys (2010). Derivation of the adaptive equations for thermal comfort in free-running buildings in European standard EN15251, *Building and Environment*, 45, pp. 11–17.

28 Humphreys, M., F. Nicol and S. Roaf (2016). *Adaptive Thermal Comfort: Foundations and Analysis*, London: Earthscan, pp. 164–167.

29 Humphreys, M., F. Nicol and S. Roaf (2016). *Adaptive Thermal Comfort: Foundations and Analysis*, Vol. 27, London: Earthscan, pp. 260–270.

30 Leaman, A., and B. Bordass (2010). Assessing building performance in use 4: The Probe occupant surveys and their implications. *Building Research & Information*, 29, pp. 129–143. DOI: 10.1080/09613210010008045

10

KILLER BUILDINGS

Heatwaves

What Does It Feel Like to Get Heatstroke?

In 1997, architect David Cox was a project manager in the Northern Territory of Australia, monitoring buildings in the 49 Aboriginal Housing Associations throughout the state. Many were very remote, including the hot desert settlement of Wave Hill some 600 km south of Darwin. He flew there early one day for a series of constant and difficult meetings and inspections. He remembers eating or drinking nothing all day and, by late afternoon, began feeling exhausted with a slight headache. He flew out to Katherine, then unwisely drove some 110 km at night on a dirt road in an air-conditioned car to Beswick Station. Trying to sleep in basic accommodation, he had developed a blinding headache that continued all night. Aspirins did not work, so he took more, and realising how stupid he had been not drinking water all day, he also started drinking a lot, which made him nauseous and feel worse. There was no medic, medical kit, or even mobile phone at the stations, and he did not want to radio the flying doctor for a call out. In the morning, he compounded the problem by going on a walk with an Aboriginal friend, who took him to a beautiful nearby waterhole, where he simply lay, often semi-conscious on the bank, doing nothing for several days, sipping water gradually, and eventually eating a little fish until he felt well enough to walk back to the station. Heatstroke is associated with making dumb decisions, and he now admits he should have contacted a doctor from the station. A trigger for his collapse was thinking he could do in the heat what he could do in his air-conditioned office. Going from an air-conditioned office to a car to a plane straight into the heat left no time for acclimatisation. His body went from a fridge to an oven in seconds. Even lying down under a tree in the outdoor heat in an isolated location was also a bad idea.

He was lucky to recover. His metabolism was already in shock earlier in the day, but his conscious mind did not realise how dangerous a situation he was in. Many people are not so lucky. Examples of apparently healthy people dying from heatstroke are increasingly reported in the news. Even a well-informed medical expert like Dr Michael Moseley, who died of heatstroke in a heatwave on the Greek island of Symi in June 2024, did not understand the dangers of walking outdoors in high temperatures he was unaccustomed to of around 40°C (104°C) at that time. Heatstroke is a growing problem.

DOI: 10.1201/9781315645070-15

Heatstroke

Heatstroke can start with nonspecific symptoms, including weakness, dizziness, and nausea, feeling of excessive heat, and restlessness or confusion lasting several hours that can turn into a full-blown heatstroke, resulting in hyperthermia (overheating) as the body's core temperature rises above 40°C (104°F), alteration of mental status, and ultimately, damage to the central nervous system, causing delirium, convulsions, and/or coma.

Classic heatstroke results from exposure to high environmental temperatures, while that resulting from a combination of environmental heat and metabolic heat generated by exercise is called *exertional heatstroke*. The onset of both can be abrupt and manifest as a collapse and/or loss of consciousness. Both can develop within hours to cause catastrophic injury to most organs of the body, including the heart, lung, liver, kidney, gastrointestinal tract, and muscles, culminating in death.[1]

Perhaps the highest-profile example of exertional heatstroke is that of heat deaths by pilgrims in Saudi Arabia. Between 2002 and 2015, more than 90,000 pilgrims had died in Saudi Arabia on the Hajj or Umrah pilgrimages to Mecca, many from heatstroke. In 2024 alone, 1.8 million people joined the June Hajj, and of them more than 1,300 people died in heatwave temperatures that soared to 51.8°C (125°F). Officials were deeply concerned that more than three-quarters of those who died were there unofficially, without the permits associated with available relief on the wayside. As part of the required rituals, pilgrims, some of whom were elderly or chronically ill, walked under direct sunlight, many without adequate water or shelter, and presumably unaware of the terrible potential for, and consequences of, heatstroke.[2]

Thermal Stress, Discomfort, and Comfort

The line between comfort, thermal stress, and serious health impacts is often surprisingly fine. People on holiday largely have the adaptive opportunities at hand to avoid overheating. They can relax beside a swimming pool, plunge into it, retreat into a cool building, fill up on cool drinks, or take a snooze indoors in the heat of the day. Of course, there are some who still fall victim to a human version of the boiled frog syndrome, not understanding they are approaching heat stress conditions. If a frog is plunged into a pan of boiling water, it will try to jump out, but if it is immersed in water that gradually heats up, it will fail to notice the creeping change in temperature until it is eventually boiled alive.[3] The thermoregulatory system of the body has no good way of alerting the body to the fact that it is dangerously overheating other than by copious sweating. People do not easily sense the transition from feeling hot when sunbathing to full-blown heat stroke on holiday. Core temperatures rising over 40°C seriously affect health, and deaths may occur when core temperatures exceed 42°C.

People are often unaware of how seriously they are affected by temperature increases during heatwaves and so fail to alter their circumstances in time to avoid fatal conditions. Hyperthermia, overheating, can also result from taking recreational drugs, including ecstasy, which when combined with physical exercise, like dancing at raves, can trigger potentially fatal rises in body temperatures, which is why some clubs provide 'chilling out' spaces, hence the term. Some employers were historically cavalier in their willingness to have their employees die under their watch, such as the mining and construction industries, as we have seen from the appalling conditions that many of the millions of South-East Asian workers endure in six oil-rich Gulf states of Bahrain, Kuwait, Oman, Qatar, Saudi Arabia, and the United Arab Emirates, where

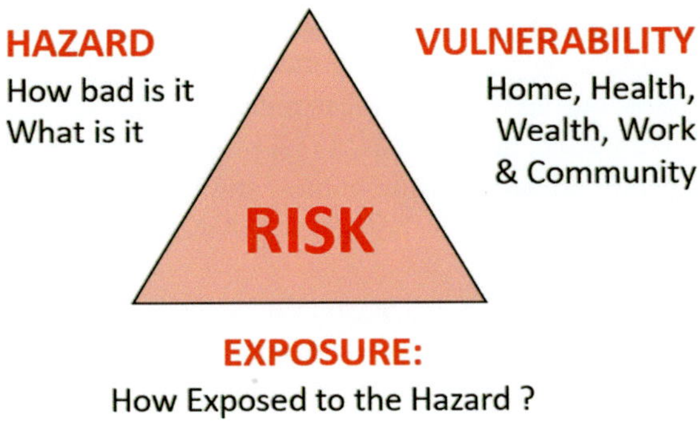

FIGURE 10.1 Risk is dependent on three variables: how bad the hazard is, how exposed a site is to it, and how vulnerable a person is.

Source: Roaf et al. (2009).

over 10,000 die every year, many resulting from being made to work in outdoor temperatures up to 50°C (122°C).[4]

Risk is affected by three main factors, as shown in Figure 10.1.[5] We have touched on the growing hazard of a heating climate and the common vulnerabilities in populations, but the exposure of a person is dependent not only on where they live but also on what building they live in.

In Heatwaves People Die in Buildings

During recent heatwaves in France and the United States, more than half of the patients were found dead in their home or hospitalised within one day of the reported onset of illness, and for those who reached the hospital and were admitted to intensive care, over 60% died and a third of survivors may sustain permanent neurologic damage.

Analyses of the 2003 European heatwave, in which over 70,000 people died, highlighted how mortality was affected by the interaction between natural and social factors.[6] Unusually high temperatures, as well as socio-economic vulnerability, led to an excess of 14,947 deaths in France, between August 4 and 18, 2003. The greatest increase in mortality was due to causes directly attributable to heat: dehydration, hyperthermia, and heat stroke. In addition to age and gender, other factors included pre-existing diseases, medication, urban residence, isolation, poverty, and probably pollution.[7] In buildings, major risk factors included a lack of mobility, as many of the elderly died while their younger relatives were on holiday away from home. Housing characteristics associated with death were lack of thermal insulation and sleeping on the top floor, right under uninsulated roofs, particularly in areas where heat built up around a building in urban areas.[8]

In the July 1995 Chicago heatwave, 739 heat-related deaths were recorded over a five-day period. Most of the victims of the heatwave were elderly poor residents of the city, who could not afford air conditioning and would not open their windows or doors simply out of fear of being robbed or murdered.[9] In the North American Heat Dome of July 2021, many all-time maximum temperatures were recorded, and even in Canada, a national temperature record of 49.6°C was set.[10] Again, as in Chicago and France, the risk of death during the event was associated with deprivation, lower neighbourhood greenness, older age, and gender, with more women dying. High indoor temperatures played an important role, as did low air conditioning prevalence and also grid failures, during which no mechanical cooling was available.[11] In America, in more urban areas with high urban heat islands, a real problem is with people working outdoors and the homeless and people in cars, mobile homes, and caravans that they cannot keep cool. So acute has the problem become that the hot states in America now keep detailed records and reports of heat death–related data.

In 2021, it was estimated that in the apparently temperate climate of the UK, some 3.6 million (15%) of homes had living rooms that overheated and 4.6 million (19%) of homes had bedrooms that overheated in the then heatwave summer of 2018. The majority of UK homes are two story, with the bedrooms more exposed to both solar gain through the roofs and heat rising from the living areas below. Two-thirds of the overheated rooms were in the London area, also impacted by the higher temperatures in its urban heat island. Again richer people who live in larger homes were found to be much less affected than the disadvantaged poor in crowded social housing. The level of exposure to the problem now bodes badly for many homeowners in a predicted much hotter future.[12]

Heat Death Reporting: Lessons for Future Climates

Because of the growing scale of the problem of heat deaths, many of the most affected regions had stringent reporting protocols for heat deaths. In Arizona, rising temperatures and numbers of vulnerable people have resulted in a huge growth in heat-related deaths.[13] In 2023, 987 people died from heat-related causes, and in the greater Phoenix area, that figure was 645, a 52% increase from 2022, peaking during heatwaves, as shown in Figure 10.2.[14] Around 75% heat-related deaths were male, around 60% were aged 50 years or older, 80% were Maricopa County residents, and 75% were exposed outdoors, of which 67% were in urban areas. A surprising 6% of outdoor heat–related deaths involved some type of physical or recreational activity, like walking, running, bicycling. About 50% of those who died had a medical history of physical/mental health conditions, 45% were experiencing homelessness, and 71% died on a day where the Weather Service issued an excessive heat warning. Nearly 60% of heat-related deaths occurred among non-Hispanic White individuals, compared to African Americans (6.8%) and, finally, American Indians (2.9%), who are overrepresented among heat deaths by population proportion.

Sadly, 95% of all indoor heat–related deaths occurred among individuals aged 50 years and over, and 75% were found when neighbour, family member, or health official called in to check on how they were. Indoor deaths occurred mainly in houses (62%) and other living spaces, in RVs/trailers (16%), apartments (14%), and mobile homes (8%). In most homes, an air conditioning unit was present on-site and, in 85% of cases, was functioning. Despite the low electricity prices in Arizona, many still cannot afford to pay for cooling, and in 2022, Arizona's regulated utilities have been banned from cutting off power during the summer, following the 2018 death of a 72-year-old woman after Arizona Public Service disconnected her electricity over a $51 US debt. Certain health conditions put individuals at increased risk of heat-related illness, and in Maricopa County in 2023, these included, cardiopulmonary disease (46%), obesity (7%), diabetes (8%), kidney disease (2%), dementia (2%), and injuries or falls (3%). The collection of such data is vital to understanding the health risks of populations in increasingly extreme heat events, to inform current action plans and new design drivers to combat overheating and

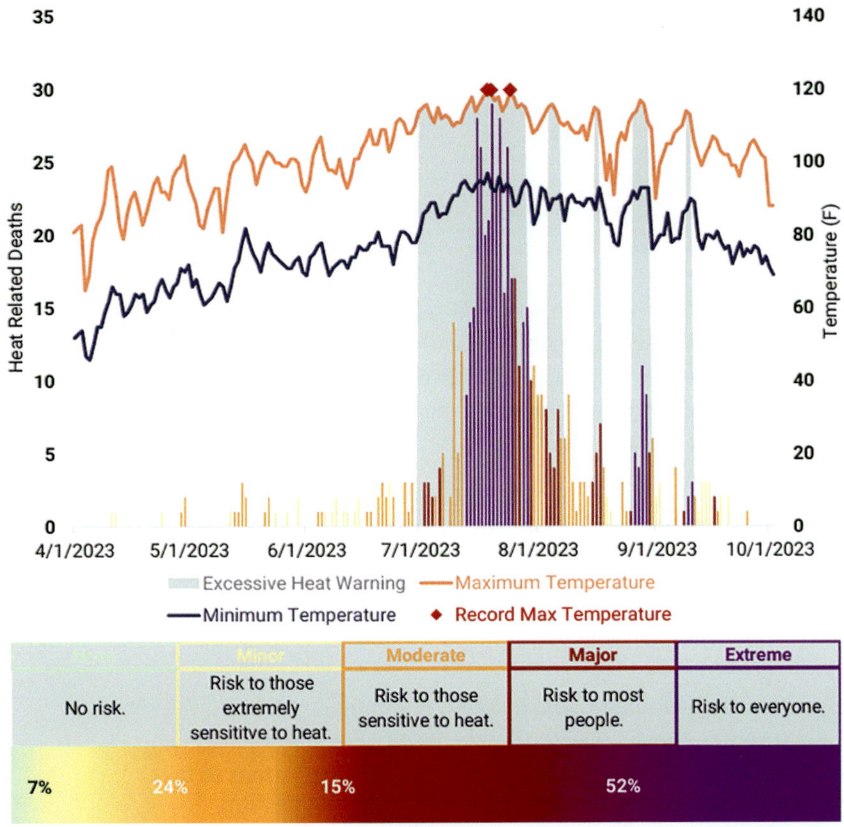

FIGURE 10.2 Temperatures at which people died from heat-related deaths in Arizona in 2023 for five levels of health risk warnings, showing the way heat deaths increase in relation to the maximum and minimum temperatures and the duration of the heat event.

Source: The Maricopa County Department of Public Health.

to scope out potential effective solutions for providing safe thermal environments during future superheating events.

Across the USA, similar exercises are being undertaken and used to experiment with, test, and share the effectiveness of overheating solutions. During the 2021 Heat Dome in Portland, Oregon, the highest risk of dying was from homeless, largely White men.[15] Currently, the sending out of mobile weather stations is being trialled in Arizona to climate data deserts in urban heat island locations where high numbers of heat-related deaths occur. Poorer neighbourhoods were traditionally less well served by temperature monitors, like temporary camps and caravan parks, where so many die in inadequate shelters.

In 2023, in Arizona, daily temperatures did not fall below 91°F (33°C) from the 10th to the 25th of July during two weeks, in which 303 people died in Maricopa County alone, and the temperature record was broken when 119°F (48°C) was exceeded five times. On 7 July 2024, the overnight temperatures in Phoenix did not fall below 92°F (33.3°C), which is 8°F (4.4°C) above normal for the city. Why do such studies matter? Why do such temperatures matter? Perhaps because conditions like this will spread around the world as the climate heats, and we should be using such temperatures to inform today how we design for the future, to develop a holistic understanding of the problems of overheating, including their social and economic drivers too.

As Marcel Schweiker explains in his work on Resilience:[16]

> *Distinctive aspects of resilience like 'toughness', 'ability to cope' or 'capacity to recover' are linked to definitions of thermal comfort which go beyond comfort as thermal relief provided by thermoneutral conditions.*

Who knows when conditions such as those now found in Arizona will spread to now cooler regions. In the UK, a maximum temperature of 40°C (104°F) was recorded in July 2022. Tragedies too often occur when people do not have the time or opportunity to adapt behaviourally and physiologically to rising temperatures in an acute heatwave. We should all now be preparing to adapt to very high temperatures.

Does Your Neighbourhood Overheat?

There are many new tools available for people to identify the areas most vulnerable to the impacts of climate change. In the UK, for instance, Climate Justice has produced a free web tool for public service providers designed to:

- Identify who is vulnerable to climate change and fuel poverty
- Highlight neighbourhoods where climate disadvantage is highest
- Explain the factors involved and help you decide what actions to take

Globally, emissions are now tracking high-emissions projections, and if ordinary citizens really understood the potential for runaway climate trends resulting in the extreme temperatures shown in Figure 10.3 on the Climate Justice website, would they still make the same purchase and lifestyle decisions they do today?[17] Many people now check whether the home they want to buy is on the floodplain, but now it is possible to also review the heat stress metrics for a location, like the predicted daily maximum and minimum temperature, temperature variability, and vapour pressure for future dates under different emissions scenarios. Such maps are available for the UK that show qualitatively similar spatial temperature patterns for the recent past, highlighting the most pronounced UK heat hazards areas that are found in the south-east English regions, especially in Greater London with its entrenched urban heat island.[18] When combined with socio-economic factors, hotspots of high heat stress risk also emerge in the Midlands and eastern England, along with the southern and eastern coastal regions. Different heat risk factors subjectively weighted on an interactive app are provided for heat stress vulnerability, which allows users to assess sensitivities and uncertainties in the projected UK heat risk.[19] All local authorities in the UK and many around the world have rapidly developing climate adaptation plans that increasingly focus on reducing vulnerabilities related to rising temperatures.

A key factor in whether a neighbourhood, street, or building overheats is also the amount of shade in and around it, whether it be from adjacent buildings, shading devices, or importantly, planting, as the following case study shows.[20]

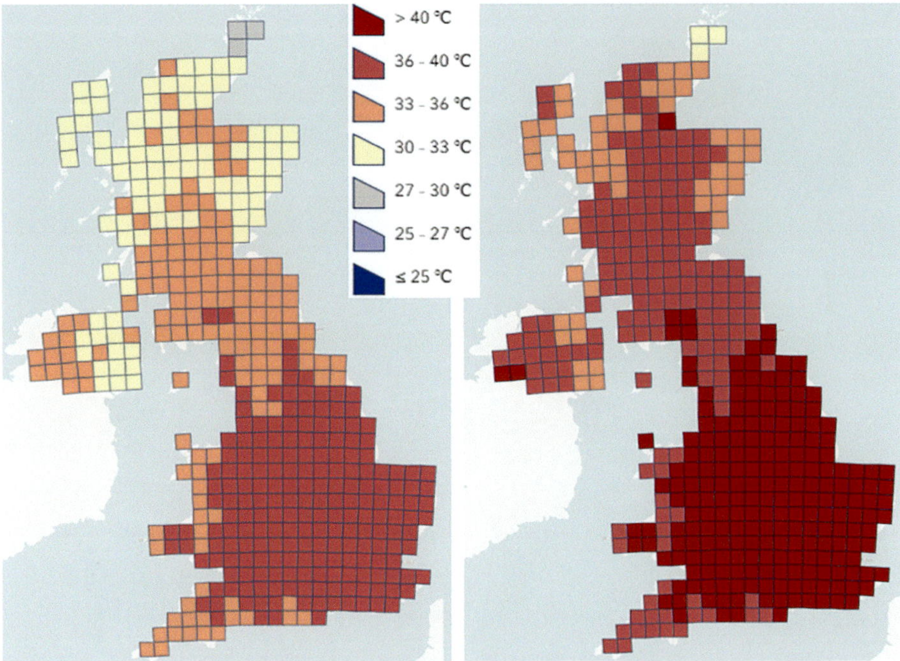

FIGURE 10.3 Heat exposure maps for 2080 modelled from 2018 data showing predicted summer extreme maximum temperatures (°C) for 100-year return periods, for two high-emission scenarios: (a) Central estimate (50th percentile). (b) High estimate (90th percentile).

Source: Climate Justice.

Experiences From the USA

People who either live in a good local climate or are wealthy enough to be able to choose their location are the lucky ones because micro-climates do make a difference. In the North American heat dome of July 2021, the street temperature in the Lents, one of the poorest neighbourhoods in Portland, Oregon, measured 51°C (124°F), while across the city in Willamette Heights, a verdant, exclusive neighbourhood with lots of trees and parks, the temperature was recorded at the same time as 37°C (99°F). A few blocks can make the difference between comfort and survival, not least when the grid fails. Figure 10.4 is a map of modelled temperatures in downtown Portland, Oregon, showing the distribution of relative hot/cold temperatures through the city. The downtown area, which is one of the hottest areas in the city overnight, with its tight-packed buildings, becomes one of the cooler areas during the day, with its narrower streets now shaded. Heavily forested areas, including major parks and certain residential areas, show a tendency towards cooler temperatures. Areas with lower canopy cover, such as those at the northern edge of the city in the industrial, port, and airport areas, eastern freeway corridors, and train yards, consistently appear hotter in the afternoon urban heat island model. Additionally, areas around freeways and arterial roadways show small, tightly knit local 'pockets' of heat.[21]

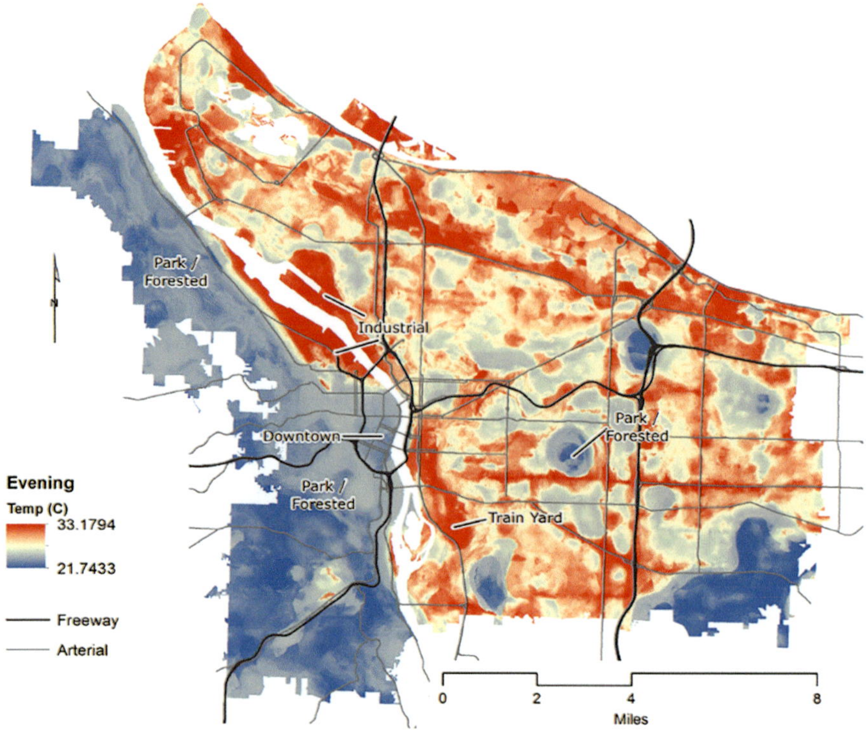

FIGURE 10.4 Modelled map of the afternoon urban heat island of Portland, Oregon, based on data collected on 25 August 2024 during a heatwave. Significantly lower temperatures exist in leafy suburbs, parks, and hilly areas, in contrast with the high temperatures around unshaded concrete, bitumen, and buildings' features.

Source: Jackson Voelkel.

What Buildings Overheat?

In any climate, buildings can become dangerous for health and life during excess heat events. In the temperate English climate, excess heat–related mortality is first detectable above the seemingly low temperature threshold of around 24.5°C (75°F). In a 2024 study of the impact of hot weather, defined as an average of 24.5°C or more for five or more days over the summer period, it was predicted that at 1.5°C of warming, more than 6 million people in around 3,000 local regions will be exposed to 'very hot weather' of 27.5°C for five or more days during the summer months. If mean global temperatures rise to +3°C, then the same areas will be regularly exposed to dangerously hot temperatures of over 30°C (86°F), as shown in Figure 10.3.[22]

Some buildings suffer from overheating more than others. In the UK, smaller dwellings under around 50 m^2, single-storey flats with poor ventilation, and bungalows with limited protection from the overhead sun tend to overheat more than larger homes, and particularly those with more than two storeys.[23] In many hot countries, even middle-class families can spend almost as much as the family earns in summer just trying to keep the family cool. Upper floors can become uninhabitable, so people move downstairs to sleep.

Ever-higher temperatures are challenging to traditional approaches to keeping buildings cool in the heat. Roof insulation in the UK does reduce overheating in the UK, although wall insulation appears to have less of an impact. This is true for two reasons: The strongest summer heat largely falls on roofs that are typically insulated with *resistive insulation*, including many bulky products, like mineral wools, strawboard, wood–wool slabs, glass fibre products, kapok, wool, and cellulose fibre. The problem with these is that air can travel through them, and once it is hot, it will rise because of the stack effect, moving through the materials where it can and, if not sealed, then escaping at the top. In roofs where the angle of pitch is flat or low and the apex of the insulation gap is sealed, then this air gap in the insulation can be effective. In walls, it is almost impossible to retain fully sealed cold air in the insulation, and so it heats up.

Isaac Meir, in the very hot Negev desert, reported that even with high-quality climatic design and construction, in the record heatwave summer of 2024, his house, which had a high thermal mass core construction with external resistive insulation around the internal thermal mass, did not keep cool enough for comfort without having the solar-powered air conditioning (AC) on all day. They turned the AC off at night, when ambient minima stayed well above the upper thermal comfort limit and occupants slept only with ceiling fans and closed windows due to the internal thermal mass having been chilled mechanically during the day, providing a heat sink during the night. New thinking on keeping occupants cool is urgently needed.

The problem is exacerbated by the compound nature of the impacts of cooling approached. Air conditioning systems can require very high levels of water usages, a problem in an increasingly water-challenged future everywhere. Fire exacerbated by extremely high temperatures is also a problem. Other resistive insulations, like expanded and extruded polystyrene, polyurethane, urea-formaldehyde, vermiculite, and perlite, perform better in the extreme heat as they are structured to contain air pockets within them. However, these materials have poor fire ratings, often giving off toxic fumes when burnt. Building, site, and regional fires are a growing problem during heatwaves, and lightweight timber buildings burn more noticeably more quickly. Reductions in the fire vulnerability of buildings are attempted by constructing them with metal frames, but this results in structures riddled with hot bridges, very effective pathways for the heat to travel into and out of buildings.

Fires spread in four ways, by spontaneous combustion, resulting from extreme radiant heat from a local fire; by creeping fire across vegetation and adjacent surfaces; by windblown sparks; and in very high wind, by heavier flying firebrands, all of which threaten ignition, particularly of external cladding and insulation. So conditions of extreme heat and high winds are particularly dangerous and must be included in design thinking for cooling solutions was seen in the Los Angeles Palisades Fires of 2025.

In very hot countries around the world, walls were typically, traditionally designed to withstand the heat using *capacitive insulation*, often described as 'thermal mass' and found as thick, heavy walls made of stone, concrete, brick, or mud constructions. Heat is conducted through the solid material of the wall, and the rate of the heat flow depends on the conductivity of the material the wall is made of, its thickness, and the difference in temperature between the air on the two different sides of the wall. Load-bearing insulation materials do benefit by largely avoiding the hot and cold bridging problem through outer walls, as covered in Chapter 16 following.

Who Overheats? The Rental Sector

A recent study in Australia across all states shows that some of the people who are worst affected by trying to survive in high temperatures are those who rent. People in poor-quality housing are afraid to complain to owners about not only uncomfortable but also at times unliveable conditions for fear of being evicted in fragile rental markets. Table 10.1 shows the maximum

TABLE 10.1 Summary of maximum temperatures and humidity and average temperatures and humidity recorded in rental accommodation in the eight Australian states

Area	*Max T (°C)*	*Max RH (%)*	*Avg median T (°C)*	*Avg median RH (%)*
ACT	34.2	91.4	23.3	60.2
NSW	40.5	94.6	25.2	64.6
NT	36.5	90.3	28.9	59.4
QLD	41.4	91.8	28.2	64.4
SA	45.3	89.7	23.2	55.9
TAS	34.2	75.6	21.7	53.1
VIC	37.3	92.2	23	60.1
WA	37.1	82.8	26.3	48.9
National	45.3	94.6	24.9	59.5

Source: Better Renting (2023–2024).

temperatures and humidity recorded in rented accommodation by the Better Renting programme in Canberra, with one resident attempting to survive in temperatures above 45°C (113°F) in their sleeping area. Their *Cruel Summer's* report exposed shocking levels of overheating in this sector, a pattern no doubt echoed in most countries across the globe for many of the poorest in society who are unable to afford to buy or run air conditioning or often even fans.[24]

Killer Spaces in Killer Buildings: Lessons From Saudi Arabia

Abdulrahman Al Shaikh looked at the summer and winter performance of 18 homes in Damman, Saudi Arabia, and found clear evidence of which design features do reduce overheating indoors.[25] The residences in his study are all built of concrete, as nearly all buildings in the Gulf, and a clear pattern emerges. High-energy buildings are uncomfortable buildings, are over-glazed buildings with large windows facing west, towards the afternoon sun. The cooling loads shown in Figure 10.5 for homes 13 and 15 are well above the average and the lower-energy homes, and those are the houses with nearly 50% of their external wall area as windows, as shown in Table 10.2.

The standout lesson from this and many other studies is that the higher the glazing area, the higher the risk of severe or catastrophic overheating at higher temperatures that preclude the effectiveness of natural ventilation for cooling people, normally above 35°C (95°F). It is madness to build all- or largely glass buildings not only in the hottest regions of the world, like the Gulf, but also anywhere nowadays, because everywhere will be increasingly subject to extreme heat events in our rapidly heating future, and during them, building cooling systems and the grids that supply them is increasingly tending to fail.

Another major lesson from this study is that not the whole building might chronically overheat but just the most exposed rooms in them, not least on the top floors. In house #15 shown in Figures 10.6 and 10.7, which was the least comfortable of the Dammam study, the east-facing living room maintains a steady, fairly acceptable temperature over the day, but the bedroom temperatures, with two windows facing south-west and north-west, soar in the afternoon. Again, it is direct solar exposure that is the killer factor in creating killer spaces in buildings in heatwaves. If vulnerable individuals occupy them, with no way of leaving them for reasons ranging from infirmity to disability or interpersonal reasons, heat in these rooms can become fatal.

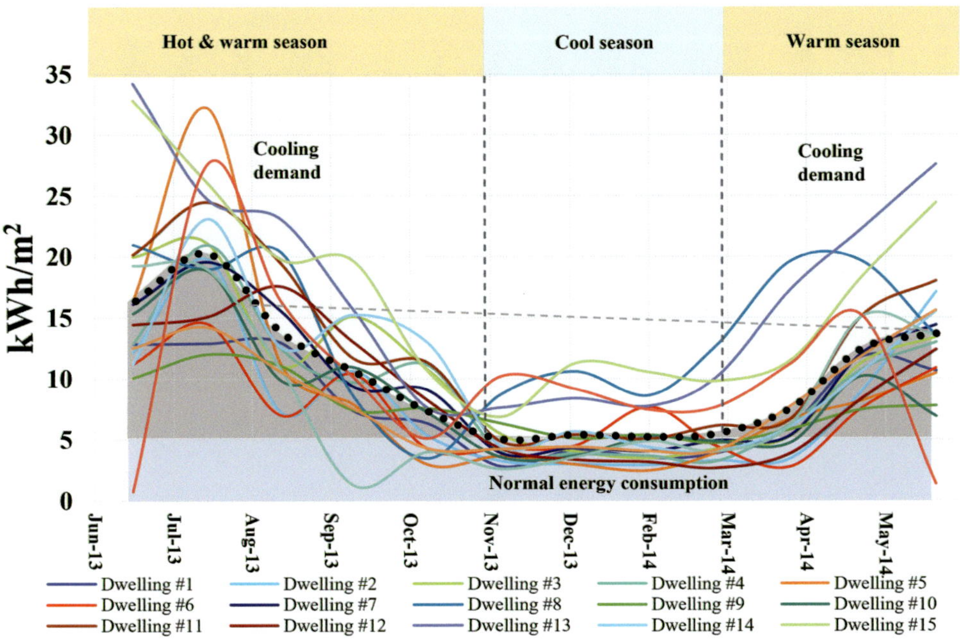

FIGURE 10.5 The electricity consumption in kWh/m^2 per month for the 18 studied Dammam dwellings, with the mean monthly kWh/m^2 from July 2013 till May 2014.

Source: Abdulrahman Alshaikh.

Unfit Legislation for Unfit Buildings in Heatwaves

Amazingly, UK legislation covering unfit housing was replaced as far back as April 2006 under the Housing Act of 2004.[26] The Housing Health and Safety Rating System (HHSRS) is a risk-based evaluation tool to help local authorities identify and protect against potential risks and hazards to health and safety from any deficiencies identified in dwellings. It includes no mention of overheating. The document, not a standard, claims that 'consideration must be given to the likelihood of occurrences that could cause harm and the probable severity of that harm to the vulnerable groups likely to occupy the dwelling'. Dwellings are deemed fit for human habitation unless they fail to meet one of ten fairly crude requirements covering (1) structural stability; (2) freedom from serious disrepair; (3) dampness prejudicial to the health of the occupants; (4) adequateness of lighting, heating, and ventilation; (5) piped supply of wholesome water; (6) satisfactory facilities for cooking; (7) WC for the occupants; (8) bath or shower and a washbasin; and (9) foul and surface drainage systems.

If a local authority deems that a property does not meet these requirements and finds it 'unfit', it can do one of four things: (1) issue a repair notice, (2) get a closing order or (3) a demolition order, or (4) include the property in a clearance area designated for demolition. This legislation was rooted in a past of slum dwellings that were excessively cold and unhealthily damp in winter. It does not deal with issues of overheating or the adequate provision of coolth in hot inner-city summers. It does not designate the condition that ONE room in a house may be unfit for human habitation while others are habitable. There is wriggle room in the *spirit of the law* to demonstrate that such a room, perhaps with no external or opening window, has inadequate

TABLE 10.2 The dwellings characteristics, the comfort survey result, and electricity consumption details of the 17 dwellings collected in Dammam during summertime 2013

	Dwelling number	*Dwelling type*	*Dwelling age*	*Building material*	*HVAC system*	*Dwelling orientation*	*Double\ single windows*	*Percentage of openings*	*Insulation*	*Mean T (°C)*	*ASHR.E*	*Annual bill £/m²*	*Energy kWh/m²*
Well behaviour homes ↔ bad behaviour homes	6	House	11	Concrete block	Fans, central HVAC, and AC unit	NE	D	18.70%	Yes	30.4	1.4	1.9	89.3
	9	Apartment	20	Concrete block	Central HVAC	NE	D	7.70%	Yes	22.4	−0.6	1.1	90.9
	1	Apartment	10	Concrete block	Window AC and AC unit	NE	D	10.80%	Yes	25.7	−1.1	1.1	95.1
	2	Apartment	10	Concrete block	Fans, window AC, and AC unit	SE	D	N/A	Yes	27.2	−0.9	1.4	97.1
	17	House	2	Red bricks	Central HVAC	E	S	N/A	Yes	27.5	−0.6	1.5	101.2
	10	Apartment	5	Concrete block	AC unit	N	S	N/A	Yes	26.7	0.5	1.1	101.4
	16	Apartment	12	Concrete block	Fans, window AC, and AC unit	NE	D	19.90%	Yes	27.6	−0.4	1.3	104.1
	12	Apartment	15	Concrete block	AC unit	SW	S	9.70%	Yes	23.1	0.4	1.6	106.5
	5	House	22	Concrete block	Central HVAC and AC unit	SE	S	11.90%	Yes	27.5	−0.1	2.8	108.8
	4	House	12	Concrete block	AC unit	NW	S	N/A	No	28.7	0.4	1.8	111.7
	7	House	20	Red bricks	Central HVAC	SW	S	N/A	Yes	24.1	−0.3	2.3	118.7
	3	House	25	Concrete block	Window AC and AC unit	NW	S	31.30%	No	26.8	−0.5	2.1	125.2
	14	House	4	Concrete block	Window AC and AC unit	NW	S	N/A	No	28.1	0	1.7	125.9
	11	House	18	Concrete block	AC unit	SE	S	N/A	No	28.6	0.4	2.9	148.3
	8	Apartment	15	Concrete block	Window AC	SW	S	45.80%	No	26.8	0.4	1.9	165.8
	15	House	28	Concrete block	Window AC and AC unit	W	S	48.60%	No	31.5	1.8	4	201.7
	13	House	27	Concrete block	Window AC and AC unit	W	S	31.10%	No	28.6	−0.1	3.9	206.4

Note: Arranged in ascending order of good house performance recorded in terms of energy consumption per meter square.

Source: Abdulrahman Alshaikh.

FIGURE 10.6 Plan of house 15 in Dammam showing the north point and the locations of the bedroom and the living room in which temperatures were measured.

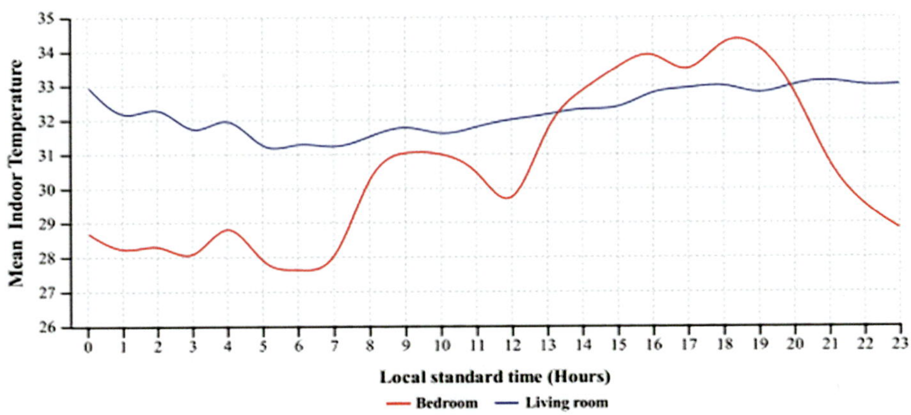

FIGURE 10.7 The performance of the average indoor temperatures of the bedroom and the living room of dwelling 15 in the hot season within a 24-hour time series.

Source: Abdulrahman Alshaikh.

ventilation to cool occupants during overheating periods, but no legal precedents have been set to clarify the Act here. Such legislation needs urgent updating in light of the rapidly increasing health and mortality risks in overheating rooms and buildings during heatwaves.

One place that the law has been successful in punishing architects for designing overheating buildings occurred in 2016. HKS Inc. and Skidmore, Owings & Merrill LLP architects settled out of court for millions of dollars in a suit brought by the Beacon Residential Community Association, occupants of a 595-unit luxury condominium complex in San Francisco where some units were so over-glazed with insufficient opening windows that they became uninhabitable in hot weather. The suit deemed that the architects were responsible, were culpable, and had a responsibility to provide fit-for-purpose accommodation both now and for the future.[27] This success was due to the persistence of one heroic Oakland solicitor alone who was not cowed by facing up to major corporate design companies in the courts.

Defining a Heatwave

Historically, heatwaves have been responsible for more deaths in Australia, Europe, and the United States of America than any other natural hazard, including bushfires, storms, tropical cyclones, and floods, second only to the COVID pandemic in mortality rates. *Heatwaves* are defined differently in different countries, usually in terms of a period of unusually or exceptionally hot weather, using a minimum sequence of consecutive days where daily maximum temperatures reach a designated threshold. Because populations are adapted to living in very different climates, the maximum temperatures for a heatwave in different regions vary considerably. Information on what constitutes a heatwave in your region should be easy to find online. It is obviously important for both legal reasons and also to trigger timely actions to have clear and appropriate definitions of what a *heatwave* is locally.

In regions with very high temperatures and high urban heat island temperatures, the extent to which heat is dissipated overnight following a very hot day dictates the accumulating thermal load impacting vulnerable people and systems. In Australia, they are using a new index, the excess heat factor (EHF), for heatwave monitoring and forecasting.[28] The index is based on a three-day-averaged daily mean temperature (DMT) that captures heatwave intensity as it applies to human health outcomes, as it takes three days of very hot weather for the mortality rate to rise significantly above its antecedent rate. High humidity tends to result in high minimum temperatures in the index, and low humidity in low minima, and this is reflected in their DMT calculation.[29] The EHF method has enabled the Australian Bureau of Meteorology to publicly release seven-day heatwave severity maps since 2014. National meteorological agencies in the UK and the United States are evaluating global seven-day and multi-week EHF heatwave severity probability forecasts.[30]

In South Australia, temperature limits are set by taking into account daily maximum and minimum temperatures, daily mortality, ambulance call-outs, emergency department presentations, and hospital admissions, obtained for Adelaide between 1993 and 2009.

Heat thresholds for health outcomes were estimated using an observed/expected analysis, and for Adelaide, heat-related mortality and morbidity are associated with high temperatures, becoming apparent above maximum and minimum temperature thresholds of 30°C (86°F) and 16°C (61°F) for mortality, 26°C (79°F) and 18°C (64.4°F) for ambulance call-outs, and 34°C (93°F) and 22°C (71.6°F) for heat-related emergency department presentations.[31]

Of all the recorded heat deaths in Australia between 2001 and 2018, at least 473 (nearly 70%) occurred in buildings, and mainly of men with increased risk with age, socio-economic disadvantage, social isolation, geographical remoteness, the presence of disabilities (physical or mental, including heart and kidney conditions, asthma, obesity, diabetes), some prescribed medications, and the absence or non-use of air conditioning or other building heat protection. In the Adelaide heatwaves, then, nearly 40% were simply going about 'business as usual' when they died, as shown in Table 10.3.

Surviving a Heatwave

But on an everyday level, we are all constrained to some extent from being comfortable. Studying secondary school behaviours, Humphreys found that clothing change was the principal means of adaptation to the thermal environment in classrooms, but it was subject to various 'constraints', which meant that some students would sacrifice a certain amount of thermal

TABLE 10.3 Heatwave fatalities by behaviour at time of/just prior to incident in Australia, FY 2001 to FY 2018

Behaviour at/prior to heatwave incident	*Deaths*	*%*
'Business as usual' – no data on any precautions taken	114	31
'Business as usual' – with no precautions taken	38	12
'Business as usual' – with some precautions taken	18	6
Dehydration – insufficient water drank	56	16
History of alcoholism	53	15
No/limited action	39	11
Attempted to cool themselves – without leaving building	38	11
Attempted to cool themselves – actively sought out cooler area (e.g. shopping mall, public building . . .)	<5	N/A
Social isolation	29	8
N/A (if unable to make behavioural choices, for example, child in vehicle) adult with dementia)	25	7
Unknown	71	20
Total	**354**	**100**

Source: Coates et al. (2022).

comfort for what we might call social comfort. The urge to conform to clothing conventions and fashions, and personal preferences, could act to prevent complete adaptation. The urge to look 'cool' can dampen the urge to be as thermally warm or cool as possible,[32] just as the fear of opening a window in Chicago in 1995 was a major restraint to action. In a heating world, the idea that some people have the power to kill others by simply not allowing them to change their clothes or activities may well become an issue of *human rights.*

People who can do nothing under orders about altering their thermal condition can suffer most. A case in point is the fact that five healthy young soldiers standing guard on the steps of St Paul's Cathedral in London for the Queen's Thanksgiving Jubilee service on 7 June 2022 fainted during the service in temperatures as apparently mild as 20°C. That the military still lose service personnel to heat stress on parade or during field exercises seems inexcusable, but perhaps it reflects the thermal histories of those in charge of such events. People only tend to consciously 'thermally know' what they have previously experienced, regardless of theoretical arguments.

In the case of those fainting soldiers, future parades will have to be shaded, and impossibly hot uniforms will have to be discarded. We no longer live in a world in which it is acceptable to kill the lower orders through ignorance, and inevitably, the military will have to get wise or be sued. A big constraint for military thinkers is the fact that their own imaginaries are built from their own un-interrogated life stories. 'If I survived it, anyone can'. No.

The SAS Survival Guide by Lefty Williams[33] confidently asserts that people can withstand forced marches of many miles in temperatures of between 40°C and 48°C, based largely on the experience of him, one man. Such apparently plausible advice may well lead to the deaths of many people in the future, for three reasons: (1) The book was based on the experience of one 'well-hard' man who was exposed and adapted to high-temperature environments; (2) people reading the book may think the extreme experience of 'Lefty' actually applies to everyone else, for instance, young untrained soldiers in Wales in much lower temperatures, and (3) people reading the book may be blinded by the toxic masculinity implied and encouraged to push themselves through thermally health-threatening undertakings. When the commanders of aspirants to become the next 'Lefty' goad the applicants on, they may not be intelligent or empathetic enough to understand the nuances of thermal comfort, distress, and how death operates in reality. This example highlights the fact that perhaps the major constraints we face to safer and more affordable comfort involve understanding that one's own experience does not provide answers for others who are adapted to very different cultures and climates.

Ignorance of what to do in the heat is a major constraint.

Heat Relief Programmes

Education of all involved is a first line of defence to prevent deaths in heatwaves. In Arizona, the Maricopa County Department of Public Health hired a Heat Relief programme coordinator to plan, organise, and coordinate heat relief efforts within the Heat Relief Network, in addition to four other full-time staff. The department is currently expanding the county's heat relief activities to strengthen the multi-agency Heat Relief Network to:

- Increase the number of Heat Relief locations and lengthening their hours and ensuring they have air-conditioned space, hydration, and food to help people recover from heat exposure
- Establish a hotline service for those affected by heat by triaging calls, answering questions, providing connection to services, and providing transportation to and from cooling or respite centres or hospitals as needed
- Widely advertise the Heat Relief Network to widen accessibility to heat relief and train centre staff and volunteers

This is emergency response planning to heatwaves, but in parallel, whole new programmes should also be instigated to:

- Provide financial support for upgrading buildings to prevent overheating, with roof insulation, shading, room-level climate refuges indoors, etc.
- Educate design professionals and the construction industry on how to both avoid overheating and provide climate safe spaces in the buildings they work on
- Implement locality cooling programmes with appropriate hard and soft landscaping

The private sector is now gearing up to ensure that people who occupy their properties are protected from the worst of the heat with programmes. Oregon Energy Trust, for instance, is supporting a community cooling spaces programme being rolled out to provided safe cool areas in multi-occupancy buildings, a project developed after the high mortality rates experienced under the North-West US heat dome of 2021. From Californian heatwaves we heard of older residents moving to occupy different homes along a street so only one of the group had to pay for air conditioning on that day of the week. There are enormous energy savings to be had from sharing spaces for both heating and cooling seasons.[34] Eric Klinenberg points out in his 2018 book *Palaces for the People* that community-based solutions are multiply complex, and to use them effectively requires us to aim to build more equal and united societies.[35] Real wisdom will be needed to build the social and physical infrastructures necessary to future-proof all citizens against a heating climate.

Timely Adaptation

Timely adaptation requires people to anticipate, to the best of their ability, what thermal conditions will be like in their buildings in the near or medium term. Extremely hot or cold killer spaces and/or killer buildings must be eliminated at the design stage or enabled to be made safe during extreme weather over the lifetime of the building. The extent to which a society acts to protect the most exposed and vulnerable of their populations from growing incidences of extreme heat is a measure of the value they place on equality. Because hot people become angry people, it may also prove to be a measure, in the long term, of the stability of the society itself.

Notes

1 Al Mahri, S., and A. Bouchama (2018). Heatstroke, Chapter 32, in *Handbook of Clinical Neurology*, Edited by A.A. Romanovsky, Elsevier, Vol. 157, pp. 531–545, ISSN: 0072-9752. https://doi.org/10.1016/B978-0-444-64074-1.00032-X

2 Spender, T. (2024). *At Least 1,301 People Died during Hajj – Saudi Arabia*, 23rd June, Accessed on 27/07/2024 at: https://www.bbc.co.uk/news/articles/c8000dgk9gzo

3 Tickell, C. (1990). Human effects of climate change, excerpts from a lecture given to the royal geographical society, *The Geographical Journal*, 156(3)(March 26), pp. 325–329. DOI: 10.2307/635534

4 Vital Signs (2022). *Deaths of Migrants in the Gulf, Report 1*, Accessed on 27/08/2024 at: https://vitalsignsproject.org/wp-content/uploads/2023/03/Vital_signs-report-1.pdf

5 Roaf, S., D. Crichton and F. Nicol (2009). *Adapting Buildings and Cities for Climate Change*, Routledge.

6 Robine, J.-M., S. Lan, K. Cheung, et al. (2008). Death toll exceeded 70,000 in Europe during the summer of 2003, *Comptes Rendus Biologies*, 331(2), pp. 171–178, ISSN: 1631-0691. https://doi.org/10.1016/j.crvi.2007.12.001

7 Poumadère, M., C. Mays, S. Le Mer and R. Blong (2005). The 2003 heat wave in France: Dangerous climate change here and now. *Risk Analysis*, (25), pp. 1483–1494.

8 Vandentorren, S., P. Bretin, A. Zeghnoun, et al. (2006). August 2003 heat wave in France: Risk factors for death of elderly people living at home. *European Journal of Public Health*, 16(6)(December), pp. 583–591. https://doi.org/10.1093/eurpub/ckl063

9 Klinenberg, E. (2002). *Heat Wave: A Social Autopsy of Disaster in Chicago*, Chicago: Chicago University Press.

10 White, R.H., S. Anderson, J.F. Booth, et al. (2023). The unprecedented Pacific Northwest heatwave of June 2021, *Nature Communications*, 14, p. 727. https://doi.org/10.1038/s41467-023-36289-3

11 Henderson, S.B., K.E. McLean, M.J. Lee and T. Kosatsky (2022). Analysis of community deaths during the catastrophic 2021 heat dome: Early evidence to inform the public health response during subsequent events in greater Vancouver, Canada, *Environmental Epidemiology*, 19/6(1), p. e189. https://doi.org/10.1097/EE9.0000000000000189

12 Lomas, K.J., S. Watson, D. Allinson, et al. (2021). Dwelling and household characteristics' influence on reported and measured summertime overheating: A glimpse of a mild climate in the 2050's, *Building and Environment*, 201(107986), ISSN: 0360-1323. https://doi.org/10.1016/j.buildenv.2021.107986

13 Smith, S., K. Goff and S. Kaufman (2024). *Arizona's Heat Related Death White Paper*, Published by the University of Arizona, Accessed on 09/02/2025 at: https://mapazdashboard.arizona.edu/article/arizonas-heat-related-death-white-paper-full-report#

14 The Maricopa County Department of Public Health (2024). *2023 Heat Related Deaths Report*, Published by The Maricopa County Department of Public Health, Accessed on 09/08/2024 at: https://www.maricopa.gov/ArchiveCenter/ViewFile/Item/5820

15 Multnomah County (2022). *Final Report: Health Impacts from Excessive Heat Events in Multnomah County, Oregon, 2021*, Oregon: Published by Multnomah County, Accessed on 8/08/2024 at: https://multco-web7-psh-files-usw2.s3-us-west-2.amazonaws.com/s3fs-public/20220624_final-heat-report-2021_SmallFile-2.pdf

16 Schweiker, M. (2022). Rethinking resilient thermal comfort within the context of human-building resilience, in *The Routledge Handbook of Resilient Thermal Comfort*, Edited by F. Nicol, H. Bahadur Rijal and S. Roaf, London: Routledge, ISBN: 978-1-032-15597-5

17 https://www.climatejust.org.uk/

18 Kennedy-Asser, A.T., G. Owen, G. J. Griffith, et al. (2022). Projected risks associated with heat stress in the UK Climate Projections (UKCP18), *Environmental Research Letters*, 17(3). DOI: 10.1088/1748-9326/ac541a

19 https://akaresearch.shinyapps.io/heatstressvulnerability/

20 Bloch, S. (2025). *Shade*, New York: Random House, ISBN 978-0-59-324276-6

21 Voelkel, J., and V. Shandas (2017). Towards systematic prediction of urban heat Islands: Grounding measurements, assessing modeling techniques, *Climate*, 5(2), p. 41. https://doi.org/10.3390/cli5020041

22 Friends of the Earth (2024). Millions at risk from extreme heat unless climate goals met, *Press Release*, Accessed on 27/08/2024 at: https://friendsoftheearth.uk/climate/millions-risk-extreme-heat-unless-climate-goals-met

23 Lomas, K.J., S. Watson, D. Allinson, et al. (2021). Dwelling and household characteristics' influence on reported and measured summertime overheating: A glimpse of a mild climate in the 2050's, *Building and Environment*, 201(107986). ISSN: 0360-1323. https://doi.org/10.1016/j.buildenv.2021.107986

24 Better Renting (2024). *Cruel Summers Renters' Diverse Experiences of Summer 2023–2024*, Published on line by Better Renting. Accesses 20/05/2025 at: https://www.betterrenting.org.au/renter_researcher_summer_24

25 Al-Shaikh, A. (2016). *Design Principles for Thermally Comfortable and Low Energy Homes in the Extreme Hot-Humid Climatic Gulf Region, with Reference to Dammam, Saudi Arabia*, PhD Heriot Watt University, Edinburgh. ISNI: 0000 0004 6347 5564, Accessed on 03/09/2024 at: https://www.ros.hw.ac.uk/handle/10399/3158

26 UK Government (2006). Housing health and safety rating system: Guidance for landlords and property-related professionals, Ref: ISBN: 978-1-85-112856-3, Accessed on 06/09/2024 at: https://www.gov.uk/government/publications/housing-health-and-safety-rating-system-guidance-for-landlords-and-property-related-professionals

27 http://scocal.stanford.edu/opinion/beacon-residential-etc-assn-v-skidmore-owings-merrill-34333

28 Coates, L., J. van Leeuwen, S. Browning, et al. (2022). Heatwave fatalities in Australia, 2001–2018: An analysis of coronial records, *International Journal of Disaster Risk Reduction*, 67, p. 102671, ISSN: 2212-4209. https://doi.org/10.1016/j.ijdrr.2021.102671

29 Nairn, J.R., and R.J. Fawcett (2014). The excess heat factor: A metric for heatwave intensity and its use in classifying heatwave severity, *International Journal of Environmental Research and Public Health*, 23(12)(1), pp. 227–253. https://doi.org/10.3390/ijerph120100227

30 Nairn, J., B. Ostendorf and P. Bi (2018). Performance of excess heat factor severity as a global heatwave health impact index, *International Journal of Environ Research and Public Health*, 8(15)(11), p. 2494. https://doi.org/10.3390/ijerph15112494

31 http://earthobservatory.nasa.gov/IOTD/view.php?id=3714 (image).

32 Humphreys, M., F. Nicol and S. Roaf (2016). *Adaptive Thermal Comfort: Foundations and Analysis*, London: Earthscan, pp. 164–167.

33 Wiseman, J. (2018). *SAS Survival Guide: How to Survive in the Wild, on Land or Sea*, Collins Gem.

34 Hawasley, M., D. Corne and S. Roaf (2010). Social networks save energy: Optimising energy consumption in an eco-village via agent-based simulation, in Transforming Markets in the Built Environment: Adapting for Climate Change, (Ed.) Sue Roaf, *Architectural Science Review*, 53(1)(February).

35 Kleinenberg, E. (2018). *Palaces for the People: How to Build a More Equal and United Society*, London: The Bodley Head.

11

KILLER BUILDINGS

Cold

Winter Is Coming: Be Prepared

The Texas ice storm took everyone by surprise. Who guessed that even in very hot places like Texas, cold can kill? On 11–20 February 2021, an unprecedented winter storm hit the USA, settling over the whole of Texas, with a deep-freeze dropping temperatures across the state to below zero. Precipitation began as sleet and freezing rain and, by the 13th, had spread throughout Texas. Temperature records for the longest freezing streak in the state's recorded history broke after nine consecutive days. Power grids failed across Texas, with at one time nearly 10 million people plunged into the dark, left without warmth and the ability to cook food. Water pipes burst; roadways iced over, causing pile-ups; and food shortages caused chaotic scenes in grocery stores, where long queues formed outdoors to collect survival provisions.[1] At least 246 people were killed directly or indirectly, with some estimates that 702 people died as a result of the crisis. Some died of carbon monoxide poisoning, in car crashes, drownings, and fires, but almost two-thirds, including infants and the elderly, died of hypothermia. A sudden warming brought the storm to an abrupt end, causing widespread flooding and additional cold deaths, particularly in areas where the power was not restored for days.[2]

Great storms like the Texas Ice Storm or the Sacramento flood of 1862 shown in Figure 11.1 have always happened, only now they are happening more frequently and systematically breaking weather records, that warns us to be prepared, not just for the heat, but for extreme winter storms as well.

Hypothermia

Hypothermia is technically defined as an *involuntary drop* in body temperature below 35°C. The body works rather like a battery, continually topping up its heat and energy levels with metabolic activity, warm food and drinks, and external heat sources to maintain a stable thermal state, *homeostasis*, keeping the core body temperature of around 37°C (98°F). The body's thermal inputs and outputs in the cold are largely controlled with three conscious *voluntary actions*, by adjusting clothing, buildings, and heat supplies.

DOI: 10.1201/9781315645070-16

FIGURE 11.1 The abrupt melting winter snows and the dumping of torrential rain over the Western US states in January 1862 led to catastrophic flooding that added deaths to the fatal cold spell experienced across the whole region, as this A lithograph of K Street in Sacramento during the Great Flood shows.

Source: A. Rosenfield/Wikimedia Commons.

When the body loses more heat than it absorbs, it chills, causing more or less severe hypothermia, defined according to core temperatures as:[3]

Mild: 32–35°C (89–95°F). Symptoms are often subtle and nonspecific, including hunger, nausea, fatigue, shivering, and pale-dry skin. Patients often shiver and may suffer from memory and judgment impairment and need to urinate more often.

Moderate: 28– 32°C (82– 90°F). Cognitive decline and lethargy are common. Normal bodily reflexes slow down, and the body shivers gradually. Occasionally, 'paradoxical undressing' may occur when the body attempts to survive by putting one last flush of warm core blood to the surface, making the person suddenly hot, prompting them to strip to cool off. Heart functions become chaotic.

Severe: under 28°C (82°F). Flow of blood to the head and body declines, and patients become unresponsive as blood pressure, heart rate, and cardiac output also continue to decrease, and heart and lung functions collapse. In some extreme cases, cold victims have been found indoors, hidden behind wardrobes or under beds, enacting a *hide-and-die syndrome*. Outdoors, people in the last stages of hyperthermia can go into *terminal burrowing* mode to survive, attempting to dig themselves into snow or earth holes.

Profound: under 24°C (75°F). Few people survive, but surprisingly some do.[4]

How Many People Die From the Cold?

Cold kills thousands every winter, from the homeless sheltering on the streets to the elderly who fall in their homes and find themselves helplessly fading in the cold. If temperatures are low enough, it does not take too long to morbidly or fatally chill.

In the USA, about 700 to 1,500 hypothermia-related fatalities are reported each year, and usually, of adults between the ages of 30 and 49, largely people who die of exposure outdoors, on the streets, where ten times more men die than women. The true health and mortality impacts of hypothermia in the USA are unknown. Of those actually diagnosed with it who are treated in hospital for moderate or severe hypothermia over 50% die.

In the UK only 39 people were reported as dying from Hypothermia from 'natural cold' in 2021. For the broader category of cold-related deaths thousands dies every year in Britain, mostly in their own homes. Over the 35 years from 1988 to 2022 almost 200,000 deaths in England and 16,500 deaths in Wales were associated with the coldest days. As with deaths during heatwaves it is mainly the most vulnerable, socially deprived, poorest and least healthy people who do die, in the poorest housing.[5] Figure 11.2 shows clearly that populations are particularly vulnerable to the health and mortality impacts of the most extreme cold weather events with over 20,000 excess cold deaths being recorded for the record freezing winter of 2010 in England. Figure 11.3 shows the relative risk of dying when outdoor temperatures are high or low, indicating that the Welsh population is hardier in the cold but more vulnerable to succumbing to higher temperatures than the English. The Scots are the hardiest of all British populations, as Figure 11.4 shows, only showing signs of death rates rising due to the cold when outdoor temperatures fall below around 2°C (36°F), but they start to feel the impacts of heat as temperatures rise towards 20°C (68°F). In Scotland, it is estimated that there is an increase of all-cause mortality risk of 9% under extreme cold conditions. Most deaths are from respiratory diseases, while cardiovascular disease risk is also heightened, particularly in the elderly in Scotland.

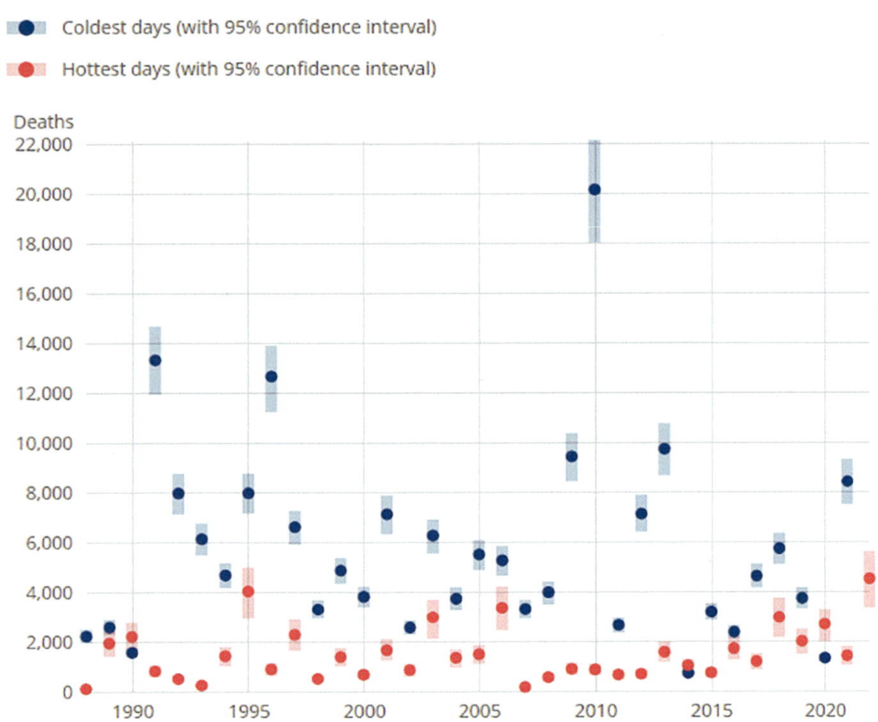

FIGURE 11.2 Estimated temperature-related deaths, England, 1988 to 2022.

Source: UK government (2023).

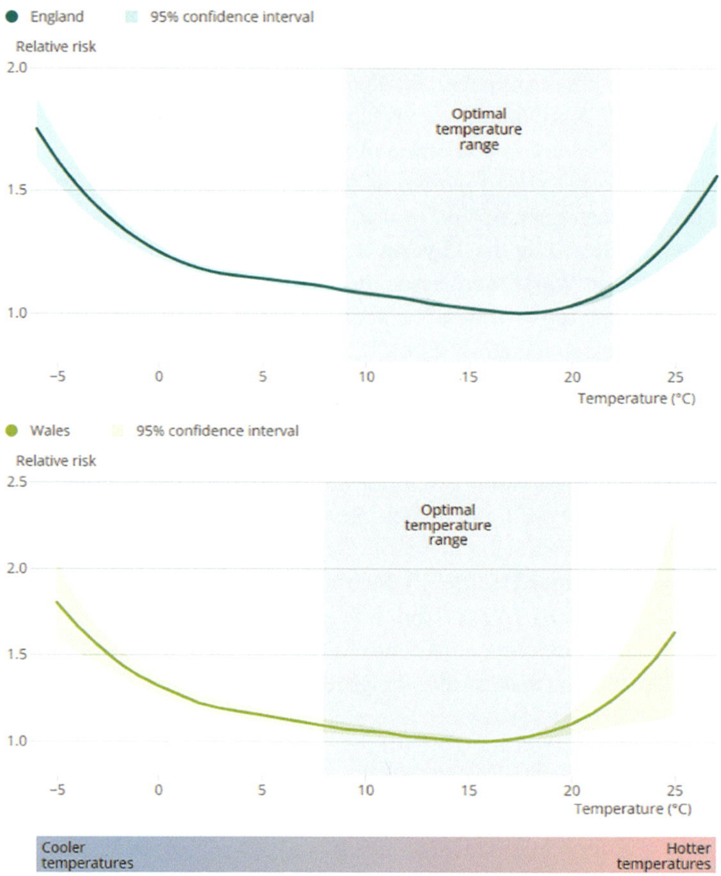

FIGURE 11.3 Temperature-related relative mortality risk, England and Wales, using data from 1988 to 2022.

Source: UK government (2023).

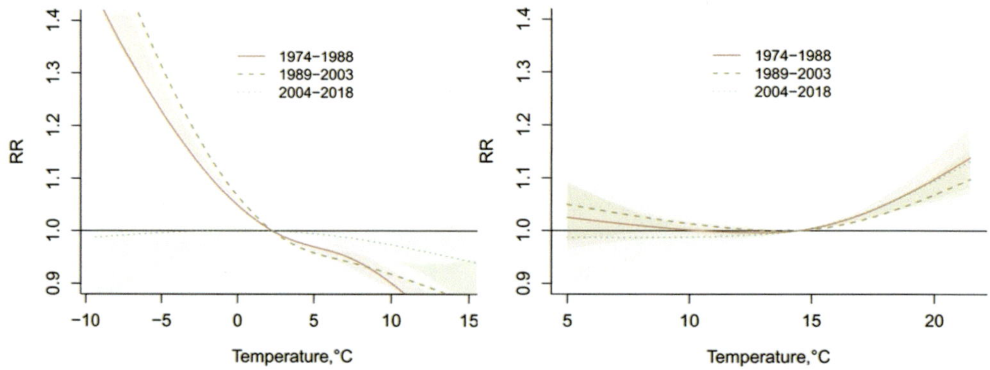

FIGURE 11.4 Meta-estimation of cumulative relative risk under daily mean temperatures in Scotland in October to April (next year) for cold effects (*left*) and in June, July, and August for heat effects (*right*) in three 15-year periods.

Source: Wan et al. (2022).

In the last 15 years, a lower cold effect on death rates might be due to a trend towards warmer winters and/or linked to policies ranging from energy and living cost subsidies to building improvement schemes to protect vulnerable populations from the effects of cold.[6]

Cold Weather Warnings

As with heatwaves, meteorological organisations around the world have developed advanced weather warnings for extreme cold weather events to warm people to hunker down at home and prepare for a period of extreme cold and possible power outages or to pack up and leave the worst affected regions. In the USA, which is particularly exposed to winter storms, those who can often move to stay relatives or in hotels with electricity in adjacent counties, or even to a warm holiday resort like Cancun in Mexico, which was the choice destination for Ted Cruz, the then senator for Texas during the 2021 ice storm. Cities are set up with warm refuges during such, as they are for cool refuges in heatwaves, but the same problem exists that many people are simply unable to reach them due to their location, infirmity, disability, or poverty. Details of cold weather bombs and where to access help to survive them will be on most local or central government websites in affected areas but that, of course, depends on having (1) internet connectivity and (2) power to run it. In the USA, detailed maps for each region are provided by the National Oceanic and Atmospheric Administration, National Weather Service, Storm Prediction Center – Watch, Warning, Advisory Display pages.[7] This service is provided by the Meteorological organisations in the UK[8] and Australia.[9]

Cold Tolerance, Sensitivity, and Impacts

Human bodies have a limited ability to adapt to cold environments. One person might be very differently predisposed to tolerate cold, or sensitive to it, depending on how acclimatised they are to the cold, their body form, thermal experiences, and attitudes. Vulnerability to the health impacts of cold, including increased susceptibility to infection, is also affected by the health of an individual with underlying conditions, such as heart, lung, or kidney disease, diabetes, Parkinson's or dementia, which may be worsened by cold exposure. Housing conditions, nutrition, malnutrition, and lifestyle choices, like the use of alcohol, tobacco, and other drug-related substances, exacerbate vulnerability to cold. People who are cold also have a higher risk of suffering cold weather injuries, potentially further increasing their own exposure to the impacts of the cold around them.[10]

Treating Very Cold People

Learning from how people with hypothermia are treated provides some clues to surviving in, and designing for, the cold. People who are very cold tend to have a warmer core around their heart and trunk and very cold peripheral limbs and skin. In moderate and severe hypothermia, a core temperature *afterdrop* can be caused when blood flows out from the warm core to the colder peripheral tissues, and the return flow floods much colder blood from the limbs to chill the central circulation and heart. In treating hypothermia, an increased blood flow to cold limbs is avoided, so a cold a person is best left resting in a flat position to reduce rates of blood flow around the body.

Medical advice is not to use warm showers or baths for rewarming as it can increase the flow of the cold peripheral blood to the warmer heart and may cause cardiovascular collapse. Anyone who has experienced really cold hands in freezing conditions and puts them under a warm tap to heat them will also know how very painful it can be. Instead, a hypothermic (very cold) person should be kept horizontal and inactive while being constantly verbally stimulated to keep them attentive, focussed, and awake. It is essential to keep them insulated with dry coats, blankets, etc., and especially from the cold ground or floor, and minimise heat loss from the head and neck by covering these areas as effectively as possible with a cap, hat, or hood to also protect them from draughts and winds. A vapour barrier around them can also protect a body from convective heat loss.[11] For the elderly, prone to falling in the bedroom, really warm bedside rugs that can be wrapped around a body can help prolong the onset of hypothermia until help arrives.

Battlefield medicine often needs to be innovative, and the Thermarmour company developed a four-layer multifunctional material that is wind- and waterproof and highly insulating. As shown in Figure 11.5(a), it is used in human carrier bags to maintain normal body temperatures (normothermia) and prevent hypothermia in wounded soldiers during transportation in extremely cold environments. What is also carried by the Finnish army on winter exercises was used by Roaf et al.[12] to great effect in a triple-skinned tent at Collins Bay in Antarctica based on a Mongolian yurt. Built in 2019 by the Polar Lodge team of the Portuguese Propolar organisation, it is still providing shelter for climate researchers on an exposed rock promontory on King George Island.

What can we learn from understanding how to treat hypothermia? How to avoid it! Even if working in a cold room, in a cold house, basic precautions can be taken to prevent mild hypothermia. Keep the peripheries warm with insulating footwear, hats, and gloves if it is really cold. Avoid draughts and cold surfaces that draw heat from the body by convection, conduction, and radiation. Keep away from, or mask radiation from, a cold window with thick curtains and carpet a cold floor. The smaller the space that is being warmed by a local heater, the easier it is to keep that space warm. Many people who cannot afford to heat their homes in winter spend a lot of time in bed, in a smaller room, avoiding the larger open spaces in a home. Even people

FIGURE 11.5 (a) A hypothermic victim on an army exercise in Ukraine is lain flat and wrapped in a super-insulating carrier bag to help maintain user normothermia prior to evacuation. (b) The Polar Lodge in Antarctica, constructed with an external skin of Thermarmour over a yurt structure, with a second internal lining of Thermarmour suspended within the timber ribs

Source: (a) https://thermarmour.com/. (b) Sue Roaf.

FIGURE 11.6 Panelled bedroom, shuttered windows, and four-poster bed in the Jeremiah Lee Mansion, Marblehead, Massachusetts, with a warming pan by the fire.

Source: Susan Roaf.

living in 'Grand Designs' houses with huge open-plan spaces with cold glass walls overlooking the spectacular view in the depths of winter often try to avoid wasting money and energy by beating a retreat into the snuggest room around like the one shown in Figure 11.6. In extremis, like the kings of old in their freezing stone castles, cover the walls in hangings or timber panelling, get into a four-poster bed, warm nightwear, a nightcap, and a warming pan.

Clothing and Shelter

Survival expert Jessie Krebs claimed of people in the wild, 'Clothing or equipment is the first line of defence, shelter is second and third is fire', but pointed out that 'people will skip right to fire after clothing has been compromised, which is a mistake. If they are not successful, they die trying to start a fire'.[13] When really cold, people start to flee from cold by burrowing into the snow or earth or hiding from cold indoors in cupboards or under beds. During the Texas ice storm, the first line of defence was clothing, to protect the core and limbs of the body with warm trousers and jumpers, as well as hat, gloves, and shoes. The second line of defence concerns the quality of the shelter on offer in a building, primarily what places in the building, or in nearby buildings, are there to which occupants can escape and hide in from the cold. Is there a snug space in the home? Many modern 'energy-efficient' homes rely on the building envelope to keep out the cold air that penetrated via cracks in openings or walls, and the creeping cold that enters across cold bridges in the envelope, particularly metal ones. Emphasis is placed on high levels of insulation for the external, but often, large open-plan living areas dominate a house plan, with one or more walls of glass, often without curtains, which lose heat quickly and radiantly chill the space and its occupants.

Reverse Engineering Cold Defences – Icehouses

Reversing the challenge is illuminating. How did people store in winter ice and preserve it for use in summer weather? In many countries ice was collected off lakes and ponds during freezing spells and stored in underground chambers for months so it could be used in the hottest months to cool drinks, food, people and for a myriad of other uses including for exotic table ice-sculptures.[14] Figure 11.7. Shows a design of an icehouse with four airlocks, the inner one filled with insulation. The structure was insulated from the air temperature by the ground and from the ground temperature by the cavity walls that also kept ground water out of the structure because water was seen as the 'enemy of ice' at those temperatures. Great debates raged in the mid-19th century as to whether trees over these structures were good to shade the earth, or bad because they dripped water onto it rather than allow drying air to access and dry out the ground. An exceptional icehouse in a good location shaded from hot winds and sun could keep ice frozen for 18 months.

The further away from the ambient air the longer a protected temperature can be preserved. During blackouts in the Texas ice storm the only heat to be found was either from burning wood or retreating into the warmer, steadier temperature of inner rooms or the earth. Ground-source heat pumps proved unusable then as they required electricity to operate.

Just as in hot weather, when occupants hide from the sun behind pergolas, shutters, and shades outdoors and buffer spaces indoors so one of the most effective ways of keeping heat in and cold out is to use air-locks. The value of a super-insulated envelope is reduced if every time the outside door is opened, the heat from the inside is lost. So in extreme cold, particularly when

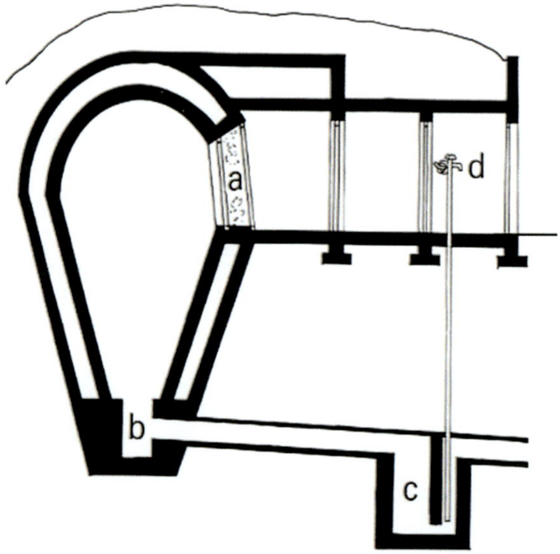

FIGURE 11.7 A five-door underground icehouse designed in 1833 by Loudon with (a) space between inner ground-level doors filled with a cushion of barley straw; (b) the sump covered by a grill beneath the ice; (c) a trap in the drain; (d) Siebe's rotary pump to draw up iced drinking water.

Source: Susan Roaf.

no fire is available, RETREAT, hide from the ingress of cold with heavy curtains across the windows, draft lobbies, and when if all else fails, burrow, possibly with a torch, into a four-poster bed, or if in a home with a basement, this may ultimately be the warmest, safest place for a day or two. If really pushed, possibly do what many nomads or refugees do in tents in winter: Sleep in the same bed to keep all the family warm in a communal warm bubble.

A good icehouse stored ice in a compacted ovoid form that, as it melted, it reverted to a ball-like shape with minimum surface area to maximum stored volume of cold. The larger the ice stored, the longer it lasted. This is a proxy for the warmth that is stored in the mass of the internal walls of a home. The more well-designed mass there is that starts off warm at the beginning of a power outage, the longer it will retain its warmth.

Fire and Cold

Fire is the heat source of last resort when mechanical systems fail. In 2022, the Scottish Buildings Regulations banned wood burning stoves in new homes in rural Scotland from 1 April 2024. Scottish Power had warned that people should be prepared for 72 power cuts that year. The coldest places in Britain are in Scotland, and historically, power outages have lasted in places up to three weeks there. That decree would inevitably have led to death by hypothermia of residents in remote settlements in Scotland during extreme cold storms. People revolted. The Regulations team had to add a fudge clause to enable people to install an ‘emergency heating system’. The Building Regulations system was overseen by politicians with apparently little empathy for the individual people and populations who are impacted by the drive to achieve ‘net zero’. The problem is the targets of the net-zero departments in governments relate simply to decarbonising the grid. Nowhere is there a department whose remit is to ensure that buildings and their occupants are, and will be, safe in evermore extreme climate trends and weather events even when the grid energy supplies fail.

Cold Adaptation in Humans

As modern humans emerged from the warmer climates of Africa, those who moved north into colder lands developed new habits, clothes, and buildings and also became more physiologically adapted to cold. The change took place over millennia from simple clothes like cloaks of feathers, skin, and fur to the more complex outfits needed to occupy frozen zones or freezing periods. Early tools evolved over time to scrape, cut, and pierce skins, including blades, awls, and later, needles, to sew together fitted garments, such as shirts with sleeves and trousers with separate legs. Examples of hide-working scrapers date from around 370,000 years ago, blades later from 300,000, and even later, the earliest examples of bone needles with eyes from after 40,000 years ago. One can only guess at when clothes became a fashion statement amongst our ancestors.[15]

Humans physiologically adapted to the cold on their long migration north. Black African populations have been shown to still have a reduced ability to produce heat by shivering and have poor cold induced vasodilation in fingers and toes compared to Caucasians and the Inuit. Repeated whole-body exposure of mainly Caucasian volunteers to severe cold undertaken in the University of Maastricht laboratories resulted in subjects adapting by feeling less cold over time to low temperatures but no major physiological changes were exhibited aside from some slightly reduced production of metabolic heat. However, during repeated exposure to milder cold conditions subjects showed increased non-shivering metabolic heat production to stay warm. The inference of this is that indigenous populations have indeed metabolically and physiologically adapted to cold climates since they migrated north into colder climates some 45,000 years ago, but individuals within one generation can only make minor adaptions to colder conditions over their lifetime. To survive in the extreme cold, people must then rely largely on preparing to mentally and behaviourally develop new skills to live and survive in the extreme cold.[16]

A major barrier to building on traditional systems of staying warm enough in buildings has been erected by existing thermal comfort regulations and standards. Developed by and for engineers they mandate narrow comfort zones of temperatures that can only be met by mechanically supplied heating and cooling systems. 'Flick of a Switch Comfort' wears away the knowledge of how comfort was achieved before those systems. Historically thermal comfort was provided in colder climates with background lower-level heating and topped up with local supplementary warmth, usually from radiant heat from fires. Comfort was achieved then without grid energy supplies and did not rely on technical upgrades of the building fabric and services to keep people thermally safe. Today, such low-cost strategies of *heating the person, not the building,* are regaining popularity and complement fabric upgrades to achieve even better savings in operation. Michael Humphries detailed how many in Britain traditionally lived in cold homes in his useful technical note for Historic Scotland on 'Staying Warm in a Cooler House'.[17]

Building Resilience to the Cold

It is extraordinary to live in a world of wild weather phenomena most of us never knew existed: atmospheric rivers, thundersnow, ice tsunamis, the Pineapple Express, and bomb cyclones. When increasingly intense cold or hot weather events occur, they normally now come with power outages. In the following chapters we detail how we might design climate-safe spaces into our own homes. An understanding of the physiology of hypothermia is useful to inform how we act during very cold weather and what clothes and other accoutrements we may need to keep our bodies warm enough. Advice on surviving cold storms often includes the need to store vital provisions, like potable water, non-perishable foods, and if possible, fuel to operate vehicles or local heating systems, portable battery systems for lighting and communications, and firewood for stoves or fireplaces. But also storing heat, in the earth or the fabric of a building, can provide days or weeks of safer internal temperatures.

Cold need not be our enemy. We now know that spending time outside the narrow engineering definition of what the thermal comfort zone can positively enhance human thermoregulatory capacities and thermal resilience, while also improving metabolic and cardiovascular health. Allowing more thermal variation over a day and year indoors can also stimulate physiological and psychological adaptation to a wider range of temperatures in humans, which can lead to improved resilience and health.[18] It is just a short step from that wider comfort understanding to being able to prepare for even more extreme variations of temperature, becoming more attuned to the natural thermal cycles of the climate and ecosystems we occupy. Recent generations before us lived with the extreme cold of the Little Ice Age of the Middle Ages, and so can we.

A key lesson is not to be reliant solely on grid energy during extreme weather events. Get a plan to change from the normal thermal modus operandi of a building into an emergency cold weather mode of occupation. Winter is coming. Be prepared.

Notes

1 https://www.ncei.noaa.gov/news/great-texas-freeze-february-2021
2 https://en.wikipedia.org/wiki/2021_Texas_power_crisis
3 Duong, H., G. Patel and C.A. Holt (2024). Hypothermia, StatPearls [Internet], *Treasure Island (FL): StatPearls Publishing*, Accessed on 09/09/2024 at: https://www.ncbi.nlm.nih.gov/books/NBK568789/
4 Duong, H., G. Patel and C.A. Holt (2024). Hypothermia, StatPearls [Internet], *Treasure Island (FL): StatPearls Publishing*, Accessed on 09/09/2024 at: https://www.ncbi.nlm.nih.gov/books/NBK568789/
5 UK Government (2023). *Climate-Related Mortality, England and Wales: 1988 to 2022: Relative Risk of Death Associated with Temperature for England and Wales from 1988 to 2022*, Published by the UK Office for National Statistics, Accessed on 09/09/2024 at: https://www.ons.gov.uk/peoplepopulationandcommunity/birthsdeathsandmarriages/deaths/articles/climaterelatedmortalityandhospitaladmissionsenglandandwales/1988to2022
6 Wan, K., Z. Feng, S. Hajat and R.M. Doherty (2022). Temperature-related mortality and associated vulnerabilities: Evidence from Scotland using extended time-series datasets, *Environmental Health*, 25(21)(1), p. 99. https://doi.org/10.1186/s12940-022-00912-5
7 https://www.spc.noaa.gov/products/wwa/
8 https://www.metoffice.gov.uk/weather/warnings-and-advice/
9 http://www.bom.gov.au/
10 Haman, F., S.C. Souza, J.W. Castellani, et al. (2022). Human vulnerability and variability in the cold: Establishing individual risks for cold weather injuries. *Temperature* (Austin), 29(9)(2), pp. 158–195. https://doi.org/10.1080/23328940.2022.2044740
11 Dow, J., G.G. Giesbrecht, D.F. Danzl, et al. (2019). Wilderness medical society clinical practice guidelines for the out-of-hospital evaluation and treatment of accidental hypothermia, *Wilderness & Environmental Medicine*, 30(4_supplement), pp. S47–S69. DOI: 10.1016/j.wem.2019.10.002
12 Roaf, S., J.P. Silva and M.C. Guedes (2020). Extreme design: Preparing for a different future, in *Green Buildings and Renewable Energy. Innovative Renewable Energy*, Edited by A. Sayigh, Cham: Springer. https://doi.org/10.1007/978-3-030-30841-4_45
13 https://www.bbc.com/future/article/20200226-how-to-survive-in-the-extreme-cold
14 Beamon, S., and S. Roaf (1990). *The Ice-Houses of Britain*, London: Routledge.
15 Gilligan, I. (2019). *The Technology of Paleolithic Clothes. In Climate Clothing and Agriculture in Prehistory: Linking Causes and Effects*, Cambridge University Press, pp. 66–79. DOI: 10.1017/9781108555883
16 Hein, A.M.D., and W.D. Van Marken Lichtenbelt (2016). Human whole body cold adaptation, *Temperature*, 3(1), pp. 104–118. https://doi.org/10.1080/23328940.2015.1135688
17 Humphreys, M., F. Nicol and S. Roaf (2011). *Keeping Warm in a Cooler House*, Historic Scotland Technical Paper 14, Edinburgh. http://www.historic-scotland.gov.uk/technicalpaper14.pdf
18 Pallubinsky, H., R.P. Kramer and W.D. van Marken Lichtenbelt (2023). Establishing resilience in times of climate change – a perspective on humans and buildings. *Climatic Change*, 176(135). https://doi.org/10.1007/s10584-023-03614-0

Ecology and Comfort

12

ECOLOGY AND COMFORT

Introduction

A building is not a sculpture or a box full of stuff floating in space. Buildings exist in time and space and within their own ecosystems within Earth's larger ecosystem. A building does not end at its envelope but is woven into landscapes and climates with living and inert threads of material and energetic flows.

In our battle to both survive in and salvage our climate, we must first cure our addiction to fossil fuels. This requires us to urgently reconnect the inside of our buildings to the world around them. We need to be kept safely comfortable in buildings, depending as far as possible on local resources, water, food, and natural energy from the sun, earth, and wind. To do that, the next generation must become eco-architects.[1]

Ecology

The word *ecology* comes from the Greek language, in which *Oikonos* means 'house' and *Logos* means 'knowledge'.[2]

The term *oekologie* was first coined in 1866 by the German scientist Ernst Haeckel and translated into English in 1873.[3] He defined it as the comprehensive science of the relationship of the organism to the environment. This is a rather recent branch of science, with its first textbook only published in 1909 by the Danish botanist Eugenius Warming, who is considered by many as the founder of ecology.[4]

The Biosphere

All animate and inanimate life cycles are linked to the biosphere, the life space of the planet, and its three main components:

- Atmosphere (air envelope)
- Hydrosphere (water systems)
- Lithosphere (the earth, soil, minerals, etc.)

DOI: 10.1201/9781315645070-18

The biota, or living communities within a building niche, include all the organic populations, from plants to people, dust mites, viruses, mice, and moulds,[5] which all function within and around the inanimate building structure and its resource streams.[6] A more frequently used term for this whole thing is environment, but the *environment* is generally seen as a more animated version of the ecology of the physical world, including also the social ecology of human relations and the cultural ecology of human creativity.

Ecological Flows

Buildings are constructed from many different inanimate components and over their lives play host to innumerable organic flows through visible species like people, pets, and pests, to the millions of invisible species and substances. At different times a building might host flows of red dust from the Sahara, or smoke plumes from distant forest fires or the COVID-19 virus spread around the world by visiting human vectors. Animate and inanimate matter may flow through buildings over a short time like tap water, or over very long cycles like in the minerals in a rock wall or the carbon embedded in a timber roof.

Ecologies sound, and are, complex but are often studied in a simple flow, as shown in Figure 12.1. Each individual property of the system moves through it, propelled by a force like energy, gravity, or pressure and, as it encounters other properties of the system, is changed by them through interactions or feedback systems. No ecosystem component flows through it are unaffected by other components that inevitably change the properties of the whole system on their way through.

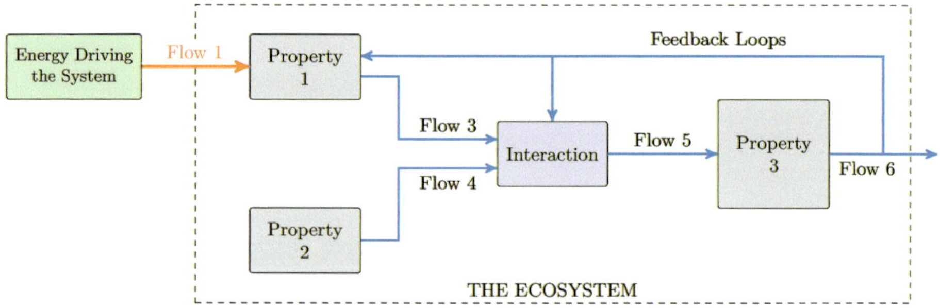

FIGURE 12.1 A systems diagram showing the five primary components of ecosystem models.

The Carbon Cycle

Perhaps the currently most studied flow through the built environment is the carbon cycle (Figure 12.2). This cycle connects the earth to the atmosphere and oceans, powered by heat, pressure, gravity, and importantly, the solar energy used by photosynthesis to convert atmospheric carbon dioxide into the carbon and oxygen that together provide the energy to keep our planet 'alive' for us.

> Carbon is the backbone of life on Earth. We are made of carbon, we eat carbon, and our civilizations – our economies, our homes, our means of transport – are built on carbon. We need carbon, but that need is also entwined with one of the most serious problems facing us today: global climate change.[7]

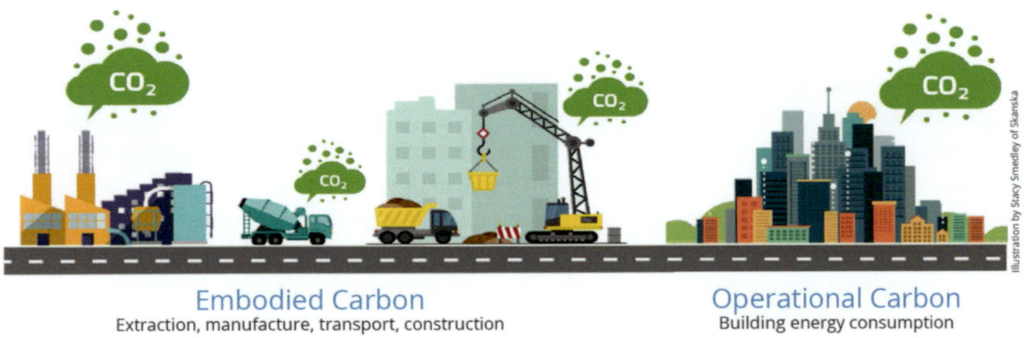

FIGURE 12.2 Buildings emit carbon into the atmosphere at all stages of their life cycles, from the embodied energy embedded in the mining and manufacture of materials for construction, from the energy they use over their lifetime to operate, and that emitted during their demolition.

Source: Stacy Smedley.

The slow carbon cycle moves carbon through the lithosphere (rocks and soil that make up the Earth's crust) over millennia, moving atmospheric carbon via rain, though the oceans, and locking it into rocks deposited on the sea floor. These strata rise, fall and fold over time, and the carbon in them is gradually re-emitted back into the atmosphere occurring via volcanic eruptions. The amount of carbon in the oceans, land, and atmosphere is regulated chemically. If increased in volcanic activity leads to increased atmospheric CO_2, for example, global temperatures rise, leading to more rain, which dissolves more rock, creating more ions that will eventually deposit more carbon on the ocean floor. It takes a few hundred thousand years to rebalance the slow carbon cycle through chemical weathering.

The *fast carbon cycle* denotes the movement of carbon through life forms on Earth, or the biosphere. It is a biogeochemical cycle in which carbon is exchanged between the biosphere, pedosphere (soil layer above the rock crust), hydrosphere, and the atmosphere of the Earth. At its heart is the process of photosynthesis taking place in plants on land and phytoplankton in the ocean that uses energy from the sun to combine carbon dioxide (CO_2) and water to form sugar (CH_2O) and oxygen. The chemical reaction looks like this:

CO_2	+	$2H_2O$	+	photons	$\rightarrow$	$[CH_2O]$	+	O_2	+	H_2O
Carbon dioxide		water		light energy		carbohydrate		oxygen		water

Carbon takes many forms and is the main component of biological compounds like wood that buildings are made of, as well as a major component of many minerals such as limestone that is also a building material, embedded in its structure over its lifetime. Carbon flows through homes over hours or days, in the food we eat, and the energy used to operate buildings. The greatest damage done to the planet's atmosphere by a single sector is by the built environment. Also, countries vary wildly in the amount of the greenhouse gasses they emit.[8] In the European Union in 2020, building construction consumed 40% of materials and 40% of primary energy and generated 40% of waste annually. Globally, in developed and developing countries, buildings contributed 33% of the greenhouse gas (GHG) emissions and 40% of the global energy consumption stemming from the usage of the equipment, the manufacturing of building materials, and transportation,[9] but again, a large proportion of this particular global emissions tally relates to high-end, high-energy buildings, such as those found in the Gulf, where all construction materials and labour have to be imported and many buildings are tall, requiring much higher levels of energy to build and operate. The amount of carbon emitted per capita results not only from the buildings they occupy but also the wealth of populations, their climate, and their national comfort cultures.[10,11] Despite over 30 years of endeavour and billions of investment in reducing greenhouse gas (GHG) emissions, they are still increasing from all sectors,[12] and fastest from the built environment.[13]

Because global temperatures are closely linked in the climate record to levels of CO_2 in the atmosphere, carbon dioxide is used as the litmus test of how far and how fast the climate is changing. In March 1958, C. David Keeling, at the Scripps Institution of Oceanography on Hawaii, started measuring the levels of CO_2 in the relatively clean environment on Mt. Mauna Loa in the Pacific Ocean.[14] Figure 12.3 gives a clear warning not only that global heating is already baked into the planet's atmosphere because of the direct relationship between manmade CO_2 emissions which continue to rise steadily in tandem but also the interconnectedness of the responses of Earth's ecosystem evidences by multiple changes to its climate systems. If we are to reduce the impacts of humans on this process, it will require not only improving building

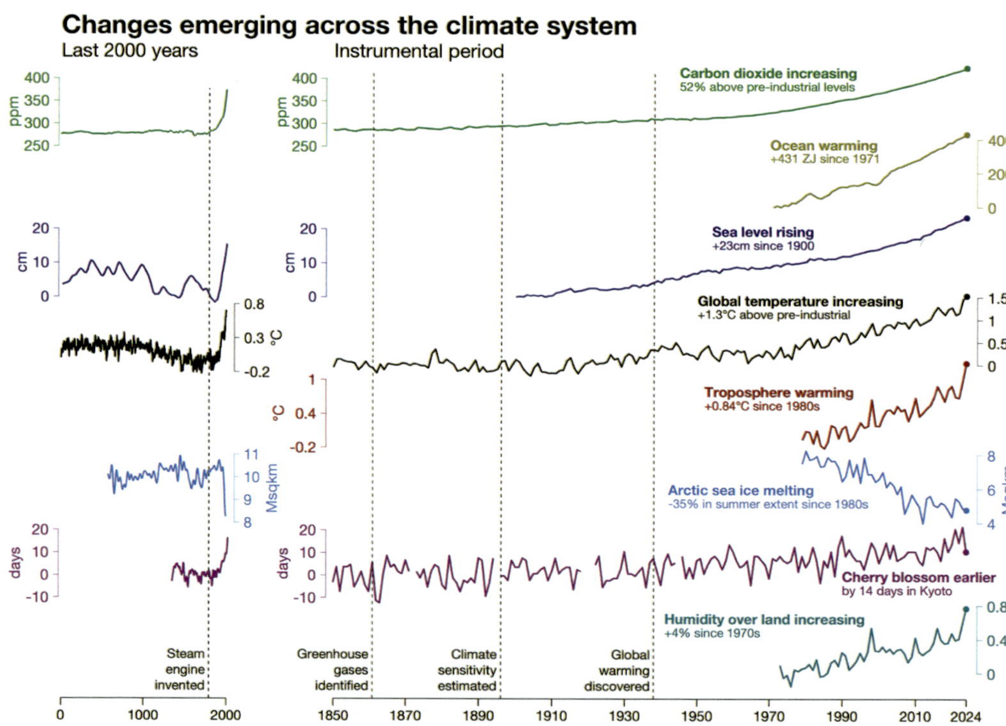

FIGURE 12.3 Multiple indicators across 1850 to the present day and over the last 2,000 years show the interconnectedness of Earth's ecosystems' responses to its changing climate systems.

Source: Ed Hawkins at https://ed-hawkins.github.io/climate-visuals/.

performance but also limiting, and perhaps even eliminating, high-energy buildings and also the changing comfort expectations and cultures.

The rapid escalation in rates and magnitude of changes in the global climate have escalated resulting in the breaking of weather records around the world on an almost-daily basis. World leaders at COP21 in Paris in 2015 promised to try to limit the long-term Earth's temperature rise to 1.5°C, which was key to avoiding the most damaging impacts of the heating climate. From February 2023 to January 2024 the global temperature was already above 1.5°C. 2023 was the warmest year on record. The past nine years have been the warmest nine ever on record. Higher temperatures have been accompanied around the world by intense heat, extreme precipitation, droughts, floods, wildfires, and between 2000 and 2019, over 5 million people globally have been estimated to have died from non-optimal temperatures exacerbated by climate change.[15]

On 6 October 2024, Jim Skea, Chair of the IPCC, warned that 1.5°C is officially out of reach, 2.0°C demands heroic action, 3.0°C is now the likely outcome, and Western Europe could face a shocking 5.0°C by 2100.[16] Some now suggest that this will pose an existential threat to humanity, a literal threat to humanity's existence as a result of the dangers that unchecked climate change can pose to our ways of life and place in the natural world. Armstrong McKay et al. in 2022 analysed nine 'core' and seven 'impact' climate tipping points (CTPs) affecting Earth system functioning and human welfare respectively. They estimated the crossing of six CTPs becomes likely, with a further four possible within the Paris Agreement range of 1.5°C. A significant likelihood of passing multiple climate tipping points exists above 1.5°C. In the greater connected Ecosystems of our planet the crossing of one CTP in one system can cascade into other tipping points leading to severe, potentially catastrophic impacts, for instance, ice loss in West Antarctica and Greenland will significantly alter ocean circulation,[17] not only raising sea levels, but also altering global climate patterns.

The lives we lead in the buildings we live and work in will over the coming decades have to change radically, including the food we eat, the cars we drive, the buildings we occupy, and the ways in which we heat, cool, and light buildings, because they are all connected via the flows between ecosystems to the ecologies around them.

The Ozone Cycle

An underrated but highly damaging ecological cycle affects the ozone layer in the upper atmosphere shown in Figure 12.4. As the planet heats, we need and use more air conditioning, but the machines involved in cooling buildings are not only major contributors to that heating through the electricity they use but also the largest source of ozone-destroying substances globally. Ozone is a chemically active colourless gas consisting of three bonded oxygen atoms (the usual oxygen molecule is two). Terrestrial ozone affects the health of many species at ground level, including sensitive vegetation and ecosystems like forests and wilderness areas. It can be very harmful to humans, causing breathing difficulties, particularly for people near busy roads during hot sunny weather[18] and in a warming climate.

In the upper atmosphere, ozone forms a thin protective layer around the upper stratosphere, some 50 miles above the Earth. This vital layer shields the Earth against harmful ultraviolet B, or UV-B, rays in the incoming sunlight that can also seriously, and fatally, affect the health of many plants and animals, not least by causing cancers in humans.

Chlorine is a major constituent of the fluorinated gasses – chlorofluorocarbons (CFCs) and hydrochlorofluorocarbons (HCFCs) – that are gases used worldwide in domestic and commercial

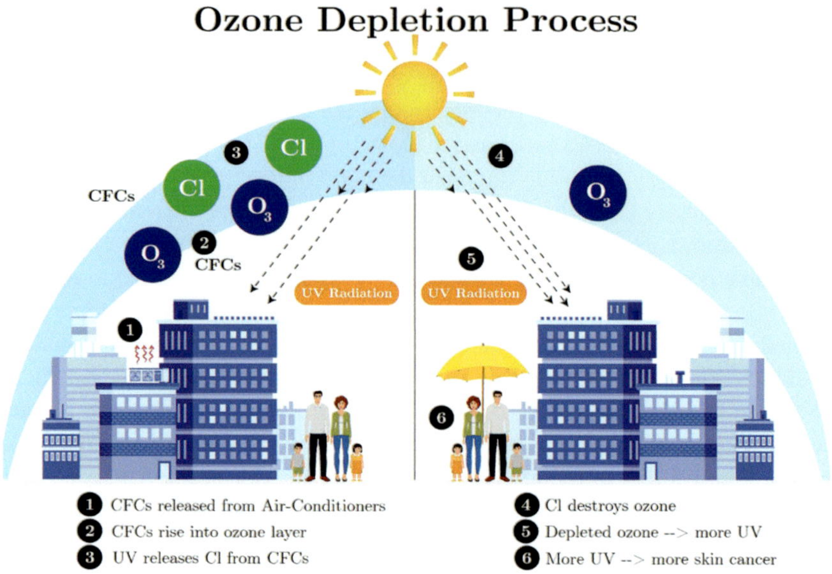

FIGURE 12.4 Simplified diagram of the impacts of fluorinated gas flows (chlorines) in the upper atmosphere in the presence of sunlight and freezing temperatures acting to thin the ozone layers above the poles, causing the ozone holes that can be so detrimental to the health of the populations below them.

Source: Jim Eveson.

refrigeration, cold storage, food processing, transport, industrial refrigeration, air conditioning, and heat pumps and chillers. CFCs and HCFCs are the main ozone-depleting substances that accumulate in the upper atmosphere, and in the presence of extreme cold and sunlight, they react with the ozone to destroy it rapidly and severely. This happens almost exclusively over the poles, and particularly in the Southern Hemisphere, where they have created the ozone holes. Most HCFCs and CFCs used in these processes have global warming potentials (GWPs) in the range of 1,000 to 4,000 times more powerful as greenhouse gases than CO_2.

Measurement of the ozone hole began in Antarctica in 1956, but it was not until the paper by Farmer et al. in 1985 made clear the scale and implications of the ozone depletion problem.[19] Immediate and effective action triggered the unilaterally agreed Montreal Protocol in 1987 that began a process of CFC and HCFC phase-out and the move to low-GWP alternatives, like ammonia, hydrocarbons, and carbon dioxide. Unfortunately, almost every alternative has more or less of the associated problems, like higher cost, flammability, toxicity, and pressure requirements in their processes.

The carbon and the ozone cycles are examples of how directly connected the individual comfort habits of people who use mechanical heating and cooling are to driving the heating of the planet. Higher levels of incoming solar radiation entering through the ozone holes over the poles melt the sea ice leaving the darker oceans to heat more rapidly. Despite international agreements and significant steps taken by industry emissions from fluorinated gases increased threefold between 1990 and 2019.[20] Comfort conditioning systems are driving global heating and the hotter it gets the more air conditioning systems will be needed in a vicious circle.

Ecological Design

It is easy to be overwhelmed by the sheer scale of the ecological implications and impacts of the buildings we design, build, or occupy, but the task of understanding how to do so has really been enabled by the development of the recent science of ecology. Leading pioneers in understanding and quantifying the material flows through the ecosystems included Eugene P. Odum, who did much to develop the science of ecological modelling from the 1950s onwards.[21] Donella H. Meadows, the environmental scientist who was lead author on *Limits to Growth*, written for the Club of Rome in 1972, shocked the world by demonstrating with credible models that unchecked human growth on our finite planet was taking the world to ecological 'overshoot'.[22]

Mathis Wackernagel understood that 'overshoot' will ultimately liquidate the planet's ecological assets' and, in his 1994 PhD thesis, developed the hugely influential ecological footprinting method to account for the major ecological flows driving climate change, biodiversity loss, and resource shortages. He linked them with a uniform metric system, referenced back to land to calculate the impacts of humans on our shared planetary metabolism.[23] These ecologists did much to inform and educate the later generation of carbon accounting systems that have been so important in helping us understand the impacts of design decisions on the global climate and later those working in the circular economy movements.[24]

Architects are seldom environmental scientists and tend to be an image-driven profession, responding to drawings and maps more than formulae. This may explain why one of the most influential books for designers linked people to their animate and inanimate environment was by the ecological planner Ian McHarg. In his 1969 book *Design with Nature*, he used images and maps overlain on each to explain the connections between the different material planes on a site covering its geology, hydrology, biology, and human settlements.[25]

Some readers may want to explore ecological design ideas further. For many others, the key message is simple: to de-couple buildings from the flows of fossil fuels and climate damaging emissions through and from buildings requires designers to reconnect the environments within buildings to the flows of free, clean and renewable energy and resources around them. Self-sufficient buildings can be as comfortable or more so than mechanically conditioned ones for much of the year if they are intelligently designed by people who understand the forces that link energy and resource flows between the micro-climates and ecologies outside a building and those inside it.

A building is not a sculpture or a box full of stuff floating in space. A building exists in time and space and, within its own ecosystems, connect directly to Earth's larger ecosystems. Everything is connected, and for mankind to survive, we need to re-learn that simple truth and work with the forces of nature, not just to needlessly destroy it.

Notes

1 Roaf, S., M. Fuentes and S. Thomas (2012). *Ecohouse: A Design Guide*, 4th Edition, Routledge.
2 Odum, E.P. (1971). *The Fundamentals of Ecology*, Philadelphia: W.B. Saunders Company.
3 McHale, J. (1970). *The Ecological Context*, New York: George Brazziler, p. 1.
4 Warming, E. (1909). *Oecology of Plants – an Introduction to the Study of Plant-Communities*, Oxford: Clarendon Press.
5 Annesi-Maesano, I. (2019). The exposome paradigm, in *Built Environments and Health*, Edited by E. de Oliviera Fernandes, Porto, Portugal: Luso Impress.
6 Odum, E.P. (1997). *Ecology: A Bridge between Science and Society*, 1st published 1913, Sunderland M.A. USA: Sinauer Assocs. Ltd.

7 https://earthobservatory.nasa.gov/features/CarbonCycle

8 Ritchie, H., and M. Roser (2024). *CO_2 emissions*, Published online at OurWorldinData.org. https://ourworldindata.org/co2-emissions

9 Sizirici, B., Y. Fseha, C.S. Cho, I. Yildiz and Y.J. Byon (2021). A review of carbon footprint reduction in construction industry, from design to operation, *Materials* (Basel), 14(20), p. 6094. https://doi.org/10.3390/ma14206094

10 Valancius, K., M. Grineviciut and G. Streckiene (2022). Heating and cooling primary energy demand and CO2 emissions: Lithuanian buildings in different European locations, *Buildings*, 12(5), p. 570. https://doi.org/10.3390/buildings12050570

11 Treloar, G.J., R. Fay, B. Ilozor, P.E.D. Love (2001). An analysis of the embodied energy of office buildings by height, *Facilities*, 19(5/6), pp. 204–214.

12 www.icarb.org contains a range of papers and presentations covering emissions from many different sectors of the economy and may provide a useful introduction for different disciplines on carbon emissions.

13 IPCC (2022). Climate change 2022: Mitigation of climate change, *Summary for Policy Makers*, Accessed on 09/06/2022 at: https://www.ipcc.ch/report/ar6/wg3/

14 Keeling, C.D., R.B. Bacastow, A.E. Bainbridge, C.A. Ekdahl, Jr., P.R. Guenther, L.S. Waterman and J.F.S. Chin (1976). Atmospheric carbon dioxide variations at Mauna Loa Observatory, Hawaii, *Tellus*, 28, pp. 538–551. https://doi.org/10.1111/j.2153-3490.1976.tb00701.x

15 Zhao, Q., Y. Goa, T.T. Ye, et al. (2021). Global, regional, and national burden of mortality associated with non-optimal ambient temperatures from 2000 to 2019: A three-stage modelling study, *The Lancet Planetary Health*, 5(7), pp. e415–e425.

16 Skea, J. (2024). It's too late to save Britain from overheating, says UN climate chief, *The Telegraph*, 6th October. https://www.telegraph.co.uk/business/2024/10/06/too-late-save-britain-overheating-climate-chief-jim-skea/

17 https://en.wikipedia.org/wiki/Tipping_points_in_the_climate_system#:~:text=In%20climate%20science%2C%20a%20tipping,and%20may%20accelerate%20global%20warming

18 Zhang, J., Y. Wei and Z. Fang (2019). Ozone pollution: A major health hazard worldwide, *Frontiers in Immunology*, 10. https://doi.org/10.3389/fimmu.2019.02518

19 Farman, J., B. Gardiner and J. Shanklin (1985). Large losses of total ozone in Antarctica reveal seasonal ClOx/NOx interaction, *Nature*, 315, pp. 207–210, DOI: 10.1038/315207a0

20 IPCC (2022). *Climate Change 2022: Mitigation of Climate Change: The Working Group III Technical Summary*, p. 142, Accessed on 12/06/2022 at: https://www.ipcc.ch/report/ar6/wg3/

21 Hall, C.A.S. (Ed) (1995). *Maximum Power: The Ideas and Applications of H. T. Odum*, University Press of Colorado, ISBN: 0-87081-362-5

22 Meadows, D., J. Randers and D. Meadows (2008). *Limits to Growth: The 30-Year Update*, 3rd Edition, London: Earthscan.

23 Wackernagle, M., W.E. Rees and P. Testemale (1995). *Our Ecological Footprint: Reducing Human Impact on the Earth*, Canada: New Society Publishers, ISBN: 0-86571-312-X

24 Aarikka-Stenroos, L., P. Ritala and L.D.W. Thomas (2021). Circular economy ecosystems: A typology, definitions, and implications, in *Handbook of Sustainability Agency*, Camberly, UK: Edward Elgar.

25 McHarg, I.L. (1971). *Design with Nature*, New York: Doubleday & Company.

13

HEAT FLOWS AND ECOLOGICAL ENGINEERING

Introduction

A primary purpose of a basic shelter is to protect occupants from the thermal dangers surrounding it and discomfort in it. The shelter works in tandem with the clothes worn and activities undertaken indoors with other available adaptive opportunities, including the use of heat and cold sources, to keep body temperatures of occupants at around 37°C. A secondary purpose might be seen as promoting feelings of thermal neutrality and, better still, pleasurably comfortable conditions indoors. Both require a shelter to be designed, constructed, and managed to control heat flows in and out of it and around the structure. Important questions, then, are what is heat, and how does it flow around buildings?

Heat is a complex phenomenon, with an entire branch of physics established to explain its relationship with other energies. In thermodynamics, heat is energy in transfer between one thermodynamic (thermally active) system and its surroundings by other than mechanical, electromagnetic, or gravitational means.[1] This transfer usually involves the movement of heat from a hotter to a colder body or state. Many of us are not physicists, so this chapter covers a few basics that might help a designer develop their own palette of heat management strategies to optimise the comfort value of natural heat transfers within a building to its occupants.

The History of Heat

Speculation on what *heat* actually is as a separate form of matter began centuries ago. In the 17th century, some thought that heat consists of a self-repellent fluid called *caloric* that flows from hotter bodies to colder bodies, or a weightless gas that could pass in and out of pores in solids and liquids. Explanations became more confused as explanations of combustion were mixed with the caloric theory. In 1667, it was proposed a matter existed that was thought to be the 'substance of heat'. It was named *phlogiston,* meaning 'burning up', from the Greek word *phlox*, a flame. The theory posited that this fire-like element was contained within combustible bodies and released during combustion. In general, substances that burned in the air were said to be rich in phlogiston.[2]

DOI: 10.1201/9781315645070-19

A charming example of how heat systems were understood to work in the late 18th century comes from Marcellin Ducarla-Bonifas, a Parisian scientist who published *The Complete Book of Fire* in 1774.[3]

Calorific Cascades

Ducarla often amused himself and friends of an evening in Paris when the fire had died down. Others would poke in vain at the smouldering wood in the hearth, adding more kindling and coal, producing nothing but thick smoke. Ducarla would eventually get hold of the fire tongs (a feat he claimed required patience, audacity, and luck), call a halt to the chatter, and pretend to cast a spell over the fire. Then elevating the glowing faces of a few half-burned logs towards each other, often unnoticed, he would sit back and wait till flames took hold of the logs. His friends would accuse him of throwing some flash powder on it or simply having made the most of the draught. Ducarla noted they did not inquire into '*the effluent (phlogiston?), the radiant kinds of heat, pyrosheres, translative speeds and calorific series*'.[4]

The useful lesson for designers here is his description of the existence of a 'calorific series' of energy flows where the interactions between the different heat flows given off by each logs and the surrounding oxygen combined to create a different, hotter thermal system that soon produced flames. Less important than the need to explain the detailed science of heat to his friends was the fun he could have playing with the calorific cascades in the fire. Like Ducarla's fire, a building is a dynamic and complex thermal environment that can be played with to manage heat flows within it across a day or year to maximise occupant comfort over time, while also maximising the amount of free, natural, local energy required to make that happen.

Mechanical Heat

The *caloric theory* was superseded by the mid-19th century in favour of the mechanical theory of heat but nevertheless persisted in some popular scientific literature until the end of the 19th century. The mechanical theory of heat begins with the premise that there is a physical quantity, which is not identical with the mechanical quantities (mass, force, pressure, etc.), whose changes can be determined by calorimetric measurements, and that is called heat. A *calorimeter* is a device used to measure the heat of chemical reactions or physical changes as well as the heat capacity of a material, and it was developed first by the Scottish scientist Joseph Black from his research on measuring heat. *Thermoscopes* were the earliest devices for detecting temperature set against a simple qualitative scale, and Galileo was thought to have built the first one in around 1592. For some authors, a *thermometer* is just a thermoscope with an exact temperature scale on it. It was not until the 1760s that Joseph Black, the Scottish scientist, first employed a device developed decades before by Daniel Gabriel Fahrenheit to grade temperatures between freezing and boiling points to quantify the intensity of heat. Black's scientific measurements of temperatures enabled the development of a theoretical understanding of thermal phenomena through the application of mathematics to the readings, and the science of thermal dynamics was born.[5] But to become a designer of natural energy buildings, mechanical and mathematical understandings of heat must also take its rightful place as only one of the ecological flows of energy and living and inert matter through a building over time.

Powering Ecosystems

Forms of energy that power life on this planet and comfort in buildings include:

Heat energy. Heat flows through buildings as:

Conduction is the process by which heat is transferred from the hotter end to the colder end of an entity, be it through solids, through liquids, and even through gasses. Heat is continually being conducted through our own bodies from the skin inwards or outwards and from the core outwards or inwards via veins and arteries along the temperature gradient from hot to cold. Heat from the skin is exchanged with surroundings via physical contact between the surfaces of different thermal systems, be it between a person and a chair or bed, or also continually between the thermal systems of different materials in a house.

Convection is driven by heat differentials between different masses of liquids or gasses that cause movement as the hotter, less-dense areas rise, and the cooler, denser areas sink in the system. Two forces are acting on each adjacent mass, the downward force of gravity and an upward force called buoyancy. The density or weight of a volume or mass of water or air measures the downward force of gravity that acts on it. If one mass is heavier or denser than an adjacent mass, it sinks, and if the reverse is true, it will rise. Convection can be very powerfully used to push heat and air around a building either by buoyancy or wind pressure.

Buoyancy acts rather like water does by gravity, but because it is lighter than its surroundings, instead of flowing downwards, it flows upwards. If you use a bucket in this as proxy for a room, if you put holes in the room (bucket upside down), their size and height will influence how much water is lost and at what rate, driven by pressure. With buoyancy, the temperature difference between the floor and the ceiling is also key to the rate at which air flows out of windows at the top of the wall. The relationship between pressure, temperature, and volume of a gas was defined in Boyle's law of 1662, which states thus:

> *When the temperature of a given mass of confined gas is constant, the product of its pressure and volume is also constant.*

In 1878 Charles's law of volumes that described how gases tend to expand and rise when heated and cooled and contract in volume as they sink.

Radiation is electromagnetic radiation measured on a spectrum from visible radiation in the ultraviolet, short wavelengths, to the infrared longer wavelengths, which are less visible to human eyes. Radiant heat results from the thermal motion of particles in matter; the hotter the body, the greater the motion of particles within it, and the direction of flow is always from a hotter body or surface to a cooler one. In physics, radiant energy is emitted or transmitted in the form of waves or particles through space or through a material medium. We mostly encounter it in the form of solar rays, microwaves, and radio waves in our everyday lives. As heat moves through various thermal systems in a building, it continuously radiates energy from a hotter to a cooler material across the intermediate spaces. Figures 13.1 to 13.3 show how we are constantly surrounded by flows of heat from and through different building elements at different temperatures, motivated by conduction, radiation, convection, and in the case of infrared radiation, also reflection.

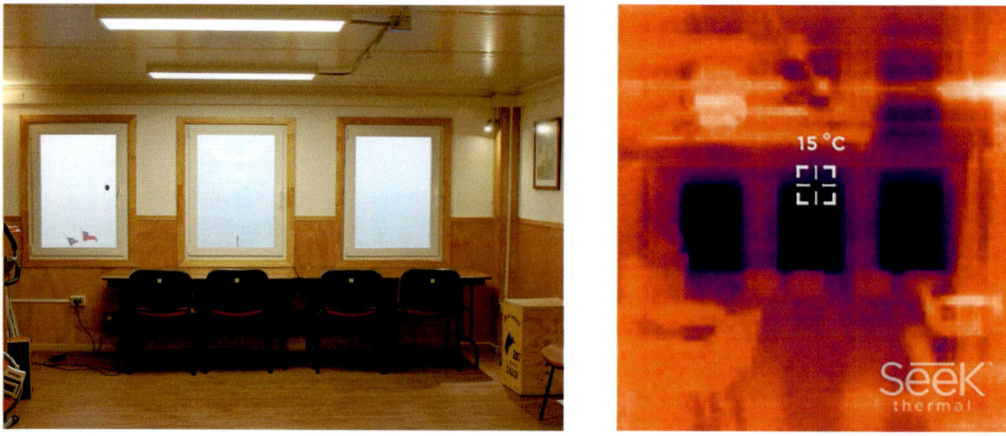

FIGURE 13.1 (a) The workroom at the Chilean research station of Escudero in Antarctica showing the very cold double-glazed glass set in metal frames, both forming more or less efficient cold bridges channelling external cold into the room by conduction. (b) The thermal image of the windows shows those cold surfaces are also radiating cold up to the metal ceiling on which they are reflected.

Source: Susan Roaf.

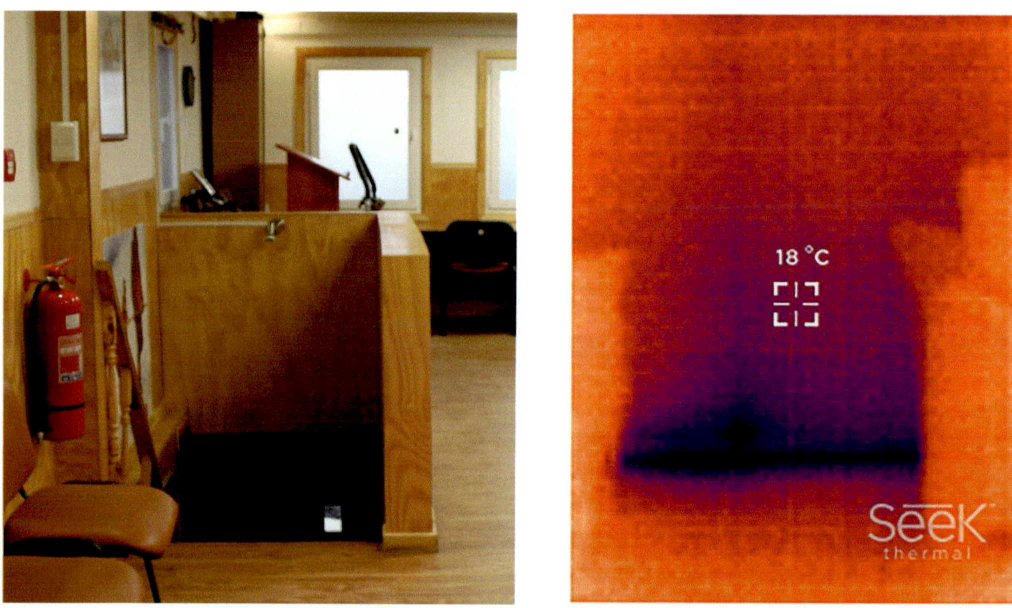

FIGURE 13.2 (a) The staircase down from the workroom to the living room below at Escudero has the door at the bottom closed. (b) The thermal images show a significant cold air inversion puddling in the lower space.

Source: Susan Roaf.

FIGURE 13.3 (a) The workroom at Escudero is heated with the electric wall radiators, and (b) the thermal image shows that as well as radiating heat to those adjacent to it, heat from it flows upwards convectively and puddles under the metal ceiling that warms to form a radiant heat source for those working below.

Source: Susan Roaf.

Reflection is usually thought of as the throwing back by a body or surface of light, heat, or sound without absorbing it. Because visible light and heat travel together, and since visible light can be reflected off a mirror, heat also can be reflected to some extent, usually only in much small concentrations than of visible light that is being reflected. In Figure 13.1, the cold windows are reflected on the metal ceiling. Polished metal surfaces or metallic coatings not only reflect light but are, as a result of their electrical conductivity, also excellent heat reflectors right into the far-infrared. Silver achieves reflection values of 98–99% at infrared wavelengths of more than 1,000 nm, which is why unglazed silvered mirrors are used to concentrate solar energy on solar collectors.

Emissivity of the surface of a material is its effectiveness in emitting energy as thermal radiation. It is measured as a factor of the radiant energy emitted by a surface to that emitted by a blackbody at the same temperature. Highly reflective surfaces like polished silver emit little radiant energy and so are good at retaining heat, which explains why polished silver teapots are so good at keeping water hotter for longer than a blackened unpolished pot. Many modern windows are treated with low emissivity coatings treated often with speckled silver or aluminium coatings between panes to retain the heat inside a space. Standard clear glass has a thermal emissivity of 0.9, while for metals such as silver and aluminium, it is less than 0.05. Unfortunately, because the emissivity of a material surface depends on many chemical and physical properties, it is often difficult to estimate, but it is a property of building surfaces that has gained increasing traction with the growth in importance of low-emissivity roof paints that have been heavily marketed as a cheap and effective way of reducing heat gain into buildings via cool roof programmes around the world.[6]

Evaporation (water). The change of water from its liquid to its gaseous form (water vapour) requires a lot of energy, and it is this energy which (for instance) cools our skin when our sweat dries. The heat required to change the sweat from a liquid to water vapour (known as the latent heat of vaporisation) is more than five times that needed to heat the water from cold to hot. Think of a pan of water on a stove. It heats quite quickly to boiling point, but then it will take far longer for the water to boil away. Additionally, water in its liquid form has a high 'specific heat', which means that for a given rise in temperature, it will take more heat than for most other materials. While the latent heat of fusion needed to change ice to water is less than that for vaporisation, it is useful for cooling drinks on a hot day.

People in hot climates traditionally used the evaporation of water from floors in and around buildings to drop air temperatures by using the heat absorbed when heating and evaporating the water to powerfully cool the surfaces of the courtyard. Evaporative coolers can provide some of the simplest and lowest-cost cooling techniques around to improve thermal comfort in both hot-dry and hot-humid climates.[7] Since antiquity, porous jugs of water cooled air moving over them, and since then many different systems have been used around the world, from evaporative air conditioner, swamp coolers, swamp box, desert cooler, and wet air cooler, to a simple jug of water thrown over a bamboo screen in the indoor outlet of a windcatcher in the Gulf. Misting systems, some using fans, are increasingly used around the world to lower temperatures around buildings and in urban spaces and at sports events, where they can significantly increase comfort and lower air temperatures by up to 5°C (9°F).[8]

Gravity underlies the power generated by the interaction between all things with mass or energy. It is the force that determines the motion of planets, stars, galaxies, and even light. It influences many different flows around the planet, including those of the oceans and the tides and the weather above them. Gravity actually anchors the atmosphere to the mass of the planet, providing us with the air that we breathe. It holds the world together and lends weight to objects that can be transformed into the mechanical energy, which is the ability of an object to do work. Mechanical energy can be either kinetic (energy in motion) or potential (energy that is stored) as shown in Table 13.1. It is gravity that produces the kinetic energy to turn water turbines and to power energy storage systems, like the regenerative braking systems used in many modern electric vehicles.

Chemical energy is released by chemical processes. It is the energy of chemical substances that is released or absorbed when they undergo a chemical reaction and transform into other substances. Some examples of storage media of chemical energy include batteries, food, and gasoline. Breaking and making chemical bonds involves energy (heat), which may be either absorbed by

TABLE 13.1 Different forms of kinetic and potential energy

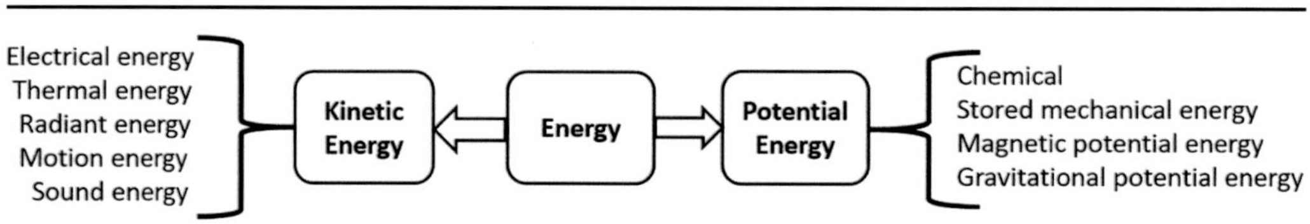

(endothermic reaction) or produced from (exothermic reaction) a chemical system. The source of all this energy is the sun, which triggers the chain reactions of life. Some of the incoming energy from the sun is absorbed by pigments in leaves and converted to chemical energy during photosynthesis, the process that converts water, carbon dioxide, and minerals into oxygen *and* energy-rich organic compounds. These compounds, when burnt, or decomposed, create heat energy and, if used in a high-density form, like wood or coal, can be burnt at a high temperature to create the mechanical, kinetic energy (associated with movement) that powers machines. The energy stored in plants is also what provides the energy necessary to power the metabolisms of all living animals in energy cascades that follows the laws of thermodynamics.

Potential energy is the energy stored in a system (say, a battery or a water tank) which can be used when required. For instance, a water tank or reservoir connected by a pipe or channel to a lower water tank or other sink, through which water flows by gravity to a lower level and is converted into kinetic energy (energy in motion) to turn, perhaps, a grinding millstone, or power a turbine, a rotary mechanical device that extracts energy from a fluid flow and converts it into useful work, like producing electrical energy in a hydroelectric plant. In a world of unreliable or intermittent energy availability, stored energy in all forms will become increasingly critical to maintain basic functions in the built environment.

Sound energy results when a force, either sound or pressure, makes an object or substance vibrate. That energy moves through the substance in waves. Those sound waves are called kinetic mechanical energy. Solids, liquids, and gases all transmit sound as energy waves.

Carried energy. Every time a person or a material flow passes into, out of, and through a building, they typically carry heat or cold with them. This is an under-considered energy flow through buildings but can cause considerable problems if unaddressed. For instance, postal sorting offices in the Midwest of America sort millions of letters and parcels from hot or cold postal vans that can overheat or overcool a premises quickly in the heights of summer or the depths of winter. Metro systems around the world often overheat simply because of the difficulties of dissipating heat from the streams of hot people and trains that flow through them. Much work is being done on overheating in the London Underground,[9] and new ways are being sought to exploit the heat from the tunnels to provide heat for buildings on the surface. The new desert city of Masdar in the UAE was built on a plinth that was supposed to contain a cool semi-underground basement car park. The desert-heated cars that flowed through and stayed in the space heated the raised-ground huge floor slab of the whole city above. In cold climates, draft lobbies for external doors, where cold outdoor clothing can be shed, are essential to reduce the carried cold through a building in winter. As hot and cold weather extremes become more intense and frequent, much more emphasis must be given to the thermal pre-conditioning of people and things before they enter buildings.

The Laws of Thermodynamics

The laws of thermodynamics are the set of rules largely developed in the 19th and 20th centuries that govern the way that energy, and in particular, thermal energy, can be seen to act in a system which is 'isolated' from the natural world. This is sometimes difficult to imagine because a truly isolated system does not actually exist in the natural world. The first established thermodynamic principle, which eventually became the second law of thermodynamics, was formulated by Sadi Carnot in 1824.[10] The third law was formulated by Walther Nernst over the period 1906–1912. Not only are they important fundamental laws of physics in general, but they are also particularly important for designers constructing heat and coolth pathways and stores in buildings.

The first law of thermodynamics, the law of conservation of energy, states that the total energy of an isolated system remains constant and can neither be created nor destroyed; rather, it can only be transformed or transferred from one form to another. Why is this important? Because a building as a structure is a bounded (though not an isolated) system, through which energy is continually flowing in one form or another and, importantly for comfort, as heat. The ability to understand and imagine the forms and pathways energy takes and stores in and through a building is a fundamentally important skill for designers looking to create buildings that use minimal energy to keep people comfortable inside them.

The second law of thermodynamics deals with thermodynamic processes but is similar to the law of conservation; meaning, that once the energy is in the building, it does not disappear but may transfer into different forms. Sunlight entering a window may hit a wall and be stored in it as heat that is gradually re-radiated back out to warm other surfaces it can 'see'. Wood might be brought into a building and, when burnt, heat the adjacent surfaces or persons by radiation or conduction and, by radiantly heating particles in the air around it, cause them to rise in warmer, less dense air, so heating the ceiling above it. The total heat in the room is the same but has just changed its properties. Heat can pass through the system as work, heat, or matter.

Heat is associated with the random motion of particles (entropy), while *work* is the energy of ordered motion in one direction (exergy). Take a smouldering log in a fireplace that has grown cool; it provides low-quality, low-temperature energy that can heat little around it. Contrast that with a white-hot burning log that can heat much of the matter around it and is working to do so, producing high-quality energy that can be used to perform work. *Entropy* is the term used to describe how much energy is in a system that is NOT available for doing useful work. Exergy is the useful part, the part that can perform work. It is a measure of energy quality.

The third law of thermodynamics states that heat always travels in one direction; meaning, that hot things always cool unless you do something to stop them. It expresses a fundamental and simple truth about the universe, that disorder, characterised as the quantity known as entropy, always increases, because of the irreversibility of natural processes that lead towards the even distribution of matter and energy in a system. Put a drop of dye in a liquid, or a blast of hot air into a cold room, and the dye will dilute to a uniform weaker colour and the room temperature will tend to even out across the space.

These three laws can inform the way we design thermal pathways through buildings, and even landscapes, where the placing of adjacent functions requiring lower temperatures in sequence can help run otherwise energy-demanding functions off the waste energy of processes further up the heat chain, as we saw in the London Underground.

Low-Exergy Design

Pioneered by Professor van den Dobbelsteen et al. in Delft,[11] an approach termed low-exergy design limits the loss of the ability to do work between and during the different steps in a heat cascade along its route. It is a horses-for-courses theory. Each step down in the level of heat intensity in a cascade design places within the system only tasks that require the appropriate intensity of work that matches what is available at that particular point in the cascade within a building, process, or landscape. This way, heat is reused several times and not wasted between each step. The waste heat created in high-temperature heavy industrial processes can then be used in gradually lower-grade functions, such as manufacturing processes, horticulture, and subsequently, residential heating. Residences and agriculture can eventually again 'feed' the power plant with biomass and waste, as shown in Figure 13.4.

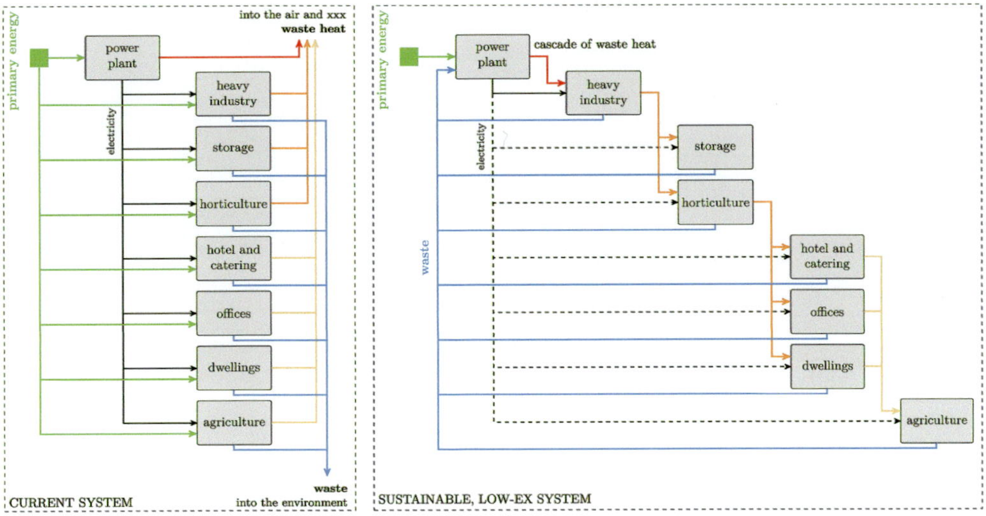

FIGURE 13.4 Graphical example of energy provision in (a) the current system and (b) in a more sustainable, low-exergy system that uses heat cascades (right) to maximise the work done by heat flowing through a planned landscape.

Source: Based on van den Dobbelsteen et al. (2008).

The use of cascading heat flows through buildings is part of the vernacular vocabularies of thermal design strategies, for instance, where main bedrooms were placed above the kitchen or stove below and masonry chimney breasts and flues were used to weave heat through a building, so requiring fewer fires. In many agricultural cultures, heat from sheep or cattle which were kept over winter in stalls on the ground floor was used to heat the family living areas above or next to it. Elsewhere on the remote island of Ulleung off South Korea, high-energy heat from kitchen fires was piped through hypocausts under adjacent high-thermal-mass living room floor areas, requiring lower-temperature heat to keep families comfortable wisely, using all the heat available to keep them safe through the long, severe winters. By contrast, in the Mongolian yurt in Chapter 7, the walls of the yurt are lightweight and retain little of the heat generated from the fire which is simply thrown away up the chimney.

Sources and Sinks

The second law of thermodynamics states that all energy naturally moves from a higher-quality to a lower-quality state as heat flows from a hotter to a colder state. The flows of particular interest to comfort designers are those that are the sources of heat around the house, via thermal stores to the lower-temperature heat sink regions. In the hydrological cycle, mountain lakes or aquifers can be a source of water that is moved by gravity flows down to the sea, which is its natural sink. In its passage the water loses potential energy, and some of that water is heated, by the sun and wind, gaining thermal energy that enables it to evaporate, forming clouds that can then rain down (gravity) onto those same mountains, replenishing the source again in a coupled ecosystem flow. At any point along the gravity-fed flow, water can be stored for different lengths

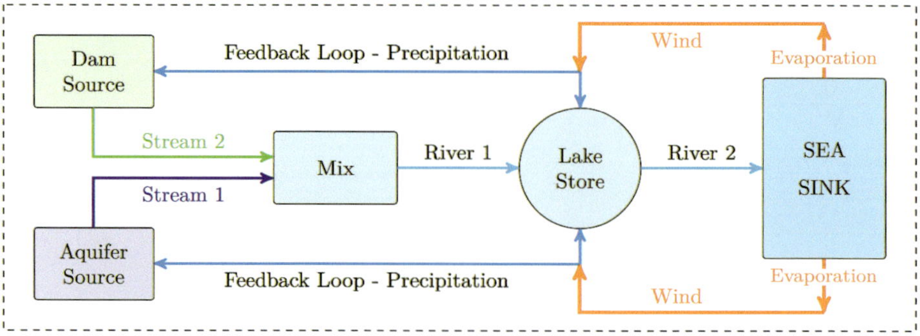

FIGURE 13.5 Flows of water through the hydrosphere are powered by gravity (blue), evaporation, and wind energy (orange).

Source: Susan Roaf.

of time. Ecological design involves the interaction between many different levels of flow and cycles of energy systems in the environment. Understanding such flows can liberate architects and engineers to move beyond the thinking of simply mechanical theories of thermal energy and design, to perhaps evolve into the different 21st-century thought systems we might need to create climate-safer buildings powered largely by local natural energy.

Ecosystem Engineering

An *ecosystem engineer* is defined as an animal species that creates, significantly modifies, maintains, or destroys a habitat by altering the flows of energy, resources, and populations through it; to a certain extent, all organisms impact their environments in some way. The term is usually used for species whose behaviour very strongly affects other organisms. Beavers are considered the classic example of an ecosystem engineer because the dams they build on waterways affect the water flows, geomorphology, and ecology of an area.

Humans are also major ecosystem engineers who have modified energy and resource flows across whole landscapes to enable people to live comfortably in some of the most hostile environments on the planet. As we move further into the much hotter and more extreme climates of the future, in which the refuge of last resort in extreme events will be the buildings we occupy, it is probably time for architects and engineers to become fairly sophisticated thermal ecosystem engineers. To do this, it is necessary to:

1. Reconnect the thermal life cycles and energy flows of people inside buildings to those in their surrounding ecosystems
2. Design and place each space and function in a building carefully to maximise the comfort value from locally available energy sources over each day in a year and to avoid energy waste and occupant discomfort

Unfortunately, many of the design tools widely used to create buildings today are really not fit for thermal ecosystem design. To create a comfortable home for a Hobbit, you do not need to be a physiologist; neither do you need to be a physicist or an engineer to design a climate-safe

TABLE 13.2 The extreme design process: a simple staged process for exploring, developing, testing, and building an extreme design used to create the resilient temporary shelter at Collins Glacier in Antarctica

	The extreme design process
1	Study the site, its environs, and its climate in great depth.
2	Source and study appropriate vernacular archetypes for inspiration and understanding.
3	Simulate improvements to traditional structural strategies for the proposed site.
4	Explore innovative material solutions building on advice from product experts.
5	Bench and field-test structures and materials in an appropriate local test facility.
6	Finalise structural and envelope design with tent makers.
7	Transport and build final design on-site.
8	Continually measure local conditions on-site, refining use of the structure.
9	Monitor structure each year: strengths and weaknesses to inform next evolved design.

Source: Susan Roaf.

building for the future. It may need a more careful design process to achieve. Innovation will be important to create comfortable buildings going forwards, requiring more development and interrogation of ideas. The process shown in Table 13.2 was used by Roaf et al. to build a temporary shelter at Collins Glacier in Antarctica in 2018–2019, relying on vernacular archetypes, product testing, expert advice, and only intermittently on design simulations to beta-test design decisions.[12]

Thinking like a thermal ecosystem engineer requires both new understandings of how and when to charge and discharge energy stores and use warmth and coolth from the earth and the air. Much can be learnt from the design skills and wisdom of Indigenous builders. The next generation of designers may also have much to learn from other species on how to survive in a heating climate.

Notes

1 https://en.wikipedia.org/wiki/Heat
2 Fox, R. (1971). *The Caloric Theory of Gases*, Oxford: Clarendon Press.
3 Ducarla-Bonifas, M. (1784). *Du Feu Complet*, Paris: Published by Moutard, Accessed from the University of Michigan Library at: https://babel.hathitrust.org/cgi/pt?id=mdp.39015035810939&seq=5
4 Bachelard, G. (1964). *The Psychoanalysis of Fire*, Boston: Beacon Press. 1st Published in French in Paris in 1938.
5 Sherry, D. (2011). Thermoscopes, thermometers, and the foundations of measurement, *Studies in History and Philosophy of Science Part A*, 42(4), pp. 509–524, ISSN: 0039-3681. https://doi.org/10.1016/j.shpsa.2011.07.001
6 Kolokotroni, M., E. Shittu, T. Santos, et al. (2018). Cool roofs: High tech low cost solution for energy efficiency and thermal comfort in low rise low income houses in high solar radiation countries, *Energy and Buildings*, 176, pp. 58–70, ISSN: 0378-7788. https://doi.org/10.1016/j.enbuild.2018.07.005
7 Kapilan, N., M. Arun and S.K. Isloor (2023). A comprehensive review on evaporative cooling systems, *Results in Engineering*, 18(101059). https://doi.org/10.1016/j.rineng.2023.101059
8 Xie, X., Z. Sun, X. Zhu, S. Zhao, Z. Wang and Y. Zhai (2024). Influence of misting system on the thermal environment and thermal comfort of seated people in semi-outdoor space in Xi'an, China, *Frontiers of Architectural Research*, 13(3), pp. 668–681. https://doi.org/10.1016/j.foar.2023.12.010
9 Greenham, S., E. Ferranti, R. Powell, K. Drayson and A. Quinn (2023). The impact of heat on London Underground infrastructure in a changing climate, *Weather*, 78(6), pp. 170–175. https://doi.org/10.1002/wea.4421

10 Carnot, S. (1824). *Réflexions sur la puissance motrice du feu et sur les machines propres à développer cette puissance* (in French), Paris: Published by Bachelier.
11 van den Dobbelsteen, A.A.J.F., L.J.J.H.M. Gommans and R.E. Roggema (2008). Smart vernacular planning: sustainable regional design based on local potentials and optimal deployment of the energy chain, *Proceedings of SB08*, pp. 1–8. https://www.irbnet.de/daten/iconda/CIB15946.pdf
12 Roaf, S., J.P. Silva and M.C. Guedes (2020). Extreme design: Preparing for a different future, in *Green Buildings and Renewable Energy. Innovative Renewable Energy*, Edited by A. Sayigh, Cham: Springer. https://doi.org/10.1007/978-3-030-30841-4_45

14

HARVESTING COMFORT FROM LANDSCAPES

Introduction: Lessons From Other Species

More extreme climate trends are already pushing many species towards regional and local extinction. Scientists now produce alarming maps of future human climate niches showing huge shifts predicted in the habitable thermal zones over time. In the absence of mass migration, a substantial part of humanity will inevitably be exposed over the coming decades to mean annual temperatures warmer than nearly everywhere today.[1]

Macro-climatic and meso-climatic approaches to mapping habitable ecological niches can obscure the lessons already playing out from extinction narratives for other species, for instance, in patterns of coral bleaching across the globe. Current studies this century of the severity, annual duration, and onset of severe bleaching risks for global coral reefs pinpoint to the existence of some vital climatic refuges for different coral species. Low-latitude coral regions are most vulnerable to thermal stress, and their move to year-round bleaching will be all but untouched by any climate mitigation actions from now on. By 2080, most reefs will suffer from bleaching earlier in the spring rather than late summer, as happens now. Earth's reef regions are currently being mapped for their accelerated bleaching risks, enabling scientists to focus early efforts to limit future loss of coral reef biodiversity in areas most at risk.[2]

The Great Barrier Reef (GBR) in Australia has already been badly, and visibly, affected (Figure 14.1). The year 2024 saw the biggest dieback of corals in the recorded history of the reef that has been growing for over 300,000 years and may well see our species out in the long run. Lessons from its current dieback patterns may help us understand how to survive comfortably even in human niches designated as uninhabitable in the future.

A 2022 study on *coral mortality in the different thermal habitats* around Heron Island and its reef shows really clearly that in the zones suffering from short-term fluctuations in extreme temperatures, like the shallow waters inside the reef lagoon, bleaching is much more and faster than in the shaded, deeper regions cooled by ocean currents beyond the reef crest that have much smaller amplitude of diurnal or seasonal temperatures.[4] As many traditional builders knew, it is the micro-climates across a landscape that are critical in maintaining comfortable/survivable conditions in a region. Alpine builders knew the benefits of siting a house on one side of a valley

DOI: 10.1201/9781315645070-20

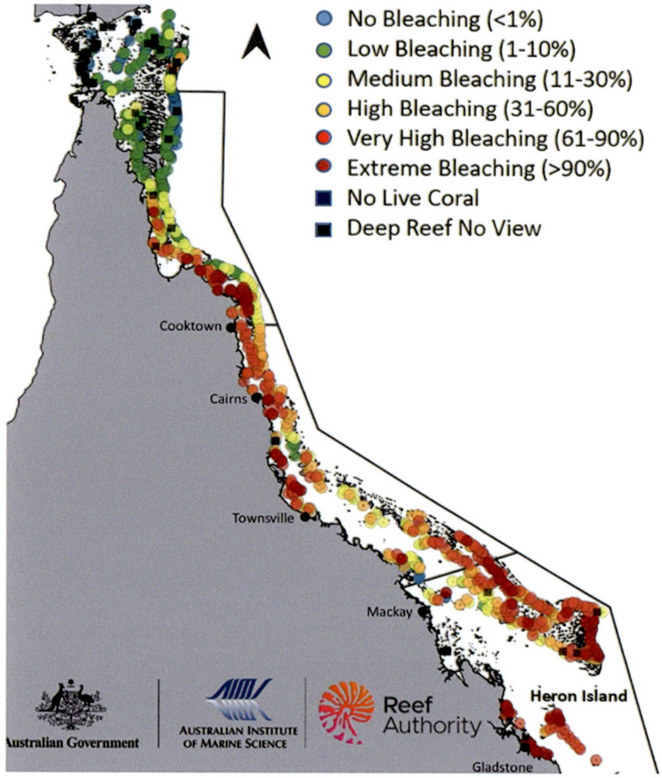

FIGURE 14.1 Aerial survey map showing the variation in bleaching prevalence observed across the Great Barrier Reef, Australia, updated in 2024.

Source: https://www2.gbrmpa.gov.au/learn/reef-health/reef-health-updates.[3]

or mountain or, in maritime climates, what a difference access to a sea breeze can make. The key factors in determining habitability over the longer term will be latitude, altitude, and the degrees of exposure to solar heating and, importantly, exposure to hot, warm, cool, or cold winds or currents in the oceans or the atmosphere.

Thermal Engineering in Landscapes

Human groups have been able to inhabit an extraordinary range of climates, from the very cold to the very hot. Unlike corals, they can move and shape the landscapes around them. They occupy extreme habitats in four ways, through:

1. Migration, seasonally occupying sites when the weather allows them to and moving between micro- and meso-climates, usually between altitudes and latitudes, to do so
2. Siting settlements in sheltered micro-climates in hostile landscapes, for example, in oases
3. Harvesting remote heat and coolth and transporting it to settlements in hostile climates, as happened with mechanical cooling in the Gulf, the importing of coal and gas where heating is needed, or the thermal engineering of landscapes
4. Occupying thermally robust structures that can protect occupants in extreme climates

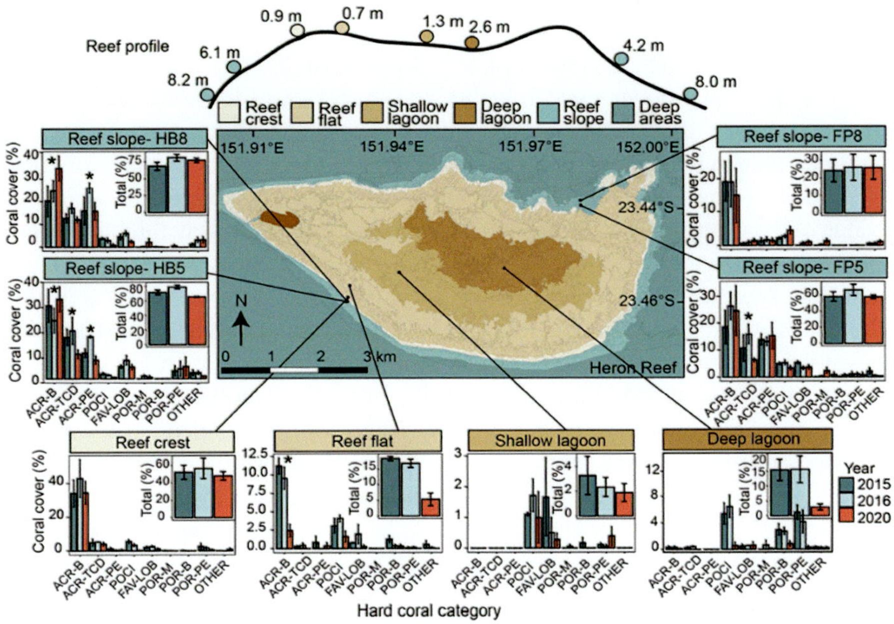

FIGURE 14.2 Cross section of Heron Reef showing different habitats and mean depth in meters. The acronyms and graphs show different coral types in the different sites. The largest relative declines in coral cover from 2016 to 2020 were observed within the thermally variable habitats of the Reef Crest (−25.6%), Reef Flat (−69.8%), Deep Lagoon (−82.6%), and Shallow Lagoon (−28.6%). At the thermally stable reef slope sites, declines occurred of lower magnitudes ranging from −0.26%) to −19.26%.

Source: Brown et al. (2022).

Dwellings in some traditional settlements are coupled to their landscapes with pre-conditioning climate systems, where ventilation air drawn from cooler caves in summer or energy and heat are drawn from warmer underground sources in winter. The *qanat* systems of Iran are perhaps the most sophisticated of all landscape-scale, climate pre-conditioning systems on Earth.

Harnessing Resource Cascades for Comfort: Qanat Landscapes

In the hot, dry deserts of Iran, different passive technologies were used to keep populations cool, fed, and watered. Settlements relied on water harvested from distant mountains and transported by *qanats*, underground water supply systems built for millennia in Iran.[5] Master craftsmen, called *Mulqani*, sank wells into distant cold mountain aquifers and transported water in *qanats* to feed farms and settlements on arid desert plains below.[6] Different uses for the water were arranged along the flow channels to take advantage of cascading water qualities in what ultimately ended up as a zero waste water system, as remnants of the supply finally tailed off to water fields.

Qanats ran for up to 50 mi long, often sunk deep beneath the desert surface, typically losing less than a degree or two centigrade in its journey from a mountain 'mother-well'. Gravity fed the water in tunnels dug at a low angle to ensure the slow flow of the water did not wear away the tunnel walls en route to the settlements.[7] Similar water supply systems were built from Morocco in the West, through the ancient Near and Middle East, right across to China and Japan.

Uncycling the Water Resource as It Flows

The usefulness of qanat water is optimised as it flows down through a landscape by arranging its functions sequentially down a water quality and energy cascade. Water quality degrades as it moves through the village, as shown in Figure 14.3, on its pathway from being a source of drinking water, then to bathing water in the hammam, to a place for the washing of kitchenware and clothes, down to turn the millstone, then on to water animals and crops.

The value of the water can also be upcycled by chilling it evaporatively through the incorporation of wind towers on a cistern's dome, and also by turning it into ice in winter. Many Iranian settlements had spectacular icehouses that had huge, shallow ice pans shaded all year by high ice walls.[8] Water was gradually turned into layers of ice a foot or two deep on the freezing winter nights of the high central plateau. Iceblocks were then shovelled down into the ice store and covered with straw and stored in a massive block until needed in the heat of summer to cool food, drinks, and fevered brows.[9] In this extremely hot, dry environment, every drop of water is used to support the man-made ecosystem in a zero waste water system powered ingeniously by many different forms of energy in pathways evolved and used over millennia in countries across the world.[10] Many traditional water systems have dried up as aquifers drop when water is extracted from them by motorised pumps, not only to supply homes and settlements, but also when extracted by powerful industrial companies, like Coca-Cola, who, in the past, destroyed whole farming ecosystems around one of their drinks bottling plants by over-extracting water from local aquifers.[11]

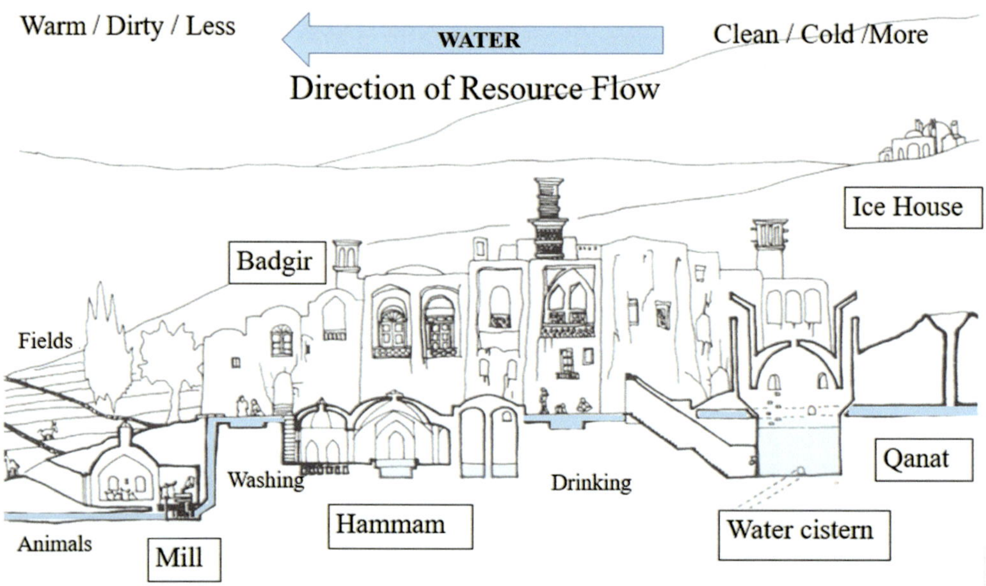

FIGURE 14.3 This cross section shows the direction of flow downhill of the pure cold mountain aquifer water from the qanat as it is used selectively through a desert village from its cleanest condition for ice formation, drinking water storage, and collection, and then in the *hammam* as bathing water, and on down a quality cascade for pot and clothes washing to flour milling, and then agricultural function for animal and field watering in a zero waste system.

Source: Susan Roaf.

Waterborne Comfort

In cities, qanats not only air-condition basements, as shown in Figure 14.4, but also provide cold clean water to homes, accessed via stairs down to a small water collection terrace by the water. Local low temperatures are also used to store and cool fruit. Ice from large icehouses was also brought and carried into homes to chill food, drinks, and water-ices, a practice also undertaken on every continent in the world, except Australia, and in the Near East for millennia.[12] Thus,

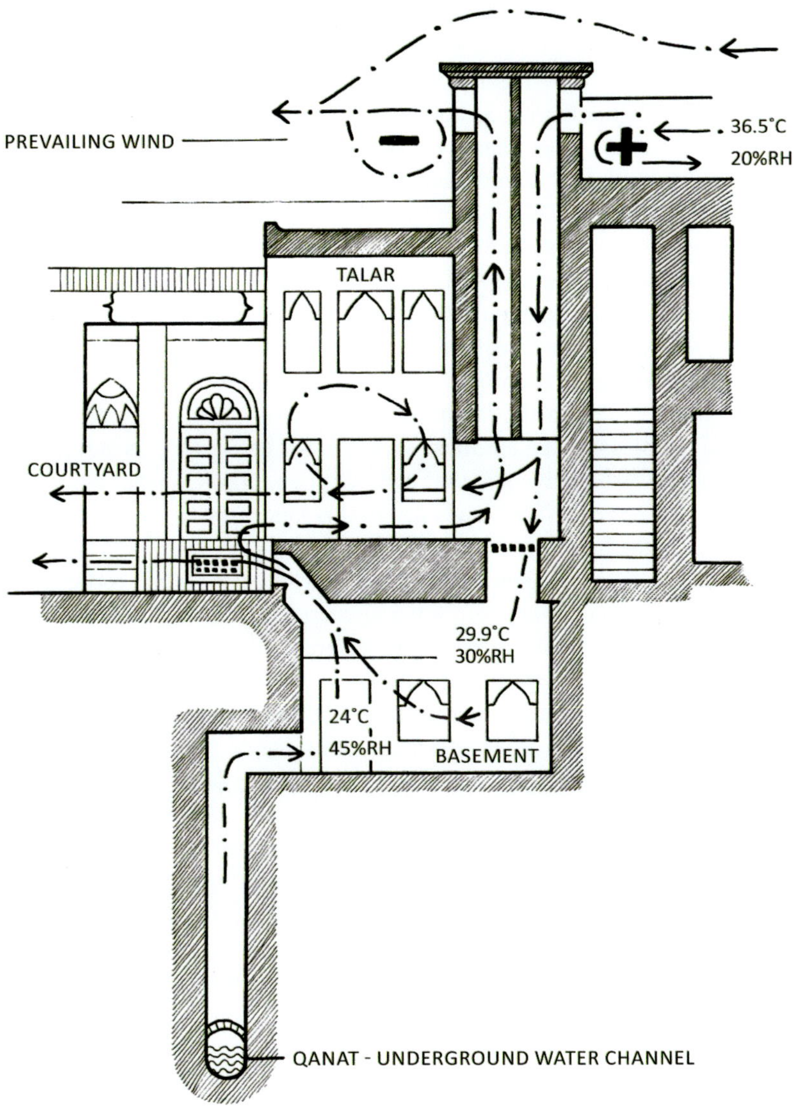

FIGURE 14.4 Qanat water can be accessed via shafts dug into the walls of basements in desert homes, where the cool mountain air drawn down the tunnel mixes with room air to provide natural air conditioning when the basements are occupied during hot afternoons.

Source: Susan Roaf.

thermal delight was experienced as well as comfort by wealthier locals who could afford to harness water quality and temperature flows from qanats in their own homes. For the poorer household, wealthy patrons built and maintained public *abanbars* (water cisterns) storing qanat water that was often further cooled evaporatively by air flows from *badgirs* (windcatchers) above them.[13]

Airborne Comfort: Villa Campolieto

Really sophisticated passive systems traditionally took a long time to evolve. Building on the brilliance of the Roman and medieval architects in the Naples region of Italy, it is still possible to visit buildings with landscape-level thermal pre-conditioning air systems, for instance, at the 18th-century Villa Campolieto in Herculaneum shown in Figure 14.5. It is one of a string of Rococo and Neoclassical villas that were built by the Bourbon aristocracy, with stunning gardens running down to the Mediterranean. The master architect Luigi Vanvitelli was one of its designers between 1763 and 1773. Apart from the use of complex natural ventilation systems in the rooms and walls and ceilings of the villa, it is a brilliant example of how air can be pre-conditioned before it enters a building. The main building, with a square plan, has four blocks organised around a central Greek cross gallery, and it is spread over five levels. The entrance is at the basement level, with a grand hall and a staircase beneath a dome topped off in a cupola, a Venturi ventilator drawing air from the basement levels up to the noble floor (piano nobile), across which air moves through the great reception rooms. The rear side of the villa opens towards the Gulf of Naples with an elliptical arcade at the entrance level surrounded by Tuscan columns that form a covered belvedere and ends with an elliptical staircase that connects the building to the gardens below. What stands out in this landscape is the pre-conditioning of the air moving up and into the building in the summer.

The strong sea breeze flows up and over the garden, where it is further cooled as it passes over a pond, fountain, and sunken garden before being channelled through tunnels around the large circular water cistern built beneath the raised belvedere platform with its colonnaded exedra or circular arcade. Pushed by the pressure of the prevailing wind, air then passes through openings in the massive walls of the basement behind, and in places, and at times passes over the subterranean ice store built to be filled seasonally with ice carried from the inland mountains to cool food and drinks for the Bourbons above. The air has been cooled by water, planting, ice, and thermal mass before it passes up the grand staircase leading from the entrance, drawn by the Venturi effect through the cupola over the stairway dome to intermingle with the air in the rooms of the piano nobile to thermally delight the nobility. Some pre-cooled air was then distributed around the building with a complex system of ventilation pathways driven by air pressure and stack and Venturi systems. The 4 m deep sunken garden crowned by a canopy of date palm fronds provided a cool haven for strollers or sitters that might be 3 to 5°C cooler than in the walkways above.

Waterscapes: Potsdamer Platz

Modern designers are beginning to play with energy and resource quality cascades in their designs in order to maximise the free heating or cooling that can be extracted from an area was done by the architects centuries ago in the gardens in Seville and Cordoba shown in Figure 14.6. Today, a modal shift has occurred to active systems and mechanical ground-source heat pumps

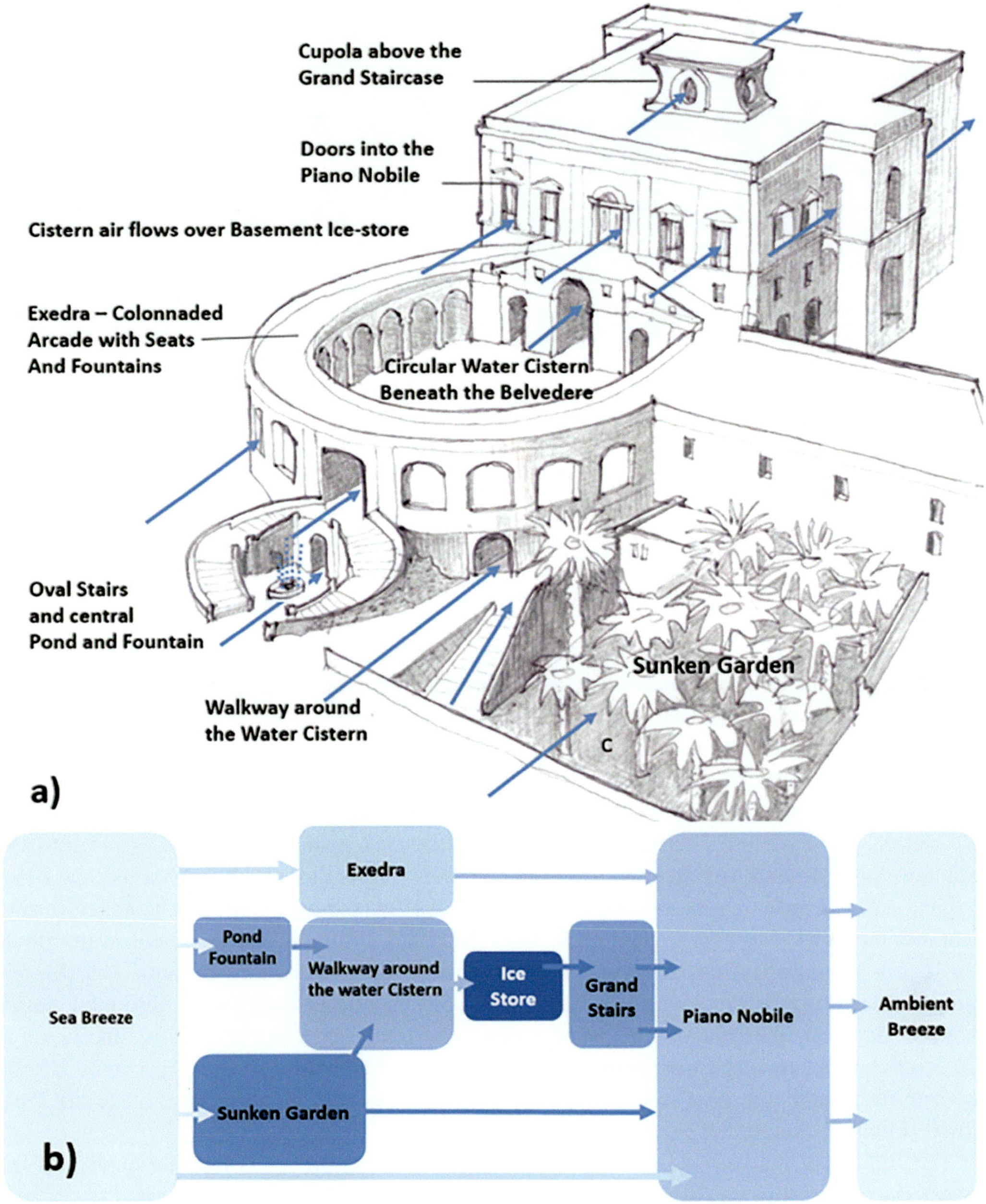

FIGURE 14.5 (a) Axonometric of the Rocco Villa Campolieto in Herculaneum near Naples showing the wind flow directions. (b) Flow diagram indicating relative temperatures of different sources of coolth interacted with as air moves through and around the villa. The darker the blue, the cooler the source.

Source: Susan Roaf.

FIGURE 14.6 Gardens at the (a) Alcazar Palace in Seville and (b) Córdoba cooled the soil, air, and pathways with planting at four different photic levels, providing shade and deep sunken gardens containing cold air inversions. Evaporative cooling of people and air occurred around fountains and by well-managed water irrigation channels.

Source: Susan Roaf.

are widely used, as are electrically driven evaporative cooling systems for open areas in streets and parks in hot climates. An example of a highly designed and managed modern waterscape is found in the famous Potsdamer Platz in Berlin. Rainwater from surrounding roofs is captured and stored in five large underground cisterns, used to top up water levels in street-level pools, for flushing toilets in offices, irrigating green areas, and cooling the area in hot spells. Water quality is key to the use of water at different stages of its flows, particularly its clarity. Solid particles settle in the cisterns. From the cisterns the water is fed into the southern pool through seepage facilities. Once there, it runs directly through a biotope (ecological habitat) with vegetation, where it is treated. If necessary, filters can be added during summer months to filter suspended algae from the water. From the northern pool, water flows through a seepage facility, along treatment biotopes with vegetation, and into the deepest point of the water features near the plaza. The water is aerated as it passes through a series of multi-layered tanks and into the showy pond and fountain of the plaza, without chemical additives and using very little energy on its pathways.[14] The whole system is powered largely by gravity to maintain flow rates and levels and by evaporative cooling to modify the local micro-climates around the pools.

The difference between waterscapes and temperature landscapes is that the quality of the water in a natural flow can be clearly visible from its colour and cloudiness. Thermal landscapes can be felt but are invisible in themselves but can leave tell-tale symptoms of temperatures around them. Just look at the ice on different surfaces of a car parked next to a building on a frosty morning, or the rate at which snow melts off differently insulated roofs in winter, or the changes in snow and ice by the road on a drive through an area with frost hollows. Temperatures across ecosystems are marked by the different plant species that will grow, or fish species in a river, by birds and insects and butterflies. Temperatures in individual ecosystems are marked by the seasonal residence of migrating species, reflecting the motion of weather cycles over landscapes across a year.

Reconnecting Populations to Thermal Landscapes

In terms of comfort, temperature can either be seen rather as a particle, a point in time experienced in one space and location. In terms of standards and regulations, it is seen as such. If the humidity of a climate is, for example, 90%, or its temperature is 40°C, the comfort temperature must, according to one model, be 23°C. In reality, it does not work like that. Comfort temperatures are actually waves, flows of thermal energy over landscapes over time, mixing and weaving with other energy streams, with gravity, pressure, and the material world around it. Few future populations will be able to survive with their old 20th-century lifestyles intact in a changing climate.

For whole populations to survive comfortably in a more extreme future, it will be necessary to rapidly innovate new ways of affordably providing comfort for the masses, which may even involve moving people between micro- and meso-climates within buildings or landscapes. An alternative of exploring new ways of managing the flows of heat, air, and water across time and space, in cities and settlements, offers exciting opportunities for new thinking. COVID taught us one thing – that no one is safe until everyone is safe. Solutions need to be found that address the challenge of keeping the many, not just the few, cool enough.

Notes

1 Chi, Z., T.A. Kohler, T.M. Lenton and M. Scheffer, et al. (2020). Future of the human climate niche, *PNAS*, 117(21), pp. 11350–11355. https://doi.org/10.1073/pnas.1910114117

2 Mellin, C., S. Brown, N. Cantin, et al. (2024). Cumulative risk of future bleaching for the world's coral reefs. *Science Advances*, 28(10)(26). https://doi.org/10.1126/sciadv.adn9660

3 Great Barrier Reef Marine Park Authority (2024). *Reef Health Updates*, Published online by the Australian Government, Accessed on 29/10/2024 at: https://www2.gbrmpa.gov.au/learn/reef-health/reef-health-updates

4 Brown, K.T., G. Eyal, S.G. Dove, S.G. et al. (2023). Fine-scale heterogeneity reveals disproportionate thermal stress and coral mortality in thermally variable reef habitats during a marine heatwave. *Coral Reefs*, 42, pp. 131–142. https://doi.org/10.1007/s00338-022-02328-6

5 Beasley, E., and M. Harverson (1982). *Living with the Desert*, Warminster, UK: Aris and Phillips.

6 Beaumont, P., M. Bonine and K. Mclachlan (Eds) (1989). *Qanat, Kariz and Khattara*, Whitstable, Kent: Menas Press Ltd.

7 Jorgensen, H. (2012). *The Ice-Houses of Iran*, Costa mesa, CA, USA: Mazda Publishers. Beazley, E. &.

8 Jorgensen, H. (2012). *Ice Houses of Iran: Where. How. Why*, Costa Meza, CA: Mazdar Publishers.

9 Dehghani-Sanji, A., and M.N. Bahadori (2021). *Ice-Houses: Energy, Architecture and Sustainability*, London: Elsevier.

10 Roaf, S. (2009). Resource quality cascades in traditional low exergy technologies: The qanats and badgirs of Yazd, *Proceedings of the Smart and Sustainable Built Environments (SASBE) Conference*. https://www.irbnet.de/daten/iconda/CIB14196.pdf

11 Roaf, S., D. Crichton and F. Nicol (2009). *Adapting Buildings and Cities for Climate Change*, Oxford: Architectural Press, pp. 89–90.

12 Beamon, S., and S. Roaf (1990). *The Ice-Houses of Britain*, London: Routledge.

13 Rahmati, H.S., M. Amiribolvardi and H. Torkshavand (2025). Ab-Anbars of Yazd: A study of geographical and historical distributions of cisterns in a desert city, in *Persian Vernacular Architecture: Lessons from Master Builders of Iran on Climate Resilient Design*, Edited by A. Cheshmehzangi and S. Roaf, Singapore: Springer Nature.

14 Dreiseitl, H., and D. Grau (2006). *Waterscapes: Wasserlandschaften – Planen, Bauen und Gestalten mit Wasser*, Basel, Switzerland: Birkhäuser.

15

MINING COMFORT FROM THE EARTH

Introduction

All buildings are more or less thermally coupled to the sky or the earth, to the continually fluctuating climate cycles above ground or the steadier temperatures to be found within it. Tall buildings are highly exposed to weather events, and deep caves are almost de-coupled from them. Urban buildings are also more or less affected by the strength of local urban heat islands (UHIs) that affect both the air and earth temperatures of cities. To successfully plan to mine for free reliable heat or coolth from the earth at the design stage, it is important to get a feel for the invisible thermal cycles and systems that drive temperatures underground.

Solar Cycles

Our lives in and out of buildings revolve around regularly repeated temperature pulses from the sun that dictate the length of days and seasons. Seasons shape global and local weather over the year in conjunction with their interaction with heat and cold stored in the land, the oceans, and the built and natural habitats of the lithosphere, Earth's outer layer.

The Earth's climate cycles are dictated over a day, or millennia, by both the changing shape of orbit of the planet around the sun and the tilt and 'wobble' of its axis. In the 1920s, a Serbian mathematician called Milutin Milankovitch began calculating the timelines for the three main variations in the distance and angle relationships between the Earth and the sun.[1] Over millennia, the orbit of the Earth around the sun changes from being elliptical to almost circular (*orbital eccentricity*), so changing the distance between our planet and the star at the heart of our solar system. When furthest away, around every 400,000 years, less solar radiation reaches the planet, leading to ice ages lasting centuries. When the orbit is more circular, the Earth is closer to the sun and warmer periods occur.

On its annual path around the sun, the Earth rotates daily around its polar axis. Because this *axial tilt* is not vertical, every six months either the Northern or the Southern Hemisphere is closer to the sun, resulting in the seasons. The Earth also wobbles slightly upon its axis during those rotations, like a slightly off-centre spinning toy top. This wobble, or *axial precession*, is due to changing tidal forces in the oceans caused by the gravitational influences of the sun and the moon that cause

DOI: 10.1201/9781315645070-21

Earth to bulge at the equator, affecting its rotation and the severity of summers in each hemisphere on roughly 13,000-year cycles. These long-term changes make little difference to the current heating of the planet that is driven by anthropogenic emissions (generated by human activities) of the greenhouse gasses that trap ever more heat within Earth's atmosphere, causing global heating. The alarming scale of that heating is clearly demonstrated in many publications by scientific organisations including the US space agency, NASA, where their climate-time-machine website brilliantly visualises historic annual increases in temperatures across Earth's lands and oceans.[2]

Mean Global Surface Temperature

The *mean global surface temperature* (MGST) is a single temperature measure for the whole planet and is difficult to calculate as, at any one time, the highest and lowest temperatures on Earth might vary by up to 55°C+ (100°F). Scientists use the concept of a global average temperature (GMST) to track changes in Earth's energy budget over time, how much sunlight it absorbs minus how much it radiates to space as heat. It is calculated now from temperatures taken in locations around the world to provide a consistent, reliable method for monitoring changes in Earth's surface temperature over time.[3]

GMST is also used to benchmark policy initiatives. For instance, the oft-cited Paris Agreement is a legally binding international treaty on climate change that was adopted by 196 countries at the UN Climate Change Conference (COP21) in Paris in December 2015. Its main goal was *to hold the increase in global average temperature to well below 2°C above pre-industrial levels* and pursue efforts to 'limit the temperature increase to 1.5°C above pre-industrial levels' but Figure 15.1 shows the impossibility of that. Coming into force in 2016 by September,

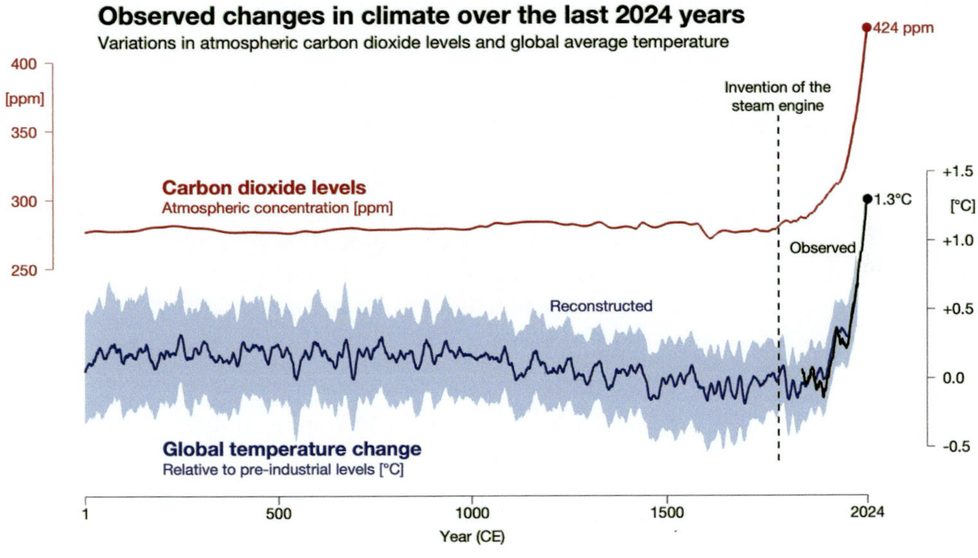

FIGURE 15.1 The man-made greenhouse gas emissions, of which c. 40% are related to the built environment driving global temperature increases, as is clearly visible in their relationship in this diagram.

Source: www.edhawkins.org.

the *1.5°C had already been exceeded by summer of 2024*. August 2024 was the joint-warmest August globally at the time of writing, with an average ERA5 surface air temperature of 16.82°C, already 0.71°C above the short-term 1991–2020 average for August,[4] and the following months continued to break records. In October 2024, Jim Skea, Chair of the IPCC, warned that 1.5°C was then officially out of reach, 2.0°C demands heroic action, 3.0°C is now the likely outcome, and Western Europe could face a shocking 5.0°C by 2100.[5]

These temperature increases are already triggering extreme weather events globally, and so there is a dual imperative now: to radically slash emissions from buildings and to also keep people safe in buildings during ever more extreme weather events. Hence, the need for climate-safe buildings that can run on local natural energy even during heat and cold waves; meaning, that designers must increasingly look to where they can mine for heat and coolth in the local environments of buildings. Many existing architectures in extremely cold or hot regions historically mined for comfort from the earth, as we saw in the last chapter. Basement rooms provide thermal inertia by burrowing into the earth, as people with extreme hypothermia also often do to survive. Canadian homes across the Prairies keep boilers in spaces below the main house, in spaces that can also serve as a climate refuge during blizzards or heatwaves. Across the deserts of the Middle East, cool basements were, and are, resorted to in traditional homes during the hottest times of day and year for cost-free comfort. But for some on the planet, the impacts of climate change will be felt much earlier than in other regions due to the large differences in the rates of heating in different regions across the globe, as shown in Figure 15.2.

Ocean Temperatures

The annual thermal pulse or heartbeat of a site results from its macro-, meso-, and micro-climates, as we saw for the coral populations of the Great Barrier Reef. As the climate heats, the

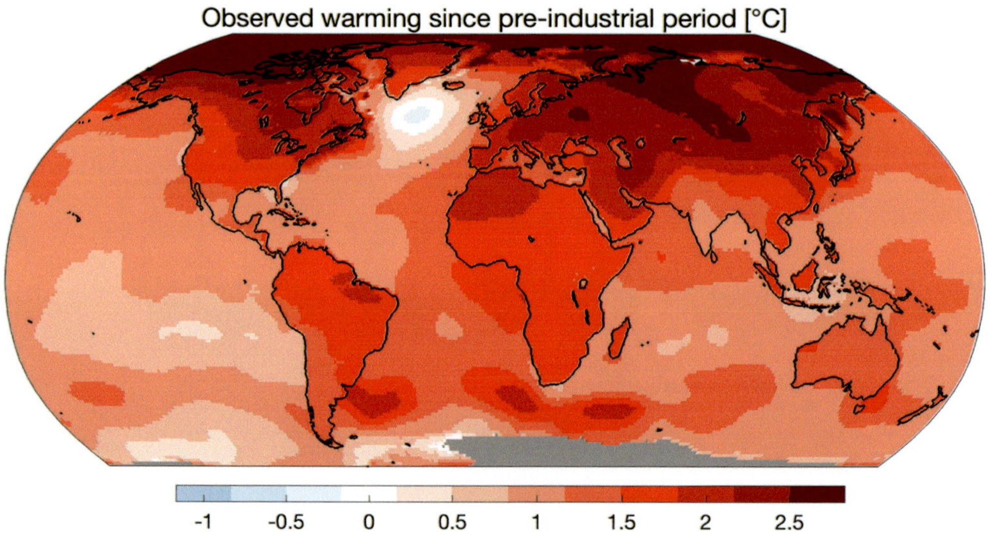

FIGURE 15.2 Some parts of the planet have heated much faster than others over the last 300 years, with the most severe warming occurring over the Arctic region.

Source: www.edhawkins.org.

oceans are warming more rapidly than land masses. The rate of ocean warming has doubled in 20 years, driving up rates of sea level rises as the warming waters expand, which have also doubled in 30 years. Coastal areas are especially impacted, with species finding themselves on the front line of a battle to breathe in shallower coastal waters with roughly 500 *dead ocean zones* identified globally, where almost no marine life remains due to a dwindling oxygen content.[6] Such a zone exists in the shallow single-entry waters of the Persian/Arabian Gulf. A study of the sea surface temperature (SST) in the waters of Kuwait Bay over five consecutive years (2016–2020) showed that sea temperature had reached a staggering 37.6°C during one event associated with heatwave, neap tides, and an extended period of desert winds, which was accompanied by large-scale intermittent fish deaths along the full length of the Kuwait coastline.[7] The Gulf water temperatures are critical for many industries, ranging from tourism and water desalination processes to the cooling of nuclear power stations sited on the seashore. The speed with which this and other estuaries and oceans are warming is alarming, and sea temperatures are increasingly affecting development planning of shore sites as well as infrastructure investment to cope with the more extreme weather events triggered by warmer water bodies.

Warmer waters lead to heavier rainfall that nations are unprepared for. For instance, many roads and developments in the Gulf cities and states have no surface water drainage systems or even sewage systems, with treatment plants' waste being hauled and dumped in landfills. Radical rethinking and planning for heavier rainfall will be needed to construct more waterproof infrastructure and repair and reinforce existing flood defences in and near buildings and settlements. In September 2023, up to 23,000 died around Derna in Libya after a dam broke. In October 2024, over 250 people died when extraordinary flooding overwhelmed whole settlements in Valencia, Spain. Warmer ocean waters are driving stronger storm and monsoon systems and the thermal impacts of ocean currents on land masses, precipitation, Earth's temperatures as fear grows over the 'domino effect' of individual tipping points on each other and the whole global climate system.[8] But for the built environment, understanding the temperatures of the Earth beneath and around sites will become increasingly important as designers seek to mine steady and reliable heat from the environment.

Seeking Stable Earth Temperatures

Terrestrial soil temperatures vary from month to month as a function of exposure to direct and indirect solar radiation, rainfall, local climate and weather, local vegetation, structures above and around it, and the type of soil and depth in the earth. Because earth is much denser than air and less exposed to the impacts of sun and wind, soil temperatures lag significantly behind seasonal and daily changes in the overlying air temperatures. Upper soil layers can vary substantially from day to day if exposed directly to solar radiation. Every site and region has its own deep earth temperature, typically around the annual average of monthly temperatures at around 9 to 10 m beneath the surface, as shown in Figure 15.3. Is it just coincidence that many historic homes, like the Sethi House in Peshawar or the grand houses of Yazd, in very hot cities in the Middle East have basements that reach down to 9 m below ground level or more?

Because temperatures at depth depend on the soil composition and moisture content, it is important to understand the geology of sites and areas and, of course, their potential levels of saturation over the year. The reason there are no basements in many traditional homes along the Gulf is that the water table was traditionally so high in settlements. Many factors impact on soil conditions and potentials for providing free heating and cooling from the earth.

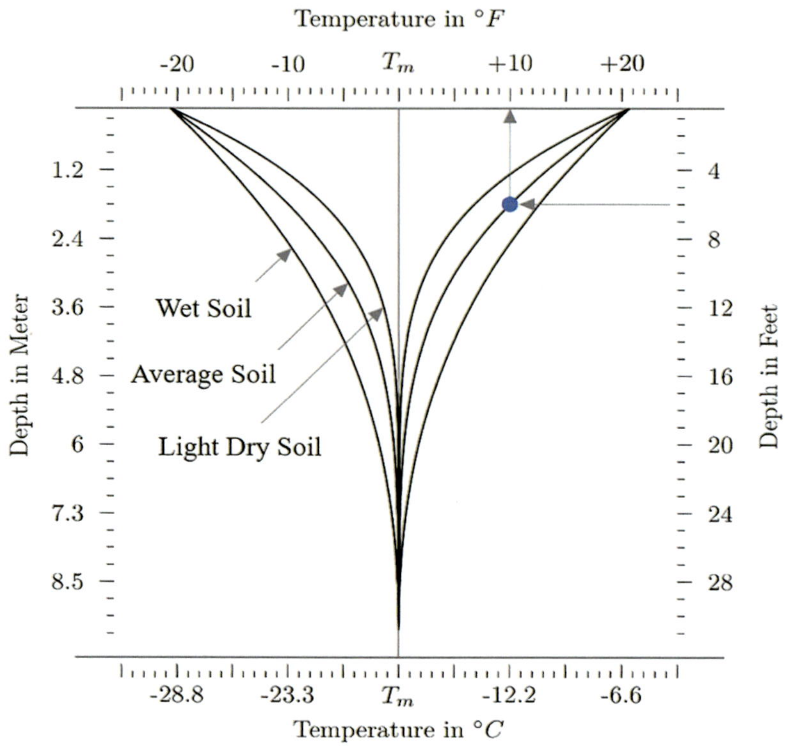

FIGURE 15.3 Amplitude of seasonal soil temperature change in winter in Virginia, USA, as a function of depth below ground surface. Depending on the soil type, at around 9 m or 30 ft below the surface, temperatures stabilise at around the local annual MST.

Source: Virginia Tech.

Urban Heat Islands

The UHI effect can result in temperatures up to 6–12°C (11–22°F) higher in city centres than their surrounding regions, often causing knock-on impacts. The city-wide rise in cooling loads is created in a vicious circle created by heat from those very mechanical cooling systems driving up the UHI, and with it, the need for more air conditioning. What began decades ago as a scientific interest in the UHI[9] has now become a critical component in planning law in many cities, not least for many high-rise city centres that are badly impacted by increasingly extreme UHI temperatures.[10]

As the GMST increases, cities like Singapore are already routinely mapping local air and earth temperatures across their urban areas to access the effectiveness of programmes to passively modify UHI temperatures with, for instance, shading and planting.[11] Singapore has ambitious plans to lower temperatures across the whole city to combat global heating and is experimenting with features like green roofs and facades, extensive urban planting programmes, and behavioural policies, like asking people to raise thermostat temperatures for air conditioning to reduce heat emitted from HVAC systems. Ultimately, key factors like population density and the morphology of a city affect how much solar energy hits the ground and impacts urban earth temperatures.[12]

Regional MSTs are rising ever faster. In the UK, mean annual temperatures at 1 m deep ranged from 12.7°C in the period 1961 and 1990 in southern England to 8.8°C in northern Scotland. By 2015–2024, Central England MST had already risen to 14°C.[13] The time has come to take earth and ocean temperatures into account when designing developments, as well as the general thermal landscaping of the areas around buildings on sites.

Thermal Energy Storage Systems for District Heating and Cooling

Geothermal energy has always been harvested around the globe, to heat bathing pools, dwellings, or settlements, for tens of thousands of years. Right through the Greek and Roman civilisations until it was more industrially utilised over the last two centuries, heat and cold (icehouses) stored underground in the insulating soil is now increasingly being used to create affordable and reliable comfort on a large scale. Heat can be either mined from the ground or aquifers below or harvested by solar panels and stored in underground tanks, as shown in Figure 15.4.[14] Heat is typically moved as liquid flows down in pipes from heat sources. The heat can either then be pumped continuously around building or district heating systems or stored in large tanks or pits for later use.[14]

The US Department of Energy published a study in 2019 outlining the huge unrealised potential for the use of geothermal energy to power residential, commercial, and district heating and cooling systems. The report also flagged its importance as a future source of constant and secure renewable electric power generation that can be flexibly used as and when needed, so providing vital stability and resiliency to power grids.[15] Power generation systems have been so dominated by the fossil fuel industries that geothermal energy has been underutilised to date. The DOE estimates the existing installed US capacity is about 16.8 GWth (gigawatts thermal), equivalent to around 2 million ground-source heat pump installations, and that by 2050 the market could be expanded to around 28 million households, or 23% of the total residential heating and

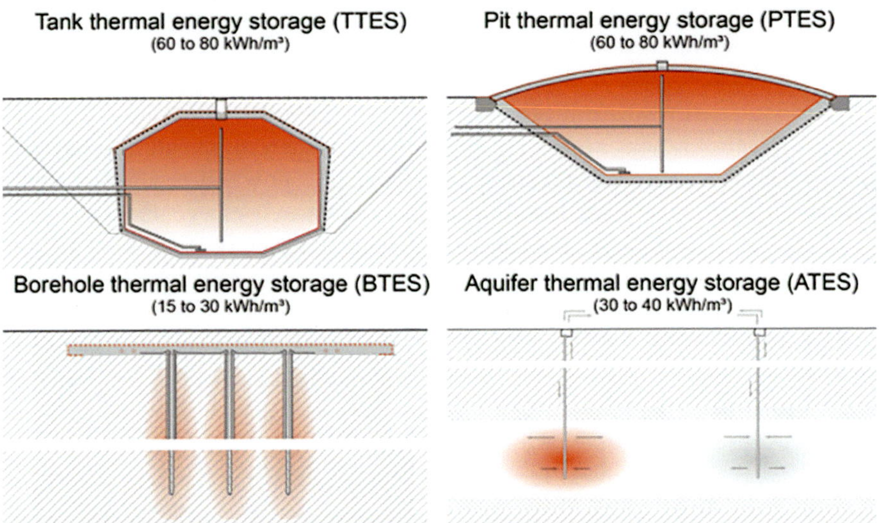

FIGURE 15.4 Overview of available underground thermal energy storage concepts that can be used with district heating systems.

Source: Schmidt et al. (2018).

cooling market share. If the generation and distribution systems can be backed up with sufficient dispatchable stored electricity to operate even when main grids fail, then even during extreme weather events, like the Texas ice storm, some buildings could continue to be supplied by free heat from the ground even during such extremes.

Geothermal heat is not always a renewable energy source, because it can be depleted. The geothermal field (RGF) of the Rotorua of North Island, New Zealand, is relatively shallow, being typically around 30–200 m deep, leading to its over-exploitation as a heat source for buildings and industries. It is riddled with features like geysers and hot springs that, in the 1960s and the 1970s, were extensively tapped, leading to their declined to critically low levels by the mid-1980s. By 1986, many of the boreholes, especially to residences, were closed, leading to some recovery of surface features and, over time, the successful recharging of the heat flows and sources, although often in ways that defied scientific explanation.[16]

The Netherlands has a country-level programme to utilise heat from both deep and shallow geothermal stores and is very aware of the need to balance energy drawn from reserves against heat recharged into them. In 2020, there were 21 deep geothermal energy projects, with an approximate total capacity of 3.6 PJ, or 300 MWth. The amount of shallow geothermal energy being utilised is also increasing, with over 60,000 ground-source heat pumps in operation there, of which over 99% of these projects involved low-temperature (< 25°C, <77°F) storage. The industry-led 'Master Plan Geothermal Energy in the Netherlands' claimed to be able to meet 5% of the total energy demand there for heat by 2030, and 23% by 2050, but this race to exploit earth temperatures is also matched by new legislation to manage over-exploitation of the resource.[17]

Digging Down for Comfort Is Expensive

It is more expensive to dig down for comfort than to simply reach up into the air for it. Campus organisations, like universities, often lead the way in long-term investments to fund large heat management infrastructures like district heating schemes to keep people warm or cool. In the former Soviet countries, towns and cities were typically powered by steam-based district heating systems. The move now is away from steam to water-based distribution networks integrating geothermal heat. In the Netherlands, many universities now use ground-sourced, pumped district heating systems. The High-Tech Campus in Eindhoven uses a geothermal energy system to heat and cool the 30 buildings, housing, c. 300 companies employing some 12,500 people. Buildings are connected to a ring pipeline and receive heat and cold from a collective system with ten cold and ten heat sources. As in the Rotorua case, protections need to be put in place to protect the heat reserves and the ecologies around them. Risk analyses for such projects are needed for geothermal permits to be issued that take into account groundwater protection areas, local seismic activity, drilling-free zones, and the use of only sustainable well designs and construction.

On the Princeton campus in New Jersey, USA, a major investment has been made to convert the traditional steam district heating system to water incorporating geothermal heat and thermal storage, with heat pumps powered largely by solar photovoltaics. Two thermal energy storage (TES) tanks are used, one for hot and one for cold water. When electricity is cheap or there is lots of available solar energy, the tank stores are topped up with heat or cooled as required in an orchestrated delivery system to both heat and cool the campus buildings by meeting energy demands with the lowest cost and carbon supply. This is a large complex project, but the rewards are huge too. Since the university's Energy Master Plan was established in 2008, energy savings of approximately $5.7 million per annum and annual CO_2 reductions of more than 25,000 t have been placed in the pipeline that is leading to the ambition of a net-zero energy campus.[18]

EARTH HEAT AS A THERMAL SAFETY NET

Winter (and summer) extreme storms are increasing in number and intensity across North America where energy supply lines are typically vulnerable being above ground, covering long distances and often suffer from poor maintenance practices. Hence the move there to using earth's heat (and coolth) as a thermal resource of last resort during power outages in novel ways. At the end of January 2019 the entire US Midwest region grid system faced near collapse during an under-reported Polar Vortex event. By declaring a Grid Maximum Emergency event they were able to import energy from adjacent regions and ask for demand reduction by consumers so preventing blackouts across the affected states. Figure 15.5 shows how Glenn Remington in Michigan has designed a Ground Source Heat (Cold) Sink system to provide year around hydronic cooling during summers, thermal resilience during extreme cold events and ice storms and year round brown-out or black-out resilience when the grid fails. The

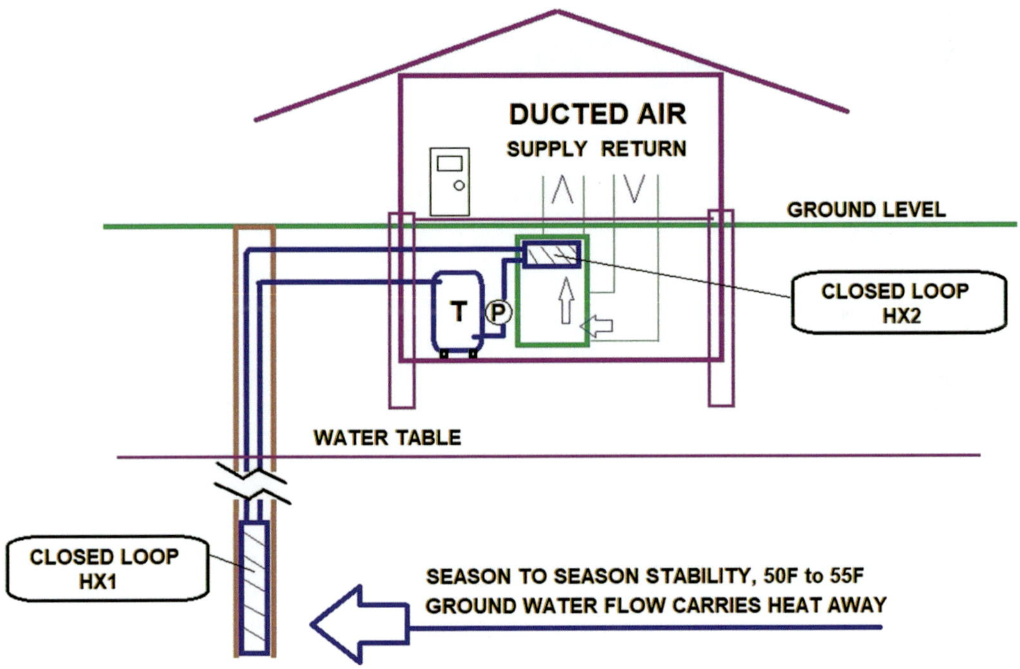

FIGURE 15.5 Section through a planned domestic Ground Source Heat Pump system in Michigan USA for seasonal water preheating and as a 550F (110C) thermal backup system during extended power outages.

Source: Glenn Remington.

pumps and controls can be powered by backup batteries and are accessible and protected from extreme wind damage being located in the basement mechanical room which can also serves as an extreme climate refuge. It is not uncommon across the US Northern winter climate zones to be making such preparations in the face of more extreme future climates and importantly to protect occupants against the unreliable local and regional electric utility capability and condition problems.[19]

Synchronizing the seasonal harvesting, dumping and storage of natural energy with solar cycles to provide zero emissions comfort indoors is possible using inter-seasonal energy stores in the earth, aquifers, water in tanks or reservoirs or even in ice-stores. Digging deep also provides potential thermal safety-nets, either cool or warm, that can be accessed in extremis when above ground energy systems fail. Temperatures within buildings are influenced by the extent to which they are exposed to the steady temperatures of the earth or the thermal vicissitudes of the sky climates and weather. Another key factor is how protected indoor climates are from climates above ground by the thermal capacity of the materials in and around the building, which is the subject of the next chapter

Notes

1 https://en.wikipedia.org/wiki/Milutin_Milankovi%C4%87
2 https://climate.nasa.gov/interactives/climate-time-machine
3 Lindsey, R., and L. Dahlman (2024). *Climate Change: Global Temperature*, Published online by the US national Oceanic and Atmospheric Administration. https://www.climate.gov/news-features/understanding-climate/climate-change-global-temperature
4 https://climate.copernicus.eu/copernicus-summer-2024-hottest-record-globally-and-europe
5 Skea, J. (2024). It's too late to save Britain from overheating, says UN climate chief, *The Telegraph*, 6th October. https://www.telegraph.co.uk/business/2024/10/06/too-late-save-britain-overheating-climate-chief-jim-skea/
6 Enevoldsen, H., K. Isensee and Y.J.L. Lee (Eds) (2024). State of the ocean report, IOC, Technical Series, 190 [212], Intergovernmental Oceanographic Commission.
7 Alosairi, Y., N. Alsulaiman, A. Rashed and D. Al-Houti (2020). World record extreme sea surface temperatures in the northwestern Arabian/Persian Gulf verified by in situ measurements, *Marine Pollution Bulletin*, 161, Part B, p. 111766, ISSN: 0025-326X. https://doi.org/10.1016/j.marpolbul.2020.111766
8 Wunderling, N., J.F. Donges, J. Kurths and R. Winkelmann (2021). Interacting tipping elements increase risk of climate domino effects under global warming, *Earth System Dynamics*, 12(2), pp. 601–619, Copernicus Publications. https://doi.org/10.5194/esd-12-601-2021
9 Rizwan, A.M., Y.C. Leung and C.L. Dennis (2008). A review on the generation, determination and mitigation of Urban Heat Island, *Journal of Environmental Sciences*, 20(1), pp. 120–128, ISSN: 1001-0742. https://doi.org/10.1016/S1001-0742(08)60019-4
10 Climate Central (2023). *Urban Heat Hot Spots*, Accessed on 30/10/2024 at: https://www.climatecentral.org/climate-matters/urban-heat-islands-2023
11 Chàfer, M., C.L. Tan, R.J. Cureau, W.N. Hien, et al. (2022). Mobile measurements of microclimatic variables through the central area of Singapore: An analysis from the pedestrian perspective, *Sustainable Cities and Society*, 83, p. 103986, ISSN: 2210-6707. https://doi.org/10.1016/j.scs.2022.103986
12 Mughal, M.O., X.-X. Li and L.K. Norford (2020). Urban heat island mitigation in Singapore: Evaluation using WRF/multilayer urban canopy model and local climate zones, *Urban Climate*, 34, p. 100714, ISSN: 2212-0955. https://doi.org/10.1016/j.uclim.2020.100714
13 UK Met Office (2024). *Monthly Weather Reports*, Accessed on 30/10/2024 at: https://www.metoffice.gov.uk/research/climate/maps-and-data/summaries/index
14 Schmidt, T., T. Pauschinger, P. Sørensen, et al. (2018). Design aspects for large-scale pit and aquifer thermal energy storage for district heating and cooling, *Energy Procedia*, 149, pp. 585–594. DOI: 10.1016/j.egypro.2018.08.223

15 U.S.DOE (2019). *GeoVision: Harnessing the Heat Beneath Our Feet*, Published by the US Department of Energy, Accessed on 09/12/2024 at: https://www.energy.gov/sites/default/files/2019/06/f63/GeoVision-full-report-opt.pdf
16 Scott, B.J., D.A. Gordon and A.D. Cody (2005). Recovery of Rotorua geothermal field, New Zealand: Progress, issues and consequences, *Geothermics*, 34(2), pp. 159–183. https://doi.org/10.1016/j.geothermics.2004.12.004
17 Békési, E., M. Struijk, D. Bonté, et al. (2020). An updated geothermal model of the Dutch subsurface based on inversion of temperature data, *Geothermics*, 88(101880), ISSN: 0375-6505. https://doi.org/10.1016/j.geothermics.2020.101880
18 https://facilities.princeton.edu/projects/path-net-zero-new-energy-facilities
19 Lund, John W. (2003). The use of downhole heat exchangers, Geothermics, Vol. 32, Issues 4–6, 2003, pp. 535–543, ISSN 0375-6505, "https://doi.org/10.1016/j.geothermics.2003.06.002" DOI: 10.1016/j.geothermics.2003.06.002.

16

THERMAL MASS AND COMFORT

Why Does Mass Matter?

Thermal mass is often described as the property of a material that enables it to absorb, store, and emit heat energy. It is considered by many as synonymous with *thermal capacitance* or *heat capacity*, the ability of a body to store thermal energy, but it really is so much more than that. In the context of a building, the presence of thermal mass can have major effects on the way the internal environment of a building responds to its local and global thermal environment. This chapter shows the variety of ways mass affects the natural thermal response of the internal environment and, by so doing, not only saves energy but also makes for safer buildings in more extreme climates and weather.

Thermal mass was part of the thermal design vocabularies of traditional builders in all climates. Thermal storage was a vital comfort and survival feature in climates from freezing Siberia, where heat is stored in the earth under the floors of cattle herders' yurts, to the tropics, where coolth accumulates below the raised floors of Kampong houses of Malaysia. In a world of more extreme weather events and climate trends, designing with thermal mass will become important again for the aforementioned uses, especially when power grids fail. It is important to understand what mass is and how it can be used, and this definitely does not involve rocket science. Thoughtful designers can go a long way using common sense and a 'feel' for temperatures, as we hope to show in what follows.

The lack of a consistent definition around *thermal mass* (Box 16.1), or an understanding of its uses, has led some to conveniently dismiss its use in building design, but their lack of understanding must be forgiven and moved beyond, as the importance of thermal mass in climate-safe design for the extreme futures we face is paramount.

The Thermal Flywheel Effect

The *thermal flywheel effect* refers to the ability of a material to maintain and extend stable indoor temperatures by intermittently absorbing, storing, and then steadily releasing heat over time, in the same way a mechanical flywheel stores and releases kinetic energy. Designers note that this is very different to the main benefit of *insulation* because the insulative value of a material relates only to the ability of a material to block heat flows into or out of a building. High-mass

DOI: 10.1201/9781315645070-22

BOX 16.1 WAYS TO BEST MEASURE MASS ARE CONTESTED

The *heat capacity* is the amount of heat that must be added one unit of volume of the material in order to cause an increase of one unit in its temperature. The International Standard unit of volumetric heat capacity is one joule per kelvin per cubic meter of the material, $J \cdot K^{-1} \cdot m^{-3}$.

The *volumetric heat capacity* of a material is the volume of a material divided by its heat capacity.

The *heat capacity* of a material can be quantified as its volume times its volumetric heat capacity.

The *specific heat capacity* of a material is the heat capacity per unit of mass, in $J \cdot K^{-1} \cdot kg^{-1}$.

The *volumetric heat capacity* can also be expressed as the specific heat capacity times its density (in kg/L or g/mL).

Density is the mass per unit of volume of a material. The symbol most often used for density is ρ, although the Latin letter *D* can also be used. Mathematically, *density* is defined as mass divided by volume, where ρ is the density, *m* is the mass, and *V* is the volume. Density alone gives no idea of how good a material is at storing heat – a lump of metal is very dense but an efficient conductor too.

Volumetric heat capacity often varies with temperature and is different for each state of a material.

The *thermal conductivity* of a material is the number of watts conducted per metre thickness of the material per degree of temperature difference between one side and the other (W/mK), also known as its lambda (λ) value.

Thermal diffusivity is defined as the ratio of heat per meter cubed to conductivity and is calculated by dividing thermal conductivity by density times the specific heat capacity at constant pressure.

While some equate thermal mass simply with volumetric heat capacity, others see its performance and value in a building, in broader but less well-defined terms, as being dependent on the relationship between the specific heat capacity, density, thickness, and conductivity of a material.

walls absorb and store heat, stabilising the amplitude of the internal thermal pulse, dissipating a fraction of it and transmitting the rest back into the conditioned space at a later time or back out to the outside environment. The thermal inertia, how much time lag it provides, of the wall is a measure of the effectiveness of the thermal flywheel effect, which becomes stronger as the heat capacity and/or volume of material of the structure increases.

Indoor temperatures of spaces are largely influenced by three types of heat load: by the amount of direct solar gain into a building, by the amount of local ambient heat transmitted into it through openings and infiltration, and also by how much heat is transferred indoors through the fabric of a building's envelope. The first two occur in sync with local weather. When all three types of heat (summer) or cold (winter) energy charging or discharging occur at the same time, indoor peak temperatures may become excessively high or low, leading to a greater need for mechanical heating or cooling.[1] The conduction of heat flux into or out of the buildings in hot and cold climates can be limited using either or both insulation or heavier structural walls with high thermal inertia, of the type shown in Figure 16.1. Mass is active within annual, as well as daily, temperature swing cycles, which makes those who suggest that only 25 mm depth of a material is thermally active claim look rather like wishful thinking. There is also an intermediate cycle that is increasingly challenging, resulting from the evermore extreme hot or cold 'storms' of a few days in length, during which mass can play a critical role in protecting occupants from the worst of the outdoor climate for a period, depending on the heat capacity of the material and how well it is designed to store and slow heat and coolth.

Where a building has no mass/thermal inertia, then any energy inputs, either from air paths from outside via air leakage or solar gain through the fabric or opening of doors and windows, will tend to drive indoor temperatures towards outdoor air conditions or lead to excess solar gain. High-mass construction is traditionally used in more extremely cold or hot countries because the thermal inertia they offer can absorb such intermittent charging or loss of heat without dramatically changing the indoor thermal environment.

The *thermal inertia* in a building describes the extent to which the mass of the building is effective in dampening indoor temperatures. The usefulness and value of the energy-storing potential of a building depends on how the building is designed.

FIGURE 16.1 In very thick outer walls, the thermal flywheel effect can mean the temperatures on the inside and outside of the wall can operate under different charging regimes, especially when the windows are closed and shaded, as shown here (a) in rural Ireland on the left and (b) at the Alcázar Palace in Córdoba, Spain. Note how the angled window reveals to optimise light ingress while controlling solar gain.

Source: Barry O'Reilly and Susan Roaf.

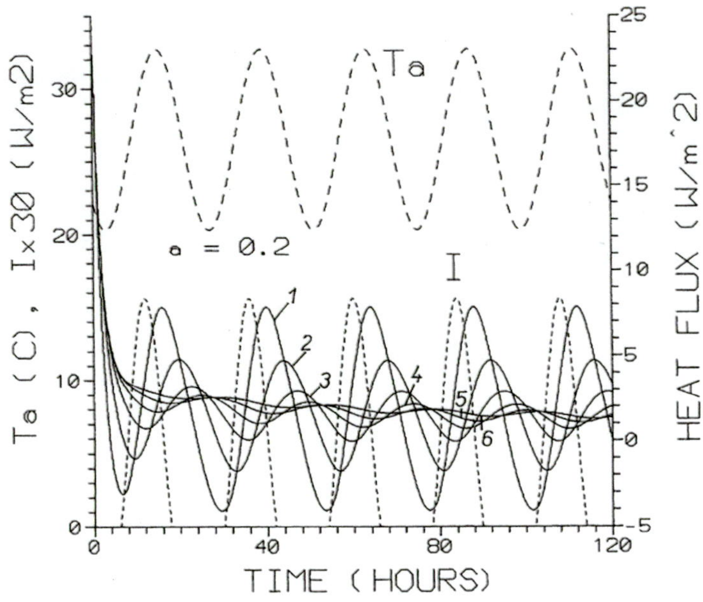

FIGURE 16.2 Comparative presentation of the calculated heat fluxes for the lightweight panel, WC1 (solid line 1), and the progressively heavier concrete walls, WC2 to WC6 (solid lines 2–6), for a succession of five sequential summer days. In the same plot, the harmonically varying ambient temperature and the time-varying solar insolation are also shown with broken and dotted lines, respectively.

Source: Tsilingiris (2003).

Reductions in energy demand also result from the fact that energy surpluses both from solar gains and heat produced from internal energy sources, such as lighting, computers, and appliances, can be reused later if and when needed. Mass was not so important historically in tropical climates with steady temperatures but vital in most traditional buildings in climates with larger diurnal and seasonal changes in temperatures. Even in very hot climates, massive wall construction could make life without air conditioners relatively comfortable when synchronised with local lifestyles.

Thermal Time Lags – Temperature Shifting

To remain thermally safe to occupy during extreme weather, building spaces generally need to be able to provide that:

1. Indoor temperatures are safely stable and not chaotic
2. Indoor air temperatures do not rapidly revert to outdoor levels during extreme weather events, whether or not the local grid has failed

In the 2021 Texas ice storm, many lightweight timber buildings got so cold so quickly that water froze inside them within a day or so. Even ground-source heat pumps failed because they require electricity to run the pumps and controls. In the absence of heat indoors from a grid supply or fire, what is needed in winter storms is a thick, heavy wall with a good time lag between outdoor and indoor temperatures, reasonably sized openings in the envelope, and little air infiltration between inside and out.

A common definition of the time lag (tL) for a building is denoted by the difference in hours between the occurrence of the peak temperature outdoors and the corresponding peak temperature indoors. The dampening of the temperature series is quantified by means of a *decrement factor*, which is the ratio of the amplitude of internal temperature wave to the amplitude of external temperature wave.

The time or thermal lag for an individual material is the time delay for heat to be conducted through that material in a wall (Figure 16.3). A material with high heat capacity and low conductivity will have a high thermal lag. Air has a very low conductivity, so does not transmit heat, but has a very low specific volumetric heat capacity, so does not store heat en route through a wall, thus providing little time lag. Concrete is much more conductive than an earth brick but can store more heat so is a good material to be used as an isolated heat store internally, but not so good as a combined capacitive insulator and thermal store for external walls, thus resulting in a lower time lag across a wall of similar dimensions. Lots of work is being done on the use of water as an internal energy store as it has a huge volumetric specific heat capacity but of little use in wall construction. Many factors will influence choices of materials, cost being one of them. A heavy hardwood wall will provide good thermal capacity at a high cost but burn less quickly. Developers' search for cost cutting leads many buildings made of low-cost timber that can perform well if well designed and insulated but burns catastrophically in a fire. It also shows

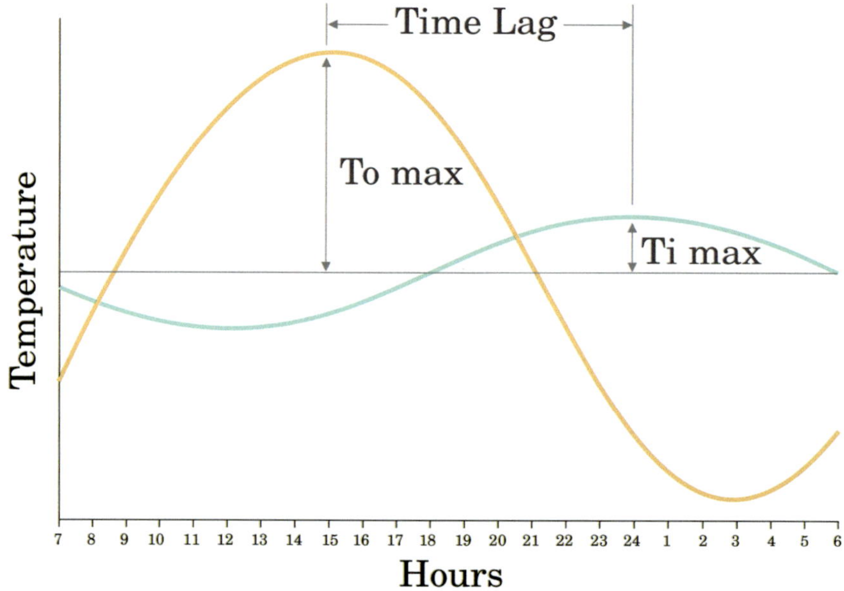

FIGURE 16.3 *Time lag* (tL) and *decrement factor* shown on standardised, generated data. The decrement factor equals to Ti/To.

Source: Jim Eveson

TABLE 16.1 The thermal conductivity, specific heat capacity, and density of selected materials

Material	*Thermal Conductivity W/m*K*	*Thermal Diffusivity mm2/s*	*Specific heat Capacity J/kg*K*	*Volumetric Specific heat MJ/m3K*	*Material Density kg/m3*
Air	0.025	19.4	1,004	0.001	1.29
Brick–kaolin	0.084	0.36	774	0.232	300
Brick–earth	0.222	0.48	795	0.477	600
Concrete 1–4 dry	0.753	0.5	657	1.511	23,00
Diamond	543.92	306	506	1.776	3,510
Glass–average	1.046	0.54	837	1.925	2,300
Glass fibre–epoxy	0.167	0.11	795	1.511	1,900
Gold	317.98	65.4	128	2.47	19,300
Plaster–building	0.431	0.32	1,088	1.36	1,250
Polystyrene–rigid	0.042	0.93	1,130	0.045	40
Rock/stone average	1.757	0.81	837	2.176	2,600
Steel–wrought	12.97	3.63	435	3.574	8,215
Water	0.60652	0.14548	4,181	4.169	997.05
Ice	2.092	0.54	4,217	3.867	917
Wood–balsa	0.05	0.17	2,301	0.299	130
Wood–spruce	0.126	0.24	1,255	0.515	410
Wood–teak	0.142	0.12	2,301	1.243	540

Source: Thermatest materials database.[2]

that you would have a very high heating bill if your external walls were of gold or diamonds, because they are such effective conductors and so useful in electrical equipment.

Three properties of a material, its specific heat capacity, density, and coefficient of conductivity, as shown in Table 16.1, are grouped together to give the diffusivity of the element, the key property that affects the rate that heat travels through building elements when exposed to fluctuations in temperature. Materials with high thermal diffusivity accumulate and move heat relatively quickly through them, thus reducing their ability to store heat over time. It is important to know what the related properties of materials are, so sensible choices can be made if a high thermal inertia is required.

Thermal lag times for a wall material are influenced by the difference in temperature between each face, that is, from outside to inside or between spaces indoors, and are difficult to consistently calculate as so many variables exist, like temperature, humidity, thermal history, etc.[3] Assessments of the thermal capacity of materials are often made experimentally using hot boxes over various exposure times to outdoor weather cycles.[4] The thermal lag is also influenced by exposure to air movement and air speed, the texture and coatings of surfaces, the thickness of the material, and its conductivity. Table 16.2 indicates how effective heavyweight materials can be in slowing down the passage of heat through walls. If heat peaks outdoors at around 1:00 p.m., it is useful if it reaches indoor rooms once the sun is setting, when occupants can access the coolth provided by the night sky and evening breezes. In coastal sites in hot climates, a temperature shift of a few hours may also tide occupants indoors over from the hot afternoon inland winds till they swing around to cooler onshore winds from the opposite direction in the early evening.

TABLE 16.2 Relative thermal lag in hours for some common 250 cm wide outer wall constructions

Concrete	6.9
Unfilled brick cavity wall	7.0
Adobe	9.2
Rammed earth	10.3
Compressed earth blocks	10.5

Source: Andrew Marsh.[5]

Coupling Thermal Mass to the Earth

The thermal time lag of an individual room within a building will depend not only on the thermal performance of its walls but also the extent to which it is coupled to the outside climate or the ground or spaces beneath and around the building. Figure 16.4 shows a section of the Hotel Suraj in Jaisalmer, subject of the doctoral study of Jane Matthews in 2000.[6] In the Hotel Suraj, there is a steep vertical time lag gradient, with longer lags on the lower floors. In some parts of the building, notably semi-underground spaces, the exposed mass is sufficient to shift the annual temperature cycle as well as the daily temperature cycles, while in areas more exposed to the weather, mass is only sufficient to shift the daily cycle, if at all.

Matthew's analysis showed that the indoor temperature time lag is not constant from season to season or even from day to day, which further compounds the problem. Previous studies presumed that such buildings had 24-hour time lags, whereas some indoor spaces in Hotel Suraj were estimated to have time lags exceeding 24 hours. It is consequently very difficult to confidently estimate the time lag by graphical means, so using two different statistical approaches (Appendix 2) in order to develop a credible method of calculating time lags from logged temperature data.

The choice of which space to occupy at any one time is determined by the climate in it at that particular time of day or year. There exists a wide variety of diurnal and seasonal climatic gradients horizontally and vertically, so the majority of rooms in a traditional house are not designed for a specific function but for a variety of specific internal environments that are used when comfortable at different times of day and year as the occupants migrate intramurally around the building over time.

The amount of heat an element is capable of storing in different areas of the building also depends on the volumetric heat capacity of the material and the size and weight of the walls. In vernacular building, a range of materials are often used for different climatic functions on one building. Some grand old Asian residences in Egypt, Iraq, Iran, and India had light timber pavilions on the roof where evening meals and entertainments could be enjoyed. Families slept beneath the cool night sky protected from winds and prying neighbours by high-roof parapet walls, above the heavily walled spaces below where the heat of the day was shedding overnight.

Properties of Mass in Materials

Thermal mass has become a victim of a circular economy conundrum. It is sidelined in modern buildings because it is perceived to cost more and take longer to construct. Research into it has often been actively blocked by lightweight construction industries, making it difficult to

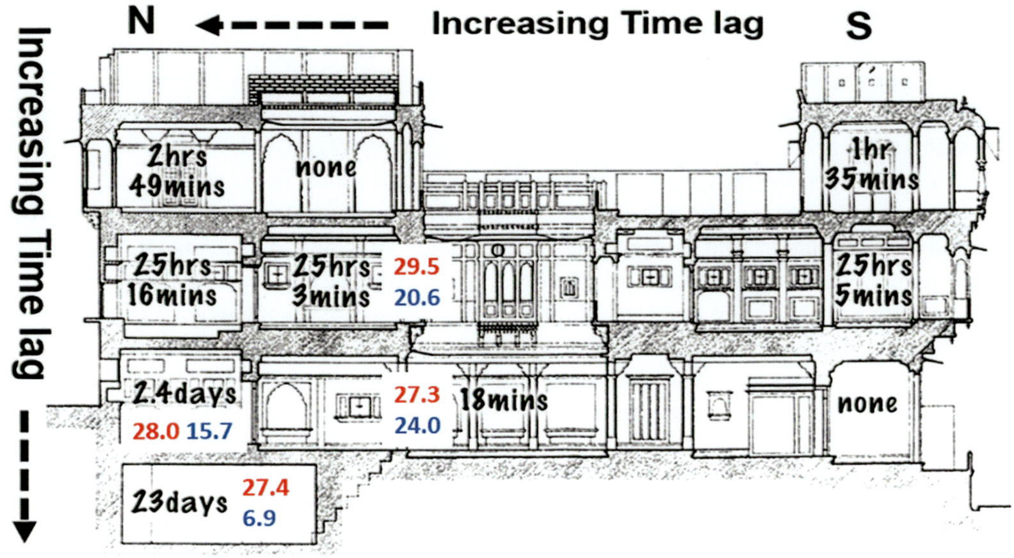

FIGURE 16.4 The varying time lags of indoor temperatures in rooms in a complex haveli in Jaisalmer, Northern India. The average annual temperatures of four spaces are shown in degrees centigrade in red, and their annual range there shown in blue.

Source: Jane Matthews (2000).

fund. Little research resulted in little progress, limiting interest in its use. This is reflected in the historic nature of the research done into it. Many of the values given to the thermo-physical properties of mass materials fed into calculations and simulations today are very dated. For any material, density, moisture content, and internal structure are the major determinants of its thermal and hygroscopic behaviour, but flaws in their measurement are rife. Factors still used were investigated decades ago. The updated compilation of solar absorptivity and emissivity factors by Penwarden at the British Building Research Establishment in the 1980s borrows extensively from an earlier review carried out at the Australian CSIRO and published in 1951 that, in turn, built on work published in the BRE 1935. In a similar way, oft-quoted emissivity values can be traced to a 1941 publication from the BRE. Clarke et al., in 2009,[7] published a review of extant data that underpins international materials standards and found the extensive discrepancies, deficiencies, and inconsistencies in all of them.

Thermal conductivity varied with moisture content and material type when timber products, inorganic lightweight insulating materials, and masonry were tested. Organic materials proved to be little affected by moisture content, so the use of a single value of conductivity for such materials is adequately accurate. Mineral fibre insulating materials showed a violent increase in thermal conductivity, with only small increases in moisture content, which may explain their poor performance in humid climates. Masonry materials' behaviour lies between these two extremes. In general, the thermal conductivity of dry materials also increases with temperature. International standards give only one value for a property regardless of the climate it relates to, so resistive insulations that perform badly in hot-dry and hot-humid climates are promoted by designers familiar with conditions in temperate conditions. Do the tenets of the European

Passive House standards really apply to design in the topics or hot deserts? Are the material property values used in the calculations for their very different climates really the same?

The *thermal inertia* of a building, its ability to dampen the amplitude of the thermal pulse of a building, depends on the *heat capacity* of its materials, as well as their mode of incorporation into a structure. If the specific heat of the construction materials is fairly uniform (e.g. around 0.2 kcal/kg°C), then the heat capacity of a building depends primarily on the mass of the materials used in it. Unfortunately, it is difficult to actually measure the *specific heat capacity* of a material – the amount of heat energy required to be added or removed to change the unit weight of a material by unit temperature – in practice. In low-conductivity materials, it is difficult to input or extract heat quickly, and test conditions for doing so are often vague.

Radiative heat transfer properties are conveniently considered together, commonly tabulated together and easier to measure. A large discrepancy can exist between measures of pristine surfaces in laboratories and those measured in the field with aged surfaces. Where possible, values appropriate to an in-use condition should be quoted to reflect how that material will perform in use.

A different problem exists for the quoted values of *emissivity* of surfaces that are dependent on the shape of a surface being flat or rounded, varying again between non-conducting or highly polished metallic surfaces. The length of a teapot full of hot water will stay hot in a slightly pitted surface of a dark pottery tea pot, which is considerably shorter, than for water in a highly polished silver teapot, which may well explain the popularity of the latter for those who could afford them.

Modelling Mass

Post-war Britain in the 1950s saw a huge amount of building as the ruins of bombed-out cities were rebuilt. By 1966, few air-conditioned (AC) buildings existed in London, and when surveyed, occupants in them found the air-conditioned offices to be cleaner and quieter, and in very hot weather, they were preferred to naturally ventilated buildings.[8] In less extreme weather, the thermal environment in the fixed-window AC buildings was, in many ways, less favoured than that in offices with opening windows, not least because the latter were of heavier, more conventional construction and had steadier indoor temperatures than the lightweight building with large areas of glazing. The growing problem in the 1950s and the 1960s of the poor thermal performance of highly glazed buildings prompted the development by the Building Research Establishment of the admittance method to assess overheating potentials in buildings by quantifying the amount of heat absorbed by and released from a particular space as the indoor temperature changes over a period of time. Appendix 2 was written by the late John Harrington-Lynne, who, among others, worked with Edward Danter and Alex Louden at the BRE to develop the tool in the 1960s as explained in Volume 2 of this trilogy on comfort.[9] The simplest type of manual, steady-state calculation method allows some insight into the flow of heat in a building, and the admittance method was the first fairly accurate method of quantifying the thermal response of building elements with mass.[10]

Subsequent, more complex computer-based models often used the admittance method to base mass calculations on. However, many now are 'black box' models, in which the underlying values, assumptions, and algorithms are obscured and cannot be modified or adapted to local circumstances by users. For teachers of designers who want to empower students to be confident enough to manage the performance of their designs, some of the pioneering work done in the United States by the passive design fraternity, led by Doug Balcomb, really helps young designers understand the issues involved in designing low-energy buildings. Appendix 3, by Murray Milne, one such pioneer, shows how to benefit from openly accessible, free design tools developed in California.

Building simulation software does have limitations, not least because much of it has been developed initially to model energy flows only and thermal flows are a secondary layer of their design. Some models do a better job than others of reflecting the dynamic potential of mass in buildings, including ESP-r (Strathclyde), TRNSys, and IDA ICE, but each country or profession will have its own preferred models into which culturally appropriate assumptions are integrated. Comfort assumptions are perhaps one of the largest stumbling blocks on the road to net-zero buildings, and the difficulty of modelling not only thermal mass but also the interactions of natural ventilation flows with mass pushes many designers to just ignore those two key building blocks of genuinely low-energy buildings. Additionally, there is a general problem with building codes globally that typically reflect the vested interests of those who write the codes and consequently limit the inputs accepted into statutory design models which not only do not reflect the thermal mass effect correctly but also lead to higher-energy buildings as a consequence.

U-Values and R-Values

Thermal building standards are typically linked primarily to the steady-state U-values or R-values, making the calculation of heating and cooling needs of a house built with high-mass walls not straightforward. For a start, many do not understand the difference between the two values.

The *R-value*, or *thermal resistance*, of a material is a measure of the performance of a single building material, dividing the thickness (in metres) by the thermal conductivity of the material, its lambda value (λ), measured in watts per meter per degree Celsius (W/m.K). The R-value can be used to compare how different thicknesses of products with different lambda values perform comparatively in transmitting heat through a material. Higher numbers are good when comparing the thermal resistance and R-values of products.

A *U-value*, or *thermal transmittance*, or overall *heat transfer coefficient*, is a measure of how much heat passes through a given area of a material or a structure. U-values measure the rate of heat transfer through matter in watts per square metre per degree Celsius (W/m^2K). U-values take into account all elements of a construction, including thermal bridges. They can only be worked out by calculation. A U-value depends on the thickness and type of the material, as well as the temperature difference between the environments on either side of the material. For example, a wall with a U-value of 0.3 W/m^2K means that for every square metre of the wall, 0.3 W (watts) of heat will pass through for every degree Celsius difference in temperature between the inside and the outside of the wall. The lower the U-value, the better the material as a heat insulator.

U-value and R-value are mathematical reciprocals of each other. Lower U-values indicate better insulating performance, and higher R-values indicate better thermal resistance. To calculate an R-value, divide 1 by the U-value figure. For example, a U-value of 0.10 equals an R-value of 10 (1 divided by 0.10). To calculate U-value, divide 1 by the R-value – a 3.45 R-value equals a U-value of 0.29. The U-value is the most accurate way to judge a material's insulating ability, taking into account all the different ways in which heat loss occurs; however, it is more difficult, and that complexity means that within the equations there is more scope to make mistakes in the assumptions on which the calculation is based.

Work done in the USA over two decades ago clearly showed that *steady-state* U and R-values do not accurately reflect the dynamic thermal behaviour of massive building envelope systems, making it difficult to convince builders, investors, code officials, etc. about the improved energy performance of massive building envelope systems. To overcome this challenge, field studies were conducted with real buildings to compare actual performance and calibrate related simulation models and methods. Early experiments were initiated in late 70s as a result of the energy crisis and focused on the application of passive solar techniques in residential buildings, with much of the work done by J. D. Balcomb and R. D. McFarland and their team at the Los Alamos laboratories, who investigated the energy performance of several passive solar wall systems and various thermal mass storage materials.[11] Many studies followed with whole buildings or hot box assemblages, some conclusively demonstrating the effect of thermal mass on space heating and cooling loads. Significant energy savings were noted for the house with a higher internal thermal mass.

Mass Works in Simple, but Sometimes Seemingly Mysterious, Ways

Thermal mass has been largely sidelined internationally by modern designers as a result of codes and regulations drafted to favour air- and water-driven mechanical systems. However, this is changing, encouraged by the emergence of a growing range of low-carbon, high-mass building materials, the chronic overheating of too much of the modern lightweight building stock, and the growing realisation that during increasingly extreme weather events, the electricity grid often fails, and mass is becoming rather fashionable amongst some architects and engineers.

The value of mass is often trivialised, with claims that only 25–50 mm of depth of wall material or plaster is all that is 'active', so more is not needed. Plaster alone, be it cement, gypsum, or clay, cannot provide 'enough' thermal mass for anything. Designing thermal mass to work well indoors involves many factors other than just the overall thermal mass of a building, including floors and ceilings and internal partitions. How much of their surfaces are exposed (i.e. not covered by carpets, cupboards, bookcases, furniture, etc.) plays a major role in their ability to absorb and release heat, as does the surface form and finish. It is essential to balance the allocation of thermal mass across all internal surfaces to make sure that the mean radiant temperature in the space is not asymmetrical. Having three heavy walls and the fourth fully glazed may lead to significant discomfort regardless of the orientation of the fourth wall, which may overheat (west-facing) or get too cold (north- or south-facing), depending on the hemisphere.

Thermal mass can also be used effectively outdoors, as shown in Figure 16.5 for shading or external building walls, pavements, and gardens, to enhance evaporative cooling and lower temperatures across in summer and warm them in sheltered, sunny winter locations in and around buildings. The appropriate use of mass in design must be case-specific and can and could eventually be usefully simulated or calculated. Thermal mass plays a number of different dynamic roles indoors, but then again, simulations in some cases may not prove as useful as thoughtful early design with a pencil, paper, and some pondering.

The Myth of Intermittency

Heavyweight components were believed to cause the average indoor temperature and difference across the building envelope to be maintained at a more elevated level. As a result, night temperature setback caused the envelope heat losses rate to be higher in massive buildings. All this supported a common belief that night temperature setbacks in massive buildings caused a

FIGURE 16.5 (a) Thermal mass also has comfort uses outside buildings. An innovative use for the low-carbon K-Briq in a jali garden wall, conditioning the air passing through while shading the space behind. (b) External wall of the c. 3,500-year-old Sumerian Temple of Inanna in Uruk, modern Iraq, using wall angles and moulded and baked surface bricks to provide shade and visual delight.

Source: (a) Sam Chapman (https://www.kenoteq.com/projects). (b) Todd Bolen.

reduction in the setback energy savings. Detailed investigation on the penalty in setback energy savings confirmed that such a reduction took place, but the magnitude of this phenomenon was very insignificant.[12] For a typical residence, the difference in setback energy savings between the massive house and an identical lightweight, wood-framed one was predicted as only 0.3%. But their study was undertaken in US homes with normal standards of construction in the 1990s. For a well-designed Ecohouse in Oxford,[13] UK, during a freezing spell in November 2024, with no heating at all, the living room temperature lost less than half a degree in temperature per day and was back up to temperature within a day, when the heating was turned on again, showing little impact of thermal intermittency due to its high levels of insulation, internal mass, and external insulation, with no cold bridges and minimal air infiltration.

Energy Savings With High Mass Homes

In the 1990s, a field investigation of the effect of thermal mass in residential buildings was completed at the Oak Ridge National Laboratory's Buildings Technology Center with a pair of test houses to compare heavy- and lightweight constructions.[14] A detailed report from the project was published in 2001, of the test of four different constructions:

- Exterior thermal insulation, interior mass (Intmass)
- Exterior mass, interior thermal insulation (Extmass)
- Exterior mass, core thermal insulation, interior mass (CIC)
- Exterior thermal insulation, core mass, interior thermal insulation (ICF)

In that study, the massive walls had about three to four times lower R-value than wood-framed walls (wood-framed wall R-value was about R-13, vs R-2 to R-5 for adobe, concrete masonry, and log walls). Final comparisons of field results and the models validated the study and showed a potential 10% savings in cooling energy and a 13% savings in heating energy. This suggests that most likely, thermal mass–related energy savings during the ICF study were in the neighbourhood of 11%. Despite clear evidence in this and many other studies of the energy benefits of higher mass construction, over 90% of all new US buildings are of lightweight timber construction. This despite also the knock-on impacts of such flammable construction in the catastrophic wiping out of communities during forest fires, increasingly fuelled by climate change.

In the Oakridge study, the clear winner was the construction with the exterior thermal insulation in terms of damping the amplitude of the internal temperatures as well as achieving the greatest energy use reductions. They demonstrated in their papers the difference in the results according to how large the temperature differences were across the wall from inside to out. Thermal mass is perhaps most useful in hot, dry climates with large diurnal swings, but in a heating world, this may apply to many regions that today still appear to have more temperate climates.[15] In colder regions and in continental climates, the use of insulated high-mass slabs beneath living areas also provides vital thermal stability for the retention of heat indoors during cold periods and the opportunity to harness daytime solar gain for evening heat where the mass is correctly insulated, to take advantage of the mass, as shown in Figure 16.6.

Thermal Mass Sufficiency

Three factors are driving the increased need for thermal mass in buildings: the need to reduce the impact of mechanical conditioning systems on the climate, the need to make comfort affordable for many, and the need to reduce overheating in modern buildings. De Toldi's 2003 paper 'Mass: More Is Good'[16] shows how effective mass can be in addressing those points now and

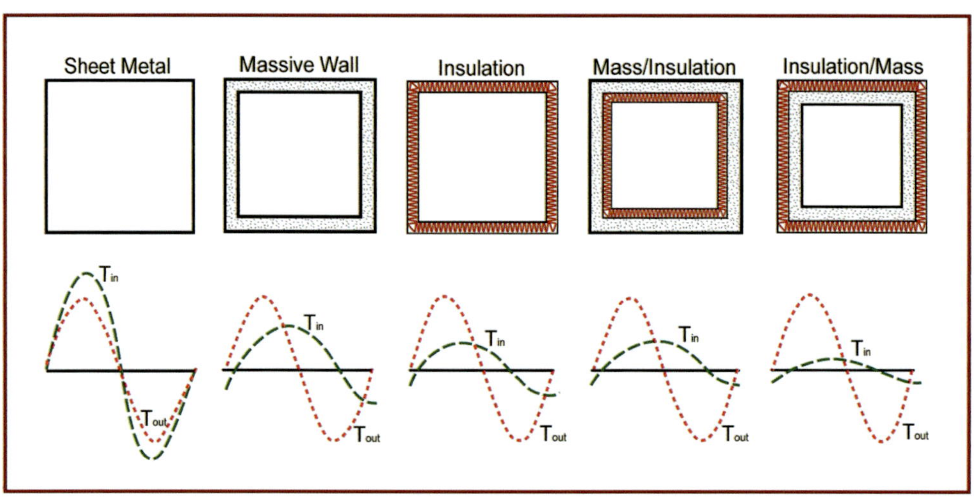

FIGURE 16.6 The best-performing walls in terms of reduction in indoor thermal amplitude are those with the thermal mass located inside the insulated envelope, regardless of the outside mean temperature, which will change over time.

Source: Isaac Meir.

can be in a hotter future in France. Using high-resolution climate projections from the recent EURO-CORDEX ensemble, he demonstrated the effectiveness of passive cooling using high levels of internal thermal mass to consistently meet over 90% of comfort temperature demand in French homes for the rest of this century, with huge energy and carbon savings possible over lightweight buildings, as shown in Figures 16.7 and 16.8.

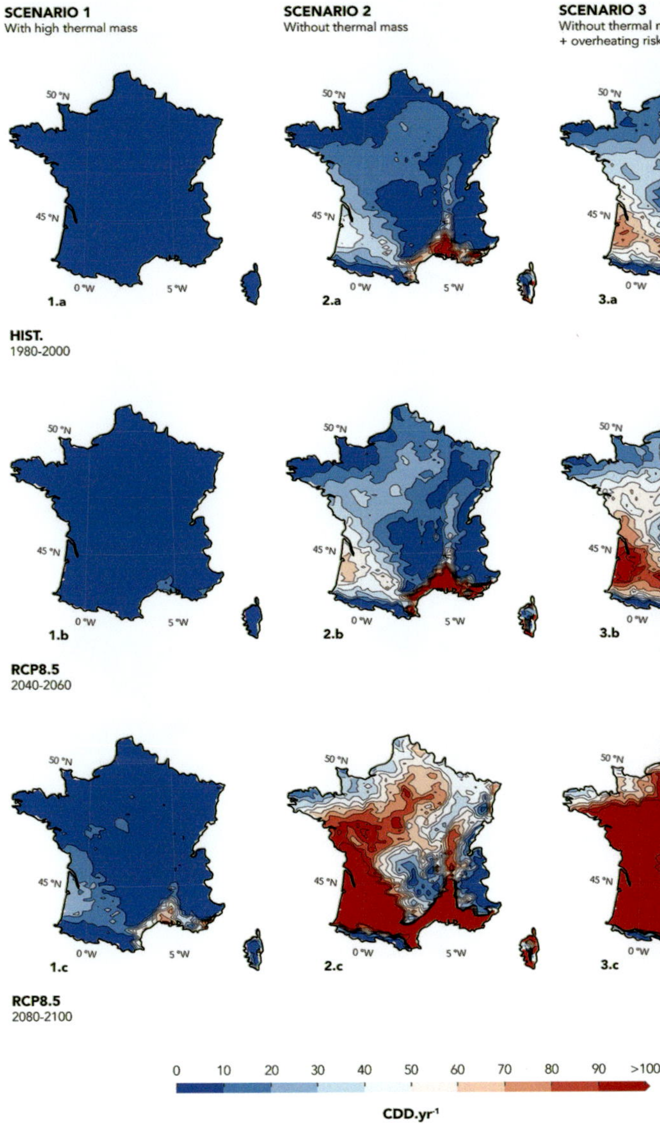

FIGURE 16.7 Cooling degree days above the adaptive comfort threshold (100%) in residential buildings in France modelled for the IPCC future climate 3.7°C global warming scenario (RCP8.5) for (a) heavyweight building, (b) lightweight building, and (c) lightweight building with a risk of overheating for using (*top row*) historic data, (*middle row*), 2040 to 2060 data, and (low row), 2080 to 2100 data

Source: De Toldi (2023).

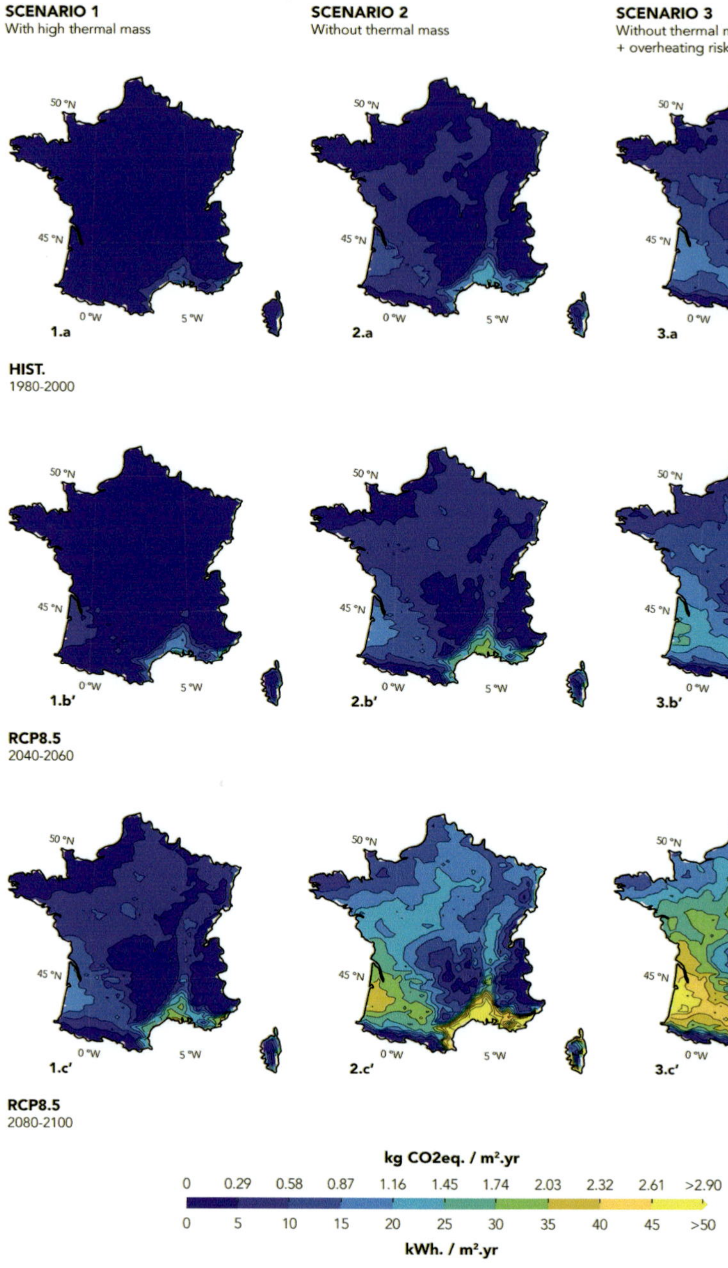

FIGURE 16.8 Comfort and cooling energy requirements in residential buildings in France modelled for the IPCC future climate 3.7°C global warming scenario (RCP8.5) showing specific energy consumption in kWh/m^2.year and indirect emissions for space cooling for (a) heavyweight buildings, (b) lightweight buildings, and (c) lightweight buildings with a risk of overheating for using (*top row*) historic data, (middle row) 2040 to 2060 data, and (low row) 2080 to 2100 data.

Source: De Toldi (2023).

His study also highlighted the limited suitability of current new construction practices, specifically lightweight buildings, favoured by green building standards, and the substantial lack of thermal adaptive capacity in the existing post-war building stock in France means millions of homes are now vulnerable to overheating. During the 2003 European heatwave, over 15,000 people died from the heat, mostly in buildings.

The question is, How much thermal mass is enough? This is like the comfort question: What is a comfortable temperature? The answer, as always, is, it all depends! Just as every region, nation, and probably city should establish an understanding of what are comfortable, acceptable, and survivable indoor temperatures, so should they have a good idea of what constitutes an adequate or reasonable level of thermal mass for local construction systems. Regulations should re-focus on both preventing overheating and excessive heat loss and as well as the storage and use of free local heat and coolth through good design incorporating well-managed thermal mass.

Delta T

The *delta T*, or ΔT, here the difference between the outside and the inside temperatures of the external wall of a building, will dictate how useful different levels of mass are in ameliorating the climate. A model of the *dynamic benefit for massive systems* (DBMS), a function of the material configuration and climatic conditions developed by Kosny et al. in the Oakridge Laboratory, showing extra importance of mass in hotter climates in the USA, in Arizona, Florida, and California.[17] Figure 16.9 shows the relative benefits of mass in ten different US climates for an externally insulated mass wall and an ICF home. ICFs are lightweight, hollow panels or blocks that are filled with concrete and left in place to insulate the structure, so the wall consists of

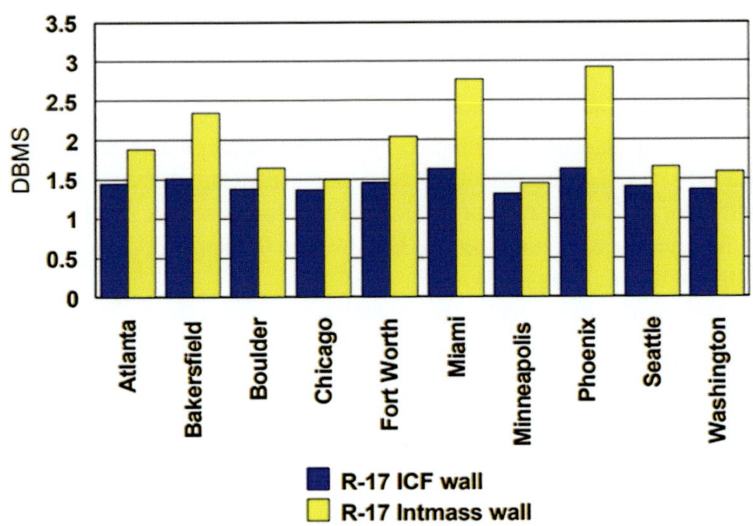

FIGURE 16.9 Comparison of the dynamic benefits of thermal mass (DBMS) in an externally insulated mass wall and an ICF construction in ten different cities across America.

Source: Kosny et al. (2000).

TABLE 16.3 Recommended levels of insulation and thermal mass for upland areas of Israel

Office, educational, public	*Residential*	
External wall R >1.6	External wall R >1.6	Thermal insulation (W/m.K)
Roof R >2.8	Roof R >2.8	
M >300	M >200	Thermal mass (kg/m^2)(M^2 of element)

Note: Higher thermal mass is advised for building types with higher indoor heat sources.

Source: Isaac Meir.

insulation, mass in the middle, then internal insulation. The exposed mass is very beneficial in the hot climate, whereas the ICF wall does not perform as well.

Researchers at the Jacob Balustein Institute for Desert Research in the Negev desert, Israel, have extensively studied climatic design, building on pioneering work by the late Baruch Givoni. Many formal and informal building experiments have been used to validate their assumptions and models of thermal mass.[18] They have published construction industry design guidelines for appropriate levels of mass for different building types in the different climates across the country, an example of which is shown in Table 16.3.[19] Perhaps their published methodology could provide a template for use in other regions and climates to provide builders with sensible estimates of optimal levels of mass in different building types.

Conclusion

Well-designed thermal mass in construction can result in significant reductions in levels of energy used to provide comfort and related emissions, as well as improving the affordability of comfort over a day, season, and life cycle of a building. As indicated in Figure 16.10 thermal mass can:

1. Dampen the amplitude of the indoor temperature swings, so improving comfort (temperature shaving and raising)
2. Create a time lag between indoor and outdoor temperatures (temperature shifting)
3. Enable free heat to be captured and stored from solar radiation and other sources
4. Enable cold to be stored as the absence of heat that is discharged during day or night to cooler air by ventilation, or through radiant fabric heat loss to the cold night sky
5. Maximise the usefulness of every unit of heat imported, captured, or generated indoors with careful design of the building's fabric
6. Minimise the need for and/or cost of imported grid energy to heat or cool buildings by enabling the diurnal and seasonal demand for heat and cold for comfort conditioning to be synchronised either with locally available free energy within a building or with low-cost/low-carbon imported energy from the grid

Thermal mass is an essential element for both building energy thermal storage over days and months and also as effective insulation in high temperatures, when resistive insulations fail in the heat. However, it is not the thermal properties of a single material or a wall construction that will slow down the thermal pulse in a building or make it safer to occupy in extreme conditions but the way the building and its spaces are configured that can make the difference between a high- or low-energy building, even if they are made of the same materials.

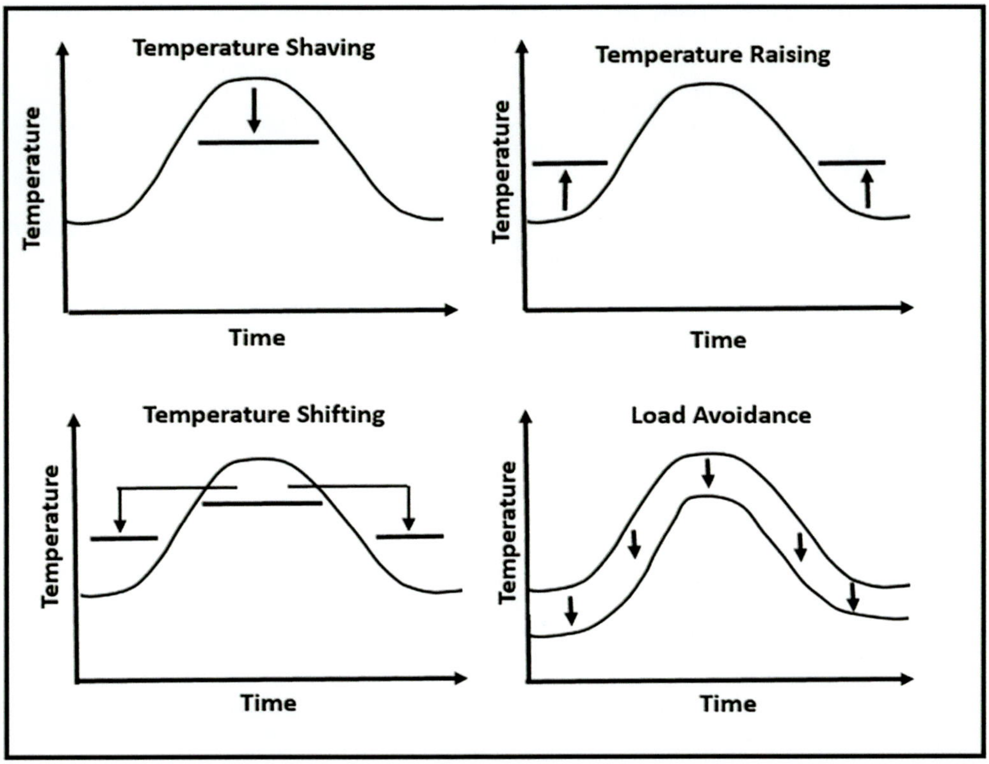

FIGURE 16.10 Temperature management using mass can translate directly into minimisation of electricity imports into buildings by temperature shaving, raising and shifting, and load avoidance, which is particularly critical at period of peak electricity use.

Source: After Alkhazraji, Salih, and Hasen (2017).[20]

Notes

1 Tsilingiris, P.T. (2003). Thermal flywheel effects on the time varying conduction heat transfer through structural walls, *Energy and Buildings*, 35(10), pp. 1037–1047. https://doi.org/10.1016/S0378-7788(03)00053-7

2 https://thermtest.com/thermal-resources/materials-database

3 Asan, H. (2006). Numerical computation of time lags and decrement factors for different building materials, *Building and Environment*, 41(5), pp. 615–620, ISSN: 0360-1323. https://doi.org/10.1016/j.buildenv.2005.02.020

4 Toure, P.M., Y. Dieye, P.M. Gueye, et al. (2019). Experimental determination of time lag and decrement factor, *Case Studies in Construction Materials*, 11, p. e00298, ISSN: 2214-5095. https://doi.org/10.1016/j.cscm.2019.e00298

5 Marsh, A. (2010). *Material Properties: Thermal Lag*, Accessed on 06/11/2024 at: https://performative-design.com/guides/material-properties/

6 Mathews, J. (2000). *Thermal Comfort in the Havelis of Jaisalmer*, PhD at London Metropolitan University, Accessed on 31/10/2024 at: https://www.researchgate.net/publication/33776485_Thermal_comfort_in_havelis_of_Jaisalmer

7 Clarke, J.A., and P.P. Yaneske (2009). A rational approach to the harmonisation of the thermal properties of building materials, *Building and Environment*, 44(10), pp. 2046–2055, ISSN: 0360-1323. DOI: 10.1016/j.buildenv.2009.02.008

8 Black, F., and E. Milroy (1996). Experience of air conditioning in offices, *Journal of the Institute of Heating and Ventilating Engineers*, pp. 188–196.
9 Humphreys, M.A., F. Nicol and S. Roaf (2016). *Adaptive Thermal Comfort: Foundations and Analysis* (Chapter 2), London: Routledge
10 Sproul, A.B. (2017). Admittance/fourier series revisited: Understanding periodic heat flows, *Procedia Engineering*, 180, pp. 292–302. https://doi.org/10.1016/j.proeng.2017.04.188
11 Balcomb, J.D., C.E. Kosiewicz, G.S. Lazarus, R.D. McFarland and W.O. Wray (1983). *Passive Solar Design Handbook*, Vol. 3, Los Alamos National Lab., ISBN: 0-89553-106-2
12 Burch, D.M., W.L. Johns, T. Jackobsen, G.N. Walton and C.P. Reeve (1984a). The effect of thermal mass on night temperature setback savings, *ASHRAE Transactions*, 90, Part 2.
13 Roaf, S., M. Fuentes and S. Thomas (2012). *Ecohouse: A Design Guide*, 4th Edition, Routledge
14 Kosney, J., T, Petrie, D. Gawin, et al. (2000). *Thermal Mass – Energy Savings Potential in Residential Buildings*, Report to the Oak Ridge National Labs and Polish Academy of Sciences, Accesses on 30/11/2024 at: https://web.archive.org/web/20040616184218/http://www.ornl.gov/sci/roofs+walls/research/detailed_papers/thermal/index.html
15 Meir, I., and S. Roaf. (2002). Thermal comfort – thermal mass: Housing in hot dry climates, *Proceedings of the Indoor Air Conference*, Monterey, USA. https://doi.org/10.13140/2.1.3449.3446
16 De Toldi, T. (2023). More is good, high thermal mass buildings for climate change adaptation and mitigation strategies in France, *Proceedings of the 8th ICARB Conference*, Edinburgh, Accessed on 16/12/24 at: https://icarb.org/2023-conference/papers/
17 Kosny, J., A.O. Desjarlais and J.E. Christian (1998). Dynamic thermal performance of concrete and masonry walls. *Thermal Envelopes VII/Wall Systems-Principles, AIVC*, (13136), pp. 629–643, Accessed on 10/01/2025 at: https://www.aivc.org/sites/default/files/airbase_13136.pdf
18 Cicelsky, A. and I.A. Meir (2014). Parametric analysis of environmentally responsive strategies for building envelopes specific for hot hyperarid regions, *Sustainable Cities and Society*, 13, pp. 279–302. https://doi.org/10.1016/j.scs.2014.02.003
19 Pearlmutter, D., E. Erell, I. Meir, Y. Etzion and Y. Rofè (2010). *Design Manual for Bio-Climatic Construction in Israel*, State of Israel, Ministry of Energy and Infrastructure, Research and Development Division – Electronic Book – 160p (in Hebrew).
20 Alkhazraji, A., M. Salih and N. Hasen (2017). Management of electricity peak load for residential sector in Baghdad city by using solar generation, *International Journal of Energy and Environment*, 8(1), pp. 63–72.

17

HARVESTING COMFORT FROM THE SKY

Introduction

Where the earth provides steady temperatures, the sky offers enormous thermal variety powered by the changing patterns of the sun, wind, and precipitation. It is easy to see why designers prefer to simulate a building as an inert, isolated box made comfortable by machines rather than to try to model how it works with real weather. Yet the builders of old created technical masterpieces without computers, not least because they had inherited an understanding of how to shape their buildings to work harmoniously with their environments. Also inherited were the skills to make that happen passed down through generations, sometimes over millennia, from the hands of master craftsmen to their apprentices.

Now, as we have to take that next vital step forwards, away from the 20th-century building ethos that pushed designers towards a total dependency on machines to keep buildings habitable, towards the next generation of mixed-mode buildings, we may have to re-learn some of those ancient skills. Rather as the sailors of ships under canvass navigated the oceans for millennia using the energy of the wind, the challenge of making safer buildings for a heating world must lie in understanding the local ways of wind and sun and applying that knowledge to the design of local buildings. Then again, sailing a ship is not as easy as simply switching on a boat's motor engine, but understanding local climates and weather really is the first vital step to creating genuinely zero (fossil fuel–free) emission buildings.

Global Climates Are Changing Fast

Macro-climates, regional climates, result from the interactions of the solar cycles with land and sea masses that create the many varied weather systems on Earth. The Köppen climate classification is the most widely used classification system, first published by German Russian climatologist Wladimir Köppen in 1884 and later modified by German climatologist Rudolf Geiger. Figure 17.1a shows the location of the three Poles in grey, North, South, and the 'High' Poles, mountain ranges like the Himalayas. The great forest basins of the Amazon and the Congo Rivers also have a huge impact on regional and global climates. Maritime climates by the sea are milder, modified by the more stable ocean temperatures. Continental temperatures in the hearts

DOI: 10.1201/9781315645070-23

of land masses, free from the stabilising influence of ocean temperatures heat up fiercely in summer and cool down dramatically in winter.[1]

The greenhouse effect is altering global weather patterns at alarming rates now due to increasing amount of heat being retained beneath the thickening overcoat of greenhouse gasses in the upper atmosphere that prevents heat, longwave radiation, escaping back out into space. Meanwhile, historically, record levels of solar energy, shortwave radiation, penetrates through the thinning ozone layers above the poles, increasing the rates of melting on land in ice caps and glaciers while shrinking sea ice further heating the sea and all these processes driving up rates of sea level rises.

Figure 17.1b shows one prediction of how much whole climate zones will shift before 2100 emphasising the shrinking of the cold, cooler, and temperate climates and the expansion of the

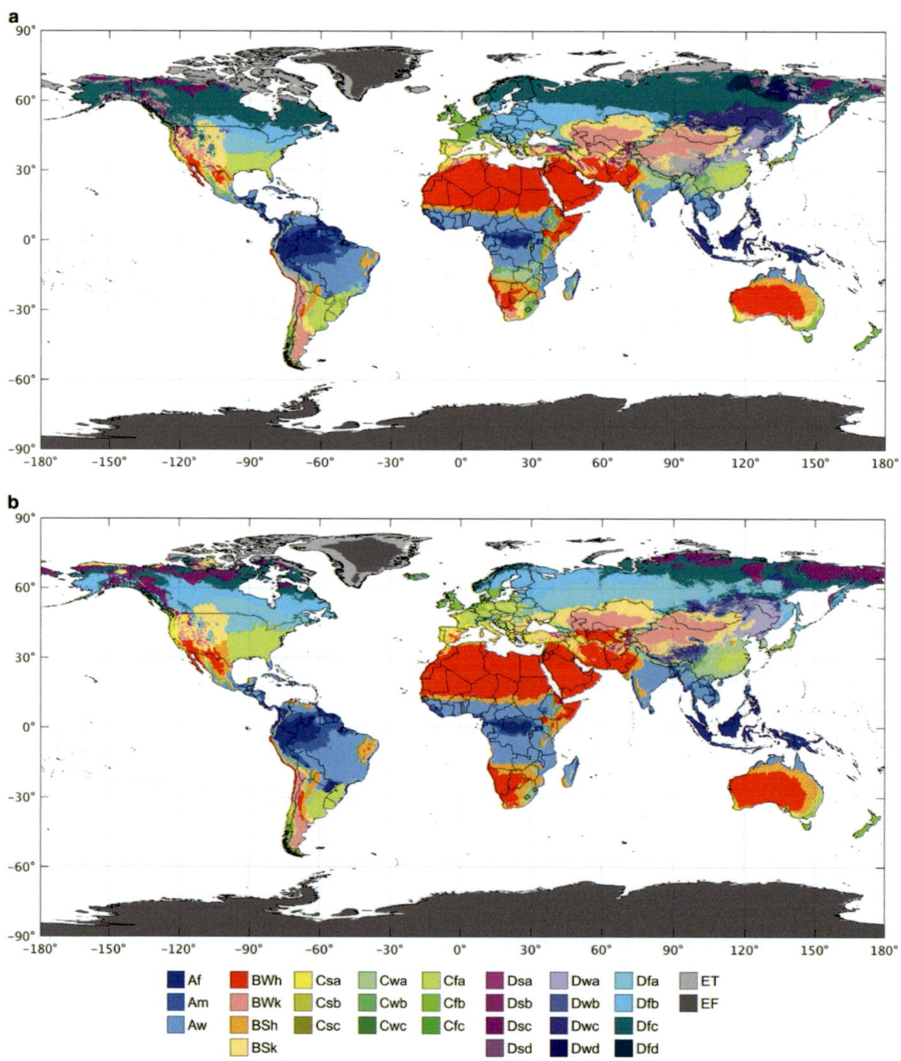

FIGURE 17.1 New and improved Köppen–Geiger classifications: (a) Present-day map (1980–2016). (b) The future map (2071–2100).

Source: Beck et al. (2000).

deserts and hotter regions, noticeably in the USA and China. Understanding how and how fast Earth's temperatures are changing will help inform how we must now design buildings for the different climates of the future.

Weather Cycles

Weather systems are driven by wind, the movement of air caused by the uneven heating of the Earth's surfaces over time by the sun. The global circulation is the world-wide system of winds that transports heat from the tropical to polar latitudes. It is divided in each hemisphere into three cells (Hadley, Ferrel cell, and polar), in which air circulates from the top to the bottom of the troposphere, which extends from the surface up to between 10 and 15 km high. This is the part of the atmosphere where most of the weather takes place.

The continued effect of the three circulation cells (Hadley cell, Ferrel cell, and polar cell), combined with the influence of the Coriolis effect, results in the global circulation. The net effect is to transfer energy from the tropics towards the poles in a gigantic conveyor belt of warm moist air from the tropics driven north by the surface winds of the Ferrel cell. This then meets cool dry air moving south in the polar cell. The polar front forms where these two contrasting air mass meet, leading to ascending air and low pressure at the surface. The atmosphere is constantly working to be in energy equilibrium. so air moves from high-energy, high-pressure areas to low-pressure areas to stabilise energy level.

Because the Earth rotates, circulating air is deflected, and instead of circulating in a straight pattern, the air deflects toward the right in the Northern Hemisphere and toward the left in the Southern Hemisphere, resulting in curved paths. This is called the Coriolis effect, and it results in the directions taken by the major wind systems on Earth, shown in Figure 17.2, that include polar easterlies, the westerlies, horse latitudes, trade winds, and the doldrums that occur in the intertropical convergence zone.

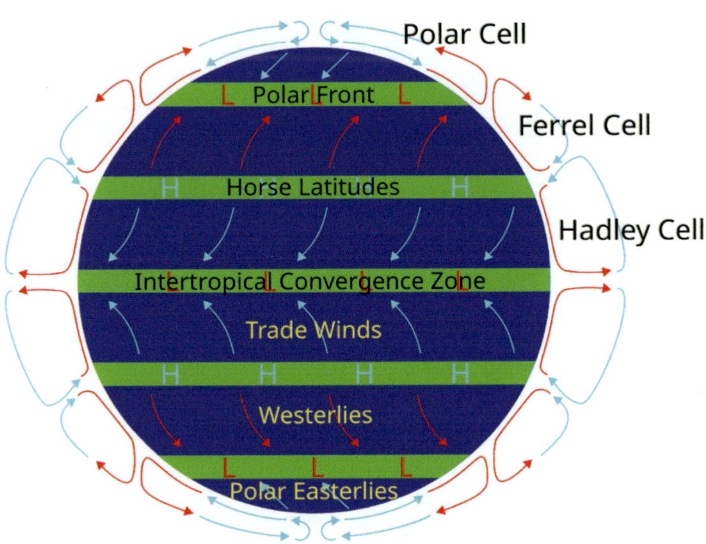

FIGURE 17.2 Diagram of the major circulation patterns of the Earth's atmospheric winds.

Source: Dwindrim~enwiki.

Differential heating of land and sea masses also drive changes in ocean and atmospheric temperatures and flows, including the influential *jet stream winds occurring between the polar and Ferrel cells.* The polar regions are now warming more quickly than the rest of the planet, causing the jet stream to intensify and meander, triggering unusually severe weather events from cold fronts to droughts. It is worth spending time updating and understanding these systems as they are changing at an increasing rate, and there are excellent websites that clearly explain how these systems work.[2] The almost-daily breaking of weather records is difficult to avoid as we become familiar with new terms for novel climate events, like *atmospheric rivers*, *heat domes*, *ice tsunamis*, *bomb cyclones*, *flash droughts*, and *firenados*.

Figure 17.3 shows how solar heating of the Earth's surface can cause convective bubbles to form which rise to the top of the atmospheric boundary layer (ABL) intruding into the global winds causing jets of very fast geostrophic wind to mix into the ABL. The shear between the ABL and the geostrophic also creates turbulence that creates eddies in the flow which are felt on the surface as gusts and lulls in the wind. The hotter the land or sea surface becomes, the stronger such effects driving the types of unprecedented weather events we are now experiences as the world heats up.[3]

Site Winds

In the planning of traditional settlements, a dominant shaping force was the wind, especially in more exposed regions and locales. So important was it to catch the sweet breezes from the sea or the mountains and to avoid the harsh air blasts from the inland regions so that

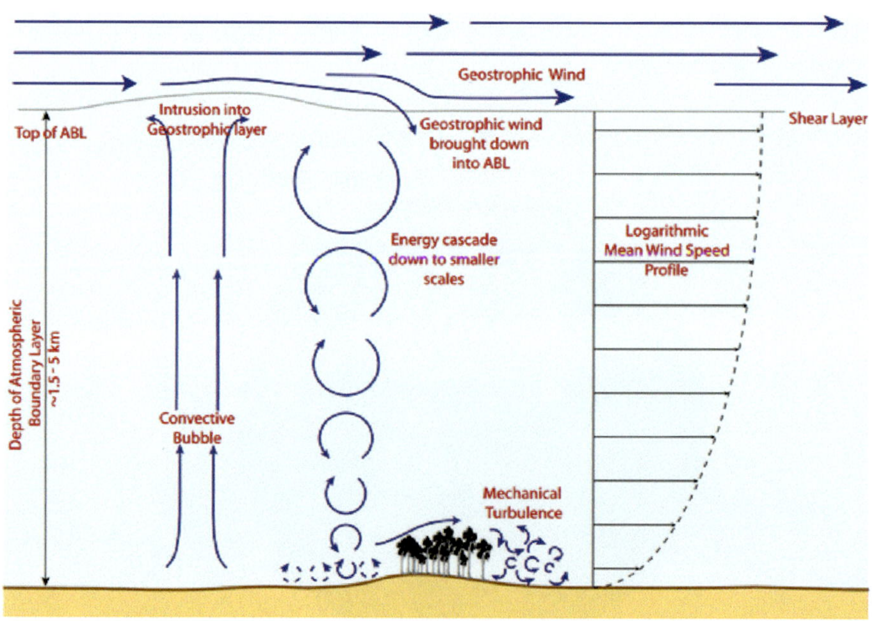

FIGURE 17.3 A diagram of forces at play in the atmospheric boundary layer between the Earth's surface and the upper-level geostrophic wind layer, where the jet stream resides.

Source: Reproduced from Sullivan (2017), with permission from CSIRO Publishing.

buildings and streets of towns and villages not only opened their doors and windows and all manner of wall forms and configurations towards the comforting winds but also turned their backs and elbows towards the hot or cold or nuisance winds.[4] There are many factors that dictate where settlements are built, like a communication route, or because of local manufacturing opportunities,[5] but once the site of a settlement came into being, other factors worked to shape it. For instance, the desert city of Yazd in central Iran is hot. Figure 17.4 shows average temperatures nudging up to 40°C (104°F) in summer with a nearly 15°C (27°F) diurnal range, while winter temperatures can fall below freezing in the following images from the Weatherspark website[6].

Yazd is located on an inland plateau *1,216 m (3,990 ft)* high and subject to strong sweet winds from the north and east in summer that were ideal for the development of the great wind catchers of Yazd (Figure 17.5). These are towers built above the once fairly level skyline of the city, with vents facing a various but designed to draw in cooling breezes down in the summer rooms of the houses below.

Michael Bonine, who studied Iranian settlements, posits that the Yazdi street orientation results from the slope of the land away from the mountains to the south, where the qanat water originates. The entire city was delineated by a grid of water channels supplying for homes and agriculture. However, the great homes of the city were designed around a courtyard, with the summer rooms facing north-east towards the prevailing winds, many with vaulted ceilings raised above the parapet line to catch the cooler breezes, as shown in Figure 17.6. When hot southerly winds blow, the vortices created by those vaults and flanking wind catchers draw cooler air up through the summer spaces from the deep basements and the courtyard gardens, enhancing the comfort of people in those spaces. Additionally, their north-easterly aspect keeps the hottest westerly and north-westerly sun out of summer spaces. Traditionally,

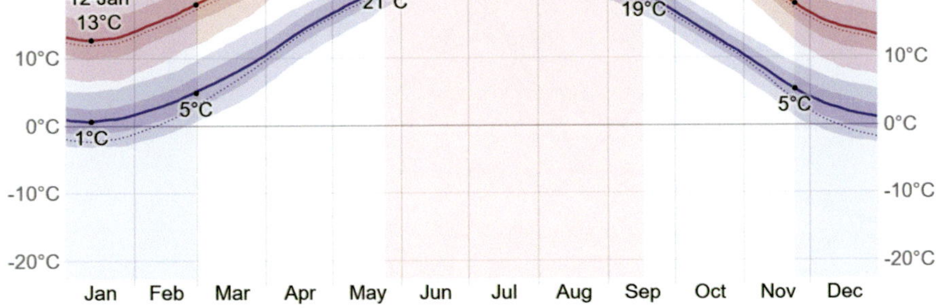

FIGURE 17.4 Average high and low temperatures in Yazd. The daily average high (red line) and low (blue line) temperature, with 25th to 75th and 10th to 90th percentile bands. The thin dotted lines are the corresponding average perceived temperatures.

Source: weatherspark.com.

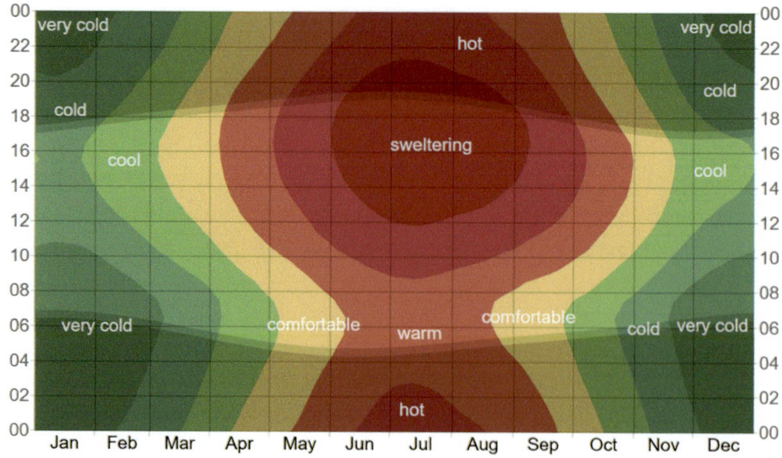

FIGURE 17.5 The percentage of hours in which the mean wind direction is from each of the four cardinal wind directions, excluding hours in which the mean wind speed is less than *1.6 kph*. The lightly tinted areas at the boundaries are the percentage of hours spent in the implied intermediate directions – north-east, south-east, south-west, and north-west.

Source: weatherspark.com.

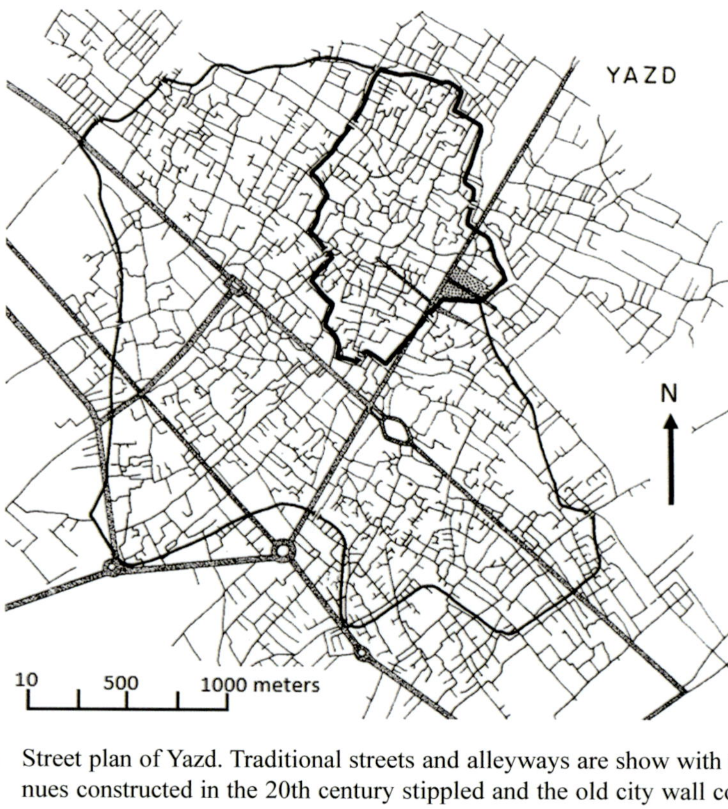

FIGURE 17.6 Street plan of Yazd. Traditional streets and alleyways are show with modern avenues constructed in the 20th century stippled and the old city wall constructed in the 14th century in black.

Source: After Bonine (1979), using maps from the Iranian National Cartographic Centre.

the forms of settlements were shaped by many cultural, historic, morphological, and climatic influences, including water supply systems and even inheritance laws and customs. In the hot desert city of Yazd, the clearest signal to be read from the orientation of streets and buildings is the need for the city to keep people cool enough in the summer in it by welcoming cooler winds, avoiding hot winds, and avoiding direct sunlight into the home.[7] So many modern buildings now overheat, and solutions typically relate to avoidance of the afternoon sun, but the harvesting of comfort from the prevailing winds is not touched on. For most locations, this needs to change.

Understanding Wind Is Important

Some classic textbooks explain how winds work rather clearly. T. R. Oke's book *Boundary Layer Climates*[8] deals with the climates that humans, animals, and plants occupy and influence. The 9th edition of *Atmosphere, Weather and Climate* by Barry and Chorley[9] comprehensibly covers the science of global climate and climate change. More accessible are online sites that use simple diagrams to explain how winds flow over landscapes and behave around hills, valleys, escarpments, and islands.[10] Wind patterns can often provide insights into why settlements are located where they are and their forms. The main driver of wind patterns are the high-level geostrophic, global winds shown in Figure 17.3, as well as the topology and location of the settlements within local wind systems, as shown in Figure 17.7.

The late Jim Wear of Perth presented a now-lost paper in 1992 called 'Air Is Stupid – It Cannot Follow the Arrows on an Architect's Drawing'. Most readers will have heard of the theory that climate chaos can be triggered by the action of a butterfly's wing.[11] Local wind effects seldom work as shown on theoretical drawings due to aerodynamic anomalies like eddies from cross draughts, objects or air introduced into or moving through spaces, and non-typical morphologies. Consequently, actual winds outdoors and indoors often act more or less atypically if not chaotically. The linear flow patterns designers imagine need to be intelligently interrogated during design to spot potential confounding factors that affect air flows.

Most regions have their *trouble winds,* the name given to strong hot winds from the north by some Aboriginal peoples in Southern Australia. Every site and continent has its good and bad winds, and when moving to design in a new location, it is important to find out about how local winds work. Ask first about the major regional winds, like the Sirocco, Harmattan, Khazri, Khamsin, Buran, Libbeccio, Loo, Sharqi, Norte, Chinook, Föhn, and Mistral. Nicol emphasises that it is important to understand how winds are felt differently by differently acclimatised populations. He found when working in Ghana that the Harmattan wind in West Africa was differently experienced by ex-patriots to locals. It is a cool dry winter wind that blows from the north-east or east in the western Sahara and is experienced by fully acclimatised West Africans as uncomfortably cold since they were used to having a moist equatorial skin but initially enjoyed by Europeans who experienced a dryer skin as being pleasant.

The significance of the correct placing of a building in relation to local winds was apparent in relation to criticality of the location of a coral on the Great Barrier Reef shown in Figure 14.2 to its chances of surviving a bleaching event. So too is this true of siting buildings in windscapes. First choose the right side of the mountain, cliff, hill, valley to avoid the trouble winds, and only then start thinking about optimising the orientation to take advantage of the sweet winds. It helps to have a feel for different what the different wind types are.

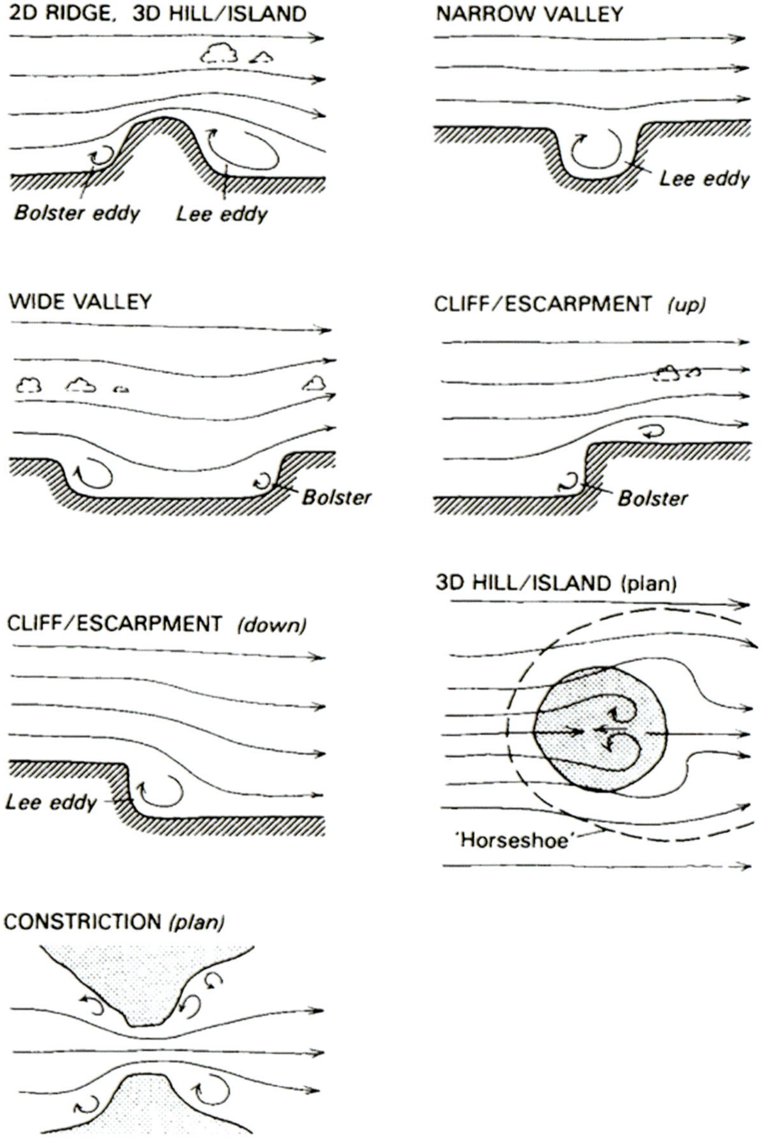

FIGURE 17.7 Typical patterns of airflow over steep and constricted topographies impact local wind speeds and directions. *Positive pressure* on the *windward* faces of features and the *negative pressure* vortices created either before or after physical obstacles.

Source: Oke (1987).

Wind Types

Regional winds are largely driven by temperature differentials, and a guess can be made about what types of winds might be encountered locally from the adjacent topography. Sea breezes and land breezes change diurnally and seasonally as air flows from cooler towards the hotter body. As the land heats over the day air rises over it drawing cooler sea breezes on-shore in the afternoons. At night and in winter this flow may be reversed with land breezes moving off-shore over the warmer sea and suddenly the local temperatures can rise or fall up to 10°C (18°F).

Anabatic winds are warm winds that blow up steep slopes or mountain sides. Driven by the solar heating of a slope causing hot air above it to rise, drawing cooler air from further down the slope below to move up the slope to replace it and be warmed in turn. These updrafts are popular with hang gliders and glider pilots but can also cause severe turbulence for aeroplanes crossing mountain ranges.

Katabatic winds are usually caused by gravity pulling higher-dense, heavier air from a higher elevation down a slope. These inversion winds often flow down from high continental interiors towards lower coasts, just as a stream flows down a mountain, constrained by adjacent morphological features or air masses. The surface of Antarctica is largely shaped by the movement of strong, persistent, and directionally constant near-surface Katabatic winds that flow from the higher interior plateau to the coast. Katabatic winds are strongest in winter, when there is no incoming solar radiation and in the right conditions Katabatic winds can rush down elevated slopes at hurricane speeds, although but most are relatively steady and mild. Perhaps the currently most famous are the Santa Ana winds, also called the devil winds, that destroyed so much of the Los Angeles region in January 2025. These are strong, extremely dry katabatic winds that flow from high-pressure, cool, dry air masses in the Great Basin and the Mojave desert down to the low-pressure zones over the Pacific, along the coasts of Southern California and northern Baja California.

Adiabatic winds are caused by the expansion and cooling of moist air, which can be triggered by winds flowing up and over a mountain, for example, so causing the water vapour pressure to exceed the saturation vapour pressure, causing droplets to form. For instance, the normal airflow over the Rocky Mountains is from the west to east, and as the air climbs up the western side of the mountains, it cools, forming rain, losing much of its water content, creating a rain shadow on the leeward slopes. When it descends on the eastern side, the increase in pressure towards lower altitudes causes its temperature to rise. In this way, many famous hot dry winds are formed including temperature to increase.

A *Foehn*, or *Föhn*, is an adiabatic wind, a dry, relatively warm downslope in the lee of a mountain range that has dropped most of its moisture on the windward slopes. Foehn winds can raise temperatures by as much as 14°C (25°F) in just a matter of hours. Switzerland, southern Germany, and Austria have a warmer climate due to the Foehn winds that have lost their moisture that was picked up over the Mediterranean Sea as they blow over the Alps.

Not only do such winds historically affect comfort and energy use, but their impacts will become greater as more extreme weather trends are manifesting with ever-stronger winds. As rain shadow belts get drier, already hot slopes get hotter, and wind flows increase in intensity, making the careful siting of settlements and buildings ever more important. When designing for new areas taking the advice of local farmers and builders, those who spend much of their life outdoors may provide life-saving and comfort-enhancing advice on the siting of new buildings.

Windscapes in and Around Buildings

A building façade is rather like a mountain slope that can rapidly heat up and cool down as the sun hits or leaves it, resulting in hot or cold surfaces. In relation to a landscape, a *façade* is typically a relatively small area, and unless the building is huge, in which case local air temperatures will play a major part in the local micro-climates, with wind speeds at street level increasing with the height of the building.[12]

Windy cities act rather like mountain ranges in themselves, so creating huge temperature and pressure differentials across them, as well as from the top to the bottom of tall towers. Already, parts of some cities are already difficult to navigate outdoors because of the strengths of winds that can prove fatal to unlucky pedestrians.[13] The taller the building, the faster the wind speed at the pavement level, which affects the wind comfort of pedestrians as well as the accumulation of pollutants at surface level.[14]

Many modern megacities have a central district with high-rise buildings which, if all are of a similar level, can form an artificial raised ground level so the city might be likened to an anthill with its own internal and surface climates. It might better be likened to a tropical rain forest with a dense tree canopy for its roof. Above and below the canopy are two completely different climates. In cities that under-canopy climate can result in reduced air movement at pedestrian level causing weak natural ventilation in some conditions causing pollutants to accumulate at street level. Such deserts have been recorded in certain weather conditions in cities including Hong Kong, Tokyo, and in New Delhi, where spikes in airborne virus causalities have even been recorded in pedestrians due to the still air conditions.[15]

Super tall buildings that stick up above the urban canopy present a particular problem for pedestrian-level safety. When wind speeds above the city increase, the tall facades form wind vertical tunnels, channelling fast air downwards to pavements from above. In such conditions, *wind comfort* around the bases of super-talls cease, and conditions occur that are recorded as being often *intolerable* and, at times, *dangerous*.[16] The super-talls heat up rapidly on hot days, acting like solar chimneys and drawing air rapidly up the heated facades, so altering weather both on the ground and above the city. City and state authorities still allow super-tall buildings to be built despite their potential threat to life and limb of ordinary pedestrians in their vicinity. In some cases they happen because of proven corruption as these uneconomical buildings have been shown to be useful vehicles for money-laundering, as noted in Chapter 25.

Tall buildings, even under the canopy, also generate their own wind systems internally, driven by the often-extreme temperature differences between the air at the top of the buildings and that at street level. Internal winds can get so strong they hinder lift door openings and make lift shafts to whistle noisily in very high towers. While the accelerated stack effect at the top of buildings can lead to *exfiltration* rates across the upper levels of the building counteracted by high levels of air infiltration in the lower floors making them often less comfortable and more expensive to heat or cool. The reverse stack effect also occurs as higher wind pressures at the top of the building can counteract the buoyancy forces. Poorly performing façades lead to higher infiltration and exfiltration rates through the building envelope with potentially major impacts on occupant comfort levels, inefficient and uneconomical use of HVAC systems, operational difficulties, and perhaps most dangerous of all, poor-performing smoke propagation control strategies, leading to ineffective fire safety controls for a building.[17] A key strategy for fire risk alleviation in tall buildings is the notion that pressure differences within escape stair will lead to the venting of smoke at a sufficient rate to keep escaping occupants safe for a while during fires. In reality, this pressure differential in fire stairs is almost impossible to maintain during real events as more than one door is opening into the escape stairs. These considerations show why it is crucial that designers also understand both the external wind impacts of building forms as well as the wind systems that exist within them.

Reducing Wind Discomfort Around Buildings

Studies show that some modifications to the form of a building can reduce wind speeds and impacted areas near buildings. For instance, Zhang et al. (2020) showed that cutting or chamfering building corners can reduce discomfort for pedestrians. Tellingly they showed that corner modifications are effective in reducing size of areas ranked as *intolerable* in terms of wind comfort and the amount of time that adjacent areas have wind speeds that are deemed as dangerous. Other minor aerodynamic modifications, such as aerodynamically shaped forms using tapered and setback levels and roofs above pedestrian areas, may also improve wind comfort, but again, orientation is critical within the context of adjacent street networks and adjacent buildings in managing wind speeds at street level.

Wind Comfort

As the world warms and high- and low-pressure climate and weather systems become more extreme, it is not only temperature records that are almost daily being broken but also those for wind speeds. A very short Google search will show many images and films of windows in modern tall buildings being blown out, and in case, whole glazed walls falling out. The loss of glass in storm events is escalating very rapidly, and not just in poor and cheap construction, but also in high-cost constructions by internationally famous architects. Although city authorities ask for aerodynamic analyses of the impact of wind inside and outside buildings, this does not seem to decrease the increasing rates of building failures as new phenomena emerge, such as the impact of *gustnadoes* or local *tornadoes* occurring at street intersections in high-rise city centres.[18]

Some of the apparently simplest decisions to be made during a design can have huge implications for the future comfort of occupants. The direction in which the main door of a property faces, and the type of door and lobby configuration used at an entrance can make all the difference to and the ability of occupants to become, stay, and pay to be comfortable.[19] For larger developments, expert wind advice is important and often obligatory, but many lessons can also be learnt from looking at the design of traditional settlements in very windy regions. Figure 17.8

FIGURE 17.8 In Punta Arenas on the southern tip of Chile, the street grid draws the katabatic winds down from the hills behind the city and draws up the on-shore sea breeze from the ocean at different seasons. Geostrophic winds in this region are so strong that entrance doors are placed in the wind shadow of the chamfered building corner at street intersections.

Source: Susan Roaf.

shows the chamfered corner door at a street intersection in Punta Arenas in Southern Chile where most doors are so configured to avoid opening them directly into the prevailing wind, a pattern that might be replicated in future city designs as the force of the wind becomes ever more important in the design of climate safe buildings.

Designers today must understand the importance of embracing sweet winds and avoiding trouble winds to maximise the comfort benefits that can be harvested from the sky as climates and weather both become more extreme.

Notes

1 Beck, H.E., N. Zimmermann, T. McVicar, et al. (2018). Present and future Köppen-Geiger climate classification maps at 1-km resolution. *Science Data*, 5, p. 180214. https://doi.org/10.1038/sdata.2018.214
2 https://www.metoffice.gov.uk/weather/learn-about/weather/types-of-weather/wind/what-is-the-jet-stream
3 Sullivan, A. (2017). Inside the inferno: Fundamental processes of wildland fire behaviour: Part 2: Heat transfer and interactions, *Current Forestry Reports*, 3, pp. 150–171. https://doi.org/10.1007/s40725-017-0058-z
4 https://weatherspark.com/s/105371/2/Average-Fall-Weather-in-Yazd-Iran#Figures-WindDirection
5 Jarrett, H.R. (1977). *The Geography of Manufacturing*, Philadelphia: Trans-Atlantic Publishing, ISBN: -978-0-71-210727-3
6 https://weatherspark.com/y/105371/Average-Weather-in-Yazd-Iran-Year-Round#Figures-Temperature
7 Bonine, M. (1979). The morphogenesis of Iranian cities, *Annals of the Association of American Geographers*, 69(2), pp. 208–224. https://doi.org/10.1111/j.1467-8306.1979.tb01252.x
8 Oke, T.R. (1987). *Boundary Layer Climates*, 2nd Edition, London, New York: Methuen Co., p. 435, Accessed on 23/02/2023 at: https://bayanbox.ir/view/6693893538424427706/T.-R.-Oke-Boundary-Layer-Climates-Second-Editio-BookFi.org.pdf
9 Barry, R.G., and R.J. Chorley (2009). *Atmosphere, Weather and Climate*, 9th Edition, London: Routledge.
10 https://www.metlink.org/resource/local-winds/
11 https://en.wikipedia.org/wiki/Butterfly_effect
12 Kim, J.-W., J.-J. Baik, B.-S. Han, et al. (2022). Tall-building effects on pedestrian-level flow and pollutant dispersion: Large-eddy simulations, *Atmospheric Pollution Research*, 13(8), p. 101500. DOI: 10.1016/j.apr.2022.101500
13 Baker, C. (2015). Risk analysis of pedestrian and vehicle safety in windy environments, *Journal of Wind Engineering and Industrial Aerodynamics*, 147, pp. 283–290. https://doi.org/10.1016/j.jweia.2015.10.001
14 Kim, J.-W., J.-J. Baik, B.-S. Han, et al. (2022). Tall-building effects on pedestrian-level flow and pollutant dispersion: Large-eddy simulations, *Atmospheric Pollution Research*, 13(8), p. 101500. DOI: 10.1016/j.apr.2022.101500
15 Mittal, H., A. Sharma and A. Gairola (2018). A review on the study of urban wind at the pedestrian level around buildings, *Journal of Building Engineering*, 18, pp. 154–163. https://doi.org/10.1016/j.jobe.2018.03.006
16 Zhang, X., A.U. Weerasuriya, K.T. Tse, et al. (2020). Pedestrian wind comfort near a super-tall building with various configurations in an urban-like setting. *Building Simulation*, 20(13)(6), pp. 1385–1408. DOI: 10.1007/s12273-020-0658-6
17 Mijorski, S., and S. Cammelli (2016). Stack effect in high-rise buildings: A review, *International Journal of High-Rise Buildings*, 5, pp. 327–338. https://doi.org/10.21022/IJHRB.2016.5.4.327
18 The Weather Channel (2024). *Why High Rise Windows Blow Out during Storms*, Accessed on 29/12/2024 at: https://weather.com/news/weather/video/why-high-rise-windows-blow-out-during-storms
19 Henderson Engineers (2024). *Stack Effect: Why Is Was So Difficult to Stay Warm This Winter*, Accessed on 29/12/2024 at: https://www.hendersonengineers.com/insight_article/stack-effect-why-it-was-so-difficult-to-stay-warm-this-winter/

18

AIR, RADIATION, AND COMFORT

Introduction – The Uses of Air for Comfort and Well-Being

The term *ventilation* covers a multitude of different uses of air in a building, including:

1. *Air to breathe.* At complete rest at sea level, an average person at rest breathes about 12 times per minute, consuming approximately half-litre per breath, around 6 L of air per minute (360 L per hour and 8.6 m^3 of air per day), of which 5% is exhaled as metabolic CO_2 (approximately 785 gm of CO_2).[1] This alters with stress, exercise, altitude, and mild disease, etc. by increasing the volume and frequency of breathing. In the 58 m^3 volume of my study, I have over six days of breathing air to use up on my own (less if the cat is in the room with me).
2. *Purging of smells.* This is best done by opening windows or directly expelling odours mechanically rather than circulating air between multiple rooms as anyone staying in a poorly ventilated *hostel* with a high dosage of sock smells will testify to. Unfortunately, supplying fresh air mechanically separately to each room is expensive, so most HVAC systems use recirculated air.
3. *Purging of pathogens.* COVID taught us that this is best done by opening windows rather than moving air between spaces mechanically.[2] Conflating air average levels of required for breathing with the purging pathogens from rooms led in cases to elderly people dying owing to surging viral loads on wards because their shallow breathing, erroneously using low CO_2 levels as a proxy for pollution/pathogen loads did not trigger the ventilation system to kick in, causing their buildup in the space.[3]
4. *Purging of pollutants.* Regulations mandate air flow rates and volumes to be delivered to, and expelled from, different rooms functions but spaces change uses frequently, and some furnishings, finishes, and equipment produce far more toxic emissions than others. A strong case can be made for air flow rates to be linked directly to annual assessments of need for pollution purging rather than the energy profligate use of one size fits all regulations[4]
5. *Directing, controlling, and containing fire and smoke.* Movement through a building in a planned or unplanned way. Advanced fire engineering uses aerodynamic modelling to de-risk buildings for fire and smoke spread.

DOI: 10.1201/9781315645070-24

6. *Managing moisture.* Unhealthy and unpleasant indoor air quality is often caused by excess moisture indoors, exacerbated in some cases by popular air and watertight constructions methods. The development of moulds interstitially or on internal surfaces results from the pooling of moisture locally. It can also be the result of changes in in the use pattern of rooms, say from store rooms to bedrooms, where the new occupancy has a higher moisture production. Regulations mandate air flow rates (not minimum percentages of fresh air) for different rooms functions but not different constructions. A heavy cavity wall house with good opening windows, plaster internal finish, and water-based paints, timber windows, and no cold bridging in the outer walls and is consequently not liable to moisture accumulation or mould growth indoors has the same required air flow rate as a steel frame building light weight structure with impermeable paint inside and out, metal windows and is riddled with cold bridges. Comfort is affected by moist internal air, but this is not improved just by simply moving more air around a space but, importantly, also by removing cold bridging from the construction and allowing the porosity of the room surfaces to enable the walls to *breathe*, moderating humidity levels and also enabling low-grade heating and cooling of the room surfaces.[5]
7. *Draughts.* This dreaded term (often spelt *drafts*) is used as a rationale for the sealing off of the indoor environment from the climate outside and seldom applied to other draught sources, like those resulting from jets of over-cool air emanating from air conditioning systems. Traditionally, draughts in outer walls in temperate climates were effective in losing some indoor pollution up and out of chimneys, but the current trend to air tightness, demonising *infiltration*, to enhance mechanical energy efficiency now mandate mechanical ventilation systems to do the same, whether they are necessary or not, again according to construction type or usage.
8. *Making occupants comfortable.* Air moving over the skin by convection can heat or cool a person with air from opening windows or internal air movement within, or between spaces, causing either comfort or discomfort depending on what the moving air temperature is in relation to skin temperature and acceptable comfort temperatures.
9. *Making a building comfortable.* Moving air that is warmer or cooler than internal surfaces can charge or and discharge heat to or from the walls, floors and ceilings of spaces and the thermal mass of the materials over hours, days, or seasons, as discussed in Chapter 17. Moving air has long been widely used to cool moist skin or building surfaces. Occupant comfort is enhanced by moving radiant temperatures away from outside air temperatures towards more acceptable indoor environmental conditions.

Perhaps a new term, like 'natural comfort cooling', should be used to differentiate between mechanical comfort conditioning of buildings and the passive cooling of buildings and people by convective and radiant energy flows to drive temperature control, requiring a new generation of passive building standards that deal either separately with the heating and cooling needs of both. These would include different values for different construction types, so a lightweight, water-impermeable building is evaluated differently from a heavyweight building with breathing walls. Regulations dealing specifically with the design of mixed-mode and naturally ventilated buildings need to be urgently reimagined to enable climate-ready buildings to be even allowed within regulatory systems, not least starting with the removal from them of the assumption that a tightly controlled, mechanically conditioned building is a 'better class' of building than its lower energy naturally ventilated equivalent.

Most of the important decisions on building performance are made in the early stages of a design. They concern the form, openings, materials and orientation of a building. A common assumption baked into simulation models is that a building is primarily mechanically conditioned and ventilated, with natural ventilation provided as an 'add on'. This assumption can be turned on its head in many new buildings, not least as people prepare to cope with surviving more extreme weather conditions and grid failures. As a minimum all buildings must be capable of operating with manual controls for the ventilation of occupied indoor spaces.

Natural Ventilation

The term *natural ventilation* commonly describes ways of bringing outdoor air into a building through openings in the outer walls and rooves that do not require mechanical systems. Moving air indoors passively typically relies on physical phenomena such as wind pressure in single sided or cross-ventilated spaces, or temperature differences creating a stack effect in a room or building where buoyant warmer air rises, as shown in Figure 18.1.

The impacts of a reverse stack effect where colder air at the top of a space sinks down through warmer air layers is often underestimated. For instance, in multi-storey homes, not only can cold downdraughts chill people in stairways, but where thermostats are placed on ground floor hall walls, it can also lead to excessive energy use and overheating of adjacent occupied rooms, in which air temperatures soar as a consequence of hall thermostats straining to reach set temperatures.

This chapter is not a ventilation design guide but raises key issues concerning how the air is measured and used in design to enhance the comfort of occupants in buildings.

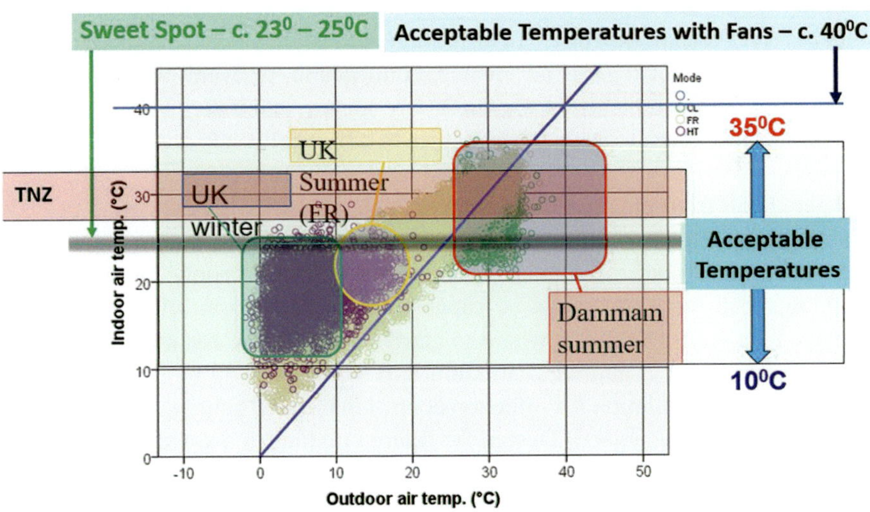

FIGURE 18.1 Building on data from Comfort Clouds for the UK (green and yellow areas) and Damman in Saudi Arabia (red area), set over data from Tokyo, Japan, physiologists and fan researchers discovered that it is possible to present approximate limits for acceptable comfort temperatures for adapted populations with air movement.

Source: Roaf, Rijal, and Nicol.

Making People Comfortable With Air Movement and Clothing

Chapter 5 on how people adapt conditions in building to make themselves more comfortable in buildings. Figure 5.9 shows that as temperatures rise above the thermal-neutral zone air movement over the skin decreases skin moisture and increases comfort as that moisture cools the skin evaporatively. It is not only that it 'feels nice' as temperatures rise to have cooling air pass over the skin, but it is a vital means of losing heat from the body, even at temperatures around 40°C (104°F) in extremes, as shown in Figure 4.8. Sourcing that cooling air through an opening window not only ensures that it is available even when energy supplies fail but also because it is 'free cooling' available to rich and poor alike.

To understand Figure 18.1, start with the need to keep the core body temperature at around 37°C, with the thermoneutral zone (TNZ) hovering around 28–32°C, the range where the shaded naked body at rest needs to do little to keep the body warmer or cooler. As the body gets cooler, the addition of clothes lowers that thermal comfort sweet spot to around 23–25°C, or much lower, depending on what is worn. Above around 32°C, heat loss from the skin becomes the key means of keeping cool, which is where air movement becomes more important. As temperatures rise, increased air speeds over the skin can remove more moisture and increase the evaporative cooling effect. Global comfort surveys outlined in Chapter 6 show that adapted populations with appropriate clothing can be acceptably comfortable indoors at temperatures as high as 35°C or more, and with higher fan air speeds and low humidity, even higher temperatures can be tolerated, up to 40°C and above. As global temperatures rise, some of the most vulnerable people around the world are having to occupy unbearable temperatures, driving them into heat stroke, as evidenced in an example in the *Cruel Summers* report on temperatures occupied in the rental sector across all Australian states.[6]

Some people who are highly adapted can learn to live in extreme heat or cold, but the usual health warning is attached to these limits in Figure 18.1. You cannot take a person who has been brought up in a very cold region or an air-conditioned lifestyle and expect them to immediately find temperatures from hot regions acceptable, comfortable, or event healthy, and vice versa. People ADAPT to those temperatures they normally occupy, and if they become uncomfortable, they learn how to take steps to return their bodies to a thermally safe and possibly comfortable state.

Cutting people off from the possibility of opening a window wide enough to provide a good level of air movement potentially threatens their ability to survive in very high temperatures if they can either not afford mechanical cooling or the energy supply to their building fails. So all buildings from now on should be capable of providing comfort cooling of occupants with air movement, so for novices, where to start? A good place for novices to start looking at the processes involved in natural ventilation in relation not only to comfort but also to the importance of natural ventilation for infection control in hospital settings is in the World Health Organisation guide on the subject.[7] Design for future buildings must also take into account the need to provide safe ventilation for occupied spaces during pandemics, so thinking about how to combine this and other air movement attributes of a building together from the start is sensible, promoting comfort, safety, and healthy.

Designing natural air flows indoors requires early decisions on what sort of a building is being designed. If the aim is that the building is capable of being efficiently naturally ventilated now, when the grid fails, or in the future, when it may have very different uses, then some simple early decisions need to be made. Later in the design refinement process, detailed are the choices

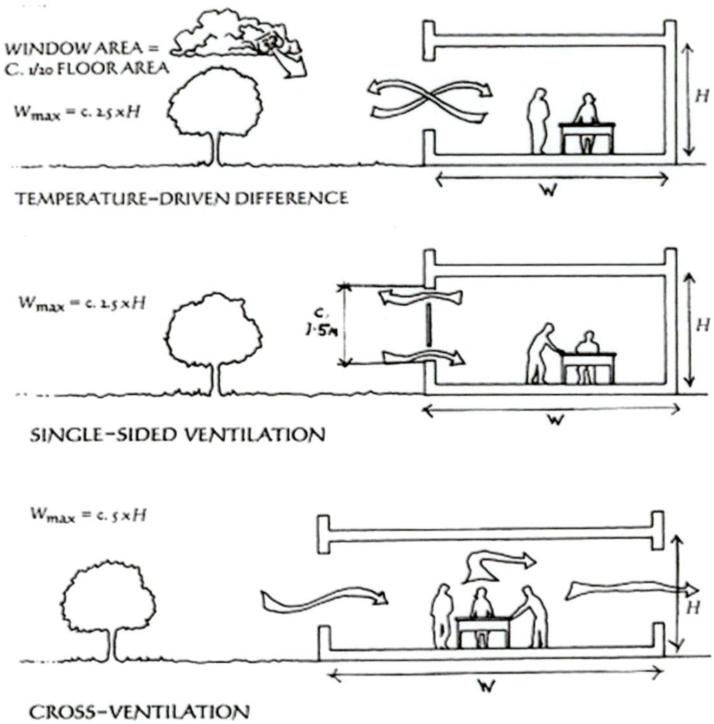

FIGURE 18.2 Natural ventilation of ordinary buildings does not involve rocket science and involves some simple rules of thumb for wind and temperature drive natural ventilation in rooms.

Source: Susan Roaf.[8]

associated with those decisions that can be developed and checked with more complex tools like those set out in the excellent guide to hybrid ventilation by Peter van den Engel published in the Klimapedia.[9] A design roadmap to follow helps.

Natural Ventilation Design

Designing natural ventilation into an ordinary building should be done from day 1 of a design, and following a simple checklist may help.

1. *Study the site climate in detail* and establish what winds affect it and how they interact with the local topography around the building. Think about how to optimise the benefits of sweet winds, and to avoid 'trouble' winds. Weigh up the extent to which the locality around the buildings can be modified to improve the quality if air coming into the building. Consider the effect of the orientation of the building (and particularly the windows) on the availability or danger of solar heating and the availability of ventilation and shading.
2. *Look at vernacular building types* on the ground and how both natural ventilation and light ingress into a building are linked and were dealt with historically for some clues.[10] For

inspiration and new design possibilities check out books on how traditional architecture follows climate,[11] how modern buildings were designed before air conditioning in the area,[12] and how the vernacular has been blended into passive design know-how over the last 60 years.[13]

3. *Start to shape strategies* to source and route outside air through the buildings at different times of day and year with and without thermal storage to heat and cool both the building itself and the people in it. Use natural ventilation design guides,[14] but also think clearly about where best to put building openings to optimise both air flows and daylight and view, a subject dealt with brilliantly in Lisa Heschong's *Visual Delight in Architecture*.[15] Take note of infection control issues indoors and fire protection advice.
4. *Choose materials* carefully to help manage energy flows and stores through the building.
5. *Use physical and computer models to test and refine designs.* This may mean testing in wind tunnels or salt baths in a laboratory, mathematical calculations, or simulations and for larger buildings, possibly expert inputs to not only ensure good internal ventilation but also avoid external wind discomfort and structural failures.

Airscaping Building Form

Welcoming the Wind

Optimising the shape of a building in a landscape to maximise the comfort of people indoors in a local climate requires a building to adopt a posture in relation to local winds (Figure 18.3). Like the human body that huddles up to retain heat, structures that need to minimise heat loss, like a yurt, maximise the usable internal space while minimising its surface area to limit heat loss. Similarly, if cold is being preserved, for instance, in an icehouse, volume-to-surface ratios need maximising. When it is hot, people sprawl out to maximise evaporative heat loss from the skin, and so do buildings with their forms extended, or perforated to enable effective cross-ventilation.

A useful early design tool developed by Nick Baker was the LT method that identified passive and non-passive zones in a building showing where natural lighting and ventilation is possibly on building floor plans. Described in Appendix 4 the method provides an idea of annual primary energy for lighting, heating, cooling, and ventilation for potentially mixed mode buildings and also was extended to evaluate the effects of shading devices on cooling loads and on lighting. The city of Berlin in Germany had a clear planning ordinance requiring buildings to have sufficiently narrow footprints so that all rooms can be naturally ventilated and light either side of a central corridor, with a plan width (c. 16 m) that is useful for either single sided or cross ventilation. This consideration is particularly important for the growing number of new hospitals being designed with natural ventilation in order to reduce levels of nosocomial, hospital acquired, infections in them.[16]

Backs and Elbows to the Wind

Buildings can open their wings and breezeways to comfortable winds to maximise the comfort potential of warm or cool winds, but avoiding damage from high winds is important. Devices like shutters block air when necessary, but on exposed sites, orienting a building corner to strong winds can significantly reduce their impacts on a structure. Wind likes to get its fingers underneath a building element to help it lift off a roof or overhang, so minimising protrusions on a façade, such as wide eaves and gutter gables, can help keep roofs on, as does the use of strong reinforced

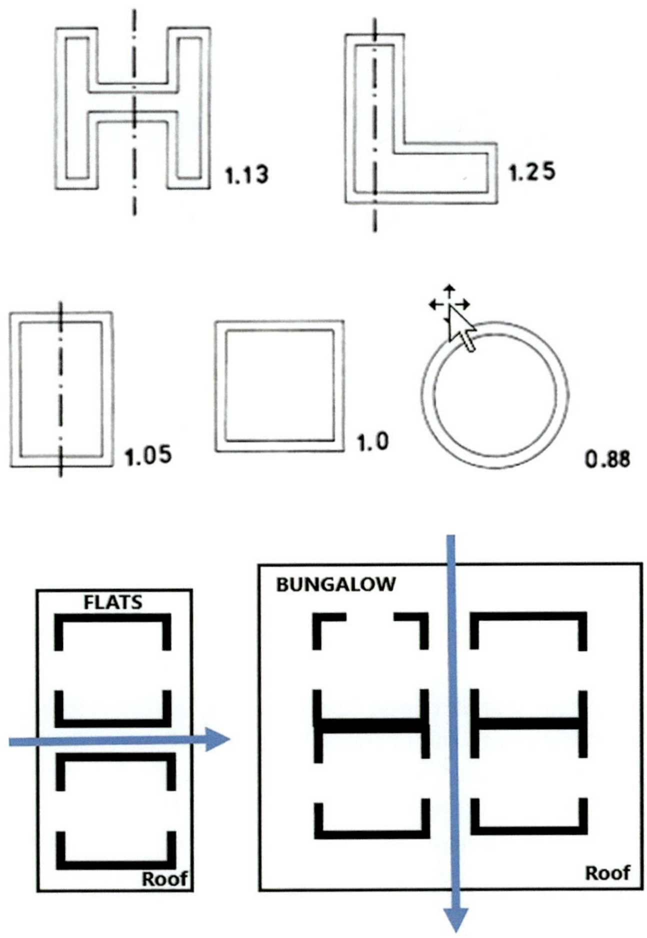

FIGURE 18.3 (a) Plans showing building footprints as a ratio of one that is a perfect square. (b) Hot climate building forms with internal breezeways showing that it is not just how the shapes of the forms and their orientation that matter but the breezeways that penetrate the structures to cool people, and the building, by convection, and also parts of the building mass en route, after sunset, and before sunrise in the tropics.

construction, like the sarking boards used in traditional Scottish rooves. Splitting the wind before it hits a building also helps keep structures stable in high winds using features like pillars, solid awnings above opening, and even flagpoles, as are used outside Swiss chalets in the Alps.

Balancing Radiant and Convective Heat for Comfort

Air temperature is not the only measure of what makes people comfortable in a space, there is also it's radiant temperature and the air movement within it.

Imagine yourself in a room with light weight timber walls with little or no insulation and one wall taken up with a large single glazed bay window, as shown in Figure 18.4. The air temperature externally is c. 3°C and the Glass is at 5°C, the internal surface temperature of the external

FIGURE 18.4 Which room is more comfortable? Air and wall temperatures in a bedroom in (a) Pucon, Chile, in winter, with heating, and (b) Oxford, in winter, with no heating on. The answer is (b).

Source: Susan Roaf.

wall around 8°C and the internal party wall 12°C and some low-level cold draughts. When on an electric air blown heater drove the air temperatures to 24°C and higher. When turned off when the local air temperature plummeted to around 8°C with heat also lost from the body to the even colder window causing bone-chilling discomfort. Fortunately, the bed had an electric blanket. This was the room Sue Roaf occupied for 10 days in Pucon in the Upper Andes of Chile in July 2024 (winter), when attending an Antarctic science conference. Many homes in Chile are similarly built with little insulation, often with high temperature local radiant and convective heat provided by low-efficiency wood-burning stoves that send a pall of brown smoke over valleys during cold air inversion conditions.

Roaf returned from the 2024 Comfort at the Extremes conference in Seville on the 25th of November after a freezing week in the UK with snow still on the ground. Her Oxford Eco-house is a super-insulated home with no cold-bridging in the outer walls, no draughts, and high-thermal-internal-mass walls and floors and triple-glazed windows. The heating had been off completely for ten days, over which time the internal air temperature had dropped only 6.6°C, making the internal air temperature 15.5°C, but the radiant temperature of the walls slightly higher. The perception was that the house was still comfortable because of the safe radiant temperature and the lack of draughts. Both spaces look similar, and the air temperature readings imply that the Pucon room should be more comfortable than the Oxford room. The reverse is true because of the overriding effect of radiant temperature on personal (dis)comfort in those situations.

Imagine yourself as a school head mistress in an office where you feel so cold it is difficult to concentrate on work. The HVAC engineer tells you that the temperature is a comfortable 24°C (75°F), but you have to call him back in because the conditions are still intolerable. He then gets out his thermal imaging camera and finds that the wall behind the desk has no insulation, so the surface you are sat next to is very cold despite the fact that the operative temperature in the room measures comfortable. Extra hot air is being poured into the space to compensate for the cold wall, which may explain why US indoor comfort standards typically call for 24°C (75°F) to compensate for the typically poor thermal performance of external walls. In Europe, standard requirements hover between 18°C and 20°C (64.4°F–68°F) in modern buildings, where high levels of insulation are mandated for making the whole thermal environment comfortable as was the case in the Oxford Ecohouse example.

Understanding radiant as well as air temperatures helps us understand where discomfort arises and also gives much broader insights into how different cultures in different climates make themselves comfortable in their buildings. While there are some excellent and interesting comfort prediction tools on-line that everyone should explore, like that from the University of California at Berkeley[17] or the ARUP's comfort tool,[18] none of them deal with the issue of difference in comfort cultures. In the 1998 SCATS project, office workers were surveyed across five countries. The data showed that in Sweden (Figure 18.5a), where the offices were air-conditioned throughout the year, indoor air and globe temperatures showed little spread.[19] In Portugal, where most offices were naturally ventilated, mean globe and air temperatures indoors spread considerably, as shown in Figure 18.5b). the difference between the air and radiant temperatures (suggested by the thickness of the line of points) was only a fraction of a degree larger than in the air-conditioned building, but the overall spread of temperatures (the length of the line of points) is very different. The offices of the two nations traditionally occupy very different building types, with different comfort regimes in very different climates.

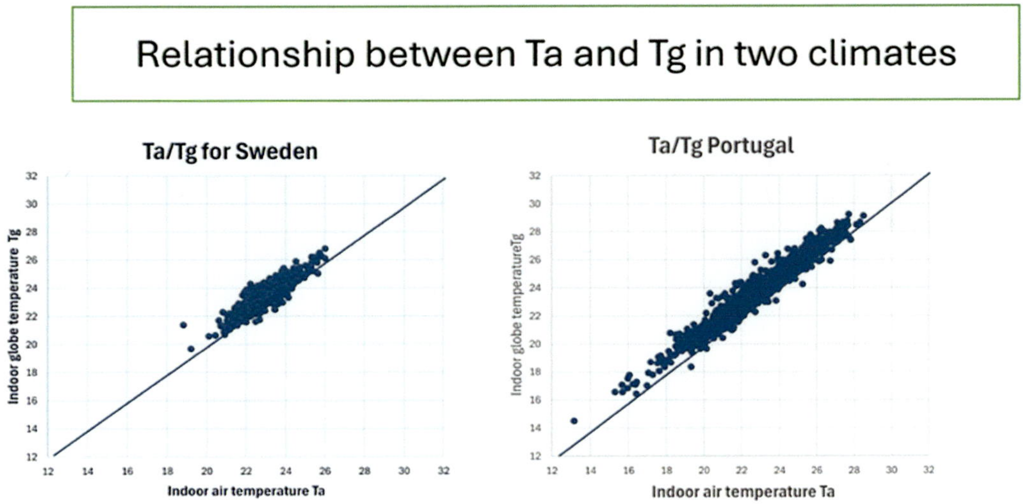

FIGURE 18.5 The relationship between globe and air temperature readings in Sweden and Portugal: (a) Mean value Tg-Ta 0.44°C. (b) Mean value Tg-Ta 0.90°C.

Source: Fergus Nicol.

The Energy Penalty of Cooling With Low-Temperature Moving Air

While high levels of air movement were traditionally discouraged in air-conditioned buildings, during the COVID pandemic, increased air flow rates were recommended in an attempt to reduce the rates of cross-infection between occupants in the same and adjacent rooms. However, increasing air speeds highlights an underlying flaw in the use of air conditioning as a comfort product that is sold as a volume of air of a certain temperature coming out of a vent at a certain speed and measured by the kWh. Air conditioning can make occupants feel colder than local air temperatures, not least because of air movement.

The first wind chill index was drawn up in 1939 by American geographer Paul Siple.[20] Figure 18.6 from the Royal Meteorological Society shows when frostbite becomes inevitable on exposed skin at low outdoor air temperatures.[21] People lose heat through their skin via conduction, convection, and radiation. Although conduction and radiation are relevant to heat transfer, the wind chill effect is caused mainly by convection. *Convection* is the process by which heat is transferred in the bulk movement of molecules within fluids, such as gases and liquids, and air is a gas. As air flows over a body, if fast it can break through the warm air layer just above the skin, replacing it with cold air. In addition, the air will extract heat when it absorbs moisture from the skin surface. Heat is lost faster to the cold than to warm air, requiring the body to increase the generation of more heat to maintain surface temperatures and minimise discomfort, sometimes not successfully, as people sat next to cold air vents or draughts may well know. The faster the wind speed, the faster the skin cools, increasing the rate at which heat is lost from the body.

Air has a thermal conductivity (TC) of 0.025 W/m*K and is not actually an efficient conductor of energy, having a low density (see Table 16.2). If you put your hand on a surface made of concrete (TC = 0.222 W/m*K) or steel (TC = 12.97 W/m*K), they will feel colder according to how conductive they are, so how quickly heat from the skin is lost to them.

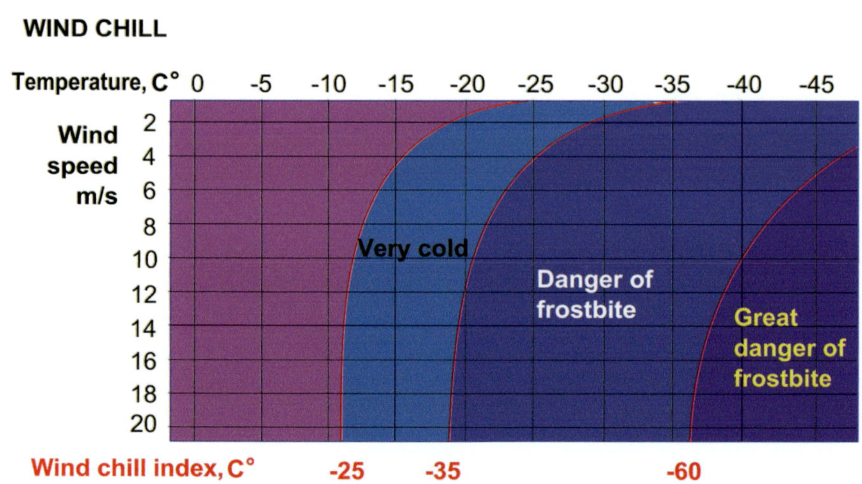

FIGURE 18.6 The wind chill index chart shows the effect of increased wind speeds on the time it takes to get frostbite in cold outdoor (or indoor) temperatures.

Source: Royal Meteorological Society.

Cold air from an air conditioner set at, say, 20°C (68°F) not only causes the skin surfaces to chill but also cools down the temperature of all the surfaces in the space to well below the TNZ, causing the body to have to work extra hard to maintain an average skin temperature at around 30°C and a core body temperature of 37°C (98.6°F). Many of the experiments in which the assumptions on what constitutes comfort temperatures in offices were done in the 1950s and the 1960s in North America with male office workers wearing business suits. Pity the poor ladies who are still expected by some companies to look attractive in skimpy dresses in such low temperatures.

Comfort standards compensate for this effect in their calculations. Take a typical Western office space that has insulated external walls, average air movement of 0.20 m/s (40 fpm) or more, with a ceiling height of 3 m or less, without skylights or active radiant surfaces, and where the external air temperature is less than 43°C (110°F). ASHRAE 55 Standard states that the mean radiant temperature of a room is assumed to be 2.8°C (5°F) higher than average air temperature for the space. This is presumably to raise the comfort temperature setting to compensate for the negative impact of the moving air on the radiant surfaces of a room.[22] A major problem is that the process of amalgamating radiant and air temperatures can result in seriously uncomfortable features or environments, like west-facing glazed walls being masked in calculations by being compensated for hotter or colder air temperatures.

The preference of many occupants for radiant heating in buildings can spring from the fact that high-temperature, hydronic (water-filled) radiators or wood-burning stoves in a space may result in the same air temperature, with some air movement, but all the surfaces in the room that are visible to the heat source, including the occupant's skin, are radiantly heated by it. So in winter, when the radiant heating is on, the chairs and tables and walls in the space feel warmer regardless of the air temperature, so the air temperature does not need to be cranked up to compensate for cold surfaces, and air and radiant temperatures are closer to each other than in air cooled spaces.[23]

Operative Temperature

Thomas Bedford in the 1940s understood that the comfort of a person depended not just on air temperature but also on the radiant temperatures of the surfaces of the occupied space and air movement around the person. He invented the globe thermometer.

Mean radiant temperature. MRT is measured in situ by a globe thermometer that theoretically exchanges a similar amount of radiant heat as a person might in that position in a non-uniform space, with little air movement.[24] Figure 18.7 shows a measurement set of the type often used in such building surveys.[25] Radiant environments can be complex, but the US ASHRAE Handbook of Fundamentals and ISO 7726[26] allow the radiant temperature (*tr*) to be calculated using the radiant values for the surrounding surface temperatures only, omitting these other sources of radiant heat exchange in the space. A 2020 study by Dawe et al. of measurements in 48 offices shows how very different radiant temperatures are when sitting close to a window than deeper in the room. Proximity to glazing can lead to cold discomfort in winter and heat discomfort in summer, as reflected in the umbrellas used in Figure 18.6.[27] MRT, measured either locally at workstations or in the centre of spaces in academic field tests, is generally now represented by a *tr* factor in HVAC calculations.

Calculating *tr* is done in a number of ways. In very complex buildings, particularly in extremely hot climates, a full analysis of radiant heat exchange may be required. For this, sometimes a few chosen spaces are calculated for, and then those results then fed back into the average figures for other spaces in buildings. A full calculation of mean radiant temperature under ASHRAE requires

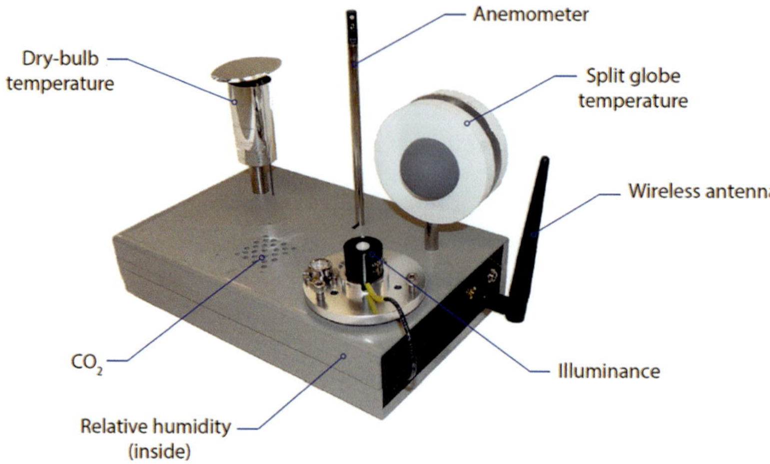

FIGURE 18.7 Typical assemblage of the traditional comfort researcher's field measurement sets with indoor climate monitors that may include dry bulb and globe thermometers, relative humidity and illuminance sensors, and more recently, CO_2 sensors have also been added.

Source: David Heinzerling.[28]

engineers to determine both the long- (trlw) and shortwave (trsw) mean radiant temperatures and assume that MRT is equal to trlw + trsw. In larger buildings, where thermal problems may arise in spaces, say, with large glazed areas on two walls, advanced simulations may be used to check the radiant discomfort potential for people seated in different locations in a particular space.

Operative temperature (OT) was the term originally given by Bedford, Danter, Humphreys, and Nicol to represent the comfort of a person in a space. It was originally given as the weighted mean of air temperature and mean radiant temperature measured by the globe thermometer, the weights being similar to the convective and radiative heat transfer coefficients of the clothed human body in the same environment.

Bedford invented a *globe thermometer* that initially took the form of a 6 in (152.4 mm) diameter metallic toilet ballcock painted matt black in the 1930s. In 1978, Michael Humphreys looked at the optimal size (and cost) for the globe of the thermometer and decided that 1.57 in (40 mm ping-pong ball size) was more appropriate for the globe, and this size of head for the thermometer was then widely adopted.[29] Today, such balls come in various sizes but are largely used in field studies to understand the thermal conditions experienced by the test kits, like the one in Figure 18.7 set next to a person in a particular position in a space. The data they collect are to measure radiant (Tr) and mean radiant temperatures (MRT) in the field.[30]

Here we have the conundrum. A globe thermometer measures not only radiant exchanges on a body but also radiant and convective impacts when there is more air movement. Air movement tends to push the globe temperature towards that of the air. In still conditions MRTs in well insulated spaces, the surface temperatures of walls, furniture, and ceilings may be close to air temperature. That is not the case where individual wall(s) behave differently from each other, for instance, where one wall is fully glazed and the MRT tracks external temperatures and direct solar incidence in particular. In rooms with open windows, or in forced-air heated or cooled spaces with brisk air movement, radiant temperatures can vary widely from air temperatures particularly where heavy mass walls have a slower thermal response than the surrounding air. Lightweight internal walls, floors, or ceilings may follow the different temperatures in spaces behind them, as anyone with an uninsulated timber ground floor over cold earth may appreciate. Any building in a city in warm or hot weather will be continually bombarded by infrared, long wave radiation, from adjacent building surfaces, the ground, and even hot passing vehicles.[31] In winter, the flow of radiation reverses as heat is lost to the street by radiation and convection, creating cold walls that are uncomfortable to sit next to despite 'comfortable' indoor operative temperatures.

Generally, commercial simulation tools will have settings that comply with national government standards specifying OT, like those from CIBSE or ASHRAE. Normal practice is to consider operative temperatures at a zone average level, and spaces can be modelled as single zones or multiple zones to capture different radiant environments, such as in 'core' vs 'peripheral conditions near a glazed facade. Few, except specialist engineers or modellers, will know what the underlying calculations are to comply or what they are actually modelling.

OT – The Comfort Design Fudge Factor

There is currently no consensus amongst many design engineers on what the term *operative temperature* actually measures, not only between comfort academics and engineers but also within the engineering profession itself. Those who design and install appropriately sized heating and cooling equipment should *theoretically* take into account the form and construction of the buildings itself. Operative temperature is calculated by engineers in at least three different ways:[32]

Method 1: With Air Movement

$$\text{Operative temperature} = (tr + (ta \times \sqrt{10v}))/(1 + \sqrt{10v})$$

Where:

ta = air temperature
tr = mean radiant temperature
v = air speed (m/s)

Method 2: Without Air Movement

$$\text{Operative temperature} = ((hr \times tr) + (hc \times ta))/(hr + hc)$$

Where:

hc = convective heat transfer coefficient
hr = radiative heat transfer coefficient

Method 3: With Low Air Movement

Where the air speed is less than 0.1 m/s, radiative and convective heat transfers may be similar, and so the equation can be simplified to:

$$\text{Operative temperature} = (ta + tr)/2$$

What could have been a complicated calculation has largely been reduced by practicing engineers to an easy step in HVAC system design. In the 20th century, air conditioning engineers fairly crudely calculated the volume of a space and how much cooling and heating air and energy would be needed to keep air temperatures within a space within the specified thermal comfort zone for that type of building in the local climate. Occupants would then have to adapt to the provided temperatures, usually by altering clothing, and possibly moving away from draughts out of air vents in a room. They were deemed to have to be comfortable in those conditions and largely had to lump it if not. Discomfort is an unfortunate result of calculating a single operative temperature for a building or a space that needs to be overcome and coped with by building users.

By the end of the 20th century, some simulation models had become increasingly sophisticated, interestingly not because of the growing complexity of buildings and expansion of western engineers to routinely working in novel climates. However, because of the dominance of 'modern design' increasingly buildings were air-conditioned and well-insulated without opening windows or the provision for natural ventilation so with little air movement, except near the forced-air vents of mechanical systems. In these cases, method 3 to calculate operative temperature is deemed acceptable in which radiant temperature is around air temperature, so air temperature sufficed as a proxy for operative temperature. In reality, many engineers understand OT or operative temperate as operational temperature, the air temperature the system should be operating at, bearing no relationship to the thermal comfort academics understanding of it at all.

Modelling Operative Temperature

REVIT is the software package widely used to create 3D models and store building information within a Building Information Management (BIM) package and database. Designing an HVAC system to achieve the desired operative temperature for a room requires a dynamic calculation to be performed that determines the amount of heat exchange required between indoor and outdoor environments, based on the specific construction details of the building and also taking into account the thermal storage within building elements. The required operative temperatures are taken from design guides for different building types and spaces in them. However, because of the low air movement levels indoors, the operative temperature is actually deemed to be the desired indoor temperature. HVAC engineers and building simulation models typically see the main purpose of ventilation as being to control indoor air quality by purging pollutants, rather than to only to regulate indoor temperature. Modellers generally consider the maintaining of a desired temperature within naturally ventilated spaces as challenging due to numerous unpredictable factors, such as wind direction, outdoor temperature, and other environmental variables.

Larger complex designs are developed using dynamic modelling that may include computational fluid dynamics (CFD) models for air circulation. These depend critically on boundary conditions, dictated by how the air, heat and cold actually move through different zones in a building, which affects the temperatures of the surrounding surfaces of any particular room. Modellers start by establishing the thermal context of the space or building and iterate from there until performance factors appear reasonable. In reality air can often move very differently in buildings from the way the models assume, and sometimes radically and even room-scale CFD grids are also often too coarse to pick up subtleties in geometry – the shape of a windowsill or the impacts of room contents.

The reality is that the real operative temperature we are after is one which mimics the thermal experiences of the occupants within an environment which is almost impossible to do in changing weather and seasons. By assuming that air temperature and radiant temperatures are similar in calculations, air-driven systems in lightweight buildings are promoted and radiant systems, especially in higher-thermal-mass buildings are denigrated and experiences of overheating can be extreme as is the case shown in Figures 18.8 and 18.9. This needs to change to

FIGURE 18.8 Working conditions in the Offices of Historic England on a sunny day, where occupants routinely use umbrellas to shade them from the radiant discomfort off a glazed wall despite the fact the room air temperatures comply with operative temperature requirements.

Source: Robyn Pender.

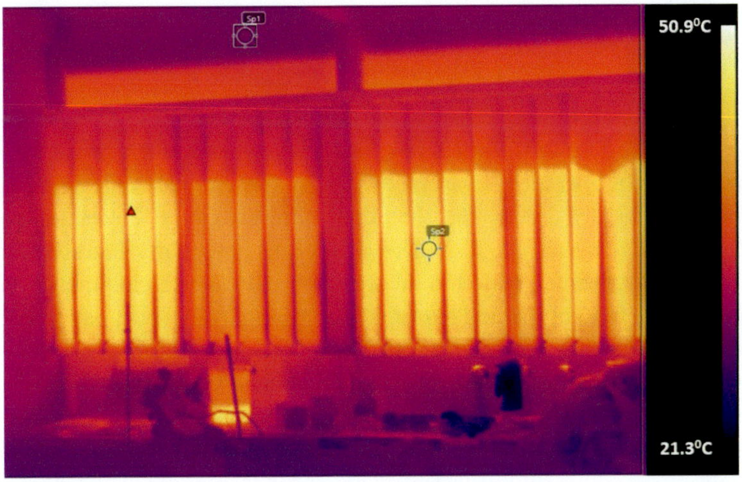

FIGURE 18.9 Ignoring the comfort implications of complex radiant environments at the design stage can lead to serious discomfort once rooms are occupied. Thermal images are useful when treating discomfort, but also to train designers on heat flows within spaces, as shown in the kitchen of a house recorded in Adelaide on 2024.

Source: Monica Oliphant.

promote the greater use of air movement for comfort cooling and lower-energy radiant heating and cooling systems using energy storage in water and mass.

HVACs Get-Out-of-Jail-Free Cards

HVAC engineers have traditionally enjoyed two get-out-of-jail-free cards to compensate for weak assumptions in design calculations.

1. *Oversizing of equipment.* By the time all the plants involved in a heating installation has been chosen, typically an engineer will add a 'resilience' factor in by oversizing the plant by typically around 30% to deal with any shortfall between air temperature and operative temperature. Unfortunately, this can come with very high-inefficiency costs and impacts.[33] This does assume reasonably accurate heat loss calculations closely operating and design temperatures match, and sadly, the practice of post-occupancy evaluation of building after construction with real people in them never really got off the ground not only because of the added costs but because designers really had no vested interest in finding out how well or poorly their buildings performed RIBA (2020).[34]
2. *The building manager*. Unless buildings are blighted with the 'no one knows how it works' syndrome common to Building Management Systems (BIMs), building managers have an important role to play in keeping most of the people not complaining for most of the time. Their tool kit includes technical solutions, like tweaking systems in use, to behavioural responses, like locating occupants who like colder conditions well away from those who like more tropical conditions.

The good news for engineers is that no design professionals to date have been prosecuted, nor not had their fees paid because they didn't calculate operative temperatures properly, but they have been for faulty systems in use.

De-Risking Design by Improving Performance Predictions

No one in the construction industry likes risk, unless they can make a profit from it. If designers move beyond current operational parameters set by fairly crude comfort standards and want their decisions to be legally defensible, then they turn initially to thermal simulations where levels of risk and budgets warrant it. Where novel and potentially lucrative innovations are being developed, prototypes are tested in laboratories and climate chambers. Today stakes are so high, not least in the field of fire and extreme heat testing, some even pay for the use of whole test House facilities to refine potentially high value design solutions. In the UK, the Salford Energy House laboratory enables whole test buildings to be built inside the laboratory that can provide robust data on, for instance, operative temperatures that will stand up in court if interrogated.[35] Unfortunately, the data from such facilities and those in the hands of industry who are very cautious about confidentiality issues which are not publicly available, but their work is hugely important in developing the next generation of climate-ready designs that work well in practice. Some data collected in such facilities are also fed into emerging standards, particularly in countries that have resisted cost-cutting drives to close national laboratories, so maintaining independence from industrial lobbying.[36] Inevitably, vested interests work to retard the evolution of standards in new directions because of the slowness of the incumbency itself to change.

However, now there is an urgent need to reduce the global impacts of over-engineered building by speeding up the move to mixed-mode buildings. The next generation of building with opening windows does not mean mechanical heating and cooling need be excluded from the but for much of the year they will be able to provide more affordable comfort, even when grids fail.

Researchers are already putting in place the evidence on which such futures standards can be based, for instance, by providing algorithms that can be fed into performance simulations to refine the ventilation strategies for natural ventilation of even larger commercial buildings in different cities and climates potentially enabling significant savings in energy consumption in offices by doing so.[37]

When Are Windows Opened?

There is nothing nicer than the feel of a fresh breeze across the skin when you need it. Air movement across the skin is particularly important for its conductive and evaporative cooling effects, when temperatures approach and exceed those around the thermoneutral zone (TNZ) above, say, 27–28°C (80–82.5°F), as discussed in Chapter 3.

When do building occupants want to open windows, and how can that be included in the design process? That question has the same old comfort answer – it all depends on what temperatures people occupy and find comfortable.

A detailed study of window opening patterns in England and Scotland by Rijal et al. in 2007 had some fairly obvious and robust results shown in Figure 18.10.[38] The proportion of open windows is lowest in winter, medium in spring and autumn and highest in summer and was a

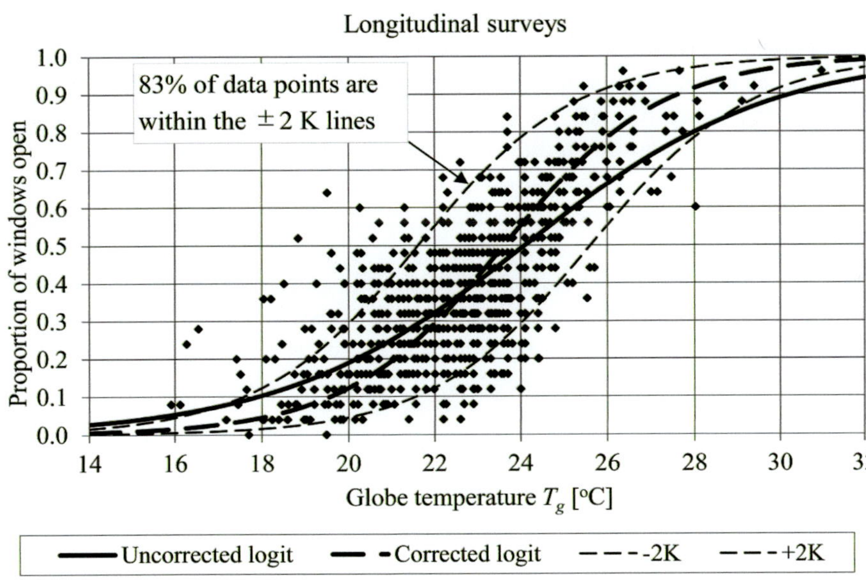

FIGURE 18.10 Logistic regression curves of windows open as a function of indoor globe temperature in all NV buildings in longitudinal surveys in the UK. Three lines (left, centre, and right) are the deadband of windows open. It is found that over 80% of the points representing the proportion of windows open occur within the deadband.

Source: Rijal et al (2007).

function of the indoor and the outdoor temperatures. An interesting finding was of a *temperature deadband* between the opening and closing of windows with a typical width of around 4°C K. (7°F K). Using the field data the team developed an algorithm for window opening that was validated in the field trials and implemented in the ESP-r simulation model. The window open behaviour as represented by the algorithm was more sensitive to changes in building design parameters than a non-adaptive approach as is typical in HVAC sizing programmes. A simulated office study showed that the resulting improved building design leads to improved comfort and a useful reduction in annual heating energy demand.

A field study led again by Rijal in Japan indicated that there are some cultural differences in window opening behaviours with the UK. Windows were seldom opened when the heating or cooling systems were on and the window opening and closing deadband was smaller than in the UK being only around 1.8°C K (c. 2°F) and most occupants either has windows opened or closed for most of the time. There are many factors that can influence why people open windows as well as comfort including cost of running conditioning systems indoors, security fears, design of windows, and ease of access or use, levels of outdoor pollution, and also building construction characteristics.[39] A similar field study in Switzerland provided robust data for offices and residences that demonstrated it was possibly to fairly accurately predict window opening behaviours in the Swiss climate and context. Both studies showed a similar trend for increased window use at moderate temperatures prior to switching on air conditioning.[40]

Ways Forward for the Natural Ventilation of Buildings

Table 18.1 highlights the problem with current comfort standards that are written by HVAC engineers. The categories in this CIBSE standard inherently imply that a highly controlled indoor environment with little temperatures range (Category I) is better than 'normal' (Category II) and particularly better than an old fashions naturally ventilated space with a wider variation in temperatures (Category III). All this based on assumption-riddled calculations that have not been validated with real comfort studies of response to those conditions by real people in real buildings. The same types of assumptions are baked into most other national and international standards, but in the face of a changing climate, perhaps it is time for fundamental changes to those standard has come.

We know that there is a lack of consensus on comfort-related measuring tools, sensors, values, and protocols, category weighting schemes and assessment class limits. From the 1960s onwards the search for temperature weightings that reflect 'comfort temperatures' has been on. From products like environmental temperature (ET), standard effective temperature (SET), dry

TABLE 18.1 Building categories for the UK Chartered Institute of Building Service Engineers in the Adaptive Thermal Comfort Standard CIBSE TM 52

Category	*Explanation*	*Suggested acceptable range (K)*
Category I	High level of expectation only used for spaces occupied by very sensitive and fragile persons	**2**
Category II	Normal expectation (for new buildings and renovations)	3
Category III	Moderate expectation (used for existing buildings)	**4**

resultant temperature (DRT), and predicted mean vote (PMV) and Operative Temperature (OT) have been promoted as the proxy for comfort.[41] Given the omission of important considerations in all of them including spatial variations, occupant behaviours, cultural factors, air movement and humidity etc. and the inability to predict in reality clothing or activity levels these are all blunt and rather discredited implements for reflecting the comfort of a person in a space. Fortunately, better simulation tools now allow the tracking of the spatial and temporal variation of all important parameters, thereby giving a more realistic insight into comfort levels even in larger buildings. Attempts to combine multiple parameters into a single index remain problematic. PMV, for example, is a steady-state concept and has no meaning if expressed as a dynamic variation. So more balanced effort must be put into standardised measurement protocols that work in the real world. As Heinzerling et al. pointed out in their 2013 paper, 'Assessment classes and categories should not be value laden and thus limited to two classes only compliance and non-compliance'.

Pragmatically, engineers believe that most well-designed and installed heating systems can usually be operated to *keep most users happy most of the time.* Pragmatic advice for architects: *Build it and they will come*. Understand the importance of *natural comfort cooling* that is hardly touched on at all in the Codes and Regulations for 'ventilation'. Understand the difference. Move away from glazed walls that prevent natural ventilation and lead to radiant discomfort. Put useful and well-sized windows into buildings that face good air and winds. Make those windows openable, accessible, and where possible, shadable, and they will be used. Air movement in well-behaved radiant environments is essential for comfort cooling.

People are not stupid and learn rapidly how to either minimise cold draughts or harvest free local energy to cool themselves indoors with moving air, and have done for millennia. One might posit that it is stupid to think they are not capable of doing so.

Notes

1 Pleil, J.D., G. Ariel, M. Wallace, et al. (2021). The physics of human breathing: Flow, timing, volume, and pressure parameters for normal, on-demand, and ventilator respiration. *Journal of Breathing Research*, 15(4). DOI: 10.1088/1752-7163/ac2589

2 Roaf, S. (2022). COVID-19: Trust, windows and the psychology of resilience, Chapter 33, in *The Routledge Handbook of Resilient Thermal Comfort*, Edited by F. Nicol, H. Bahadur Rijal and S. Roaf, London: Routledge, ISBN: 978-1-032-15597-5

3 de Man, P., S. Paltansing, D.S.Y. Ong, et al. 2020. Outbreak of coronavirus disease 2019 (COVID-19) in a nursing home associated with aerosol transmission as a result of inadequate ventilation, *Clinical Infectious Diseases*, 73, pp. 170–171. https://doi.org/10.1093/cid/ciaa1270

4 Kabele, K., Z. Veverková and M. Urban (2022). Methodology of IEQ assessment in energy efficient buildings, Chapter 23, in *The Routledge Handbook of Resilient Thermal Comfort*, Edited by F. Nicol, H. Bahadur Rijal and S. Roaf, London: Routledge, ISBN: 978-1-032-15597-5

5 Craig, S., and J. Grinham (2017). Breathing walls: The design of porous materials for heat exchange and decentralized ventilation, *Energy and Buildings*, 149, pp. 246–259. DOI: 10.1016/j.enbuild.2017.05.036

6 Better Renting (2024). *Cruel Summers Renters' Diverse Experiences of Summer 23–24*, Published by Canberra. http://www.betterrenting.org.au/

7 Atkinson, J., Y. Chartier, C.L. Pessoa-Silva, P. Jensen, Y. Li and W.H. Seto (Eds) (2009). *Natural Ventilation for Infection Control in Health-Care Settings*, Geneva: World Health Organization. PMID: 23762969.

8 Roaf, S., M. Fuentes and S. Thomas (2009). *Ecohouse: A Design Guide*, 4th Edition, London: Routledge.

9 Van den Engel, P. (2023). *Hybrid Ventilation: A design Guide, Klimapedia*, Accessed on 12/01/2025 at: https://klimapedia.nl/publicaties/hybrid-ventilation/

10 English Heritage (2014); *Practical Building Conservation: Building Environment*, London: Ashgate and Routledge.
11 Ioannou-Naoum, A. (2024). *Architecture Follows Climate: Traditional Architecture in Five Climate Zones*, Basel: Birkhäuser.
12 Barber, D.A. (2020). *Modern Architecture and Climate: Design before Air Conditioning*, Princeton and Oxford: Princeton University Press.
13 La Roche, Pablo (2024). *Carbon-Neutral Architectural Design*, 3rd Edition, London: Routledge.
14 Passe, U., and F. Battaglia (2015). *Designing Spaces for Matural Ventilation: An Architects Guide*, New York: Routledge.
15 Heschong, L. (2021). *Visual Delight in Architecture*, London: Routledge.
16 Roaf, S. (2021). Building Preparedness for Pandemics: Recommendations and Summary for Policy Makers, in *AMR and the Environment: A Global & One Health Security Issue*, Edited by G.F. Upham, Geneva: Published by the AMR Think Do Foundation, pp. 62–77, ISSN: 2673-9232. https://www.amrthinkdotank.org/book
17 https://cbe-berkeley.gitbook.io/thermal-comfort-tool/
18 https://comfort.arup.com/ACT_Tutorial.mp4
19 McCartney, K.J. and J.F. Nicol (2002). Developing an adaptive control algorithm for Europe, *Energy and Buildings*, 34(6), pp. 23–635. DOI: 10.1016/S0378-7788(02)00013-0
20 Lankford, H.V., and L.R. Fox (2021). The wind chill index, *Wilderness & Environmental Medicine*, 32(3), pp. 392–399. https://doi.org/10.1016/j.wem.2021.04.005
21 Royal Meteorological Society (2022). *Wind Chill*. https://www.rmets.org/metmatters/wind-chill. Image *by RicHard-59 – Own work, CC BY-SA 3.0*. https://commons.wikimedia.org/w/index.php?curid=18295345
22 ASHRAE ANSI (2020). Addendum to ANSI/ASHRAE Standard 55–2017: Thermal Environmental Conditions for Human Occupancy, published by ASHRAE, Atlanta, Georgia, ISSN 1041-2336
23 Dawe, M., P. Raftery, J. Woolley, et al. (2020). Comparison of mean radiant and air temperatures in mechanically-conditioned commercial buildings from over 200,000 field and laboratory measurements, *Energy and Buildings*, 206, p. 109582. DOI: 10.1016/j.enbuild.2019.109582
24 ASHRAE, ANSI (2013). *ASHRAE Standard 55: Thermal Environmental Conditions for Human Occupancy*, USA: American Society of Heating, Refrigerating and Air-Conditioning Engineers Inc.
25 Heinzerling, D., S. Schiavon, T. Webster and E. Arens (2013). Indoor environmental quality assessment models: A literature review and a proposed weighting and classification scheme, *Building and Environment*, 70, pp. 210–222. DOI: 10.1016/j.buildenv.2013.08.027
26 ASHRAE (2013). *2013 ASHRAE Handbook: Fundamentals*, USA: American Society of Heating, Refrigerating and Air-Conditioning Engineers Inc.
27 Dawe, M., P. Raftery, J. Woolley, et al. (2020). Comparison of mean radiant and air temperatures in mechanically-conditioned commercial buildings from over 200,000 field and laboratory measurements, *Energy and Buildings*, 206, p. 109582. DOI: 10.1016/j.enbuild.2019.109582
28 Heinzerling, D., S. Schiavon, T. Webster and E. Arens (2013). Indoor environmental quality assessment models: A literature review and a proposed weighting and classification scheme, *Building and Environment*, 70, pp. 210–222. DOI: 10.1016/j.buildenv.2013.08.027
29 Humphrey's Michael (1977). The Optimum Diameter for a Globe Thermometer for Use Indoors, *Annals of Occupational Hygiene*, 20(2), pp. 1355–1400.
30 Nicol, F., M. Humphreys and S. Roaf (2012). *Adaptive Thermal Comfort: Principals and Practice*, London: Routledge, ISBN: 978-0-415-69159-8
31 Kardas, A. (2023). Hot, rough and turbulent: What a city is like from an atmospheric scientist's point of view, in *Anthropocene: Towards Regenerative Architecture*, Edited by K. Kepinski and A. Krezlik, Warsaw: Published by the National Institute of Architecture and Urban Planning, pp. 225–235.
32 https://www.designingbuildings.co.uk/wiki/Operative_temperature
33 Booten, C., C. Christensen and J. Winkler (2014). *Energy Impacts of Oversized Residential Air Conditioners – Simulation Study of Retrofit Sequence Impacts*, Colorado: Published by National Renewable Energy Laboratory.
34 RIBA (2020). *Post Occupancy Evaluation: An Essential Tool to Improve the Built Environment*, London: Published by the Royal Institute of British Architects.
35 https://www.salford.ac.uk/our-facilities/energy-house-labs
36 Roaf, S. (2024). *Overhaul the Building Regulations: The Role of Research, Buildings and Cities: Commentary on the Challenges Ahead: Research Has a Role to Protect the Public Interest and Inhabitants*, Accesses on 16/01/2025 at: https://www.buildingsandcities.org/insights/commentaries/building-regulations-research.html

37 Rawal, R., T. Mehta, Y. Shukla and S. Roaf (2024). An investigation of the operational energy and carbon savings from practicing adaptive thermal comfort theory, *Proceedings of CATE 2024, Comfort at the Extremes*, Ahmadabad, India, ISBN: 978-93-83184-84-2.
38 Rijal, H., P. Tuohy, M. Humphreys, F. Nicol, et al. (2007). Using results from field surveys to predict the effect of open windows on thermal comfort and energy use in buildings, *Energy and Buildings*, 39(7), pp. 823–836. https://doi.org/10.1016/j.enbuild.2007.02.003
39 Rijal, H.B., P. Tuohy, M.A. Humphreys, J.F. Nicol, A. Samuel and J. Clarke (2007). Using results from field surveys to predict the effect of open windows on thermal comfort and energy use in buildings, *Energy and Buildings*, 39(7)(July), pp. 823–836. https://doi.org/10.1016/j.enbuild.2007.02.003
40 Schweiker, M., F. Haldi, M. Shukuya and D. Robinson (2012). Verification of stochastic models of window opening behaviour for residential buildings, *Journal of Building Performance Simulation*, 5, pp. 55–74. DOI: 10.1080/19401493.2011.567422
41 Szokolay, S. (2014). *Introduction to Architectural Science*, Routledge, ISBN: 978-0-415-824989

19

THERMAL LANDSCAPING OF BUILDINGS

Buildings as Thermal Ecosystems

Traditionally, people moved between different locations in a building, or even within a room, at different times of day or year in search of comfort. Rather than *form following function*, in fact, *function (staying comfortable) follows form*, which in a building shapes local micro-climates. Imagine a British sitting room with an open fire around which the family sits in winter. Opposite is the large bay window, open to the garden, sunny, and sat in during the shoulder seasons, but retreated from into room shade on the hottest summer days. Occupants seek out, engage with, or avoid local micro-climates according to their needs and opportunities. Breakfast rooms, winter gardens, patios, snugs, and verandas all evolved within the more complex thermal landscapes of houses to provide pleasant conditions for a particular time and season. Many simple rural homes used heat from the cattle sheds and hay stores below to heat living rooms above. Bedrooms in cold climates were placed over the kitchen to benefit from the rising heat from cooking and living areas, which were often co-located around the main fire in the house in the kitchen (Figure 19.1).

Thinking ecologically, heat in and around buildings over time moves within a dynamic landscape of energy flows from sources to sinks, along thermal pathways and between warm and cool reservoirs. Those flows and reservoirs can be optimised to reap the benefits of every molecule or kWh of available energy. Heat or coolth reservoirs can be sized to store useful amounts of external ambient energy from solar gain or winds, or internally generated heat from appliances or human activities for use when needed. Pathways between sources and sinks in a building can be fast conveyor belts of heat or cold via radiant or convective flows or slower conduits moving via conduction through most building materials, except most metals. Making thermal designs work may be easier for those who can imagine a building as a *thermal ecosystem,* but not everyone can. We all have different brains. Engineers tend to be more numerate, architects more visual. Everyone should be able to understand that by linking the very varied comfort needs of occupants to the forms, orientation and fabrics of a building through thermal ecosystem design modern passive designers can radically reduce both energy consumption in, and ghg emissions from that building. Vernacular builders did that as a matter of course, trained in skills passed on by generations workers before them.

DOI: 10.1201/9781315645070-25

FIGURE 19.1 The living and working area an early log cabin at Mount Tamborine Heritage Centre in the Queensland rainforest demonstrates the multi-purposing of energy sources by co-location of functions in a home with the high mass wood fired stove built to cook and keep the space warm overnight

Source: Susan Roaf.

Heatscape Imagining

How does one start to thermally landscape a building? Put aside the computer, forget comfort codes and standards and pick up a pencil and paper to start thinking about first about how that particular building fits into its thermal landscape. Where are the local thermal reservoirs located around a building at different seasons and times of day? What forces are acting on the site from wind, sun and rain? Most buildings have a more or less, hotter or colder side, except in the tropics where the sky is hot and the shaded building or earth beneath it is cooler. That is why traditionally tropical buildings wear great big shade hats for roofs. In hot areas prone to rainfall, buildings wear both hats and boots in the form of raised, often mud and stone or concrete plinths or a whole building may be lifted up on stilts above floodplains. Where prevailing winds are strong tight eaves help to keep rooves on. Noticing how building forms co-evolve with their weather and climates can extend the thermal design vocabulary and understanding of how comfort is achieved locally.

People who take time to measure heat and air moving around a building can often imagine it in other buildings. We can all turn on the thermal switch in our minds by making an effort to become sensitive to changes in temperatures and air movement as we move in and around buildings in our everyday lives. The thermal sense gives the designer an advanced tool with which to understand and enjoy how buildings work climatically in practice. Aleksandra Kardas expressed this *Second Site* understanding beautifully in her chapter on Urban climates in the 2023 book *Anthropocene*[1]*:* She asks readers to imagine the sun's rays falling on the ground in a city and bouncing of multiple surfaces until they are absorbed into the fabric of buildings or streets, so driving up temperatures and the urban heat island effect. Then to imagine the heat coming off every car, light, piece of equipment and air conditioning machine further adding to overheating cities. This is not difficult in places like Phoenix, Arizona named locally *Blacktopolis* because people have to live there surrounded by a sea of unshaded concrete and bitumen roads and parking lots that throb and simmer with heat in summer. She compares a city of tall buildings to a mountain range, with hills that cause incoming rivers of air to split at building faces, forcing some of the air down through the canyons to the street level below and some upwards. Downdraughts descend over different buildings onto the streets, then turn back on themselves to create sometimes dangerous unpredictable corner vortices and turbulent air blasts around pedestrians and buildings. Updraughts can create heat plumes above the city drawing cooler air from its surroundings at ground level to replace the air driven by the same forces that move the atmospheric winds shown in Figure 17.3. The windscape of a city can be a mosaic of these up and downdraughts. Cascading air flows down and around buildings, accelerate when squeezed through tight channels, and slow down in wider ones, rather like water bouncing off and navigating round rocks in rapids.

Steven Szololay compared hot still air to water in a bucket, only upside down, with the driving pressure instead of gravity in water, replaced by the uplifting power of buoyancy in warm air. The simple use of high-level windows in spaces can drain heat out of them like water flows out of a hole in the bottom of a bucket. Simple principles used for centuries in green houses and conservatories to keep spaces habitable for people and plants. The bigger the hole and the lower down in the bucket the faster the flow. It is not difficult to imagine whole thermal landscapes where heat sources drive up warm air into reservoirs, channels, rivers, and waterfalls of heat under ceilings and up staircases that can either be used to warm spaces or to expel heat out of walls or roof windows.

Every space has a dynamic thermal landscape that if understood can be used to enhance the feel of a spaces and also address other aspect of what makes us comfortable, like feeling secure in our homes. Note the high and low-level small vents beside the French doors in Figure 19.2 that enable either cooler air (lower) or warmer air (higher) to flow in or out of the room, well they small enough to be left open to securely ventilate the room when the home is unoccupied

Responding to Radiant Environments

Excessive radiant heat and cold can be very uncomfortable, and humans, like solar cats,[2] continually respond to them. They move a chair toward the fire if they want to be warmer, or away from it to be cooler. They shut the curtains to keep the cold out or turn on a radiator to be warmer. Radiant heat sources are not just point sources of heat but can also heat up particles in the air around them, causing the air that contains them to rise, forming a reservoir of hot air under the ceiling and turning the ceiling into a low-grade radiator, similar to that shown in Figure 18.2. A cat always moves to the warmest spot in the room/house to snooze in, but people

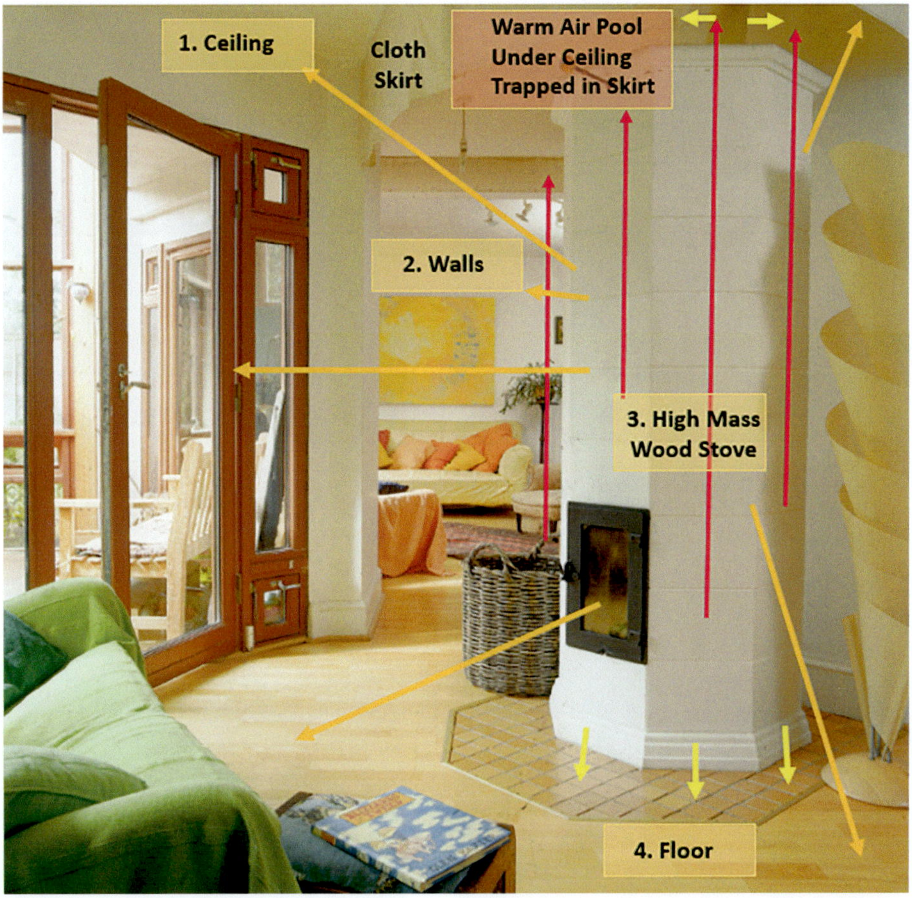

FIGURE 19.2 Optimising the usefulness of energy from a wood stove means capturing all available heat flows from the high-mass structure from convection (red arrows), radiation (orange arrows), and conduction (yellow arrows) and storing heat, here in concrete reservoirs (1–4) in walls, floor, ceiling, and the stove itself and a pool of warm air under the ceiling trapped by a cloth skirt suspended from the ceiling around the wood-burning stove.

Source: Susan Roaf.

who are also attuned to the air flows and radiant environments around themselves have the extra superpower of being able to turn on and off and manage the heat and cold flows around a building. Engineers often protest that you cannot leave the management of micro-climates to occupants, like opening a window in a bathroom after a hot shower. In reality, they do not want people to manage indoor airflow in particular because it interferes with their controlling narrative. In 1950s America, one excuse not to open a window was that 'it may let the rain in when you are not looking'.

Radiant landscapes, as in Figure 19.3, are easier to 'see' with thermal imaging cameras than air flows that respond to the slightest butterfly's wing of movement affected by draughts, open doors and windows, furniture, and people moving through spaces. Anyone who has mused on

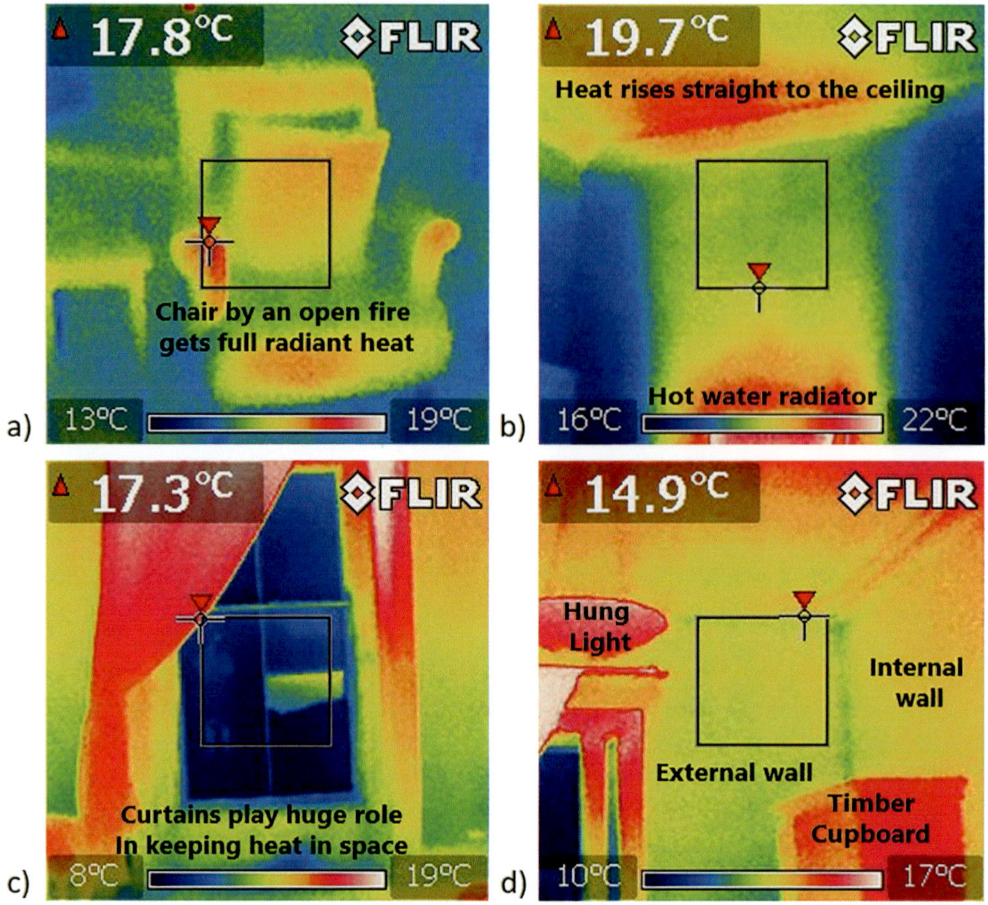

FIGURE 19.3 Thermal images in an 1867 Edinburgh flat with 4.5 m ceilings, showing how (a) a chair by a coal fire catches the radiant heat; (b) heat is lost directly upward to gather under the ceiling above a hot radiator; (c) how critical curtains are in conserving heat in a space; and (d) that the external wall is colder than the internal wall and heat collects under the ceiling.

Source: Roaf.

the patterns of visible water eddies in a hot bath will get some ideas of the complexity of the flows at play within the thermal landscapes of a room.

Horizontal Thermal Landscapes

In Chapter 9 it was clear that people like micro-climates with fairly stead temperatures, finding them typically more comfortable than those that vary widely. Two forces tend to change temperatures indoors quickly: solar gain and changes in wind direction and strength. Figure 19.4 shows the diurnal temperatures in a house in Dammam, Saudi Arabia, where the north-east-facing living room is buffered from the solar gain and varied in temperature only a couple of

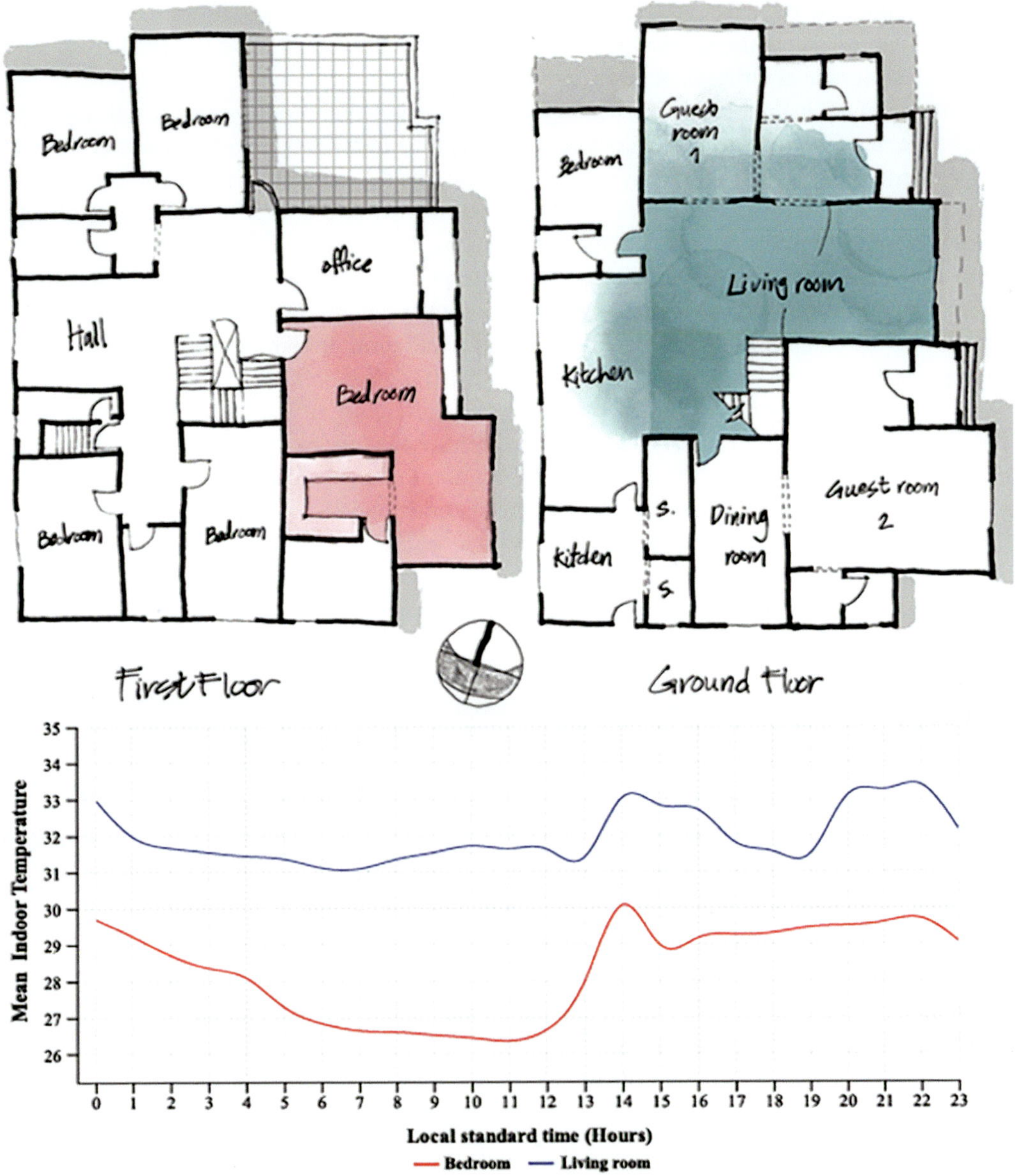

FIGURE 19.4 House #6 from Alshaikh's study of houses in Dammam demonstrates the impact of large open-plan spaces on heat across the room, and also heating rooms above connected by halls and stairs show in these plans and graphs of average indoor temperatures of the bedroom and the living room in the hot season within a 24-hour time series.

In very hot climates, like Saudi Arabia, the difference in the temperatures experienced around a simple home can be large, making some rooms adequately comfortable and others impossible to remain in at some times of the day and year.

Source: Abdul Rahman Al Shaikh.

degrees centigrade over the day. The west-facing bedroom has two windows to the south and west that get full sun at the hottest times of the day causing temperatures to fluctuate over seven degrees Centigrade creating intense discomfort.[3]

Rules of thumb in the UK around heating in its temperate climate advocated that living rooms of a house that face south can reduce winter heating bills by a third. Another was that for every degree Centigrade extra that one has to heat a building to, or from outdoor temperatures requires an extra 10% of energy. Such assumptions rely on 'standard designs'. Buildings have changed since those rules were developed, not least for housing. Living room doors into a garden became patio doors in the 1970s and within decades had morphed into whole walls of glass with Bi-Fold doors. These not only can cause significant overheating, particularly in the shoulder seasons, but do not provide small portals for safe exchange of ventilation or cooling air as do those shown in Figure 19.2.

When sketching out early plans for the thermal landscaping of a design and features within a building a good place to start is to draw coloured sketches for each season of where the main reservoirs of heat and coolth will be and the different energy pathways, radiant, convective, conductive and reflective that connect them to each other and to occupants to enhance comfort. Time and place where those reservoirs of flows may cause discomfort should be identified and dealt with early on in the design.

Vertical Thermal Landscapes

In the 2003 European Heatwave come 15,000 people died in France, many poor, ill, elderly and isolated. Many were on the top floors of buildings under uninsulated roofs. Heat built up within the building and piled up in a reservoir of air on the top floor where many died. Figure 19.5 shows a section through an idealised modern house with a high overheating potential. The light-weight timber construction offers no thermal storage opportunities that might help attenuate the

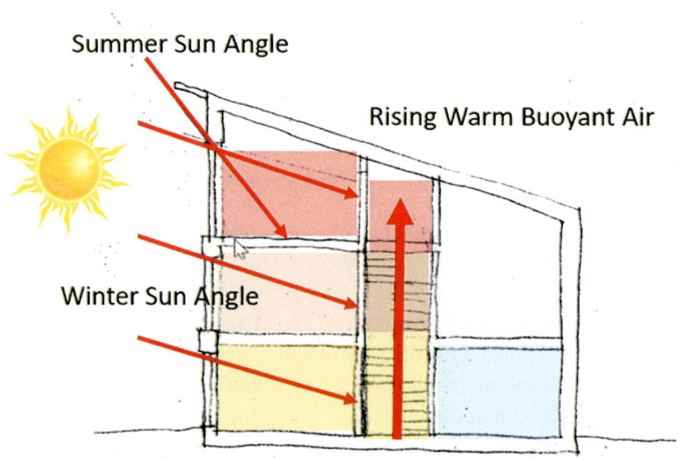

FIGURE 19.5 Sketching out a simple overview of the thermal landscaping of a building in different seasons, at an early stage, may provide pause for thought and prevent serious problems with overheating indoors once built

Source: Susan Roaf.

heat peaks and lows. Heat builds up in south-facing rooms and inevitably rises to the top floors, where they pool, being trapped in the moisture and vapour barriers within the super-insulated side and back walls. Those who rent often suffer the worst being typically constrained from making changes to their buildings.

It is not only homeowners who suffer from modern designs that expose occupants often extreme discomfort from solar gain. In the 2000s, a new trench of privately financed public schools was built in Britain, many of which suffered badly from overheating. One girl famously fainted from heat in a top-floor classroom during her final exams. Newer schools are more likely to overheat than their older counterparts, because they are more highly glazed, have fewer opening windows, more insulation and lower ceilings. School building stock retrofits form a key part of UK's commitment to net-zero carbon target by 2050 but the retrofit of school buildings for winter performance with insulation has been shown to affect the cognitive performance of children in summer.[4] Even with the benefits of increased air flow by opening windows little emphasis in our warming climate is not given to protecting classrooms from overheating. The systematic shading of school windows has not even considered in the UK. Shades and shutters seem to stop at Calais! Lobbyists push industrial agendas claiming that air conditioning will be required, where shading is not even considered first. Already struggling school budgets will have to sacrifice teaching staff to pay for what may be unnecessary mechanical systems. Schools and colleges can be effectively reorganised to take into account their thermal landscapes by moving people around inside them. Examination rooms taken off top floors, and year-round administrative staff can be put the cooler sides of buildings to avoid overheating. Gardens planted and watered on cool sides of buildings can lower temperatures significantly to enhance natural ventilation indoors and on sunny sides to reduce heat impact from reflected heat off hard surfaces.

Where poor thermal planning can be actually life-threatening is in care homes for the elderly or infirm who may be unable to move from their beds or incapable of opening windows or asking for help. A recent study showed that in London care homes indoor temperatures exceeded 30°C (86°F) in newer care homes with older buildings less prone to overheating. Night-time ventilation of spaces was found to be the single most effective passive solution for buildings in the current climate, while modelling showed that a combination of night ventilation, external shading and high-albedo external walls was the most effective packaged solution for cooling. By the 2080s, air conditioning was modelled to be the most effective solution for reducing temperature, but also had the highest cooling load, highlighting the importance of passive solutions. Nowhere was the thermal landscaping of the homes considered, moving the vulnerable elderly from the hot to the cooler side of the buildings or from the top to lower floors to save money and improve comfort.[5]

Warm and Cool Cores and Floors

Most modern developers prefer lightweight timber construction as it is easier and faster to build, particularly with off-site construction, thus increasing building profits. Governments encourage their partners in economic growth, the developers, by confusing the calculations of embodied energy with total, whole life emissions accounting which actually gives an idea of what the building will use over time including embodied and operational energy. A well-designed heavy weight building will earn back in a couple of decades the extra energy costs embodied in its construction with reduced energy running costs over time as shown in Chapter 16.

Floors constructions matter. Given the importance of getting some thermal storage into buildings its location can be rationalised into where it may provide most value for money. Even some developers of pre-Fab housing will put mass into ground floors, insulated beneath. In winter they can store solar gain to lower heating bills, and if well-shaded, will not cause too much overheating in summer. In the tropics raised ground floors can benefit from the inter-seasonal storage of coolth in the ground beneath buildings. In the Mongolian yurt, we saw heat being stored in the earth beneath the tent as was the case in many vernacular buildings. Intermediate floors can also provide varying degrees of mass and in turn thermal stability, while still enabling external walls to be of timber.

Warm cores to buildings also work well. Historically, these would be built in as heavy chimneys, in which some of the heat from fires in kitchens, living, bed, and boiler rooms would be stored, providing a very pleasant warm core to a room of people around which people were attracted to sit not just in front of an open fire but beside a warm chimney breast. In modern homes, the clever integration of heat coolth flows up and down through the mass of a warm core can make a huge difference to the comfort felt indoors. Take a central staircase that can be high mass and have hot water up and down pipes in it, also channel the heat in a thermal stack upwards to be expelled via a Velux or a dormer window above the stair or in the ceiling of an adjacent room. In the height of summer, roof and upper floor windows can be opened to not only throw out heat but also channel cold air inversions down through the mass to create a cool core indoors, with those windows closed as the day outside heats up. Traditionally, the hot heart of a building in many cold climates was the heavy kitchen range, like the old Russian stoves in Europe or the AGA kitchen ranges in Britain.

In Sweden in 1767, architect Carl Johan Cronstedt invented the *kakelugn* (kakkelovn), a high-mass stove similar in principle to the one in Figure 19.2. It was constructed of bricks clad with tiles, with a small, enclosed firebox with several flues running through the bricks in a small vertical hypocaust system. It was dramatically more efficient at retaining heat than other wood-burning stoves and transformed the way people were able to live comfortably in their homes during the winter months. It provided a steady, safe heat that lasted long after the fire in it went out and provided a very successful warm core for rooms and buildings, being built into the heavy fabric of adjacent walls, into which its heat was also conducted, warming, in turn, the rooms behind.[6]

Thermal Buffer Spaces and Ransom Strips

Thermal buffer spaces provide an intermediate zone between areas of extreme heat outdoors and occupied rooms indoors. Buildings in extremely hot regions of the Middle East and Asia include whole floors that are unoccupied in the hottest weather above the summer living quarters, as can be seen in Figure 16.4 of a large house in North India. Summer living rooms facing north over courtyards in Iranian desert houses are not only flanked by corridors with side doors to enable cross-ventilation from the courtyard doors, but also the corridors are flanked by further buffer rooms. The south-facing rear wall of rooms also have buffer spaces for storage or ancillary functions to provide multiple buffer spaces in addition to the complex roof constructions like that shown in Figure 19.6.

In old houses in Baghdad, the top one or two floors are used as ransom strips to separate the summer heat from the building occupants. In Mediterranean architecture, breezy loggias on roofs protect the spaces below, and it is painful to say, but in many hot areas on larger modern

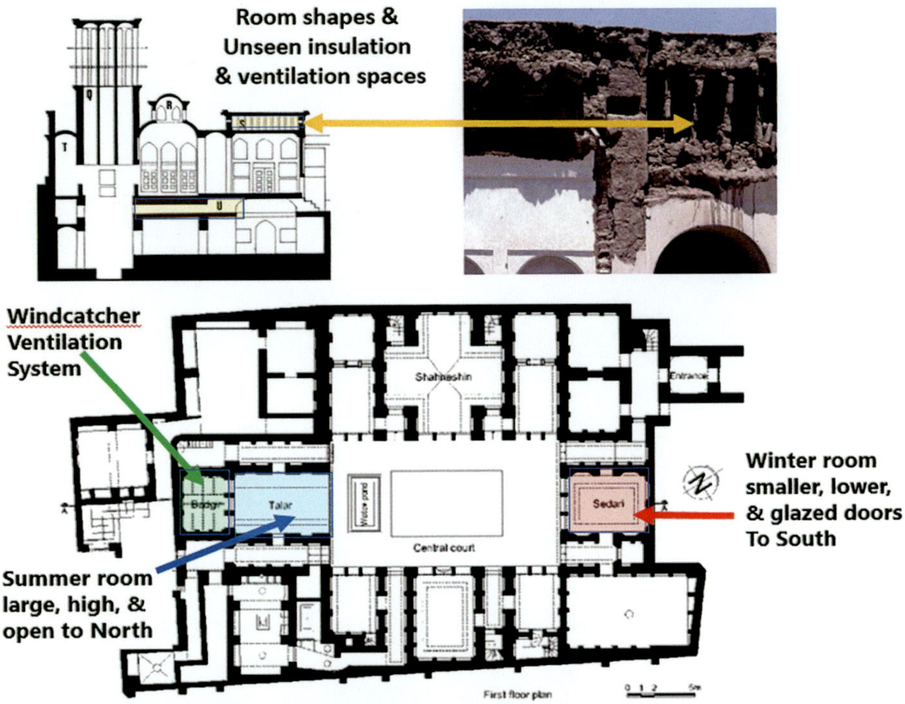

FIGURE 19.6 A large house in Yazd, Iran. A key feature of the design for house involves the use of buffer spaces around occupied summer rooms of the house and isolating the radiant surfaces of the spaces using twin wall roofs above them.

Source: Susan Roaf.

homes in African and Asia, servant quarters are placed on the hot roofs to provide buffer spaces for the wealthy owners below. In other vernacular building types, a raised and ventilated roof above ceilings also form a thermal buffer for the dwellings below.

In cold climates, critical buffer zones between outside and in are provided by front and back lobbies, porches, vestibules, halls, or conservatories. These provide airlocks in which some of the cold outdoor air can be shed in the form of coats or shoes and left in the space. The air in these transition spaces will reflect both indoor and outdoor temperatures and in extreme cold provide pre-conditioned air for internal windows which in some regions were built in as hopper window top lights above the internal door of a porch. A very easy way to slash heating bills and increase comfort indoors is to put a porch onto a front door that previously opened directly to the outside, not least because with solid floors it can also provide a thermal break for the invisible heat or cold that travels through the flood slab from out it inside the building.

Cost and Energy Benefits of Managing Thermal Landscapes

The huge operational savings available by managing thermal landscapes of buildings is clear in the example of the Porta Fira Hotel in Barcelona designed by Toyo Ito and b720 architects in Figure 19.7. Completed in 2010, this award-winning skyscraper is 113 m tall, has 345

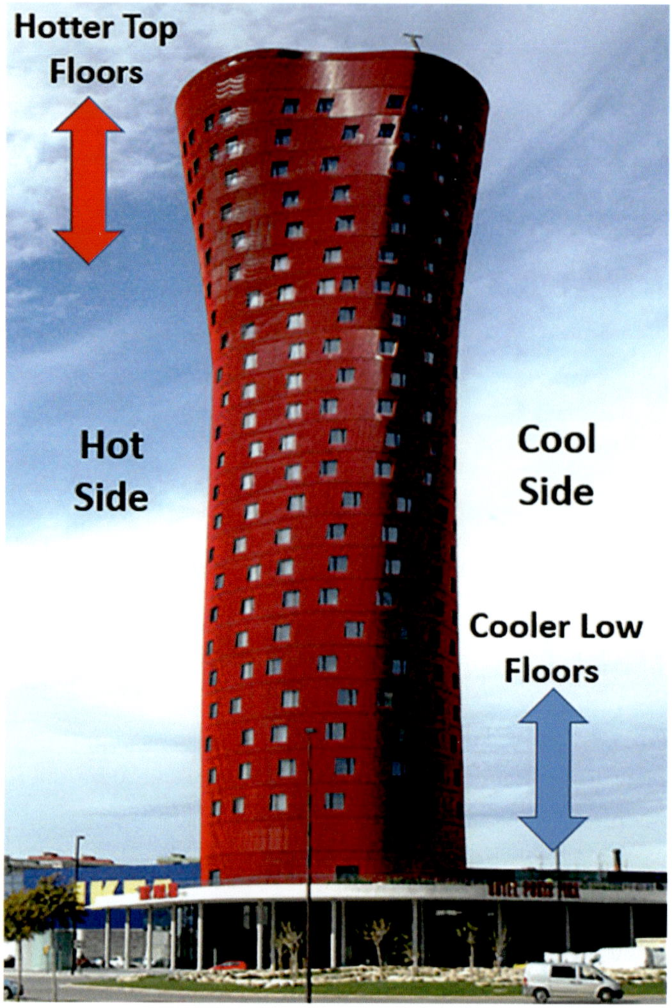

FIGURE 19.7 The Porta Fira hotel in Barcelona uses flexible provision of air conditioning in hotel rooms combined with natural ventilation and whole building shading to create an award-winning low-energy and comfortable hotel.

Source: Susan Roaf.

rooms on 24 floors each supplied with its own indoor VRV air conditioner, located externally on the roof of the hotel. Each room has an opening window which, when opened, switches off the AC. The entire building is sheathed in an exoskeleton of a red bamboo shade screen fixed some 10 cm away from the outer wall of the hotel. The use of heating and cooling systems is flexible across all floors, with heat recovery and dispersal systems that enable temperature shifting and shaving across floors and vertically up through the building. Hotel guests can control their temperatures in their own rooms, or they can be fine-tuned centrally. Real energy savings are made by careful matching of different room functions to climatically appropriate local micro-climates at different times of the day and seasons. Patterns of occupation also provide savings as in winter guests can be located in warmer, sun-facing rooms, and in summer, cooler areas of the hotel are occupied first. Seasonal efficiency is marked by exploiting the thermal landscapes of the hotel with intelligent people and equipment management using flexibly controls.

Climate Refuges

Eric Kleinberg wrote perhaps the best book ever on the subject of a heatwave, the 1995 Chicago heatwave in his book titled *Heatwave*. In his subsequent book on how to build a more equal and united society he puts his finger on the numb of the matter. Inequality drives division and division drives chaos.[7] There is no gated community in the world in which people will remain safe once the mobs become hungry, angry and hot. How does one build equality? Access to the infra structure of a decent life helps, community libraries, meeting places, sports facilities and education helps. How does one stop people from getting hot and bothered? Give them climate refuges, places they can feel same in extreme heat or cold in their own homes, workplaces, and the cities and settlements they live in.

These spaces have a range of different names: *cool retreats* in residential buildings in the Mediterranean regions.[8] *Climate refuges* is a term favoured by many, particularly for urban centres.[9] Geelong Council in Victoria, Australia has pioneered the placing of *climate–safe* rooms in homes to keep vulnerable home owners safe in the city during heatwaves and hot weather. Results have been very encouraging showing improvements in physical and mental health and quality of life for those in the refurbished homes.[10] Buffalo City, New York State is amongst many US cities already lining up to be designated as Climate Havens by scientists with a view to attracting high net worth residents to enjoy their *climate-safe* status and warm/cool amenities. A recent study has highlighted that in cities climate refuges are now critical for urban infrastructures to support adaptation to, and survival in extreme weather events. Cities open climate shelters spaces like parks, sports fields, libraries, and civic centres for citizens to take refuge in during episodes of extreme temperatures.[11] Ironically climate-safe spaces closely overlap with those spaces that Kleinberg promote for their importance in buildings socially safer societies, spaces that money hungry local Governments appear happy to dispose of for the right or wrong price.

Climate Safe Design

It only takes a pencil and a piece of paper, and a good deal of careful thinking to build climate and comfort resilience into a design. In Chapter 14 we learnt that it is not which tropical island you choose to live on, but where on that island you build. Build a climate nightmare building in a climate haven city and you will regret it.

It is crucially important that from now on buildings are thermally well-designed to both be capable of exploiting all the visible and invisible energies in and around them and to keep occupants thermally safe and comfortable during the more extreme weathers and climates of the future. Start on *day 1* of the design to create the dynamic thermal narratives for a design, as was done by Roaf for the Oxford Ecohouse shown in Figure 19.8, because in the long run they will play a key role in determining the future success of the building and the mental, physical, and emotional comfort of its occupants.

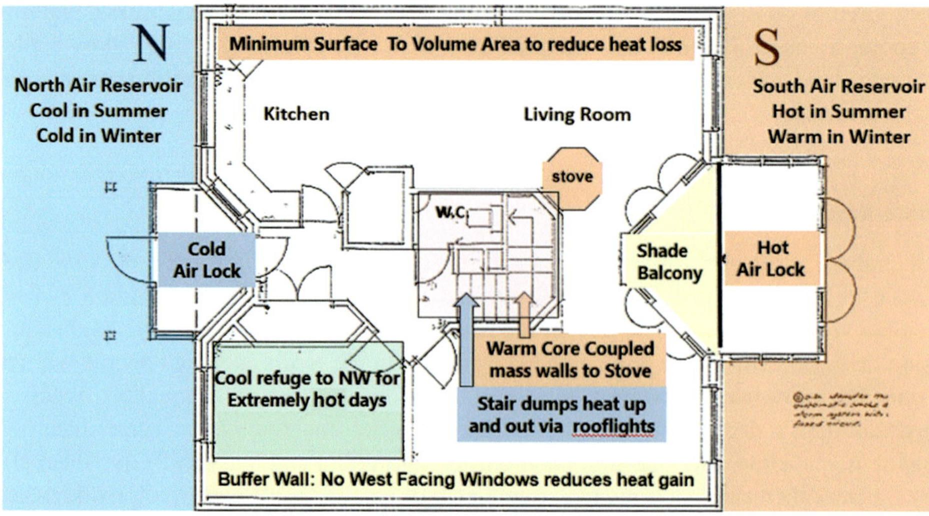

FIGURE 19.8 Early, simple heatscape imagining of the basic thermal design strategies of the Oxford Ecohouse.

Source: Susan Roaf.

Notes

1 Kardas, A. (2023). Hot, rough and turbulent: What a city is like from an atmospheric scientist's point of view, in *Anthropocene: Towards Regenerative Architecture*, Edited by K. Kepinski and A. Krezlik, Warsaw: Published by the national institute of Architecture and Urban Planning, pp. 225–235.

2 Augustyn, J. (1979). *The Solar Cat Book*, Berkeley, CA: Ten Speed Press, ISBN: 0898150183

3 Al-Shaikh, A. (2016). *Design Principles for Thermally Comfortable and Low Energy Homes in the Extreme Hot-Humid Climatic Gulf Region, with Reference to Dammam*, Saudi Arabia, PhD Heriot Watt University, Edinburgh, ISNI: 0000 0004 6347 5564. https://ethos.bl.uk/OrderDetails.do?uin=uk.bl.ethos.712727

4 Dong, J., Y. Schwartz, I. Korolija and D. Mumovic (2024). Unintended consequences of English school stock energy-efficient retrofit on cognitive performance of children under climate change, *Building and Environment*, 249, p. 111107. https://doi.org/10.1016/j.buildenv.2023.111107

5 Gupta, R., A. Howard, M. Davies, et al. (2021). Monitoring and modelling the risk of summertime overheating and passive solutions to avoid active cooling in London care homes, *Energy and Buildings*, 252, p. 111418. DOI: 10.1016/j.enbuild.2021.111418

6 Scherman, S. (2010). *The Swedish Country House*, London: Thames and Hudson, ISBN: 978-0-50-051530-3

7 Klinenberg, E. (2018). *Palaces for People: How to Build a More Equal Society*, London: The Bodley Head.

8 Monge-Barrio, A. and A.S.-O. Gutierrez (2018). *Passive Energy Strategies for Mediterranean Residential Buildings: Facing the Challenges of Climate Change and Vulnerable Populations*, Springer International Publishing.

9 Amorim-Maia, A., I. Anguelovski, J. Connolly and E. Chu (2023). Seeking refuge? The potential of urban climate shelters to address intersecting vulnerabilities, *Landscape and Urban Planning*, 238, p. 104836. DOI: 10.1016/j.landurbplan.2023.104836

10 Geelong Sustainability (2023). *Climate Safe Rooms Findings and Final Report*, Published by Geelong Sustainability, 53 pages, Accessed on 10/06/24 at: https://www.geelongsustainability.org.au/wp-content/uploads/2023/11/Climate-Safe-Rooms-Final-Report-Sep-2023.pdf

11 Marandi, A. and K.L. Main (2021). Vulnerable city, recipient city, or climate destination? Towards a typology of domestic climate migration impacts in US cities, *Journal of Environmental Studies Science*, 11(3), pp. 465–480. DOI: 10.1007/s13412-021-00712-2

Designing for a Hotter Climate

20

RECONNECTING DESIGNERS TO CLIMATES

Introduction

Architects have been largely forced to hand over responsibility for the thermal performance of their larger buildings to HVAC engineers who write most of the energy codes and are paid according to how much HVAC equipment they put into buildings. HVAC engineers are seldom taught about natural ventilation, sizing thermal mass or shading strategies to minimise overheating. The tools they use to size systems are typically blunt and inflexibly applied. HVAC engineers have no financial incentives under current contractor contracts to maximise the benefits of natural energy systems to reduce the need for mechanical systems and are typically not trained to do so in their student courses. Both architects and engineers urgently need to reconnect with the climatic challenges of building design so that with the help of 'Fit for Purpose' design appointment contracts both can work together to create genuinely low energy, at least partly passive buildings. Their training in architecturally engineered design needs to start from Year One of their courses so that good climatic design becomes part of the DNA of the next generation of building professionals.

Understanding the Scale of the Climatic Design Challenge

Working in any climate designers need to get a 'feel for' the character of the local outdoor conditions. This can be simply done by looking up the annual climate data for a location to see maximum and minimum temperatures, relatively humidity and sunshine hours over a year. A great tool to explore climate data sets for over 27,500 locations worldwide is the CBE Clima Tool that enables access to a range of climate variables.[1] Our advice is, before heading off down a rabbit hole of black box simulations tools or clever values for compounded factors, just sit back and look at the simple climate data – get a feel for it and where building performance sits in it. Baruch Givoni, a pioneering comfort researcher who particularly in hot dry regions showed that a well-designed *bio-climatic* building, with adequate and appropriately designed thermal mass and insulation, temperatures indoors can roughly equal the mean of the outdoor maximum and minimum temperatures.[2] So it is relatively clear what the climate does and how a really good

DOI: 10.1201/9781315645070-27

building may perform in it, but the key ingredient is missing. It is important to put people into the equation. People in field studies adapt to local conditions with higher comfort temperatures in hotter climates and vice versa. Roughly acceptable and/or comfortable indoor temperatures loosely equate to the adaptive comfort temperature for a location which can be readily accessed online from the CBE Adaptive Tool developed at the Centre of the Built Environment at Berkeley.[3] Design students from Year One of their studies should regularly be exploring the difference between indoor comfort temperatures and those outdoors in different climates. This information indicates the extent to which a building will need heating in winter and cooling in summer and can be simply put together using a Nicol graph.

The Nicol Graph

Fergus Nicol developed this very simple tool while helping designers visualise the bio-climatic design challenge of designing for different regions in Pakistan. The Nicol graph can be put together using climate data on a Nicol graph generator.[4] It reduces the bio-climatic challenge to three data lines:

1. *People*. The adaptive thermal comfort line which links comfort to the climate, can got from an online tool or calculated using an algorithm like:[5]

$$T_n = 13.8 + 0.53(\pm 0.02)T_{o,}$$

Where T_n = neutral temperature and T_o = outside temperature, and the relationship is explained in detail in volumes 1 and 2 of this trilogy on adaptive thermal comfort.

2. *Climate*. Shown as mean maximum and mean minimum temperatures for each month at the relevant site. Other climate features can be shown on the same graph, like sunshine hours, relative humidity, and rainfall. Monthly extreme maximum and minimum temperature can be added to give an idea of how extreme the local weather can get and as an indication of possible future temperature means.
3. *Building*. The mean outdoor temperature, the median line between the mean maximum and mean minimum lines, gives an indication of what temperatures might be achieved indoors in a well-designed building with thermal storage.

The original estimated comfort temperature line in the Nicol graph is also historic, built on data from 1936 to 1978. People around the world now have become used to air conditioning, which may change their programmed responses to temperature,[6] but the Nicol graph only gives a 'ballpark' feel for local heating and cooling requirements, as shown in Figure 1.1. The Nicol graph was originally built from 1990s data, and the comfort readings in Figure 20.1 for Yazd, Iran, using that data, are indeed different from those in Figure 17.4 with the current Yazdi climate data.

The online Nicol graph for any location in the world uses manually input or automatically generated data in its simplest form for early learners at www.nicolgraph.com that includes relative humidity as well as the temperature and mean daily sunshine hours. It is also available as the Nicol graph generator implemented for the Rhino Grasshopper environment by Abraham Yezioro.[7]

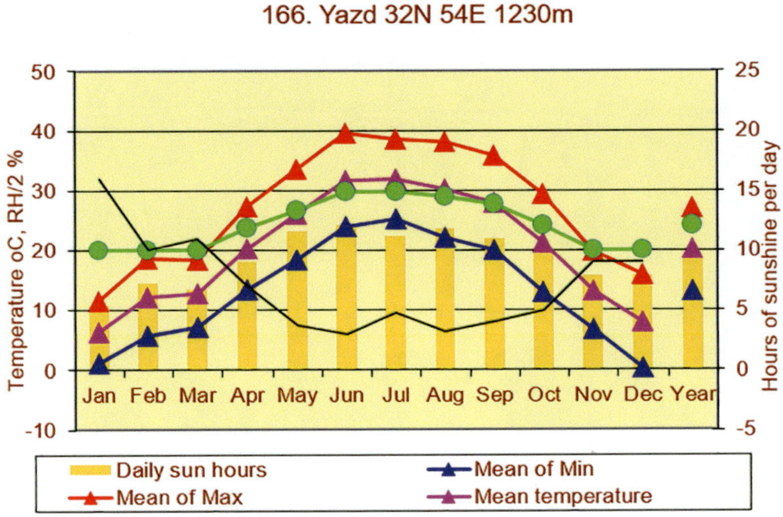

FIGURE 20.1 Nicol graph for the pre-1990 climate data for Yazd in Iran.

What can a Nicol graph in Figure 20.1 show us?

1. The monthly mean maximum temperature recorded for summer is 39°C. This is a temperature, above the TNZ, which may be comfortable with a good breeze.
2. The monthly mean minimum temperature recorded over the year is around 13°C.
3. The mean comfort temperatures are derived from the outdoor temperature and are only accurate within ± 2–3°C.
4. Annually, the mean temperature maximum is 32°C, and its minimum is 6°C, with its upper level comfortable in summer and the lower needing warmth to be added in clothing or heating.
5. The maximum adaptive comfort temperature is 30°C, compatible with the TNZ with a good breeze.
6. The sunshine hours over the year range from 5 to 12 hours per day, which shows that passive solar gain can provide free heating, even in winter, but possible not the extra 15°C needed in January and February, if indoor winter temperatures of 20°C are maintained. In reality, they are not, and people in Iran use table heaters (Koursi) beneath quilted covers in winter, like those used from Portugal to Japan (Figure 20.2).
7. The maximum diurnal range occurs in summer, when there is a 15°C difference between the outdoor mean minimum and mean maximum temperatures. In a building with thermal mass, this can translate into half of that being either free heating or cooling. Thus, the maximum temperatures were reduced to around comfort temperatures in traditional homes in mid-summer with free cooling.

So much knowledge gained already from this simple graph. Note that when scoping out the opportunities for a passive building, three design premises can be useful:

1. The adaptive thermal comfort equation line, shown in black, indicates the scale of the challenge to run the building comfortably on local natural energy sources, from the earth, the air, or the sun.

FIGURE 20.2 A kotatsu, an under-table heater in Japan, of the sort used in winter traditionally from Asia to Iran (Koursi) and Spain (Brasero).

Source: Photo by Roaf.

2. Field studies (Chapter 6) show clearly that from c. 20°C upwards to 33–35°C, opening windows are used to enhance comfort, and local fans can provide comfort at even higher temperatures (Chapter 4), so one can identify when natural ventilation can be used – in Beer Sheva, for around eight months of the year.
3. The adaptive comfort line indicates that for around three to four months of the year, some heating may be necessary, but the graph also shows that there is ample solar gain across the year, so a building can be precisely designed to run on passive solar principles during winter – taking care to avoid excess gain, even during the warmer months.

Designing for a More Extreme Future

As global temperatures are rising so fast historic data is already obsolete and buildings today should be designed for a very different future. Figure 20.3 was built from pre-1990 data for Beer Sheva. Figure 20.4 shows the Nicol graph using data from 2007 to 2021. In those thirty years

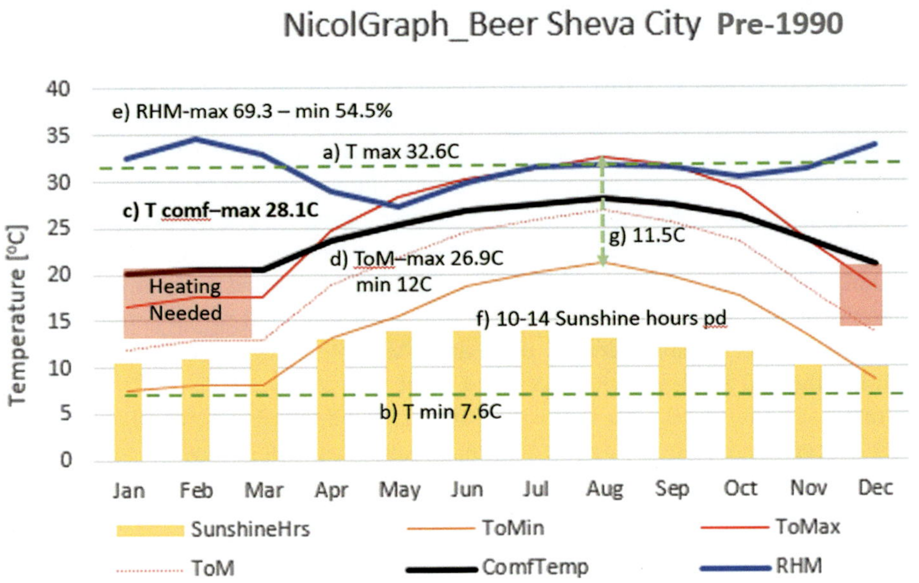

FIGURE 20.3 Nicol graph of the climate in Beer Sheva City in the Negev desert using data up to 1990. Israel Oxford showing the mean maximum and mean minimum temperatures for each month of the year.

Source: Graph by Abraham Yezioro.

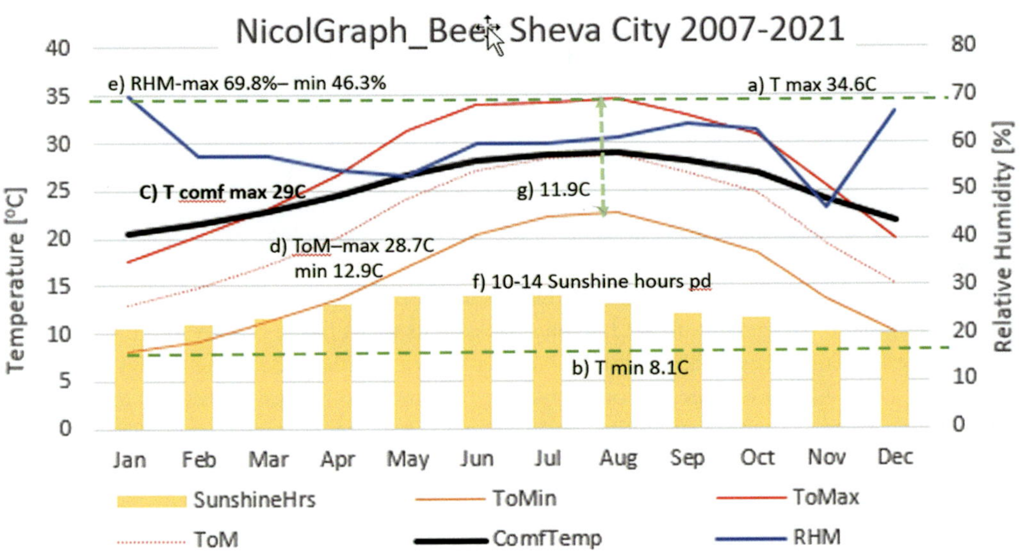

FIGURE 20.4 Nicol graph of the climate in Beer Sheva City in the Negev desert using data from 2007 to 2021, showing the scale of the changes that have occurred in the climate of the city in the past 90 years.

Source: Graph by Abraham Yezioro.

the change has been alarming. It shows us that the mean maximum temperature has increased by 2°C taking it up around 35°C, an upper level for comfort in buildings. This climate change may reduce the proportion of the year that natural ventilation can be used over a year. Then again the diurnal range of temperatures in summer has increased by up to 1°C, perhaps offering more scope for nighttime cooling. The mean internal maximum temperature in a passive building in Beer Sheva will be some 2°C warmer in summer and it is slightly warmer indoors in winter with the mean minimum temperature is winter up 1°C, so less heating may be needed then.

In both the preceding data sets, the humidity poses no real problem for the comfort of occupants. Humidity only really becomes a key comfort variable around and above upper skin temperatures, as dealt with in Chapter 3. Doing this simple exercise will reveal some interesting insights, for instance, in tropical climates there are often a few degrees of heat or coolth to be gained using thermal storage because day and night time temperatures can be four or five degrees different in some areas. In a future, as buildings need to relay increasingly to survive on their passive performance every degree in temperatures that shifts in indoor conditions can make a real different to the comfort experienced by occupants.

In the Nicol graph generator online at www.nicolgraph.com, one can not only enter the historic data that informs current climate figures but also seek out future data sets to explore the future conditions they should be designing for today for 2050, 2080, or even 2100.

Sun-Paths and Heliodons

Solar gain is free energy source availably to designer for heating buildings in winter. It can also drive overheating of buildings if poorly controlled over a year. There are a range of online sun path tools to use in the early stages of design to maximise the benefits and minimise the problems associated with solar gain. Favourites are Andrew Marsh's 2D[8] and 3D[9] Sun-Path charts that are for any latitude and longitude of a particular site. The 3D models show how adjacent buildings block or allow solar penetration onto the site for any hour or day of the year showing how high in in the sky (altitude) the sun will be and the angle from the horizon it will hit the building at (azimuth). Many simulation models will include solar design tools, but playing with it independently before the building is sketched may seed design ideas on how to optimise the solar benefits of the site.

A first step in passive design is to develop a feel for how a building will respond to the local climate. The best way to do this early on in the design is to use a heliodon to experiment with different building forms and opening orientations and sizes. Figure 20.6 Shows one being used for teaching by its' designer Norbert Lechner. Heliodons use a frame of movable lights above a flat surface to mimic the sun at all times of day or year for all latitudes. The model is placed at the centre of the heliodon table or floor, enabling the model to be easily tested to assess different building shapes, shading strategies, and impacts. They can be highly engineered products or made simply from cardboard and light bulbs for use in teaching studios and professional design offices.[10] The visualisation of exactly how the sun hits a building using an easily adapted model liberates designers to spend time thinking about how to welcome the sun into the design or shade against it, stimulating novel solutions and new thinking that is hampered by the lack of physical access to easily and repeatedly changing the model in simulations.[11]

Ideally, every school of architecture and building engineering and every office, where possible, should have a heliodon, however rudimentary, for designers to play with and get a 'feel' for how the sun moves over a site and the buildings on and around it. Playing with form and sun, connecting people to proto-buildings in their climate context at the very early design stages, is

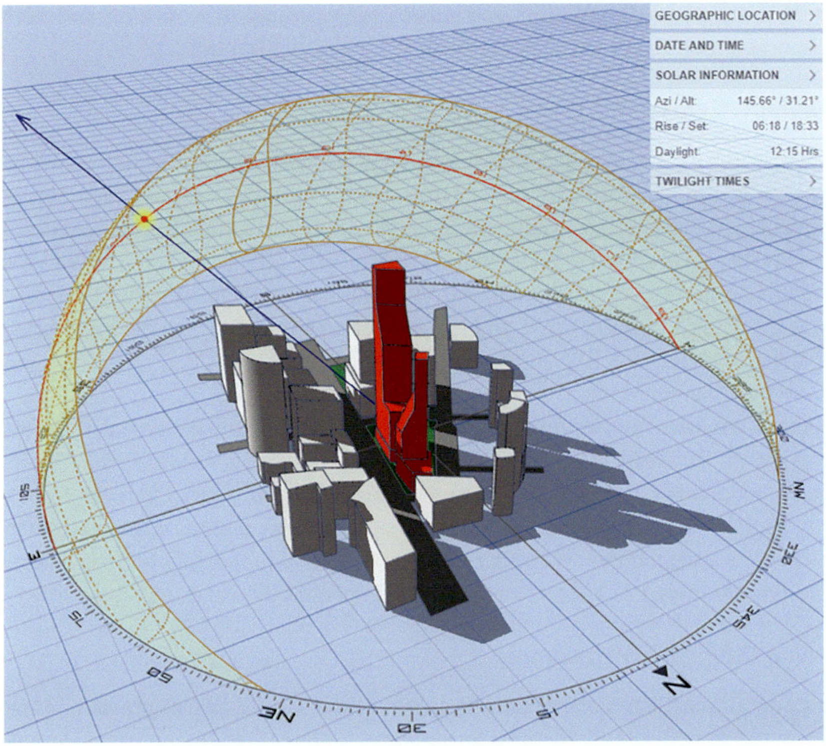

FIGURE 20.5 A three-dimensional Sun-Path diagram for a site from Andrew Marsh's website.

Source: Andrew Marsh.

FIGURE 20.6 Norbert Lechner demonstrating the ease with which a heliodon can be used and the benefits of doing so at Oxford Brookes University.

Source: Norbert Lechner.

so important to the ultimate potential and success of a project. It is so easy and such fun to do with a heliodon, much faster than testing form iterations in simulations.

The LT Method – Connecting Space, Heat, and Light

In the 1980s, commercial and residential buildings grew in size in the UK, making the challenge harder to understand and calculate energy use for lighting, heating, and cooling, and their interdependence. To ensure that designers understood the potentials for using natural energy to service spaces, the LT method was developed in the early 1990s. The aim was to make clearly understandable to architects areas of a building that can operate in natural energy modes by explicitly and visibly mapping areas and micro-climates that can exploit local heat, light, and wind and those that require full servicing, as shown in Figure 20.7. The LT method helped designers develop a generic understanding of how the main energy sources, flows, and occupant uses interact in response to the design. Educating designers to understand those meant some designers went on to apply the principles intuitively in subsequent designs, showing that the method had an educative and strategic value in addition to being just a tactical tool to test a specific design. Appendix 4 includes a short history of the development of the method that encouraged and enabled designers to wean themselves off their dependence on black box simulation models and help return to an expert, principles-led design ethos and understandings.

Reconnecting Designers With Climate

The way students are taught about design colours their whole life in practice. Many teachers around the world have been advocating for a more 'hands on' approach to introducing students into how buildings work in practice using physical models and new data collection tools. None

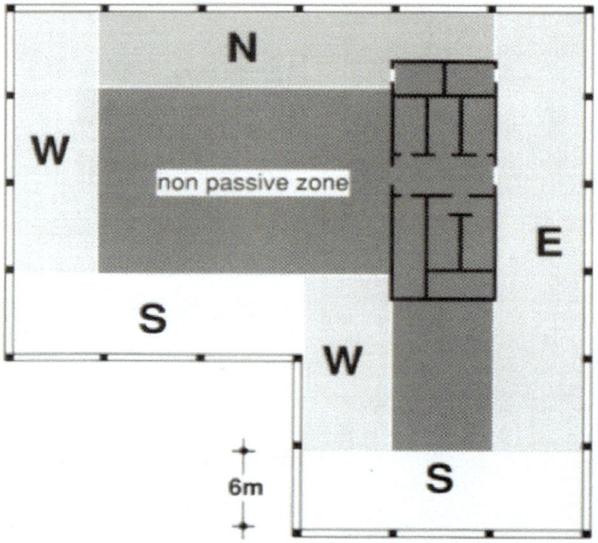

FIGURE 20.7 The LT method enables designers to differentiate between zones of the building that could run on passive energy and those that would need servicing.

Source: Nick Baker.

more so than the Society of Building Science Educators in the USA, who pioneered the use of tool sets and provided tool day training courses for their teachers and students alike so all involved could feel and sense the buildings they were analysing.[12] A whole new generation of affordable comfort data gathering monitors is being developed around the world to encourage the sensual understanding of what makes buildings comfortable.

In one Edinburgh exercise, students were required to study the thermal conditions in buildings using equipment that enabled them to feel and record buildings. Ula Odoku and the team there came up with this useful checklist of what is required for holistic teaching of comfort, including that students be:[13]

1. Taught that comfort is a multidimensional phenomenon
2. Given the tools to explore the many dimensions of comfort
3. Educated in the importance of their own sensory experience and learning processes
4. Taught to listen to building users to understand more on comfort
5. Enabled to link heart and mind and learning by playing with physical models
6. Advised that simulation is good for optimising opportunities, not defining them
7. Imbued with the need to *ground truth* solutions by exposing their designs to potential users

It is time that the power balance between architect and engineer is restructured, not least in how money is allocated in the design process through building contracts, because moving forwards, it is the building itself that will provide most protection from the extreme weather conditions of the future. This fact seems to have been forgotten. Is it time perhaps for all building designers to become architectural engineers?

Notes

1 https://clima.cbe.berkeley.edu/
2 Givoni, B. (1991). Performance and applicability of passive and low-energy cooling systems, *Energy and Buildings*, 17(3), pp. 177–199, ISSN: 0378-7788. https://doi.org/10.1016/0378-7788(91)90106-D
3 https://comfort.cbe.berkeley.edu/
4 https://nicolgraph.com/
5 Humphreys, M.A., H.B. Rijal and J.F. Nicol (2010). Examining and developing the relationship between climate and thermal comfort indoors, *Proceedings of Conference on Adapting to Change: New Thinking on Comfort*, Cumberland Lodge, Windsor UK 9–11 April.
6 Harso Karyono, T., S. Heryanto and I. Faridah (2015). Air conditioning and the neutral temperature of the Indonesian university students, *Architectural Science Review*, 58(2), pp. 174–183. https://doi.org/10.1080/00038628.2014.1002828
7 https://github.com/ayezioro/Nicol_graph
8 http://andrewmarsh.com/apps/releases/sunpath2d.html
9 http://andrewmarsh.com/apps/staging/sunpath3d.html
10 http://www.heliodons.org/
11 Lechner, N. (2015). *Heating, Cooling, lighting: Sustainable Design Methods for Architects*, 5th Edition, Wiley.
12 https://www.sbse.org/
13 Uduku, O., B.K. Satish, G. Treacy and Y. Zhao (2022). Teaching comfort: Critical approaches, digital interventions and contemporary choices, Chapter 31, in *The Routledge Handbook of Resilient Thermal Comfort*, Edited by F. Nicol, H. Bahadur Rijal and S. Roaf, London: Routledge, ISBN: 978-1-032-15597-5

21

FIRMNESS, COMMODITY, AND DELIGHT

Introduction: Order Out of Comfort Chaos?

Architecture is not an easy path to follow. Much is expected from a group of professionals who are mostly fairly poorly paid relative to many others. Architects work in a world driven by narratives of energy efficiency, net zero, industrialisation, speed of erection, and the hunger for profit. Developers and financiers increasingly rule the construction roost. With profits come risks and uncertainties as architects grapple to produce good designs despite multiple technical, regulatory, and financial constraints and influences at play. Top of the briefing requirements for most building types should be, but often isn't, the need to keep building users happy, comfortable, and healthy. Ignoring that imperative has led over time to a current priority for the building stock – to stop overheating in ever-hotter summers while keeping energy running costs affordable, especially during local power failures.

Over previous chapters it has become clear that a key factor in people's experiences of thermal comfort is the amplitude of the thermal signal over time, the thermal pulse of a building. Inspired by a chance visit to the Perfume Museum in Grasse in 2014 gifted by the Japanese firm Daikin, Roaf saw parallels with the process of designing thermally well-behaved buildings to the science of perfume design. Perfume science sets the design and composition of a scent into a musical framework based on time, and the amplitude, flavour, and intensity of the olfactory, the smell signal.

The Science of Perfume Design

One of the many sciences that was flowering around the time of the Bauhaus movement in the early decades of the 20th century is that involved in perfume production and classification. Perfume has been used as a means of stimulating sexual and social interaction since at least the dawn of recorded history in ancient Mesopotamia, some 5,000 years ago.[1]

Since the 1920s, perfumes have been classified using a musical metaphor, having three sets of *notes*, designed to result in a *harmonious scent accord,* as shown in Figure 21.1.[2] To maximise the impact of the scent, these notes reveal themselves in a sequence related to the olfactory

DOI: 10.1201/9781315645070-28

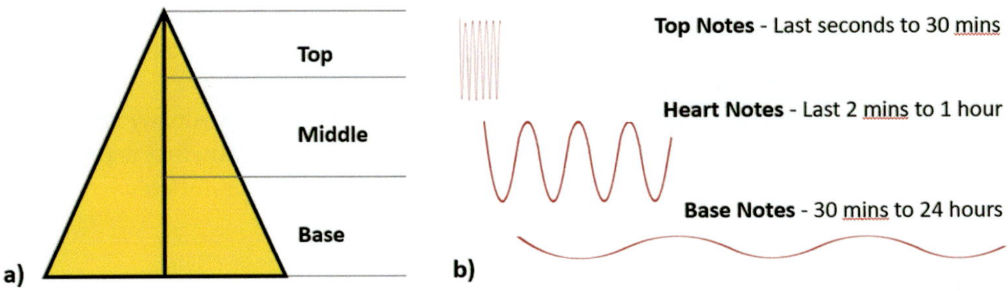

FIGURE 21.1 A perfume is designed with three 'notes' to make a 'harmonious accord'. Each note for an individual perfume is patented in a fragrance pyramid, with scents classified according to their evaporation times: (a) *Top note* wakes up the user's senses and is most volatile; (b) the big flower scent, or *middle note*, gives the perfume its personality, followed by the (c) *base note*, the least volatile ingredient that can last for hours.

strength and volatility of each note, which is dictated by the evaporation coefficients of their ingredients, how long their smell lingers in the air.

- *Top notes.* The fast 'wake up and take notice' smell that hooks the attention (e.g. lemon, bergamot, mint, lavender, and coriander).
- *Middle notes.* The slower, full-bodied smells that are sensed long enough to let users revel at the scent (e.g. jasmine, tuberose, frangipani, and sandalwood).
- *Base notes.* The long, slow tail of the scent that creates the perfume's legacy (e.g. meat, tobacco, amber, frankincense, musk, myrrh, patchouli, ylang-ylang, musk, and vanilla).

Top Notes – Short-Term First Impressions

Formed of small, light molecules that evaporate quickly, these are the 'wake-up' or *head notes*. These notes are perceived immediately to hook users from the first second they smell them. They are very important in marketing a perfume, and to grab the attention, they are fresh, assertive, or sharp. A person entering a building or space for the first time will also have formed a first and strong impression of the meaning of it to themselves within seconds of entering it. Perhaps in a 'fight or flight' reaction, they will conclude that it is safe or risky for them or, if they like it or not, for a multitude of reasons and react accordingly. Physical spaces also have top notes.

Middle Notes – Mid-Term Sensations

Just before the top note disappears, the *heart notes* emerge to form the 'personality', or the heart, of a perfume. These scents are usually more mellow and rounded and can last from two minutes to an hour. They also act to mask the often-sharp initial impression of base notes, which they induce to become more pleasant over time. They can evoke particular memories or emotions for the user with fragrances associated with themes like season, place, person, culture, purity, or seduction.

Middle notes are the big, signature scents of the fragrance and are made from a wide range of natural and synthesised substances ranging from flower oils like jasmine, tuberose, frangipani, and rose to those made from seawater or sandalwood. The reason that organic perfumes are so expensive is that these oils can be time-consuming to extract and may require large amounts of the source material to produce even small amounts of the oil.

For a person occupying a building, the first impact of the top note is gradually replaced by a deeper sensual appreciation of how the building feels over time. A stunning first impression can soon turn sour if the space is sensually, and particularly thermally, uncomfortable.

Base Notes – Long-Term Relationships

The base and middle notes of the perfume together create its main impact. Base notes bring depth, character, and solidity to a scent and become apparent close to the departure of the middle notes, usually not being perceived until around 30 minutes after application, and can last over 24 hours, lending character and body to the more transient sensual impressions left by the other two notes. Compounds of this class are typically rich and deep. The trail of scent left behind by a person wearing a perfume is called its *sillage*, after the French word for *wake*, as in the trail left by a boat in water. Base notes can fix the affections of users, who either return to the fragrance for more of the same or not. People also get fond of buildings, liking or disliking the way they feel, for a range of reasons, including thermal or other sensual delights. Buildings also leave a *sillage* of impressions on how they felt to be in.

The Composition

The signature impact of a perfume is actually patented under these three different aspects of it, but its success depends on how they work together.

Manufacturers who publish descriptions of their perfume notes typically do so with the fragrance components presented in lyrical and abstract terms for the components listed in its fragrance pyramid, often enhanced by descriptions of where they come from. Like wines, natural scents can vary in quality and cost, as shown in Figure 21.2, depending on when and where they are harvested, how they are processed, and the extraction method used. Each perfumer (also affectionately known as a *nose* – or in French, *nez*) will prefer a component from one particular place. A great perfume is born of its own time and place, just as a great building is also formed in its own times. Fashions, like climates, change.

How might perfume science relate to architectural design? A building, like a perfume, exists in time and is experienced over time. Both the initial cost and the long-term value of a perfume are related to the quality and durability of its base note. In a world of unpredictable economies, shrinking resources, growing populations, and a heating climate, it is obviously sensible to produce buildings that have a low risk of failing in uncertain times. The trick will be to design buildings with the narrative of *scents and sensibilities* too, with long-lasting base notes that will continue to keep people both thermally and emotionally safe, comfortable, and even delighted in the challenging decades ahead.

Understand What Matters Over Time

What we build now will be critical to future populations in a heating world. We have to design buildings now that will keep people safe in them in the much more extreme climates of the decades to come. Inspired by perfume science, it becomes clear what matters in adaptive and

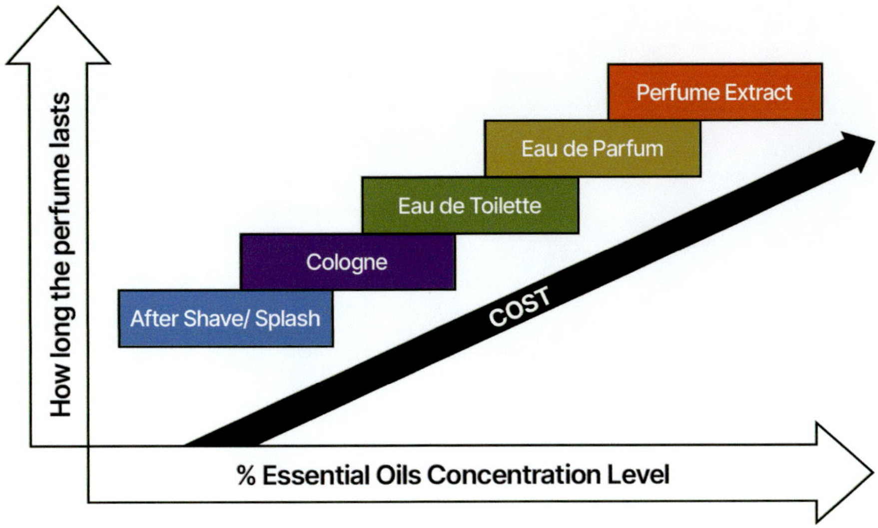

FIGURE 21.2 The price of a scent is related to the quality and concentration of the essential oils it is made of. That, in turn, dictates how long a scent lasts in the air, on the skin, and in the memory.

durable design and the building itself. Understand what does last and what is transient in the design of buildings. Compose a building to create a harmonious, durable, and high-value accord using thermal notes:

STEP 1: The Base Note: Character – Firmness

Everything about the building that cannot be changed. Start with what matters most – the building itself, its form, openings, and orientation, which dictate the way it works with the local climate on its site to ensure that the basic building has a safe, 'well-behaved' thermal character over a day and over a year and can run largely on local natural energy to ensure it uses minimal imported gas and electricity to keep people comfortable. This step will dictate the thermal amplitude of the daily or annual thermal 'heartbeat' of the building and determine its long-term comfort character. Durability adds value.

STEP 2: The Middle/Heart Note: Personality – Usefulness – Commodity

Everything about the building that can be changed. This second step designs in flexibility, enabling passive and mechanical features to be included to enable the building to adapt to changing conditions in weather, climate, economies, society, and technologies. Flexibility adds value.

STEP 3: The Top Note: Emotional Rapport – Delight

All about the mind. Designing a benign building that promotes the physical, mental, and spiritual well-being of people in and around it through the quality of its design is also necessary to provide comfort on many levels for occupants. A well-liked building will have a sound character and pleasant personality, and a great building will talk to the heart and soul as well, like that shown in Figure 21.3.

FIGURE 21.3 A building does many things as well as providing shelter. Some buildings provide spiritual peace, like the living room of the poet and artist Anthony Weir in the medieval village of Caylus in SW France

Source: Anthony Weir.

Firmness, Commodity, and Delight

These three proposed design steps resonate right back to the foundations of the earliest written architectural theory and Vitruvius's claim that good architecture is about *firmitas*, *utilitas*, and *venustas*[3] – firmness, commodity, and delight. In the 16th century, Andrea Palladio reversed the order of the first two, promoting, rather, *comodità, perpetuità, bellezza* – commodity, longevity, and beauty, in that order.[4] In both their minds, beauty comes last, after the robustness of construction and long-lasting usefulness of the design. We have simply replaced structural firmness with thermal stability, two properties of a building that often go hand in hand.

The following design approach may not appeal to everyone. Clients live with varied world views. First ask *why* they are building in the first place, to help refine the end product and also help them clarify their own thinking. Some may want to spend more money on creating a solid building, with a robust base note, that will last for 500 years and may be less worried about the wow factor. Another client may want to spend as little money as possible on the building to maximise profit when sold.

In the golden age of Western capitalism around the turn of this century, big spenders liked to be associated with trophy buildings that flaunted a *wow factor*, an overbearing top note. While some of these icons stand out as potentially being lasting monuments, many of those buildings now look and are pretty tired. Norman Foster's London City Hall, for instance, has had to be completely redesigned to make it both affordable to run and comfortable to occupy. Many others have been demolished. Single-skinned concrete Brutalist structures are often gone or abandoned, and many 1960s and 1970s glass towers, too, that are almost impossible to clean, maintain, or afford to heat, cool, or service. *Dead building syndrome* seemed to take hold as early as the last decades of the 20th century, when single-glazed tower blocks were either demolished or left empty, blighting areas often near main railway terminals in cities around the world, where the markets for offices boomed then. Nowadays, even high-end commercial real estate in business districts, like London's Canary Wharf, is now appearing dated and is astronomically expensive to upgrade, causing companies to move back into more modest, durable, comfortable, affordable office space in the climatically pleasanter Old City of London.

Some clients want the 'best of the middle of the road' solutions at every step to ensure a low-ish energy consumption and acceptably attractive buildings that will not lose its value in the short or medium term. Initially, it is important for designers to ascertain *why* a client is building to understand how to maximise the cost–benefit available to them from their investments through well-focussed design. A good designer can provide a well-performing building at most cost points.

Notes

1 https://en.wikipedia.org/wiki/Perfume
2 Teixeira, M.A., O. Rodriguez, P. Gomes, V. Mata and A. Rodrigues (2013). *Perfume Engineering: Design, Performance & Classification*, Oxford: Butterworth-Heinemann.
3 Vitruvius (1931). *On Architecture*, Translated by F. Granger, Vol. 1: Books 1–5, Loeb Classical Library 251, Cambridge, MA: Harvard University Press. See also: Vitruvius (1934). *On Architecture*, Translated by F. Granger, Vol. 2: Books 6–10, Loeb Classical Library 280, Cambridge, MA: Harvard University Press.
4 Palladio, A. (1965). *The Four Books of Architecture*, New York: Dover Publications.

22

DESIGNING THERMALLY WELL-BEHAVED BUILDINGS

Introduction: The Times They Are a-Changin'

Every generation faces new challenges. Their buildings are shaped by the availability and type of fuels on hand; the economic status and aspirations of a person, company, or population;[1] the health of the population; and the catastrophic changing of circumstances around them, like those wrought by our changing climate. These forces are clearly reflected in the rapidly evolving design mantras over the last half-century, where the ordinary citizen and their governments alike try desperately to comprehend and keep up with the global pace of change. Every decade the narrative evolves.

1. *Energy efficiency*. The key concern in the 1990s, focused on mechanical efficiency and issues around making building envelopes secure against cold penetration with more insulation and better windows, eliminating infiltration and cold bridges. A menu of heating season design actions typified by the early Passive House movement.[2]
2. *Building sustainability* dominated the 2000s with a confusion of multiple factors that were weighted in black box systems to provide a measure of how sustainable a building is.[3]
3. *Building resilience.* The hot topic of the 2010s as the devastation wrought by increasingly dangerous weather events woke many up to the need to create a generation of buildings that '*bounce forwards*' from previously failed states with resilient re-instatement and new thinking on design priorities.[4]
4. *Building sufficiency.* A top priority in the 2020s as it becomes clear that as building emissions continue to grow, extreme weather events become more extreme, associated with routinely failing energy systems. Ideally, building sufficiency is a state in which people's basic needs for energy services are met equitably and ecological limits are respected.[5]

Radical change is obviously urgently needed now, and *sufficiency* is currently seen to be the best way forward being about *long-term actions* driven by *non-technological solutions*,

DOI: 10.1201/9781315645070-29

which consume less energy in absolute terms. *Efficiency*, by contrast, is about continuous *short-term marginal technological improvements*.[6] Sufficient buildings need to keep people safe against weather and harm; be warm, cool, light, and dry where needed; be equipped for the activities undertaken in them; and provide room and space for privacy and community.[7] Sufficient buildings in most climates probably need a very well-behaved thermal base note.

The Base Note – Everything You Cannot Change

Designing a good thermal building may mean the whole building has a fairly stable acceptable temperature. Alternatively, individual room(s) in the whole building may provide safe temperatures for different seasons. The base note of a room or building is largely a consequence of all the things you cannot change in a building in relation to how it interacts with the climate and physical landscapes around it. Thickness, form, materials, ceiling heights, openings, and orientation, for a start.

When the outdoor temperatures are within an acceptable comfort range, then one may need no walls at all, but even in most climates in the tropics, the air temperature around a building and direct and indirect solar gain were modified with blocking walls and, in some places, fairly small windows. Increasing the thickness of a wall might provide increasing protection as the Delta T of the difference between outside and inside temperatures increases, as shown in Figure 22.1, but not if the form of the buildings is poorly designed.

For cooler northern European climates, it is easy to see what basic building forms do work well (Figure 22.2):

A poor base note building. A glass box building has little ability to dampen diurnal or seasonal swings in indoor temperatures over a day or year, making it virtually impossible to occupy without energy-expensive mechanical systems.

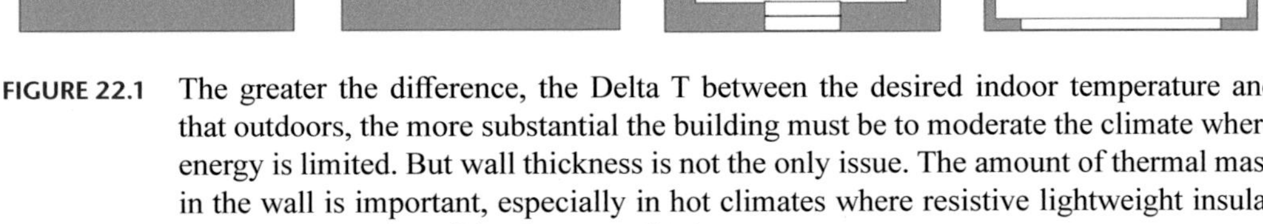

FIGURE 22.1 The greater the difference, the Delta T between the desired indoor temperature and that outdoors, the more substantial the building must be to moderate the climate where energy is limited. But wall thickness is not the only issue. The amount of thermal mass in the wall is important, especially in hot climates where resistive lightweight insulations do not work well but heavy walls do.

Source: Susan Roaf.

FIGURE 22.2 Three examples show buildings typifying the fact that if they are thermally 'well-behaved', they have very stable indoor climates; if they are generally well constructed but have too much glazing facing in the wrong direction, that will overheat in sunny weather and/or prove too cold in winter. These are deemed to have a so-so thermal performance. If they are glass boxes, temperatures inside them will swing wildly in response to weather conditions externally.

Source: Susan Roaf.

A so-so base note building. The form of a building, its orientation, thermal mass, amount of glazing, and ability to dump heat in summer or harvest coolth in summer with natural ventilation, will result in overheating potential and higher energy bills due to its inability to store heat or coolth in its walls and floors to provide free conditioning when necessary.

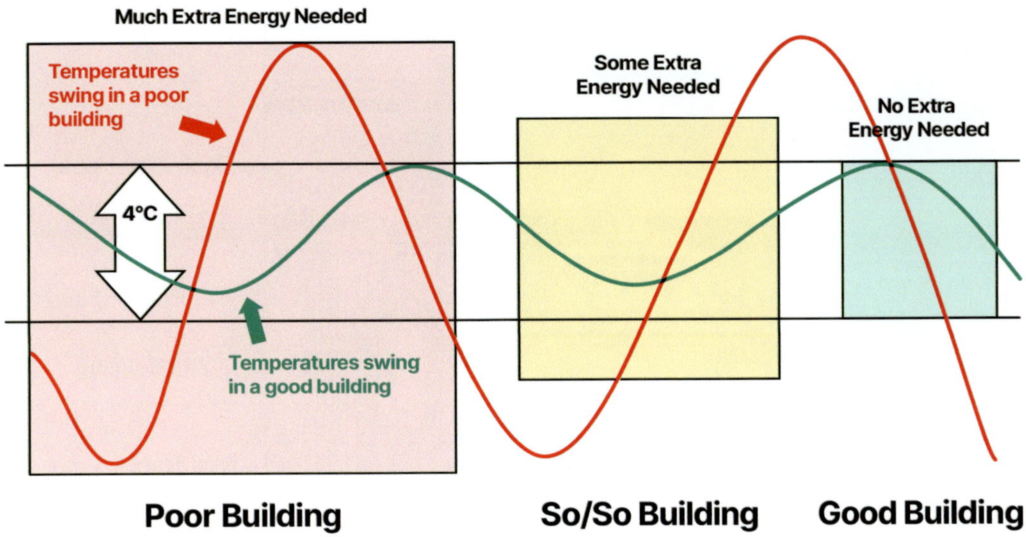

FIGURE 22.3 A daily indoor temperature swing of c. 4°C might represent an acceptable thermal amplitude over a day. Energy loads, costs, and climate impacts can vary enormously between good and poor base note buildings. In many cases, even so-so buildings may be able to reduce indoor temperature swings with good shading.

Source: Susan Roaf.

A good base note building. The Oxford Ecohouse was designed from day 1 to have a very stable base note. The design was based on the experience of ten years' living in the hot deserts of Iran and Iraq and was designed specifically not to overheat, even during the most extreme hot weather predicted under conditions of climate change by 2100, or to get too cold in winter.[8] The building's form, airlocks, lack of west-facing windows, high levels of externally insulated thermal mass, and excellent cross and stack ventilation systems all contribute to its very stable base note temperatures.

A good base note for a room or a whole building will keep occupants safe indoors in a heating world and bake in long-term value to a construction, created by all those things difficult to change over time, so get them right from the start of the design to save money and energy, and reduce GHG emissions and discomfort as suggested in Figure 22.3.

The Heart Note: Everything You Can Change

At each stage of a building's life, it may have to face change or a change of heart. The same basic building will adopt different personalities over time, both thermal and presentational. Imagine a Georgian house in London that has changed its appearance over two centuries. From the elegance of Regency decor in 1825 to the cluttered decoration of the same room in the 1875 Victorian period, shown in Figure 22.4, and on to 2025, when its floors are stripped to bare wood, the walls painted white, thick curtains are not think blinds, and the fire replaced by air from a heat pump. The internal thermal personality of the room will have changed considerably,

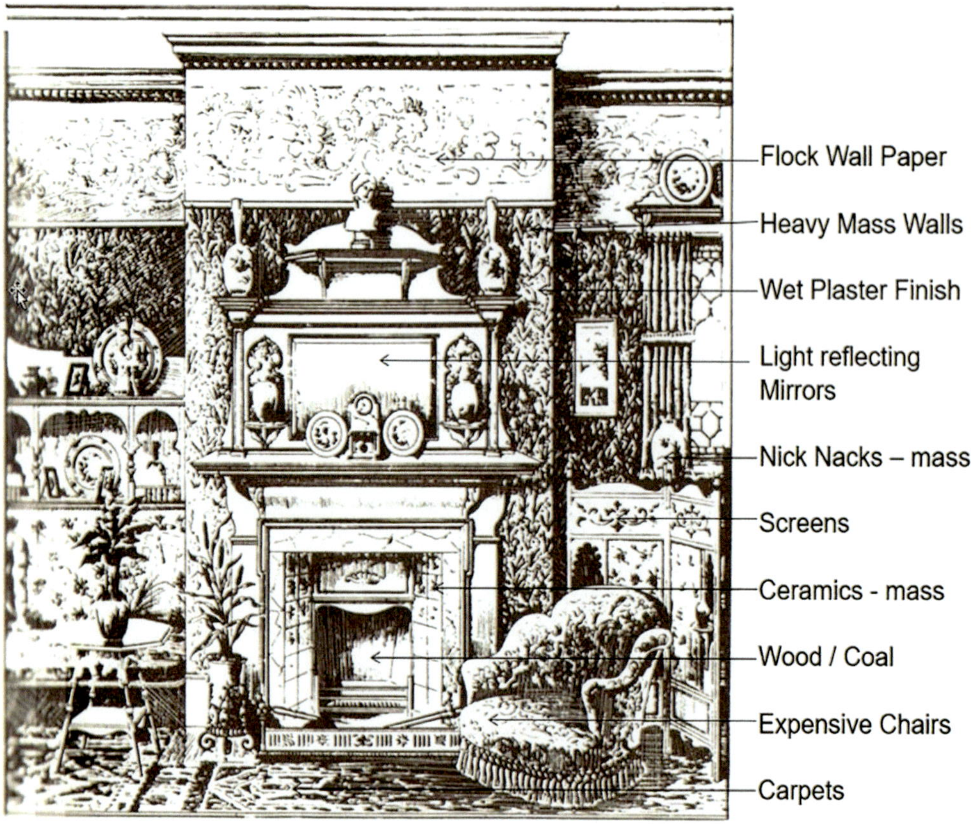

FIGURE 22.4 Interior of a sitting room in a Victorian house where every detail is contributing to making occupants comfortable while using as little energy as possible. Imagine the same room today stripped down to painted grey walls, revealed floorboards, and minimalist furniture, needing large amounts of energy to heat the air in the room at a modern comfortable temperature.

Source: Humphreys et al. (2011).[9]

but the thermal heartbeats of the sun through its windows and the modifying effect of the high-mass walls will remain similar.

While the basic building remains the same, there are many things about the building that can change. *Passive adaptive opportunities* to change the thermal performance of a building include:

- Modifying climates around the building with landscaping and vegetation
- Insulating walls, floors, and roofs
- Changing occupancy in different spaces
- Modifying window performance
- Managing airflow through the building
- Reducing draughts and infiltration

- Changing furnishings, including wallpapers and hangings
- Installing window shading systems
- Installing peripheral airlocks, sunspaces, balconies, halls, porches, and verandas

Mechanical adaptive opportunities to improve the thermal performance and environmental impacts of a building might include:

- Upgrading the efficiency of installed mechanical technologies and supply systems
- Improving controls to improve comfort while reducing energy consumption
- Changing the mode of conditioning, for instance, from air drive to radiant
- Changing levels of dependency on mechanical or natural ventilation systems
- Changing the mechanical systems
- Changing fuel type
- Replacing centralised mechanical systems with local, space-based environmental technologies

The ability of a building to adapt to social, economic, and climate changes will depend on the extent to which a building is capable of being adapted. Many modern building types are not easily adaptable. How do you attach an external shade on a glass tower block or to a stucco-finished, chipboard external wall? It may be possible, but it will be difficult. The bigger the building, the bigger the problem, and with 'innovative' building forms, many unanticipated problems can arise. The 1960s tower blocks were designed with rubbish chutes that are so small they can no longer be used for the trash bag sizes normal today, so the chutes have to be sealed off. Modifications to a building can revolutionise their thermal performance, so simply making the decision to design a building so that it can be adapted over time is an important step. A well-thought-out range of available adaptive opportunities can go on over time to provide acceptable indoor temperatures even as weather conditions become more extreme. Even if first costs do not allow it, providing the means for later additions of adaptive features may well extend the longevity of a building over time, and thus its value. Buildings like those in Figure 22.5 can last for centuries.

Designing a Climate-Safe Building

The message is pretty clear for those wanting to design climate-safe buildings. Design a robust basic building with a well-behaved base note that is fitted out with the potential to access a range of adaptive opportunities that will enable further modification of its thermal performance over time. Get the basic building right and it will provide 'sufficiency' over its lifetime by enabling occupants to continually fine-tune it to their needs in rapidly changing climates and fashions.

Everyone involved in designing resilient buildings must understand that *what must come first is the building itself*, not the mechanical services in it. Those are a secondary considerations and will inevitably be changed every decade or so, possibly radically. This means the architecture of the buildings itself and its adaptive opportunities inform the core design narratives of buildings in theory. In reality, it is the short-term commercial interests of developers and shareholders who often control the reins of many developments now and are too often not interested in the long-term value of a structure in the local built environment, just profits in their pockets at point of sale.

FIGURE 22.5 Two top buildings on the Promenade des Anglais in Nice, France, with no opportunity to open the windows or shade them. Two nearby buildings lower that enable occupants to choose the indoor conditions they prefer using many different adaptive opportunities. The top two constantly require mechanical conditioning at huge financial and environmental costs.

Notes

1 Lourdes Gutierrez, A. and E.Williams (2016). Co-alignment of comfort and energy saving objectives for U.S. office buildings and restaurants, *Sustainable Cities and Society*, 27, pp. 32–41. https://doi.org/10.1016/j.scs.2016.08.010

2 Roaf, S., and M. Hancock (Eds) (1992). *Energy Efficient Building: A Design Guide*, Blackwell Scientific Publications, ISBN: 0-632-03245-6

Roaf, S. (Ed) (2010). *Transforming Markets in the Built Environment: Adapting for Climate Change*, Earthscan Publications Ltd. April.

Roaf, S., and F. Nicol (Eds) (2018). *Running Buildings on Natural Energy: Design Thinking for a Different Future*, London: Routledge, HB ISBN: 978-0-81-539603-1

3 Roaf, S., A. Horsley and R. Gupta (2004). *Closing the Loop: Benchmarks for Sustainable Buildings*, London: RIBA Publications, ISBN: 978-1-85-946118-1

4 Roaf, S., D. Crichton and F. Nicol (2009). *Adapting Buildings and Cities for Climate Change*, Routledge.

5 Darby, S., and T. Fawcett (2018). Energy sufficiency – an introduction: A concept paper for ECEEE, *European Council for an Energy-Efficient Economy*. https://doi.org/10.13140/RG.2.2.31198.08006

6 IPCC (2022)*: Climate Change 2022: Mitigation of Climate Change. Contribution of Working Group III to the Sixth Assessment Report of the Intergovernmental Panel on Climate Change*, Edited by P.R. Shukla, J. Skea, R. Slade, et al., Cambridge, UK and New York, NY, USA: Cambridge University Press. DOI: 10.1017/9781009157926

7 Bierwirth, A. and S. Thomas (2019). *Energy Sufficiency in Buildings: Concept Paper*, European Council for an Energy Efficient Economy, Accessed on 04/06/2019 at: www.energysufficiency.org

8 Full details of how the Oxford Ecohouse was designed and performs can be found in: Roaf, S., M. Fuentes and S. Thomas (2012). *Ecohouse: A Design Guide*, 4th Edition, Oxford: Architectural Press, Elsevier, November.

9 Humphreys, M., F. Nicol and S. Roaf (2011). *Keeping Warm in a Cool House*, Edinburgh: Historic Scotland Technical Paper 14. http://www.historic-scotland.gov.uk/technicalpaper14.pdf

23

THERMAL DELIGHT IN DESIGN

Introduction – The Top Note

The top note, the first, fast glimpse we get of a building or space, creates an immediate impact on our minds and emotions and can leave in us a lasting impression about the building in its wake. A top note can sell or sink a building in the eyes of the beholder. It can draw us to it or repulse us on approaching it.

In evolutionary terms, that first impression can have a survival function if arouses fear, triggering a fight-or-flight reaction. That split-second image also has a range of adaptive functions in our everyday lives. It may make us decide if we enter a building or space at all, and when we enter or leave a building, it may inform actions to remain safe and comfortable. What is seen when opening a door to the inside or outside of a building will inform clothing choices, consciously or otherwise. That glimpse sub-consciously informs the mind about how the space will make a person feel in it, on many levels. First fleeting impressions of buildings play on the minds and emotions of people in a myriad subtle ways that have been exploited by some designers ever since spaces were consciously shaped and arranged in buildings for effect.

Impressions and Pathways

The design of the 3,000-year-old Egyptian Temple of Khonsu in Figure 23.1 shows that designers of religious experiences have been skilled in manipulating the senses and emotions of visitors to their temples for millennia. People travelling to the temple over the hot flat bright desert would first have been awestruck approaching its grand portal, then hit by coolth, shade, and blessed thermal relief on entering the first of a skilfully orchestrated series of spaces leading to the altar. Starting in the vast columns of the shaded hypostyle hall, they were drawn through ever-smaller, darker spaces, via rising floor levels, to experience the ultimate experience of being close to their deity in the innermost sanctum of the gods. Each new space presents a different impression in a ritual pathway of a type that has been evolved over millennia by global religions aiming to engage, elevate, or subjugate visitors to their mysteries using feelings of fear, awe, reverence, and metaphysical reflection.

DOI: 10.1201/9781315645070-30

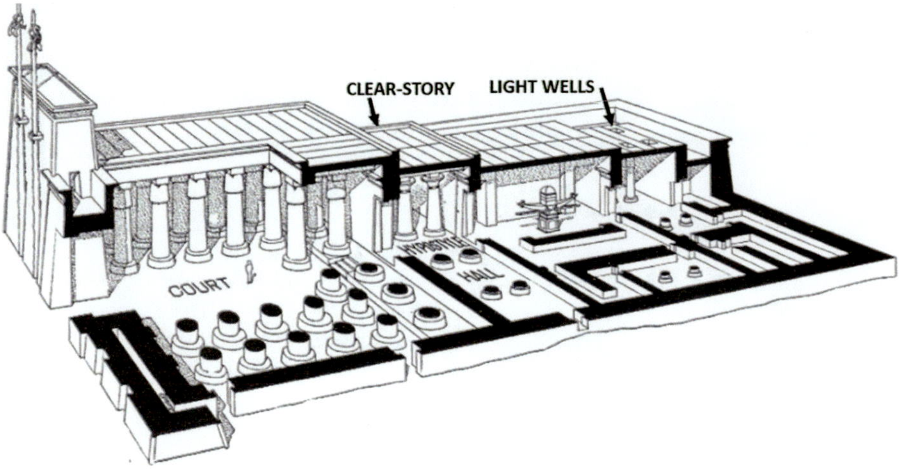

FIGURE 23.1 Temple of Khonsu, built by Ramses III, at Karnak, Egypt, in 1150 BC. The use of physical-scale, graded lightscapes and rising levels instil reverence and awe.

Source: Bannister Fletcher.[1]

How do pilgrims feel at the first sight of the humble shrine or the great Mosque–Cathedral? How are pilgrims mentally and physiologically rewarded for their arduous journeys? Alternating exposure to extremely hot or cold stressful experiences counterbalanced by periods of delightful coolth or warmth in the shelter and the comfort of a sanctuary can emotionally reward pilgrims. Similar processional sensual experiences are used in secular public buildings to impress populations with the importance, wealth, and grandeur of the building owners and the powers behind them, be they empires, industries, or individuals.

Figure 23.2 shows the hall of the great Mosque–Cathedral in Córdoba and the foyer of the Riverside Office in Brisbane designed by Harry Seidler, an architect who was a master of managing first impressions in his best buildings. Picture yourself walking into the cool halls of either of those great spaces on a very hot day in Córdoba or Brisbane. Imagine the thermal relief and delight that will be felt sensually on entering those spaces, and add to them the mental elation felt on entering those grand spaces. The temperatures in a busy air-conditioned office floor above may be similar to those in the Seidler office foyer, but the cadences of the 'Top Notes' felt by people entering the spaces will be very different. How are architects taught to understand and design with the complex playbook of experiences that can be evoked in buildings?

Patterns Inform First Impressions

Feelings of safety and security can result from what we know about buildings. Many buildings in particular places and climates are similarly designed and built and thus require less (split-second) interrogation at first sight. We are comfortable with the known. Christopher Alexander's classic books *The Timeless Way of Building* and *A Pattern Language* provide an exhaustive description of design patterns for towns, neighbourhoods, houses, gardens, and rooms and explain the rules or discipline that links them together in a building or town. Despite the fact

FIGURE 23.2 (a) Interior of the hypostyle hall in the great Mosque at Córdoba, Spain, started in 785, and (b) the entrance foyer of the 1986 Harry Seidler Riverside tower in Brisbane.

Source: Susan Roaf.

that his archetypes are of his time, culture, and countries, some of his published patterns are universal. Temperature patterns and the thermal design of buildings figure little in his books, with exceptions for things like the value of a sunny corner in winter and the role of the hearth indoors. It would be interesting to see how factors he does dwell on, like the sense of intimacy within spaces in buildings, relate in practice to the thermal landscapes in and around buildings in different environments, climates, and seasons.[2]

Rahat and Hygge – Thermal Delight in Different Climates

The problem is that we do not have a common language to describe the very different thermal environments and their patterns around buildings. Language limits our understanding of universal thermal sensations. Lisa Heschong brilliantly captured some such experiences in her classic 1979 book *Thermal Delight in Architecture*,[3] but globally we are hampered as designers by the lack of a common thermal design language.

Imagine moving outdoors on a hot day to lie beneath a vine-clad pergola with a sea breeze cooling the skin, stretched out and relaxed, the body at rest. In the Persian language, the word *rahat* (Figure 23.3b) is used to describe such sensations, of being relaxed and delightfully comfortable. What is the equivalent Danish word for that feeling?

The Danish word *Hygge* has attracted lots of attention since Meik Wiking wrote a book in 2016 on it subtitled *The Danish Art of Cosiness*.[4] This word encompasses so much more than just temperature; it not only relates to people in a cooler environment enjoying being comfortable with the good things in life and good people around them, often illustrated with the glow of fire and candlelight and 'snuggling postures', but also harks back to a more primal relationship people may have with fire (Figure 23.3a). Hygge is very much a *well-being* word rather like the Dutch word *gezellig* and the German word *gemütlich*, which have similar meanings

FIGURE 23.3 Timeless delight: (a) This reconstruction of a skin-clad, cave-dwelling Neanderthal family grilling and eating meat together reflects the primal Hygge sensations of warmth and happiness given off by some fires. (b) Foreign tourists enjoying the timeless and universal sensation of being Rahat, cool, and relaxed in a café beneath the Khaju Bridge in Isfahan, built in c. 1650.

Source: (a) Gorodenkoff. (b) Susan Roaf.

encompassing ideas of being at ease and cosy in convivial company. These terms do deal with the experience of emotional feelings of warmth, contentment, happiness, and safety and, for some, may also encompass ideas of thermal pleasure and delight. The Swedish idea of *Lagom* has also been associated with the Danish idea of *Hygge*. Literally translated, it means 'moderation', and Niki Brantmark[5] explains, in his book on living a happy, balanced life, it as 'not too little, and not too much, but just right'. It focuses on the desirability of living a life based on ideas of *sufficiency*, which in turn lead to mental peace of mind and a safe and comfortable life. Top notes associated with such word groups evoke soft and kind sensations.

Languages include wide ranges of words about thermal sensations associated with experiences in room types at different times of day and year common to their climates and cultures. Responses to invitations to enter a space may be tempered by preconceptions of what thermal feelings will be involved. For instance, 'come into the *den in winter*,' 'the *veranda in summer*', or 'the *conservatory in spring*' may be filtered through what is known about the micro-climates in the spaces concerned and the extent to which the host is known for keeping their house warm in winter or cool in summer. 'Join me for lunch in the basement' evokes very different presumptions in winter in Britain than it does in a traditional house in Yazd in the Persian desert in the summer. We are continuously revising our understanding of the built environment based not on words but on context and the experience of our sight and other senses in order to anticipate and react to stay safe and make the most of situations as they arise.

A Pattern Language for Thermal Experiences?

Mark DeKay wrote of his 1986 experience visiting Louis Khan's 1973 Yale Centre for British Art as a revelation. Walking past a busy street of shops and entering through the cave-like entrance, he emerged into the sun-lit four-story entrance court and passed on through a sequence of differently sensual spaces that he described as a *delightful symphonic multi-sensory sequence of light and space, temperature, touch, and view*. He was awestruck at the extraordinary experiential intentions consciously designed into that building by its architect. Lamenting the fact that it is difficult to excite students with knowledge around technical standardisation of buildings, he and Gail Brager set out to create a language of environmental design that was more experimental, more about pleasure and performance, a language connecting sustainability to human feelings beyond comfort. They published *Experimental Design Schemas* in 2023 to expand a designer's consciousness beyond the realms of delight and comfort to expand architecture into connecting with the peaks and potentials of the human experience of buildings.[6] In it they divide experiences in buildings into issues around properties of space and light and narratives of comfort and delight aroused in spaces and around progress routes through buildings in a thoughtful reference work for those experimenting with visitor and occupant experiences. The work does allude to a common theme for architectural theorists around the commonalities between architecture and music.

Architecture as Frozen Music?

In 1833, Johann Wolfgang Goethe first alluded to architecture as music. The German polymath and poet wrote:[7]

> *Imagine that Orpheus is building a city to the accompaniment of his lyre. A vast site is prepared; the divine soloist, after having chosen the most suitable location, takes up his lyre. Suddenly the*

rocks, obeying the irresistible charm of his harmony, detach themselves from the mountain in regularly cut and trimmed forms. Then seized with enthusiasm, they shake themselves out and move on; they organise themselves according to the rules of a learnéd architecture, arranging themselves on foundations, following the laws of the music's rhythm, to form the walls. Thus are the streets built up, one after the other. Thus is the city constructed; with defense walls forming its enclosure. Then the sounds of the lyre cease; but the harmony remains.

Goethe nurtured the late 18th-century idea of architecture as music, describing it in the German as 'mute' music, implying that architecture is not dead but is alive, communicating with us in non-verbal ways. On translation, the German mute became 'frozen' in English, implying now a lifeless form, frozen in time. Does architecture, then, become an inanimate sculpture on display, to be admired from all angles like the Piano Building in Figure 23.4 for the genius of its artistic form, impact, and of course, its designer?

Frank Lloyd Wright, a good pianist, himself (Figure 23.5) was one of the most impactful, lyrical architects of the 20th century. He claimed to compose his architecture-like musical pieces when he wrote of architecture as an arresting narrative, like the melody in music:[8]

Much like how a good designer incorporates plotting, scheming, planning, etc. into things, so does a musical composer. Just like the music from Mozart or Beethoven, which has a beginning, a middle and an end, so too should good architecture. Just like a good song, at one point or various points, the building feel and approach not only within its interior, but exterior as well.

FIGURE 23.4 The Piano Building is the Urban Planning Exhibition Hall of Shan-nan New District in Huainan City, China.

Source: Professor Zhihe Shen.

FIGURE 23.5 Frank Lloyd Wright playing the piano at Taliesin.

Source: The Frank Lloyd Foundation.[9]

When Daniel Libeskind, also a musician and an architect noted for his impactful building like that in Figure 23.6, was asked in 2014 if he thought architecture was like music. He responded that it was so, intellectually and emotionally:[10]

> *Oh, absolutely! Architecture is not just an intellectual or abstract exercise; it is an emotional experience just as music is. It has to communicate to the soul and everybody has to share it in a deep emotional way. It is always about a performance and what happens after the performance. When you leave a building, it is like leaving a piece of music. It is still in you and still with you.*

These three authors address three different aspects of architecture:

- Goethe wrote of the *rhythms* in traditional settlements, where buildings and landscapes are welded together by natural forces alone.

FIGURE 23.6 Daniel Libeskind's graduate centre at London Metropolitan University, designed to get the university noticed and to leave an impactful legacy on the minds of viewers.

Source: Susan Roaf.

- Wright talked of *melodic* compositions within architecture that should embody a positive and comprehensible narrative incorporating both building and place.
- Libeskind, himself a musician, evokes the w*ow effect* with his buildings that he sees as theatrical performances created by a maestro for visual effect and legacy.

Rhythm, melodic narrative, and audience impact are very different angles from which to interpret the relationships between architecture and music. How strong are the threads that link them? If one asked a thousand people to match an image of a building to a piece of music then played Beethoven's pastoral symphony, a rousing military march, and a cacophonic composition by Stockhausen, coupled with showing them pictures of a country cottage and landscape, the Fascist Palace of Justice in Milan and a post-modern deconstructionist building by Libeskind, how many might come up with the same matches? Might one then compare what emotions Stockhausen wanted to evoke in his music with the emotions Libeskind seeks to arouse in visitors to his buildings? Logic suggests yes! What might those emotions be, and what physiological processes arouse them?

Hormones

Human behaviours are often determined by *hormones*, the chemicals that trigger emotional functions in a body. Hormones carry messages through the blood and nervous system to organs, skin, muscles, and other tissues. They are controlled by the *endocrine system*, which is the messenger system.

External signals received by the brain, or nerves in the skin, are sent as chemical messages to the *hypothalamus* at the base of the brain, which is the neural centre for the whole endocrine system. The endocrine system links out to the nervous system via the *pituitary gland* to effect behaviours. It controls the feedback loops of hormones that are released by internal glands directly into the circulatory system to target and regulate distant organs.

Testosterone and *oestrogen* are the male and female hormones, respectively, which can, at different times, trigger very different physiological responses, such as changes of mood, levels of aggression or depression, and health issues. Many environmental factors can increase or decrease testosterone and oestrogen levels, including temperature.[11]

Adrenaline helps the body react to stressful, exciting, or dangerous situations, making the heart beat faster and increasing blood pressure. In the workplace, it can cause hearts to race and poor decision-making while under pressure or give on the extra burst of energy that can help workers meet deadlines. *Cortisol* is produced by the adrenal and endocrine glands above the kidneys, which helps regulate the stress responses, inflammatory responses, and immune functions.[12] Cortisol can be trigged by overly cluttered and harshly bright environments, while calming spaces with natural light, natural and verdant surfaces, and more spacious layouts may lead to lower cortisol levels.

Endorphins are produced by the pituitary gland and the hypothalamus and bind opioid receptors to the central nervous system to block pain and increase feelings of well-being, for instance, after exercise, eating (especially spicy food and dark chocolate), and sex. *Serotonin* and *dopamine* are known as the happy hormones, neurotransmitters that are involved in mood regulation, learning, and decision-making. Dopamine is associated with rewards and motivation and acts to control body movements and coordination. Serotonin is associated with happiness, focus, and calmness and affects digestion and appetite. Dopamine manages short-term rewards and urges immediate action, while serotonin manages long-term benefits and counsels patience.

The top note of a building, that first fleeing impression of it, may well trigger a noticeable hormonal response, but it will also be a filtered response, depending on an individual's life experiences and state of mind. Musical compositions can be manipulated for desired effects, like a military march rousing men into a fighting mood. A Fascist headquarters may inspire fear, while a bucolic melody or an image of a chocolate box cottage encourages peace of mind and happier thoughts. Complex emotional systems are triggered by sub-conscious programming and also unconscious responses aroused by primitive metabolic responses, for instance, to health-threatening temperature changes.

We accumulate expectation and response filters over time based on our experiences. They help us deal efficiently with pathways through days and landscapes. The scale of those filters becomes clearer when they are removed. A designer of age care facilities in Melbourne told of walking down a corridor with an old lady with advanced dementia holding his arm. He was surprised when, as she approached one corner in shadow, she froze, clutching his arm and crying out that she could not walk into the dark corner because she was afraid of it. They turned back, but this demonstrates that once the layers of understanding, expectation, and control which we overlay our naked emotions with over a lifetime are removed, then the body moves back to basic responses of the fear and delight of more animal responses.

The body has a palette of reward and punishment systems involving hormones that have evolved over millennia to promote the survival of an individual and the species, like alliesthesia. As Humphreys wrote in the first lines of volume 2 of this trilogy:

> 'Adaptation is at the heart of the survival of all species in the ever-changing climates of our planet' and for humans, hormones fuel the engines of our adaption.

Alliesthesia

Alliesthesia is a mind–body (psychophysiological) phenomenon where a stimulus to the brain, skin, or organs of a body can either trigger sensations of pleasure or displeasure depending on the scale of the potential good or harm that stimulus causes to the body. If someone is hungry or thirsty, food and drink may trigger a pleasure response. If more food or drink will cause damage to the body, a displeasure or pain response may be sent.

Human health is hugely dependent on maintaining a stable core body temperature to maintain homeostasis. Alliesthesia is a driving force that conditions the body to take action to maintain homeostasis. A surge of hormones that register as displeasure may be released when external temperatures are moving away from safe thermal conditions to stimulate action necessary to return to safety. When the body is moving back towards conditions conducive to thermal stability, the alliesthesia response rewards the body with a surge of pleasurable hormones to encourage the move. The endocrine system is constantly signalling to the body to return to homeostasis, like 'Get out of the hot sun', 'Move into the cool breeze', 'Put more clothes on', 'Take them off', 'Warm up in the sun'. We have all felt those wonderful hormonal pleasure rewards, sitting in the warm sun on a cold day, or in a cool breeze on a warm day.[13] In reverse, the system works by signalling displeasure, for instance, sitting next to a chillingly cold wall or beneath an overly cold air conditioning outlet. More clothes may be put on, or a person may change positions, activities, or locations in response to the cold sensation signal.

The Fifth Force

Michel Cabanac, the pioneering researcher who first built a science around alliesthesia in the 1970s, has suggested that the world is ordered by a limited number of forces. Physicists have recognised four forces that act to alter our physical world: Newtonian, gravitational, electromagnetic, and the weak and strong nuclear forces. He suggested in 2010 that the effective axis of consciousness – pleasure/displeasure and joy/distress – operates in conscious animals in a similar way to how the four forces of physics influence masses and particles; that is, it drives/influences their behaviours.[14] His team showed that alliesthesia is not limited to simple physiological function, like temperature and food sensations, and explored its function in other sensations. Experimental evidence has shown that alliesthesia may occur at the service of all physiological set points and plays a major role in decision-making, whatever the field of decision, including ethical and perhaps even religious decision-making processes. He posited that alliesthesia can now be accepted as the fundamental mental mechanism of cognition and decision-making, probably phylogenetically inherited for efficacy in producing useful behaviours and its ability to promote comfort and happiness. Alliesthesia thus does not only elicit purely sensory responses but also involves interaction with the whole 'mind-scape' of an individual.[15] Cabanac suggests that this profoundly important psychophysiological control system is *the fifth force* providing order to our biotic world. It is certainly the main force driving the adaptive thermal comfort by continually steering the body back to thermally safe and acceptable conditions. Can designers use it to create great buildings?

Emotional Pathways Through Buildings

How many psycho-physical senses are employed in good design? Vision, view, sound, and smell can all trigger pleasure or displeasure responses and are universally used in design. Middle Eastern and Mediterranean courtyard gardens like that shown in Figure 23.7 have evolved over millennia and are great examples of how to evoke a multi-sensory paradise on earth in some climates.

Great architecture plays with proportion, symmetry, and a narrative sequence of sensations through a building punctuated by endpoints and staged vistas, as anyone who has sat in front of the Taj Mahal may remember.

The Sri Lankan architect Geoffrey Bawa's buildings are visited daily by architectural pilgrims from around the world to experience and learn from some of the beautiful pathways he created through his buildings reflected here in Figure 23.8. He was a master of sequencing

FIGURE 23.7 A multi-sensory palette of coolth, colour, water sounds, perfumes, shade, and breeze combine to create thermal delight in this simple centuries-old square in Seville.

Source: Susan Roaf.

FIGURE 23.8 Three views down the passage from the front door to the garden beyond in Geoffrey Bawa's country estate house at Lunuganga.

Source: Susan Roaf.

sensual experiences, drawing people from top note to top note as they move through them and reward visitors at each leg of that journey with a visual gift of a tableau or a framed picture, a differently lit space, an artwork or landscape vista.[16] In that mellow tropical climate, the rewards he offers visitors are predominantly visual, relying heavily on light and view but spiced up with fresh breezes, scents, sounds, and the reflected colours and light of waters.

Alain de Botton, in his book *The Architecture of Happiness*, wrote:[17]

> *The places we call beautiful are . . . the work of those rare architects with the humility to interrogate themselves adequately about their desires and the tenacity to translate their fleeting apprehensions of joy into logical plans – a combination that enables them to create environments that satisfy needs we never consciously knew we even had.*

Visual Delight

It is not surprising that Lisa Heschong, who published her first edition of *Thermal Delight* in 1979, has been mulling over the idea of delight ever since and has now produced her seminal work *Visual Delight in Architecture*. In it she examines the many ways people's lives are enriched by the presence of natural daylight and window views in buildings.[18] This book should be required reading for all architects aspiring to use light, space, and view to create those top notes that elevate architecture from the mundane into high art. Her insights into the science of light impacts on people and ways to use that knowledge to create more humane, liked, and uplifting spaces, provides some of the best advice there is on how to create top notes in design.

A Masterclass in Top Note Design: The Crab-Eye Window

In the oldest civilisations in the world, buildings have been evolving for millennia, so it is no surprise that some of the most ancient cultures have the most sophisticated top notes in their buildings. A wonderful example is found in the form of a *crab-eye* light well in the traditional courtyard houses of the fishing village of Zhouzhuang in south-east China shown in Figure 23.9.[19] The light well sits behind two opening shutter windows from a corridor behind the main living area of the house. The light well has a planted garden beneath an overhanging

FIGURE 23.9 The crab-eye window in a traditional courtyard house of the fishing village of Zhouzhuang, China.

Source: Brian Ford.

roof, providing shelter from strong winds. The geometry of the light well prevents direct light penetration into the house while bouncing light into the room from the white rendered masonry walls beyond, and this light has a psychological as well as a physical effect in achieving 'light without heat'. But the genius does not stop there. The light wells are linked to the corridor via operable timber screens sprung from a raised seating bench some half a metre off the floor, called the 'beauty seat', where the ladies of the house sat in summer in the cooling drafts of rising air moving from indoors to exit via those windows, up through the light well on hot days, making that seat the freshest and most comfortable in the house. Air flow rates over the 'beauty seat' were measured as around four times faster than those across the main living area, where the men were. So the highest comfort levels were allocated to a highly valued asset – the ladies of the house. The 'crab-eye' light wells provided protection from the cold winds, an acoustic buffer between the public areas beyond and the private rooms of the house behind, and a blank wall canvas on which to reveal the beauty of the perfumed plants before it and the symphonies of light across the seasons that play from above.

Top Notes Add Value and Longevity

The 16-storey Pershing Square Building on Hill Street in Los Angeles shown in Figure 23.10 was built in 1924 in solid stone in the Italian Renaissance Revival style that features classic design elements, symmetry, arches, and ornamental stone brackets. For many years, the main office floors were unoccupied, although now recently renovated and in demand as creative spaces in the heart of downtown. What is interesting is that for many years, two micro-climates in the building have continued to be extremely popular because of their unique and delightful thermal ambiences, their top notes. On the top three floors with roof gardens, the Perch Rooftop Bar and Restaurant thrives in pleasant weather, with people flocking to sit out in the fresh breezes and well-planted and designed seating areas around those floors. By contrast, in the basement, three floors beneath the ground floor entrance are cool, dark, and interestingly light spaces now occupied by the Mrs. Fish Japanese Restaurant and previously by a very noisy and crowded nightclub.

Larger buildings can contain a range of very different types of thermal landscapes, and within them different micro-climates that can be enhanced to create the interesting and delightful types of top notes, to which people do flock to spend time pleasurable, stimulated by their environment. Such thermal experiences can add real financial value to buildings and to their longevity as attractive investments in crowded marketplaces.

From Wow to Well-Being

Alain de Botton also concluded in his book *The Architecture of Happiness* that:

> The failure of architects to create congenial environments mirrors our inability to find happiness in other areas of our lives. Bad architecture is in the end as much a failure of psychology as of design.

He argues that people are out of touch with what they really want or need, causing grief and discontent. Perhaps modern architecture does not provide the types of sensual experiences that can be translated into words we can use in dialogues about designing for thermal delight and

FIGURE 23.10 The Pershing Square Building in downtown Los Angeles that maintained two vibrant venues in very different micro-climates at the top and the bottom of the building, while many floors between were unoccupied.

Source: Susan Roaf.

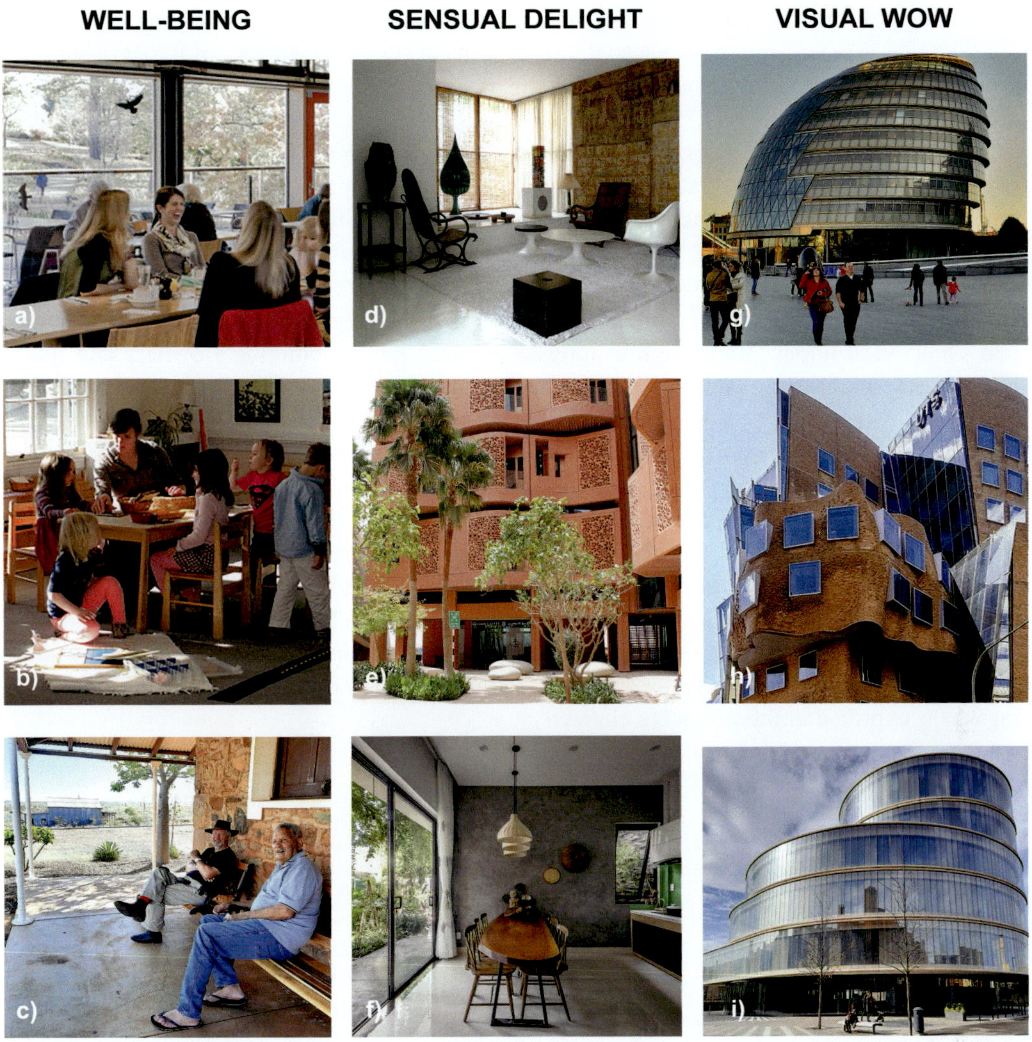

FIGURE 23.11 Sensory impressions in a space can illicit many emotions, from shock and awe to deeper feelings of well-being, delight, and happiness, as hinted at in figures (a) to (i), which play with ideas around the unique selling points of the top notes of very different buildings: (a) John Hope Gateway, Edinburgh Botanic Gardens by Edward Cullinan Architects; (b) Penobscot Primary School, Maine; (c) The Old Courthouse, Roebourne, Western Australia; (d) Reflective Interior of Geoffrey Bawa's house in Colombo, Sri Lanka; (e) apartment building in Masdar City, Abu Dhabi, by Norman Foster and Partners; (f) The Wind House, Da Nang, Vietnam, by Green Concept + Nha Cua Gio Architects; (g) Former London City Hall by Norman Foster and partners; (h) Dr Chau Chak Wing Building, UTS Sydney, by Frank Ghery; (i) Blavatnik School of Government, Oxford, designed by Herzog and de Meuron.

Source: All images from Susan Roaf, except (h) by Martin Kraft – Wiki Commons; (g) Wiki Commons pix, by Gary Knight and (f) Minq Bui Photography.

the feelings of happiness it can create. Perhaps we now lack the self-awareness to explain in our own words what we want to sensually feel, perhaps Hygge in winter and Rahat in summer, or understand the types of visual sensations that our souls seek to touch the heart or inspire the mind.

Has the age of industry, computers, and Capitalism put so much importance on firmness and commodity that the art of creating delight is largely lost from the current lexicon of architecture? Has the art of creating and explaining the top notes in buildings, be they first impressions, uplifting pathways, or genuinely beautiful, sensual features within buildings, been lost? Is the productivity of workers in buildings so much more important than their health and spiritual uplifting? Is what really matters in a building simply its transient *wow factor*, and how does that blunt force impact translate into long-term value for the building's owners and users? Figure 23.11 highlights that time spent in exploring what elements the top note of a building might be composed of, be it wow factor, sensual delight, or enhanced well-being, will inevitably raise the quality of a design and also provide clients with the confidence to hire and invest in thoughtful designers and their buildings.

Time spent in exploring, creating, explaining, and sharing the architectural ambitions of a designer and their client in terms of how the building is intended to add *top note* value to the investment must inevitably result in better buildings on so many levels. Communicating designer intent using a language of thermal experiences will definitely help refine outcomes on whatever point in the wow-to-well-being scale designers are aiming for. Issues around comfort are so interwoven with the physical, mental, and even spiritual well-being of people in buildings, as is explored in the following chapters.

Notes

1 Sir Fletcher, B., and B.F. Fletcher (1905). *A History of Architecture*, 5th edition, London: Batsford.
2 Alexander, C. (1977). *A Pattern Language: Towns, Buildings, Construction*, USA: Oxford University Press, ISBN: 0-19-501919-9
3 Heschong, L. (1979). *Thermal Delight in Architecture*, The MIT Press, ISBN: 978-0-26-258039-7
4 Wiking, M. (2016). *The Little Book of Hygge, the Danish Way to Live Well*, London: Penguin.
5 Brantmark, N. (2017). *Lagom: The Swedish Art of Living a Balanced, Happy Life,* London: Harper Thorsons, ISBN: 978-0-00-826010-1
6 Dekay, M. and G. Brager (2023). *Experimental Design Schemas*, Novato California: Oro Editions, ISBN: 978-1-957183-73-2
7 Gœthe, J.W. (1833). *Maximes et réfexions*, Translated from German by S. Sklower, 1842, Paris: Brokhauss et Avenarius, pp. 173–175, Accessed on 17/04/2019 at: https://fr.wikisource.org/wiki/Maximes_et_R%C3%A9flexions_(Goethe,_trad._Sklower)/07
8 Twombly, R.C. (1987). *Frank Lloyd Wright, His Life and Architecture*, Hoboken, New Jersey: John Wiley & Sons.
9 Frank Lloyd Wright Foundation (2022). *The Muse of Music: Wright's Legacy in Song, Frank Lloyd Wright Quarterley Magazine*, Accessed on 18/04/2025 at: https://franklloydwright.org/the-muse-of-music-wrights-legacy-in-song/
10 https://the-talks.com/interview/daniel-libeskind/
11 Husak, J.F., M.J. Fuxjager, M.A. Johnson, et al. (2021). Life history and environment predict variation in testosterone across vertebrates, *Evolution*, 75(5), pp. 1003–1010. https://doi.org/10.1111/evo.14216
12 Thau, L., J. Gandhi and S. Sharma (2023). Physiology, cortisol, in *StatPearls [Internet]*, Treasure Island (FL): StatPearls Publishing, Accessed on 08/02/2025 at: https://www.ncbi.nlm.nih.gov/books/NBK538239/
13 Parkinson, T. and R. de Dear (2015), Thermal pleasure in built environments: Physiology of alliesthesia, *Building Research and Information*, 43(3), pp. 288–301. https://doi.org/10.1080/09613218.2015.989662

14 Cabanac, M. (2010). *The Fifth Influence: Or, the Dialectics of Pleasure*, New Jersey, USA: Published by Book Whirl.
15 Cabanac, M., L. Brondel and A. Cabanac (2020). Alliesthesia. Up-Date of the Word and Concept, *American Journal of Biomedical Science & Research*, 8(4). https://doi.org/10.34297/AJBSR.2020.08.001293
16 Robson, D. (2004). *Geoffrey Bawa: The Complete Works*, Thames and Hudson, ISBN: 0-500-34187-7
17 De Botton, A. (2014). *The Architecture of Happiness*, London: Penguin Books.
18 Herschong, L. (2021). *Visual Delight in Architecture*, Routledge, ISBN: 978-0-367-56322-6
19 Lau, B., B. Ford and Z. Hongru (2014). The environmental performance of a traditional house in China, in *Lessons from Vernacular Architecture*, Edited by W. Webber and S. Yannas, Abingdon: Earthscan, pp. 92–99.

Comfort and Well-Being

24

COMFORT AND WELL-BEING

Physical Health

Introduction: What Is Well-Being?

Wikipedia in 2024 defined *well-being* as:

> Well-being, or wellbeing, also known as wellness, prudential value, prosperity or quality of life, is what is intrinsically valuable relative to someone. So the well-being of a person is what is ultimately good for this person, what is in the self-interest of this person.[1]

This seems like a very limited, egocentric, mercenary take on what constitutes human well-being. Very often it is the good of those around the individual, or community, that makes for happiness. A kinder definition of *well-being* might be 'the state of being comfortable, healthy, or happy'. Let us stick with that.

Partly in response to the growing angst experienced within many societies about the state of the world, issues of well-being have come to the fore. Making sense of the enormous number of factors that affect our own personal well-being is difficult. General tick-boxing of subjects in order to designate a building as being comfortable, or its occupants as being well, never addresses conditions in any one particular building, for any particular group or person, at any one time. Each building in its place at any one time is unique and must be individually evaluated against locally relevant measures to ensure the highest levels of well-being for its occupants that can be achieved with the means available.

To organise ideas around well-being, the framework presented in what words follows was developed by the Māoris of New Zealand, and is useful.

Hauora: Well-Being, the Māori Way

The Hauora is the Māori philosophy of health and well-being, unique to New Zealand.[2] It is a holistic system which includes issues around physical, emotional/mental, social, and spiritual caring. The Haura, shown in Figure 24.1, is taught in most primary schools there and is understood and appreciated by many New Zealanders. Well-being is, in the Māori imaginary, built

DOI: 10.1201/9781315645070-32

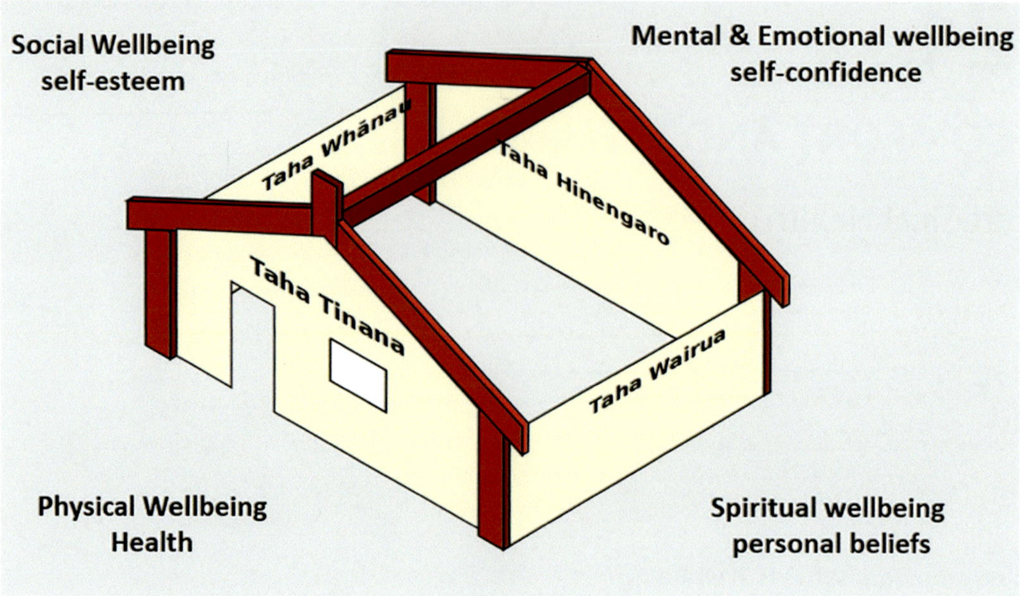

FIGURE 24.1 Diagram of 'te whare tapa whā' – the four cornerstones (or sides) of Māori health. It is imagined as a Whare (house), named with domains (walls) of Hauora (well-being).

Source: https://en.wikipedia.org/wiki/Hauora.

like a house (Whare) with four walls, each representing a different dimension of well-being, and each wall necessary for the health and well-being of the whole. Without one of the following supports, the strength and stability of the whole house fail too:

1. *Taha Tinana.* Physical well-being – health.
2. *Taha Hinengaro.* Mental and emotional well-being – self-confidence.
3. *Taha Whanau.* Social well-being – self-esteem.
4. *Taha Wairua.* Spiritual well-being – personal beliefs.

Physical Well-Being, Comfort, and Health

Well-being is often used to describe a state when an individual feels both comfortable and healthy.[3] Factors that affect perceptions of well-being may include a person's physical and mental health, as well as the buildings they live and work in, and the dynamics of their relationships at work, at home, and in the community.

Just as how comfortable a person feels is linked to their own perceptions of well-being, people are different. Arguments often arise amongst friends, family, and colleagues about what temperature to set the thermostat at in a commonly occupied space. Some like it hot, others hotter or cooler, often influenced by personal factors, like age, weight, and underlying health issues, as well as their income level, social status in a group, who pays the bills for energy, and personal thermal histories. Did they come from a warmer or cooler home to work, or vice versa, etc.?

Even in modern buildings in moderate climates, where people have access to appropriate clothing and housing with central heating and cooling, people often acclimatise to very different temperatures in different seasons.[4] Perceptions of how an actual temperature feels in the cooler north change from autumn, when the fabric of the building still has some summer warmth in it, to spring, when all the heat has been sucked out of the walls by winter. Air temperatures are the same, but the radiant temperatures from walls of a space may be very different. Staying comfortable across seasons in homes requires not only adaptation of the environment but also the expectations of occupants who have acclimatised using clothing, lifestyles, and mind-sets to actually change what they perceive they need to make themselves comfortable. People's comfort and health dictate how well they feel; their well-being and age and health will influence the extent to which ordinary behaviours can be used to avoid discomfort outside normally accepted comfort levels,[5] how much thermal stress a person can tolerate, and how likely they are to suffer ill-health as a result of the conditions they occupy.

Figure 24.2 gives an idea of the factors at play in understanding the extent of the resilience of a person or a building, based loosely on the area under a stress/strain curve for metals. People or

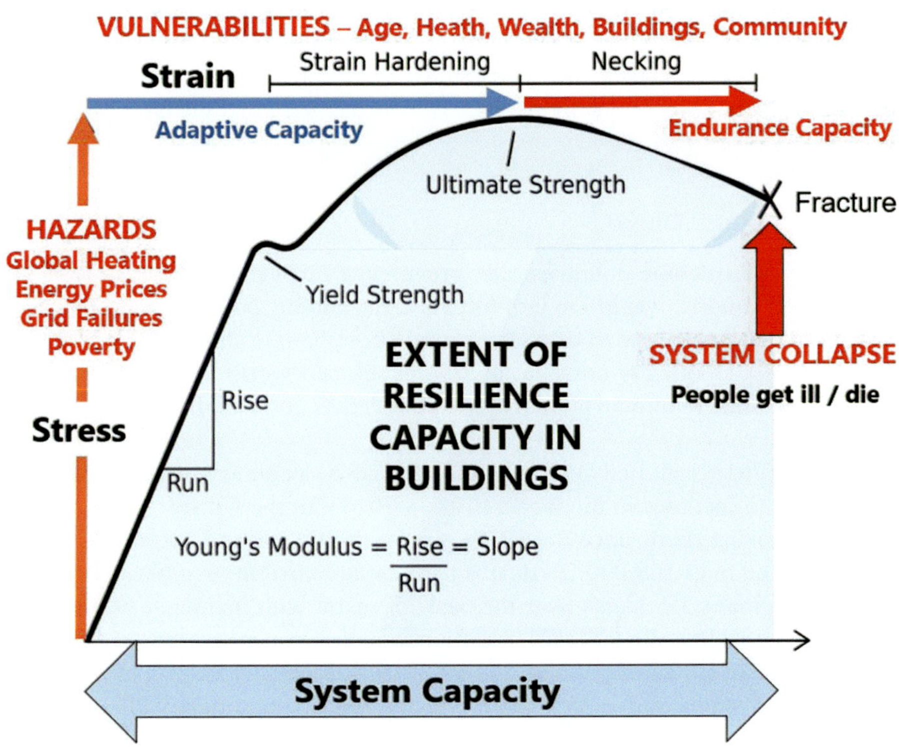

FIGURE 24.2 This diagram is based on a metallurgists stress/strain curve for a metal bar. Hazards for individuals can be seen as the stressors, and the strain people can endure depends on their vulnerabilities.

Source: Diagram by Susan Roaf based on graph in Campbell (2008).

buildings are impacted by hazards or stresses and can adapt to them to a certain point according to how vulnerable they are to those stresses. When a person can no longer adapt to conditions, they endure them, for instance, as energy prices soar during very hot or cold weather, until a point they collapse and become very ill and/or die. Physical health will significantly bolster the resilience of an individual, even under extreme conditions.

Behaviours and activities undertaken as part of the customary lifestyle of a person or population are important in maintaining steady body temperatures within a known environment. More active lifestyles in cold climates over a day are essential for comfort, in contrast to less-active daily routines in warmer climates. The Tuvan nomadic pastoralists in Siberia in Chapter 7 were physically active for 10 hours a day while exposed to temperatures below 0°C and dealt with the cold by being between 17 and 31% more active than predicted, as probably by theoretical physiological models. To maintain such levels of activity, you need to stay well and fit to both adapt to and endure extreme cold in this case.[6]

Routine physical activity has been shown to improve body composition, weight control, and bone density; reduce blood pressure and inflammation; augment cardiac functions; and so on, but also important is that it is associated with improved psychological well-being through reduced stress, anxiety, and depression. Psychological well-being is particularly important for the prevention and management of cardiovascular disease, but it also has important implications for the prevention and management of other chronic diseases, such as diabetes, osteoporosis, hypertension, obesity, cancer, and depression.[7] There appears to be a graded linear relation between the volume of physical activity and health status, such that the most physically active people are at the lowest risk.

Comfortably Rich and Obese

In low- and middle-income countries, the prevalence of overweight and obese people is generally higher among wealthier individuals than among poorer. However, as nations become richer, the percentage of overweight and obese people shifts to poorer demographics. The recent rise in obesity rates around the developed world is perhaps the most rapid population-scale shift in human phenotypes, their shapes and weights, to occur in recorded history.

Bentley et al. (2018[8]) studied obesity rates in the USA, a country that has one of the highest GDP per capita incomes in the world (fifth in 2024). In the United States, adult obesity rates have been rising there since the 1970s, and currently, around two-thirds of US adults are overweight and over 100,000 US deaths per year are attributed to obesity. Obesity rates having tripled in many US states over the past 25 years, with incidence accelerating since 1990, when around 11% of a typical US state population was obese and no state had more than 12% obesity in its adult population. By 2015, US obesity rates had more than doubled, with several states above 35% adult obesity and no state below 20% obesity in the population.

Medical research on the causes of obesity has highlighted the impacts of diet and, in particular, drinks with processed sugars, food deserts, and the positive correlation in the United States between reported lack of activity and exercise (leisure hours) and obesity. The health consequences of obesity, as shown in Figure 24.3, can be many and severe, and theories for its rise range, including attributing it to social transmission within population clusters, via

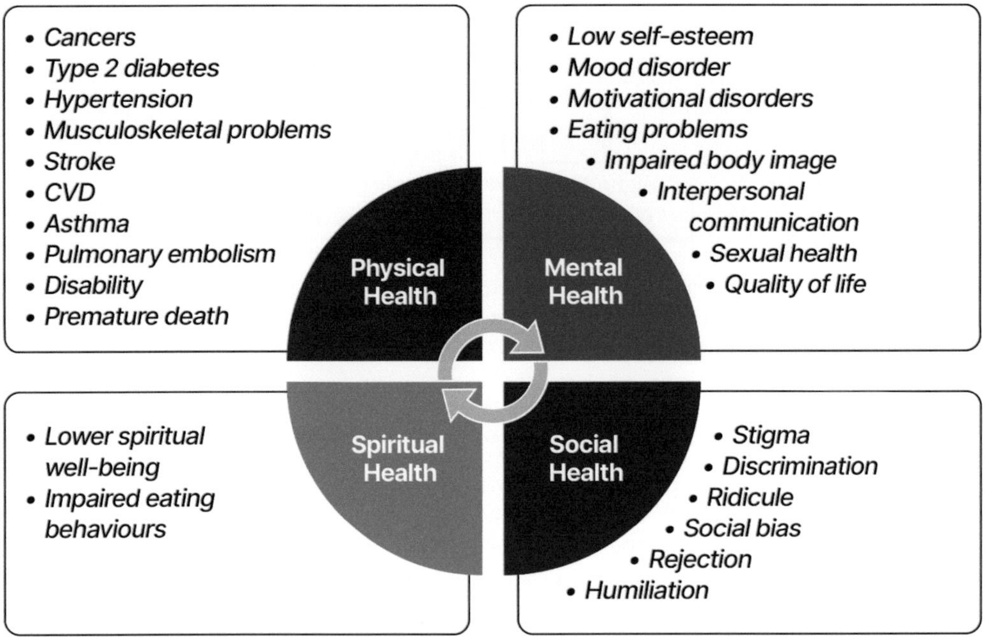

FIGURE 24.3 The most common consequences of obesity on the main domains of health.

Source: Djalalinia et al. (2020).[9]

learned behaviours at a societal level. It has even been proposed that in high-income countries, heavier women tend to marry into poorer households due to through an 'anti-fat discrimination' in marriage.[10] By contrast, promote the idea that it is where you live that has social learning (SL) explanations emphasise if obesity clusters in social networks merely because those clusters are similar with a shared cultural ancestry, with similar habits while adapting to similar environments.

Technological rationales for the growth in obesity are rare, despite a plausible link between indoor comfort temperatures and obesity existing. Some people who can afford to stay indoors more are less active because they have become habituated to cooler indoor environments in warmer climates, and warmer indoor climates when it is cold out, not liking the thermal shock experienced when travelling from in to out, the Dislike of Going Out (DOGO) noted in Figure 24.4. A resulting lack of activity and more time spent watching television or on computers can lead to weight gain. One study done in southern Brazil found that occupants in four mixed-mode and one air-conditioned offices that overweight occupants are more likely to register 'warm' discomfort than normal or underweight occupants, whereas normal or underweight occupants are more likely to experience 'cold' discomfort. Air conditioning use was more frequent amongst those who reported being warm.[11]

In cold weather, having a uniform indoor thermal environment all day every day reduces exposure to seasonal cold indoors and out, to the environments that enable the body to learn how

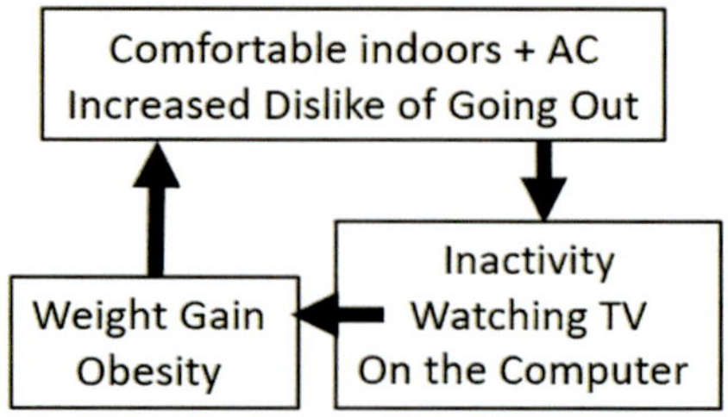

FIGURE 24.4 DOGO: The Dislike of Going Out. A link between artificially cooled environments discouraging activity and exercise, leading to weight gain, which further leads to increased DOGO.

Source: Susan Roaf.

to adapt to cold, not least by developing brown fat tissue in the body that stores energy (heat) in winter. There is growing interest in the causal link between increased time spent in 'thermal comfort' and increased white body fat in a population, despite gaps in understanding of the pathways involved. Decreased energy expenditure in keeping the body thermally comfortable in heated buildings should be matched by decreased energy intake, but in a food environment in which energy-dense and tasty foods are widely available, that often does not happen. There is no doubt that more work needs to be done on both short-term and long-term thermal exposures of 'comfort regimes' on body weight that could lead to the development of novel therapies to address obesity at an individual and a population level.[12]

Comfort Foods

In stressed, or depressed, humans, chronic stress either induces increased comfort food intake and body weight gain or decreased intake and body weight loss. Many people in fuel and food poverty in cooler climates gain weight not least because a common mechanism used to cope with cold is to eat *comfort foods.* Eating cheap, high-fat, and high-carbohydrate comfort food does cheer people up and may make them feel and function better, by dampening stress and anxiety responses. However, it often produces abdominal obesity and affects other hormonal functions that can lead to dysphoria, a mental state in which a person has a profound sense of unease or dissatisfaction and an obesity that is strongly associated with type II diabetes, cardiovascular disease, and stroke. Eating comfort foods may temporarily improve mood but is likely to be bad for long-term health.[13]

Personal Comfort and Resilience

Whether rich or poor, the well-being of populations is intimately tied to their own health and comfort experiences, and also the health of the planet. Around 40% of the annual global CO_2 emissions come from the built environment, and much of that from the burning of fossil fuel to provide pleasant and healthy indoor spaces and protecting people from weather extremes. It is mainly the rich who are helping drive climate change with their disregard for the global impacts of their energy use in energy-hungry buildings or over-sized cars. Energy is too often unnecessarily squandered to provide tightly controlled thermal environments indoors, with some buildings having thermostats set to the same target temperature all year round. However, it is increasingly clear that exposure to constant, comfortable indoor temperatures can have significant negative health impacts on health and deteriorate people's physiological capacity to deal with thermal challenges.

Lack of access to affordable heat or coolth is a chronic and growing problem, not least during extreme weather events, where we regularly see people living on the streets in temperatures above 50°C or in freezing conditions. Importantly, for those who can afford comfort conditioning, spending time outside the thermal comfort zone is known to enhance human thermoregulatory capacities and thermal resilience, while also improving metabolic and cardiovascular health. Allowing temperatures to vary diurnally and seasonally indoors, and using passive adaptive features like natural ventilation and shading, was shown for one new house in the Netherlands to save well over half of all energy used in buildings,[14] slashing energy costs and CO_2 emissions while designing in the potential to stimulate physiological and psychological adaptation to occupied climates for individual that can lead to improved personal resilience, health and well-being.[15]

In a heating world, the well-being of the rich, or the poor, will increasingly depend ultimately on how good a building they occupy is and how they stay comfortable.

Buildings and Well-Being

Many different physical attributes of a space or building can affect the health and comfort of occupants, including its thermal, visual, and acoustic performance, as well as the quality of the air and the presence or absence of chemicals and pathogens indoors. There are a number of useful, classic texts on the subject, like the *Rosehaugh Guide to Healthy Buildings*,[16] or practical guides, like *Healthy Housing*,[17] from the 1990s, when issues around sick building syndrome were researched as problems grew, associated from the 1990s onwards on the increasing popularity of mechanically serviced buildings without opening windows.

In parallel with this trend came the increasing focus of laboratory studies on individual elements and attributes of building systems that affect health. Noise, chemicals, pathogens, light, solar gain, ventilation, etc. were all studied in isolation. Philomena Bluyssen makes the point on the health of indoor environments that unhealthy conditions in buildings are often caused by several factors acting together, causing multiplier effects.

A good example of the need to think holistically related to materials and furnishings and fittings in buildings that can have a huge impact on the health of occupants and, in turn, on their well-being.[18] Apparently benign materials can become hazardous as indoor temperatures change. As a room gets warmer and drier, off-gassing of chemicals from furniture and fittings in a space can exacerbate health issues for the skin, ears, nose, throat and headaches and tiredness. As indoor air in a building gets colder and moister, due perhaps to poor construction, inadequate services, or leaking fabric, respiratory and skin illnesses can be triggered, affecting the general health in individuals. Choice of healthy, more natural materials becomes essential for the health, well-being, and comfort of occupants in the face of evermore extreme weather events occurring.

It is the rising temperatures that are the greatest cause for concern, affecting materials, the performance of services, and the challenge of not only staying thermally safe in and around buildings during extreme weather events but also remaining adequately comfortable between them, in everyday lives. For those who actually have a choice of the type of building to occupy, issues around comfort inside it and the cost and health impacts of their choice will increasingly drive purchasing decisions for many. It is important to get some perspective on what really matters for a person's well-being over time. Which region, climate, part of a city, side of the street, floor, side, or type of a building chosen to settle in may dictate the comfort and well-being of a person over their whole lifetime. Figure 24.5 shows the land use cover for the state of Singapore with the cluster of mid- and high-rise buildings in the centre of its southern shore.[19] Figure 24.6,

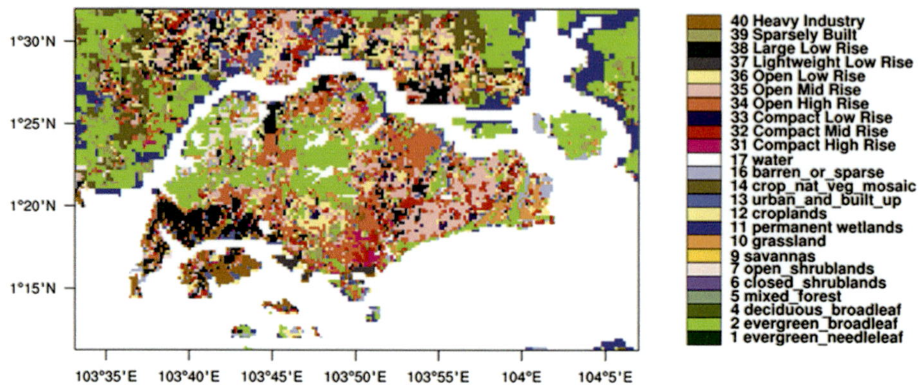

FIGURE 24.5 Land use and land cover map for Singapore showing the clusters of high-rise buildings and open spaces.

Source: Mughal et al. (2020).

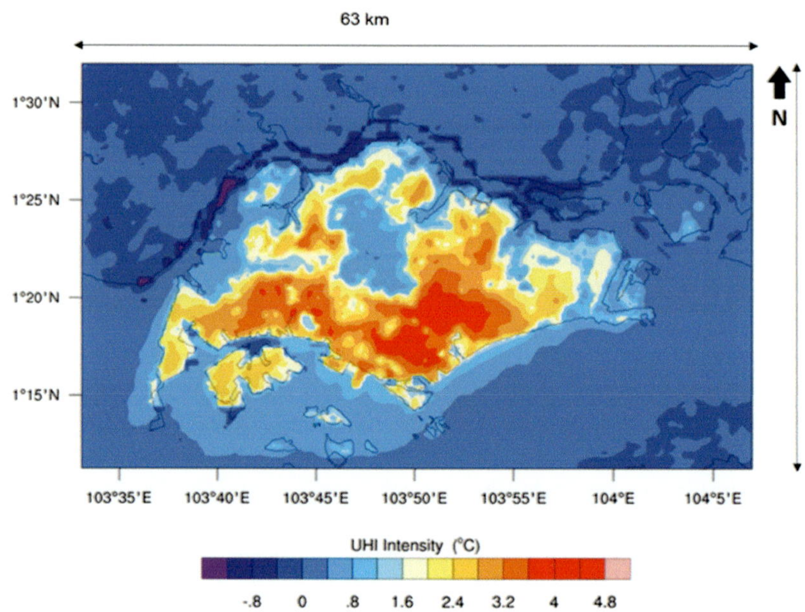

FIGURE 24.6 Map of the urban heat island (UHI) in Singapore in 2016 showing the overlap between the clusters of high-rise buildings and the maximum temperatures of the UHI.

Source: Borzino et al. (2020).

of the urban heat island (UHI) effect for the state, clearly indicates the temperature differences between the rural areas of the island and the high-density heart of the city.[20]

In the future, UHI maps may influence where well-informed people choose to buy homes, just as many now avoid buying homes on floodplains. Many cities globally have UHI satellite maps available online that may show that some areas of a city can be up to 5°C (9°F) to 10°C (18°F) cooler or warmer than adjacent zones. People in hotter areas or buildings are more vulnerable during heatwaves to associated illness and can pay significantly more for annual cooling bills. The converse is that people may, in cold regions, flock to warmer cities.

Three Ways to Improve Occupant Well-Being

1. Regular health checks on individuals, buildings, and cities
2. Accessible adaptive opportunities in buildings
3. Education for all on dealing with heat and cold stress

Regular Building Health Checks

As the climate changes, the well-being of some people will be adversely affected more than others'. The vulnerable, who may be old, ill, young, poor, disabled, or just unable to cope, must be regularly assessed to maximise their well-being and that of those around them, particularly during extreme weather events, as temperature records and comfort thresholds are exceeded. This can be done at a city level, too, as the technology exists to constantly monitor different areas of the city so that actions to keep people thermally safe can be targeted at the areas where the hazard is greatest and, within those areas, at the most vulnerable people, to minimise health and mortality impacts.

The greatest impacts of more climate heat on well-being will be on those in the worst-performing buildings and also in buildings where individuals have no control over their own indoor environment. In our 2022 book on resilient comfort,[21] Karel Kabel points out that it is impossible to create buildings as an environment where all users are satisfied, and that users are significantly more critical and less tolerant of buildings with artificial environments, unconnected to the external climate.[22] He also found that indoor environments that are fixed and maintained by centralised control systems do not always bring a feeling of well-being. There is much evidence that lack of control over their occupants' environments of temperature, window opening for ventilation, shading, and lighting increases discomfort in many offices in particular[23] controls. Kabel stresses the need, and provides a methodology for, undertaking regular health checks on buildings to fine-tune the quality of indoor environments over time and enhance comfort satisfaction within them, to ensure that all the building services are working optimally and occupants are enabled to incrementally adapt both the building and themselves to stay comfortable in changing internal and external conditions. Just like having an annual MOT service on a car or a regular health check, checks avoid failures and can increase comfort and well-being, as well as saving energy use in and emissions from buildings. There are great books on post-occupancy evaluation,[24] but not only does the title imply it is done once after construction, but also many of the indicators of what constitutes a good building treat the occupants as problems, measuring absenteeism, staff churn, accident rates, incident rates measured against revenue, profits, staff utilisation and performance, and customer feedback. Well-being, comfort, and happiness do not even figure in some indexes as a process is described promoting the corporate interests rather than those of the employees. Regular building health checks need to be done on the building environments and the people who occupy them.

Accessible Adaptive Opportunities in Buildings

In an increasingly unpredictable climate, people's need to feel they are able to change the building when they become uncomfortable is extremely important. Many readers may have experienced the feeling in their own home of rooms getting too hot, or cold, and there is nothing they can easily do about it. Many modern buildings have no or inadequately opening windows to comfort and cool occupants, or shades in places in which to escape the heat or cold. The controls of larger buildings are universally out of the occupants' hands, yet control over personal comfort is highly important for people's well-being. This realisation, as well as the need to drastically reduce the cost of comfort and emissions from buildings, is driving the move back to *mixed-mode* buildings, in which for as much of the year as possible, local, passive steps can be taken to improve local comfort and, consequently, health and well-being. Heschong and Day[25] point out that the challenge is not just to design the building to include a full range of adaptive opportunities but also to give occupants the competence and information needed to operate the adaptive opportunities. They sensibly propose seven ways to do that by making the opportunities:

Logical, durable, reliable, accessible, responsive, adjustable, and repairable

How simple is that! No complex modelling needed to do that in a design, just a lot of clear thought and hard work. The rewards are high, and studies show that indoor temperatures of the mixed-mode buildings were more dynamic and connected to outdoor temperatures and seasonal conditions; people are much happier when they can control a variety of adaptive opportunities, particularly openable windows, pedestal fans, and local heaters; and individuals within groups are enabled to adjust their local conditions to meet their perceived thermal preferences.[26] Providing and enabling the use of a range of local adaptive opportunities can improve the health and, in turn, the well-being of individuals and employees.

Education for All on Dealing With Heat and Cold Stress

The first line of defence against more people being unwell, suffering discomfort, and dying during increasingly extreme weather events is to educate people on how they might act or design to make buildings thermally safer, and their occupants healthier. Start with the home, which is where people most often become thermally distressed, unhealthy, as die during heatwaves and cold snaps. Education should cover how to provide a passive means to heat or cool buildings; what range of adaptive furniture, fittings, and behaviours will make a difference; and importantly, how to use them. Everyone should be educated, starting with the vulnerable and those who have care over them, the population at large, and importantly, the people who shape our built environments.

Educating Primary School Children

Comfort education should start in primary schools. Domínguez-Amarillo and Fernández-Agüera studied the comfort perceptions of a cohort of primary school children and used the reported wide range of issues that the children brought up themselves to educate themselves and the children on heat adaptation in Seville. Many factors were declared and shared that should cost no money, including the positions of tables in classroom, the study hours for different subjects, access to and operation of windows, blinds and lights to improve comfort. Children came to us with hugely innovative ideas, like changing hairstyles and clothes, tree planting, and a range of classroom ventilation strategies. They hoped that raising awareness of actions that the children could take to avoid discomfort they hoped that the thinking and the lessons would then be taken home to improve the comfort and well-being of the families of those children.[27]

Educating University Students

Every university design and building course should now include modules on comfort and well-being, introducing students to their own thermal environments and teaching them to 'feel' temperatures and humidities in the buildings they occupy, in their neighbourhood and city, so they start to re-learn how to learn from the world around them and trust in their own judgements, based on that learning, as touched on in Chapter 20.

In the 21st century, comfort rules and approaches designed in the 20th century for totally air-conditioned buildings really no longer apply, and the best way to educate designers of thermally safe buildings in an unpredictable future is to train them to understand thermal environments and trust in their own judgement in designing for them. The well-being of populations around the world is dependent on legislators and designers understanding that the comfort requirements for people in the regions, cities, buildings, and cultures in the myriad different places around the world are not the same, as assumed in many modern comfort standards. The role of the 21st-century building designer is to use their innate understanding and intelligence to provide buildings that will keep people safe in the very different and unpredictable climates of the future. To provide a solid foundation for the well-being of future generations, the design course should now be teaching students that innate understanding of temperature and comfort to give building occupants in the future a chance of surviving comfortably and safely indoors.

Educating the General Public

As a heatwave or cold spell appears on the horizon, emergency communications are sent out across all media. Wear a hat, put on sunscreen, avoid the sun, put on more clothes, and information on the location of cold or hot weather comfort refuges. There are many excellent sites, for example, from the like of the Red Cross,[28] but the actions have to be achievable. The US government advice on surviving extreme heat if you don't have air conditioning is to go to a local climate shelter.[29] However, many cannot get to a shelter; they don't know where one is or cannot afford to get there, with no one to help them either make suggested changes to their homes or get to shelters.

As we saw in the Geelong Council study on building climate shelters into homes, small interventions like making one room a cool or warm refuge can really make a difference to people's physical and mental well-being, but the information given out must be correct and un-biased to be really useful. US government advice states:

> *Do not rely on a fan as your primary cooling device. Fans create air flow and a false sense of comfort, but do not reduce body temperature or prevent heat-related illnesses.*

This is factually inaccurate and biased towards discouraging fan sales in favour of air conditioning systems. In fact, fans do lower body temperature as they result in more evaporative cooling of sweat from the skin, so lowering its temperature, and in so doing cools the body, and so in many cases preventing heat-related illness. What they do not do is lower room temperatures. A guide, 'How to Properly Cool Your Home With a Fan', was published in *Time* magazine on 5 June 2024, full of correct and useful advice on the importance of using fans for providing low-energy, low-carbon, effective cooling in a variety of different ways, was published of the type that could genuinely save the lives of people in their own homes.[30] The authors pointed out that to get the most out of these devices, it helps to experiment with several ways of using them. Every home is different, and comfort is subjective, so it isn't one-size-fits-all. They go on to outline how to tailor fan usage with homes and lifestyles to save money and energy and increase comfort, health, and well-being for a broad range of users.

For education of the general public through any medium, only reliable sources should be listened to. For governments and larger organisations, experts providing advice should be filtered for affiliation bias, and advice taken not just from one profession or industry to ensure that the aims of the funders and the providers are aligned.

Goldilocks Homes: Educating the Elderly

Older people are often overrepresented in morbidity and mortality statistics associated with hot and cold weather, despite spending most of their time indoors in their houses. People adapt to very different thermal conditions not only because of the climate they occupy but also because of the culture they are part of. One size will not fit all, and in a heating world, not only is education vital to keep people thermally safe in their own homes, but also, local councils in different regions must develop bespoke programmes to enhance the health and well-being of their own populations, particularly their older citizens. The economic case for adapting behaviours to avoid discomfort is clear, not just to save money by keeping the older people fit and well in their own homes, but also reduce the burden of their care on the state by keeping them out of hospitals and care homes.

After an increasing numbers of heatwaves in South Australia, the government pioneered a programme designed to keep older people thermally safe, healthy, and alive even during extreme weather events. Research on *improving thermal environment of housing for older Australians* looked at the relationships between the indoor environment, building characteristics, thermal comfort, and the perceived health/well-being of older South Australians over a study period that included the warmest summer on record and a cold period, when temperatures in some homes reached over 35°C in summer, dropping to around 10°C in winter. The cost of energy was a key factor in making occupants more often turn to adaptive behaviours, such as changing clothes, using fans and local heaters to achieve thermal comfort instead of turning on the home air conditioning or heating systems, even though they perceived themselves to be less comfortable and less healthy by doing so. Occupants felt that their health and well-being worsened as temperatures rose above 28°C or fell below 15°C. During hot weather, they experienced tiredness, shortness of breath, sleeplessness and dizziness, coughs and colds, painful joints, shortness of

FIGURE 24.7 Many elderly resort to using adaptive behaviours rather than turning on the home heating or cooling systems to save money.

Source: Soebarto et al. (2021).

breath. In cold weather, influenza was their main reported health problem. A health/well-being perception of 'very good' was achieved with indoor operative temperatures between 18.4°C and 24.3°C and a 55% relative humidity for this group of mainly White middle-class people from European stock.[31]

The University of Adelaide researchers then developed the concept of 'personas', putting the older people into a number of discrete 'thermal personalities'. Hierarchical cluster analysis (HCA) was implemented to analyse the features of research participants, resulting in six distinct clusters. Quantitative and qualitative data from the project were then used to develop the thermal personalities of each cluster. The thermal personalities represented different approaches to achieving thermal comfort, taking into account a wide range of factors, including personal characteristics, ideas, beliefs and knowledge, house type, and location.[32]

They then produced a guide, *Thermal Comfort at Home: A Guide for Older South Australians*, designed to be user-friendly and useful, because different individuals could choose which of the thermal personalities and which particular home type and lifestyle related most closely to them.[33] The guide provides a brief explanation about what thermal comfort is and why it matters, as well as the key factors that influence thermal comfort. To assist older readers to find the most relevant strategies for them, they are asked to review their own space and find which comfort profiles fit them the best, to then explore options to improve thermal comfort in and around the home. These strategies vary – from adopting certain personal behaviours or actions as in Figure 24.7, making either smaller or bigger changes to the design and operation of their current dwelling environment, to using equipment or appliances. Readers are assisted in their thinking and choices by six bespoke case studies and practical solutions to help them get a plan to improve their own comfort.

An important step in the development of the guide document was that a draft was sent to a group of research participants so that they could provide detailed comments not only on the contents but also on the entire format of the document, images, and illustrations, and even the font type and sizes. Workshops were subsequently held to discuss their feedback and detailed comments so that the final format was relevant to their circumstances. This user feedback is essential if an education programme is to have meaningful impact.

The guide aims to help occupants achieve a Goldilocks home environment, not too hot and not too cold: comfortable and pleasant specifically for them, because thermal comfort feels different for every individual. Maintaining and improving personal comfort conditions will also improve their overall health and well-being.

Right at the heart of the challenge of keeping populations well and healthy is the issue of thermal comfort. Some of us live in a miracle world, where work has been made easy by machines and where comfort can be created at the flick of a switch or the setting of a thermostat, but at what cost? Many modern buildings have artificial 'ideal' indoor climates that lock nature out and cause people to become dependent on the temperatures defined as comfortable by experts who know nothing about their lives, cultures, or climates, with many overt and opaque impacts on their physical, social, emotional, and mental health.

Notes

1 https://en.wikipedia.org/wiki/Well-being

2 https://en.wikipedia.org/wiki/Hauora

3 Ortiz, M., S. Kurvers and P. Bluyssen (2017). A review of comfort, health, and energy use: Understanding daily energy use and wellbeing for the development of a new approach to study comfort, *Energy and Buildings*, 152, pp. 323–335, ISSN: 0378-7788. https://doi.org/10.1016/j.enbuild.2017.07.060

4 van Ooijen, A.M.J., W.D. van Marken Lichtenbelt, A.A. van Steenhoven and K.R. Westerterp (2004). Seasonal changes in metabolic and temperature responses to cold air in humans, *Physiology & Behavior*, 82(2–3), pp. 545–553, ISSN: 0031-9384. https://doi.org/10.1016/j.physbeh.2004.05.001

5 van Marken Lichtenbelt, W., Hanssen, M., Pallubinsky, H., Kingma, B. & Schellen, L. (2017). Healthy excursions outside the thermal comfort zone, *Building Research & Information*, 45(7), pp. 819–827. https://doi.org/10.1080/09613218.2017.1307647

6 Sellers, A.J., D. Khovalyg, G. Plasqui, et al. (2022). High daily energy expenditure of Tuvan nomadic pastoralists living in an extreme cold environment. *Scientific Reports*, 12, p. 20127. https://doi.org/10.1038/s41598-022-23975-3

7 Warburton, D.E., C.W. Nicol, S.S. Bredin (2006). Health benefits of physical activity: The evidence. *CMAJ*, 174(6)(March 14), pp. 801–809. https://doi.org/10.1503/cmaj.051351. PMID: 16534088; PMCID: PMC1402378.

8 Bentley, R.A., P. Ormerod and D.J. Ruck (2018). Recent origin and evolution of obesity-income correlation across the United States, *Palgrave Commun*, 4, p. 146. https://doi.org/10.1057/s41599-018-0201-x

9 Djalalinia, S., M. Qorbani, N. Peykari, R. Kelishadi. (2015). Health impacts of obesity, *Pakistan Journal of Medical Sciences*, 31(1)(January–February), pp. 239–242. https://doi.org/10.12669/pjms.311.7033. PMID: 25878654; PMCID: PMC4386197.

10 Hruschka, D.J., and S.Y. Han (2017). Anti-fat discrimination in marriage more clearly explains the poverty-obesity paradox, *Behavioral and Brain Sciences*, 40(January), p. e120. https://doi.org/10.1017/S0140525X1600145X. PMID: 29342583.

11 Forgiarini Rupp, R., J. Kim, R. de Dear and E. Ghisi (2018). Associations of occupant demographics, thermal history and obesity variables with their thermal comfort in air-conditioned and mixed-mode ventilation office buildings, *Building and Environment*, 135, pp. 1–9, ISSN: 0360-1323

12 Johnson, F., A. Mavrogianni, M. Ucci, A. Vidal-Puig and J. Wardle (2011). Could increased time spent in a thermal comfort zone contribute to population increases in obesity?, *Obesity Review*, 12(7), pp. 543–551. https://doi.org/10.1111/j.1467-789X.2010.00851.x
13 Dallman, M.F., N. Pecoraro, S.F. Akana, S.E. La Fleur, F. Gomez, H. Houshyar, M.E. Bell, S. Bhatnagar, K.D. Laugero, S. Manalo (2003). Chronic stress and obesity: A new view of "comfort food". *Proceedings of the National Academy of Sciences of the United State of America*, 100(20) (September 30), pp. 11696–11701. https://doi.org/10.1073/pnas.1934666100. Epub 2003 September 15. PMID: 12975524; PMCID: PMC208820.
14 Alders, E.E. (2017). Adaptive heating, ventilation and solar shading for dwellings, *Architectural Science Review*, 60(3), pp. 150–166. https://doi.org/10.1080/00038628.2017.1300132
15 Pallubinsky, H., R.P. Kramer and W.D. van Marken Lichtenbelt (2023). Establishing resilience in times of climate change – a perspective on humans and buildings. *Climatic Change* 176, p. 135. https://doi.org/10.1007/s10584-023-03614-0
16 Curwell, S., C. March and R. Venables (1990). *Buildings and Health*, London: RIBA Publications, ISBN: 0947877789
17 Ranson, R. (1990). *Healthy Housing: A Practical Guide*, London: E & F N Spon. ISBN: 0419154000
18 Woolley, T. (2017). *Building Materials, Health and Indoor Air Quality*, Routledge. ISBN: 978-1-138-93447
19 Mughal, M.O., X.-X. Li and L.K. Norford (2020). Urban heat island mitigation in Singapore: Evaluation using WRF/multilayer urban canopy model and local climate zones, *Urban Climate*, 34(100714), ISSN: 2212-0955. https://doi.org/10.1016/j.uclim.2020.100714
20 natalia.borzino@frs.ethz.ch Borzino, N., S. Chng, M. Mughal and R. Schubert (2020). Willingness to pay for urban heat Island mitigation: A case study of Singapore, *Climate*, 8. https://doi.org/10.3390/cli8070082
21 Nicol, F., H. Rijal and S. Roaf (Eds) (2022). Resilient comfort standards, in *The Routledge Handbook of Resilient Thermal Comfort*, London: Routledge, ISBN: 978-1-032-15597-5
22 Kabele, K., Z. Veverková and M. Urban (2022). Methodology of IEQ assessment in energy efficient buildings, in *The Routledge Handbook of Resilient Thermal Comfort*, Edited by F. Nicol, H. Bahadur Rijal and S. Roaf, London: Routledge, ISBN: 978-1-032-15597-5
23 Rupp, R.F., J. Toftum and E. Ghisi (2022). Thermal comfort and occupant disposition in mixed-mode offices in a Brazilian subtropical climate, in *The Routledge Handbook of Resilient Thermal Comfort*, Edited by F. Nicol, H. Bahadur Rijal and S. Roaf, London: Routledge, ISBN: 978-1-032-15597-5-4
24 See for instance: Oseland, N. (2022). *Beyond the Human Zoo: Humanising the Office*, London: Routledge, ISBN: 978-0-367-65533-4 and Baird, G. (2010). *Sustainable Buildings in Practice: What the Users Think*, London: Routledge, ISBN: 978-1-13-847147-4
25 Heschong, L., and J. Day (2022). Why occupants need a role in building operation: A framework for resilient design, in *The Routledge Handbook of Resilient Thermal Comfort*, Edited by F. Nicol, H. Bahadur Rijal and S. Roaf, London: Routledge, ISBN: 978-1-032-15597-5
26 Shahzad, S., and H.B. Rijal (2022). Mixed mode is better than air-conditioned offices for resilient comfort: Adaptive behaviour and visual thermal landscaping, in *The Routledge Handbook of Resilient Thermal Comfort*, Edited by F. Nicol, H. Bahadur Rijal and S. Roaf, London: Routledge, ISBN: 978-1-032-15597-5
27 Domínguez-Amarillo, S., J. Fernandez-Aguera, M. Gonzalez and T. Cuerdo (2020). Overheating in schools: Factors determining children's perceptions of overall comfort indoors, *Sustainability*, 12, p. 5772. DOI: 10.3390/su12145772
28 https://www.redcross.org.uk/get-help/prepare-for-emergencies/heatwaves-uk/advice-and-tips
29 https://www.ready.gov/heat
30 https://time.com/6985139/how-to-use-fan-air-conditioning/
31 Hansen, A., T. Williamson, D. Pisaniello, et al. (2022). The thermal environment of housing and its implications for the health of older people in South Australia: A mixed-methods study, *Atmosphere*, 13(1), p. 96. https://doi.org/10.3390/atmos13010096
32 Bennetts, H., L. Arakawa Martins, J. van Hoof and V. Soebarto (2020). Thermal personalities of older people in South Australia: A personas-based approach to develop thermal comfort guidelines, *International Journal of Environmental Research and Public Health*, 17(22)(November), p. 8402. https://doi.org/10.3390/ijerph17228402
33 Soebarto, V., H. Bennetts, L. Arakawa Martins, et al. (2021). *Thermal Comfort at Home: A Guide for Older South Australians*, The University of Adelaide, ISBN: 978064-6851655

25

WELL-BEING

All About the Mind

Mental Well-Being, Health, and Comfort

People exist in many realities, including:

- *Physical ecosystems*. Their own body systems linked to the physical world around them.
- *Socioeconomic ecosystems*. Encompassing family, community, work and play contacts, society.
- *Mental, intellectual, and spiritual dimensions*. Attitudes, propensities, and beliefs.

The cognitive, bodily, and behavioural mechanisms linking these realities are infinitely complex and explored by many disciplines in medical, psychosocial, and pure science research beyond the scope of this book.

To understand how mental well-being, health, and comfort are linked, we start with looking at the basic need to feel *safe,* not only in the pursuit of living a satisfying life, but also to provide the sort of environment in which people can creatively and productively flourish.

When threatened, the brain and body work together to deal with stressors. In a dangerous situation, the 'fight-or-flight' response is activated and a person is *mobilised* to act. If a person feels overwhelmed or powerless, a body can shut down, disconnected from surroundings and *immobilised*. When feeling safe in social and physical contexts, a third response state can be that of social engagement, when a person feels secure enough to openly communicate with those around them. But everyone is different, and features of an environment that feel comfortable and enjoyable, to some, may be unsettling or frightening to others.[1]

In the following sections are case studies of how experiences of feeling safe, or not, in and around buildings, climates, society, or groups of people contribute to the everyday experiences of a person.

Comfort and Fear

Fear is a feature of nearly all clinical mental health problems and is a root cause of some of the most common ones. It is strongly associated with anxiety disorders and depression and also

DOI: 10.1201/9781315645070-33

strongly linked to poor physical health, as evidenced by the links between obesity and mental psychopathologies. Personal vulnerabilities are also affected by people's ability to cope mentally and behaviourally with events and situations.[2] An extraordinary range of ways in which the built environment affects our mental comfort, health, and well-being is touched on with the following exemplars that highlight some challenges and solutions.

Fear of Crime: The 1995 Chicago Heatwave

During heatwaves, often the most vulnerable group are the elderly poor, many of whom live in buildings prone to overheating. Many fear for their own well-being, and even survival, as their homes warm, then overheat, particularly those who are socially isolated and who are, or simply feel, helpless when dealing with heat stress.

From 13 July 1995, Chicago experienced a week-long heatwave in which temperatures reached over 41°C (106°F) and 700 people died. Soaring electricity use for air conditioning caused power grids to fail, leaving many residents without electricity for up to two days. In his classic book *Heat Wave*, Eric Klinenberg conducted a 'social autopsy', examining the social, political, and institutional factors that made this urban disaster so much worse than it should have been.[3] He described how the literal and social isolation of seniors, the institutional abandonment of poor neighbourhoods, and the cutbacks in public assistance programmes contributed to the high fatality rates. Hundreds of people died behind locked doors and sealed windows, out of contact with friends, family, community groups, and public agencies, in a compound catastrophe that implicated everyone in their deaths.

In a city that one old lady described as 'a shooting gallery, and [she is] a moving target because [she] walk[s] so slowly', she preferred living on the third floor because it was much safer than the ground floor, where she would be even more vulnerable to break-ins, even if it was hotter than the lower floors. Fear of crime was a leading reason that so many vulnerable citizens, many of whom could not afford to buy or run air conditioners, were found dead in their apartments with the doors and windows locked, even when they might have survived the heat if they had had more air movement, if they had opened a window and/or a door. Across America today and in most cities across the world, people lock their doors and windows to prevent thieves and murderers from entering.

For little extra cost, security grilles can be fitted to windows and doors on any floors prone to being accessed by criminals. This is true for detached homes and high-rise apartments as well.

Fear of Falling

During heatwaves, it is common to read reports of people trapped in apartments that overheat. In particular, flats or homes without cross ventilation and that face west will have to cope with low direct sun during the hottest time of day and year. Even flats with opening windows on one side only could find relief with the comfort cooling from a window breeze at temperatures at or below the thermo-neural zone (c. 26–32°C) but are prevented from doing so because the windows only open around 10–17 cm due to limitations of the window design/window restriction regulations to prevent people falling from them.

There are occasional reports of people falling from windows, and the reasons for this are varied, surprisingly, by culture, and almost never linked to outdoor weather conditions. One study from a children's hospital in Zurich analysed data over seven years of children suffering

FIGURE 25.1 A well-ventilated ward in King's College Hospital in 1908, where cross-infection between beds was reduced by cubicle curtains and pathogens purged in strong updraft in still conditions and in high cross-ventilation in windier conditions.

Source: King's College Hospital Archives.

from head injuries/multiple trauma, of which 31 fell out of a building, mainly at home and from the third floor or lower. This study identified young age, an immigrant family setting, low socio-professional category of the parents, dangerous house constructions, inappropriate furniture placement, and summertime evenings as risk factors for serious building falls in children.[4] Of that cohort, sadly, two of the children attempted suicide.

UK advice on preventing suicidal jumps from high-rise buildings covers a range of locations that people can jump from, including stairs, ramps, floors, balconies, roofs, light wells, or atria. Preventative measures cover falling barriers that, in the UK, need only to be a minimum height of 1.1 m above floor level. If a mandatory risk assessment considers there is a significant risk of people attempting suicide, then the barrier height should be higher, 1.4 m or 1.6 m, for instance, where cycle access is adjacent, and 2.1 m height beside railway bridges. Barrier planting is also suggested to keep people away from dangerous edges.[5]

Many flats without opening windows have no access to natural ventilation to alleviate thermal stress during hot weather, while others have very limited access due to opening restrictors. But many do have balconies with relatively low barriers that can be accessed by children and prove fatal if fallen from. Schools with poor ventilation from limited-opening windows in the classrooms often have central atria with surmountable barriers within them. The same is true of many modern hospitals with multi-story 'feature' atria at their heart but no opening windows in their wards and staff rooms that could be opened to effectively purge the spaces of pathogens, as has been in a number of studies during the COVID pandemic.

The optimal situation would be to have openable windows in occupied spaces protected by security grilles that can provide much-needed cross- or side-ventilation while preventing anxiety in occupants about being robbed, assaulted, or falling out.

Fear of Going to Hospital

Many people fear that if they go to hospital, they will get sick, with some reason. Hospital acquired (nosocomial) infections are caused by common pathogens including bacteria, viruses, and fungi, a common one being the bacterium *Staphylococcus aureus*. Others common pathogens include *Escherichia coli* (*E. coli*), *Enterococci*, and *Candida*, which are common culprits, and all can be normally found on the skin and mucous membranes. Recent worrying outbreaks of antibiotic-resistant strains, such as methicillin-resistant *Staphylococcus aureus* (MRSA), that can be especially dangerous are difficult to treat and are increasing.

Healthcare-associated infections (HCAIs) are defined in the USA as infections that occur while receiving health care, developed in a hospital or other healthcare facility that first appears 48 hours or more after hospital admission, or within 30 days after having received healthcare. The scale of the problems they cause are huge. The US Centers for Disease Control and Prevention identified that nearly 1.7 million hospitalised patients annually acquire HCAIs while being treated for other health issues and that more than 98,000 patients (1 in 17) die due to these.[6] Advice on reducing the impacts centres simply on handwashing, more often and better, completely ignoring the aerial transmission routes of pathogens. This also happened before 6 July 2020 with SARS CoV-19 when a group of international scientists complained the World Health Organisation, who was forced to admit that aerosols were the major transmission route for the pathogen.

By May 2020, 15.8% of people in the UK who were being treated for COVID-19 had actually acquired it in a hospital, and over a quarter of those died from the disease (Read et al., 2021). Some 20% of symptomatic COVID-19 patients in hospitals in England in the first wave had acquired their infection in hospital settings, and it was proposed that increased awareness and testing, especially of patients on discharge, was needed to prevent hospitals from becoming vehicles for SARSCoV-2 transmission (Knight et al., 2022). Tens of thousands of medical staff died globally in the pandemic, having been exposed to the virus in the hospitals they worked in, and must be universally praised for their dedication to service when they knew full well of the infection risks, they faced at work.[7]

Evidence that SARS-CoV-2 can be transmitted by aerosols via HVAC ventilation systems emerged early on in the pandemic, but at that point, there was nothing to be done about it. In the deep plan Uppsala Hospital, Sweden, SARS-CoV-2 was detected in the central ventilation systems in that hospital already by April and May 2020, distant from infected patient areas, showing clearly that the virus was transported long distances in airborne form, even with relatively low air change rates in these wards.[8]

Clear evidence exists now that by simply having a regular window opening regime in wards to regularly purge pathogens from the air indoors, rates of cross-infection and deaths can be significantly reduced.[9] Why then were the original principles of healthy hospital design instigated by Florence Nightingale as illustrated in Figure 25.1, using sunlight and fresh air to purge pathogens largely abandoned?[10] With the post-war advent of antibiotics and improved aseptic practices, the medical establishment and their architects began to believe that patient healthiness could be maintained regardless of room design. This is far from true, argues Heschong in writings on healthy daylight.[11]

Hospitals in the future should be planned with shallow footplates and many spaces in them to be naturally ventilated when useful.[12] They should have easily accessible opening windows safeguarded against people falling in occupant accidents and incidents, as has been beautifully achieved in the Ng Teng Fong hospital in Singapore shown in Figure 25.2. Of

FIGURE 25.2 (a) Floor plan and (b) external elevation of the Ng Teng Fong hospital in Singapore. Every patient has their own window close to their bed, with a view to a garden, designed with CPG Healthcare by studio505 and HOK to provide a comfortable, healthy, and soothing environments to help patients heal.

Source: Lip Chuan Lim.

course, hospitals need heating and cooling as well, particularly in some specialised spaces, like operating theatres, but to protect them from catastrophic failure of the sort we first saw in Hurricane Katrina involving HVAC, computer, and grid failures, windows should be enabled with manual overrides.

Fear of Returning to Work

A sense of well-being is inevitably related to issues of trust. Residual concerns about sick building syndrome have been brushed under the research carpet, as we saw in Chapter 8. However, the fear of catching COVID during the pandemic often resulted in a trust wedge developing between employers and employees.[13] Even with advice from experts, many employees did not, and still do not, feel safe about returning to high-end offices for work, not without reason, as many solutions for reducing pathogen transmission in them, like increasing air speeds in room air supply or using high-end filters, are either proven to not always work or are unaffordable.[14,15]

In the aftermath of the COVID pandemic, many people realised their own health and well-being had improved considerably by working largely from home. Employers tried many different ways to lure people to come back to the office. Some were very happy to do so, having been stuck unhappily alone or with others in their own home, where they found it depressing and/or difficult to work in and missed the office culture. Others were much happier working from home, avoiding onerous commute and wasted time in non-productive meetings. Attitudes varied from reluctance to return for health reasons to physical and mental ones, as some were happy to avoid contact with toxic colleagues. Other employees found being around colleagues and friends who understood their industry gave them a sense of belonging and fulfilment, so a range of strategies emerged from redesigning the work culture to enhance camaraderie with increased snack and beverage budgets, free in-house lunches, and gym facilities, to providing collaboration opportunities in shared kitchens and water cooler corners. Some retain high-end employees by adding more benefits to employment packages, like parking spaces, day-care stipends, and flexible work schedules.[16]

Fear of High Winds

People who live in regions that are impacted by very high winds experienced during storms, hurricanes, cyclones, and tornadoes have a primary fear of their homes being blown away or collapsing in situ. In higher-income countries, another growing fear is of being unable to afford the soaring insurance premiums for wind and water damage that can enable buildings to be rebuilt once they are damaged or destroyed.

This explains the move in the Southern States of the USA, including the Florida Panhandle, Alabama, and South Carolina, to build wind-resistant homes. Some are now being built with reinforced concrete walls, heavier roof decking, more fasteners per foot, impact-resistant windows, and doors anchored directly to the concrete, as well as standing-seam metal roofs designed to withstand hurricane-force winds.

The US Insurance Institute for Business Home and Safety (IBHS) has, over recent decades, developed the FORTIFIED homes standards. These are voluntary construction and re-roofing standards developed to strengthen structures against severe weather, particularly strong winds and hurricanes. The standards go beyond basic building codes to enhance a building's resilience and protect it from potential damage during storms, and perhaps the biggest difference is that the standard's FORTIFIED-approved homes must be inspected more thoroughly, more often, and then re-inspected over time.

While concrete-walled structures in Alabama can cost about 50% more than conventional homes, their insurance premiums are considerably less. In one case in 2025 in Pensacola, a 1,500 ft^2 (square feet, or 150 m^2) home valued at around $400,000 had an average quoted annual insurance premium of $1,600 a year. Local, comparable homes built with conventional wood framing and shingle roofs saw house owner premiums as high as $7,000 a year. A study by the Alabama Department of Insurance and the University of Alabama's Center for Risk and Insurance Research found that FORTIFIED-built homes fared significantly better in Hurricane Sally, which hit southern Alabama and north-west Florida in 2020. Insurance company payouts would have been about 75% less if all impacted homes in two coastal Alabama counties had been built to the standards.

There are three comfort issues at play here. There is the mental peace of mind premiums for homeowners who know they have a significantly higher chance of not losing their homes in a windstorm, and also, they can afford their insurance premiums if some damage does occur. There is the added bonus that the FORTIFIED home standard also creates potentially far more thermally comfortable homes with their high-thermal-mass external walls and reinforced, better-insulated roofs. Calculations of embodied and whole life energy and carbon calculations completely ignore these vital aspects of building design in a heating world, resulting in an increasingly fragile built environment and communities that are ever more exposed to catastrophic emotional and physical collapse because of it.[17]

Fear of Fires

In the USA, homeowner insurance premiums are linked to average catastrophe losses per policy at state level and reflect exposures in areas of high natural perils. In Florida, wind and flood damage from hurricanes push up annual premiums, and in California, the highest premiums are localised in areas with the greatest exposures to wildfire risk.

In January 2025, the Los Angeles area wildfires were among the most destructive and costly in history. At least 29 people died and thousands more lost their homes or businesses in the Palisades and Eaton districts. Total insured losses are estimated to be as high as 45 billion US dollars. Many living in those affected areas had little idea of the fire risks they were living with. With increasingly active fire seasons across the globe in hotter summers, the fear of fire is a growing one, not just for those in the paths of large fires, but also those millions of households and people in them whose every day comfortable lives were impacted by smoke around them. The personal and emotional damage done to those in exposed fire and smoke zones were well described in *Fire Weather*, the book by John Valliant on the devastating 2016 wildfires in Canada. He highlighted the fact that for people in the path of that extreme fire, there is little or nothing that can be done to protect homes, but changes of mind-set, preparedness, and behaviours offer avenues for impact reduction and escape strategies for many.[18]

Wildfire smoke is a complex mixture of fine particulate matter, called PM2.5, and gases, such as volatile organic compounds, nitrogen oxides, and carbon monoxide. The composition of the mixture depends on many variables, including the fuels that are burning, the temperature of combustion, the weather, and the distance from the fire, which all affect the severity of the health damage potential indoors, as well as the design and construction of a building itself.

The first line of defence against smoke ingress is to minimise air leakage through the external envelope and its openings. Some are not advocating that at-risk homes have a clean room set up to minimise levels of smoke and other particles during wildfire smoke events. It must be free from activities that create particles, such as cooking with doors and windows that can be kept tight shut to prevent smoke from getting in. It may also contain a portable air cleaner that makes the air in the room cleaner than the rest of the home. Some engineers suggest using expensive HEPA filters in whole building HVAC systems but any ducted air system with air inlet vents to the outside risk sucking fire sparks indoors via intakes that can set the building itself alight. Wet sheets hung over openings in use and the use of a bucket and mop to clean ash off surfaces can reduce fire particles and residue in spaces. Much work is going on now to try to reduce the health and fire dangers of smoke ingress into buildings, but as with other climate risks, the most effective responses come from personal actions and community support.

People must be educated to understand their own susceptibility to the risk and steps taken to support the most vulnerable. Once smoke invades a home, simple advice, like 'Do not do strenuous exercise in smoky environments', kicks in. The harder a person breathes, the more smoke they inhale. The average resting adult breathes about 6 or 7 L of air per minute, but this can easily increase to 60 L during intense exercise. Wear face masks, if necessary, closely fitted to the face. Drink plenty of water. Portable air cleaners with HEPA filters can significantly reduce indoor PM2.5 concentrations when used properly but can also work poorly, so care must be taken with them.

Local community support must be clear and accessible. Libraries, community centres, and shopping malls often have large air filtration systems and relatively good indoor air quality, but they often only provide relief during opening hours. If possible, people must have a plan in place to stay with friends or family or move to a hotel in an area not affected by the smoke.

In areas around major fire, day-to-day comfort can be compromised by smoke ingress for days or weeks on end. As with other weather-related risks, like wind, education of populations in ways to minimise those impacts is important. Advice on good event planning, better and stronger buildings, safe rooms, and behavioural action plans can all significantly contribute to both the mental and the physical comfort of building occupants and communities that live under the growing pall of fire and smoke risk around the world.

Fears About Health and Mortality From High Temperatures

Much current debate on comfort centres on what is a safe temperature to occupy and be included in standards and regulations. The answer is the same as for most other comfort questions: *It all depends*. The adaptive principle holds here too: People adapt to those temperatures they normally occupy, and when they become uncomfortable, they change themselves or their circumstances to restore comfort. There are three different situations to consider:

1. *Mortality risks resulting from heat generated indoors.* The heat source can be managed and will likely be set to a level that is necessary for a particular process or the requirements of a particular population. Workers in steel and glass factories may be exposed to temperatures up to 85°C. They deal with it by standing sideways to the heat, clothing including masks and hat, and regular breaks in cooler places. Process by process and building by building, temperatures can be agreed by occupants and management set accordingly. A major problem does occur with building management systems that are remotely controlled and non-responsive

to room-level variations in demand, particularly in buildings for the vulnerable who are ill, elderly, isolated, and disabled.[19]

2. *Thermal stress resulting from hot weather affecting indoor climates*. Although weather and climate change are predominantly large-scale phenomena, resilience and vulnerability to their effects are highly localised, depending on many factors, including demographics of age, wealth, and the health of occupants.
3. *Thermal stress resulting from hot weather outdoors*. Between the first and thirtieth of March 2024, in India around 25,000 suspected heat stroke cases and nearly a hundred fatalities were confirmed during the hottest month of the year, when temperatures in New Delhi hovered around 50°C (122°F). The European summer of 2023 saw an estimated 61,672 nearly reaching the 71,449 excess deaths registered during the months of June, July, August, and September of 2003. In 2022, mortality rates (deaths per million) were highest in the southern Mediterranean countries: Italy (295), Greece (280), Spain (237), and Portugal (211).[20] Fear is triggered by the fight-and-flight response used by humans to protect themselves from danger. Perhaps we should be educating people to fear intense heat, not least since the high-profile death of health expert Dr Michael Mosely on the Greek Island of Symi on 5 June 2024, from heatstroke on an afternoon walk. Eight other tourists also died of heatstroke in June 2024 in an early heatwave in Greece, including walkers experienced in the local terrain. As temperature records are broken globally, the need for more heat warning and awareness is universal, not just for already-hot regions.

It is not inconceivable that, in the future, people seeking out climate-safe homes will scour available heat and mortality maps, such as those in Figure 25.3, before making their choice of where to settle, just as they now look first at flood maps before contemplating buying a house. The distribution of excess deaths due to heat and cold is also linked to areas of rural and urban deprivation. However, during extreme weather events, even wealth cannot be guaranteed to protect against extreme heat and cold.[21] The best protection a person has in those circumstances is to live in a climate-safe building in a strong and supportive community.

Fear of Zombies

Perhaps the most potent and fear-inducing predator of all is unconstrained capitalism. 'Form follows finance' is an old architectural maxim that related well to property markets prior to the COVID pandemic. The year 2020 saw the confluence of various powerful forces that have re-shaped the prospects for a 20th-century business-as-usual model of property capitalism to continue unabated. The collapse of the Chinese construction market, the Russian invasion of Ukraine, soaring energy prices, and the realisation that high-emissions buildings are a major force in driving climate change all contributed to a step change moment for architecture. One victim of the collapse of conventional development economics may be what Matthew Soules called 'zombie urbanism', which he describes as the rise of under-occupied secondary homes acting as investment properties manifested as 'ghost urbanism'. *Zombies* are houses and apartment units left vacant by absentee speculators, who buy them not to live in or rent out but to hold, as long-term investments. Two extreme forms include, at one end, 'ultra-thins', a term that applies to buildings whose height is proportionally much greater than their footprint, yielding a pencil-thin tower of penthouses. The inverse of that is the 'iceberg', created by those who,

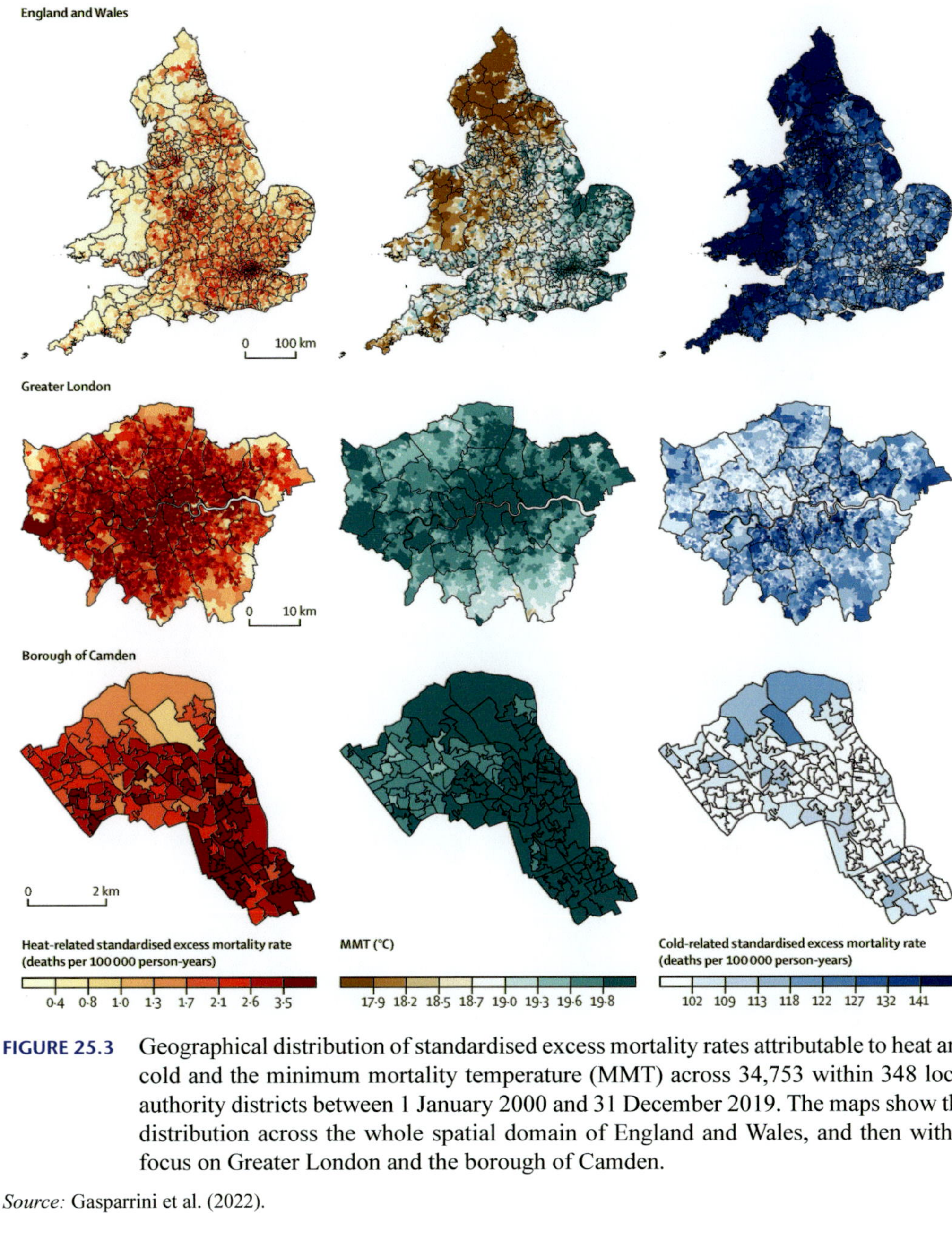

FIGURE 25.3 Geographical distribution of standardised excess mortality rates attributable to heat and cold and the minimum mortality temperature (MMT) across 34,753 within 348 local authority districts between 1 January 2000 and 31 December 2019. The maps show the distribution across the whole spatial domain of England and Wales, and then with a focus on Greater London and the borough of Camden.

Source: Gasparrini et al. (2022).

having maxed out their allowable height and footprint, extend several storeys below ground level, a particular feature of London's Mayfair district, for instance.[22]

Zombie residences and offices, constructed but unoccupied, or only partly occupied, have emerged from landscapes around the world from China to Canary Wharf.[23] A classic example in Australia is No 1. Barangaroo Point, the Crown Casino in Sydney shown in Figure 25.4. Designed for Chinese gamblers, accepted for planning permission by State Premier Barry

FIGURE 25.4 The largely unoccupied hotel and residential apartments No. 1. Barangaroo Point, Sydney, built with laundered money.

Source: Sue Roaf (1 May 2024).

O'Farrel in 2012 despite almost-universal disapproval from Sydney City authorities and civic groups. Why should one fear them? Apart from the fact that ordinary citizens cannot afford to live in them (e.g. in Astana, the new capital of Kazakhstan) and others baulk at living in sealed buildings, letting in laundered money is corrupting to building markets, with knock-on effects to the local population. For instance, the Sydney Casino was built by Crown Resorts but denied a gambling licence after the company was convicted of paying for its construction with money laundered across other casinos. Topped out in 2020, it was sold for A$8.9 billion to Blackstone investment in 2022. Under new management, it gained its gambling licence in April

2024.[24] Only the super-rich can afford a residence in this tower, not least because it is centrally conditioned and apartments need to be artificially conditioned 24/7/365 to avoid developing unhealthy indoor environments. So occasionally occupied by the very few with huge impacts on some of the poorest residents of the city.

The casino's impacts on many vulnerable people in Sydney will likely be devastating. According to researchers from the University of Sydney Gambling Treatment Clinic, strategies designed to protect Sydney citizens from the impacts of problem gambling, like minimum bet size, do not prevent real damage from being done to cohorts like students, new immigrants, and working-class individuals, who are more likely to have their lives and mental health blighted by losing large amounts of money at the casino's tables and electronic games.[25]

The middle classes also suffered in the development of this holiday home for the super-rich. No 1. Barangaroo Point has affected the quality of life of many ordinary Sydney citizens, who see the monster building and its neighbours as blighting the historic headland of Darling Harbour Point and Observatory Hill in Sydney, built so high so that its largely absent, foreign owners can get a good view of the Opera House. We live in a world where the well-being of the many can be trashed by such buildings for the fleeting benefits of the very few.

The Walkie-Talkie building in London, designed by Rafael Viñoly, is another monster structure that blotted out the skyscape of the City of London when built in 2013. It was voted the ugliest and most hated building in Britain in 2015 and described as a 'thuggish comedy villain of a building, which has melted cars and caused winds strong enough to knock people over'.[26] Many were emotionally upset at its destructive visual impact on London City, angered at the flawed planning process, where permission was granted by a committee peopled with some who make a living out of dealing in commercial office space in the city, overseen by city authorities and politicians who put profit above the well-being, mental health, and street-level comfort of the ordinary Londoners. Zombie buildings too often result from short-term thinking and greed by the self-interested and result in long-term mental health problems for the well-being and comfort of the many. Perhaps they should be feared more because of their many negative impacts on populations who understand them as symbols of lack of control over their own futures and those of their fellow citizens, visual symbols of the lack of any semblance of democracy in planning and political systems.

Luring People Back Into Buildings – Lipstick on a Hippopotamus?

The stampede out of multi-storey glass office blocks with large open plan offices during COVID lockdowns in 2020 was almost universal and marked the advent of a new era of *working from home* (WFH), with major consequences for global commercial property markets. At that point, the power balance changed as understanding of the role and importance of physical and mental health and well-being of employees grew. Initially, the move toward WFH was to avoid serious illness or death from COVID. Then, many people realised that they were more comfortable, happier, and largely more productive WFH, not having to tolerate difficult and costly commutes. There are obviously exceptions, not least people lonely or unhappy at home, but for many their sense of well-being was significantly enhanced. It was no longer possible to pretend otherwise, and en masse employees made their feeling known, and employers and building markets responded.

Key drivers for the WFH model are seen as offering an improved work-life balance, promoting better health and well-being for employees, and a reduction in stressful and expensive commuting. Time-richer people have been shown to be more productive and in a 2024 survey three in four workers felt both more productive and motivated and 85% said it had improved their job satisfaction.[27] Hybrid working has also been shown to help talent attraction and retention, reducing recruitment costs, particularly for women who substantially benefit from working from home not least by improving their promotion prospects. There are also huge environmental benefits as people working from or close to home in reduced commuting and office servicing costs and emissions, so WFH practices enables companies to significantly reduce their operational carbon footprint as well as the size of their main office.

Some employers have worked hard to develop a WFH model that suits them and also their employees, depending on the type of work being done, for instance, routine and administrative or conceptual and intellectual. Secondly, the scope of the work – Does it involve individual activities, or is it team-based? And thirdly, what temporal patterns work best for the company, such as hours worked in a day, days in a week, or weeks in a year. A wide range of bespoke agreements are emerging to enrich both workers and employers in both senses of the word.

Some employers mandated that workers must return to full time work or lose their jobs, noticeably during COVID those who own their own headquarter buildings and have to keep them fully operational even when sparsely occupied or empty. While this proved unpopular with many workers, other companies tried to lure employees back to the workplace by making it feel more 'homely' with comfy sofas in cosy and relaxing spaces, encouraging office events and friendships, providing gyms, childcare, and free lunches. Such cosmetic strategies often did not work in practice because the real problem was the buildings itself, its location, and the type of comfort, health and well-bring it actually provides with commuting added to the working days with its rising costs.

Ultimately, it is in employers' interests to listen to what both research says about creating happier and more productive workplaces and their own employees about what work/life balances suit them. Losing workers is expensive, and better staff can move to firms that offer more well-being at work. Some research shows that communications between staff is impaired by a lack of person-to-person contact in the workplace, but to counter this, more advanced means of remote working and hybrid working patters have emerged.[28]

The market impact of the WFH movement was reflected in the 2023 announcement that the HSBC bank will move from its iconic 45-storey Canary Wharf headquarters building back to the City of London when the building lease expires after 2026. Twenty-five years after leaving the City it will moves back into offices around a third of the HQ size because many of its 8,000 workers still work from home for two or three days in a week slashing the floor space needed to house them.[29] In June 2024 KPF had won the architectural competition to reimagine the 45-storey building with seductive images of great chunks of the building chopped out to transform it into a unique destination with workspaces, leisure, entertainment, education, and cultural attractions. With a price tag of hundreds of millions of pounds to complete the project its future remains uncertain.

In Canary Wharf in June 2024, at least 11 glass tower blocks were either partially or fully vacant (c. 5), slated to be fully vacated (c. 4), or seeking tenants to sublease space to (c. 2). In post-pandemic cities worldwide, companies are downsizing and rationalising office space because of hybrid working, soaring energy costs, and to enhance the well-being of their employees and business models. The outlook for such buildings has become so bad that some investors were buying up the buildings, betting that the land they sit on is now worth more than the structure itself, with plans to demolish and redevelop or refurbish the structure for high-rise housing. To facilitate this change, for instance, New York City is re-zoning downtown Manhattan to allow more offices to be converted to apartments.

In 2023, Helen Chandler-Wilde in Blomberg.com pointed out that the façades of office buildings can be an issue as the once 'modern' glass exteriors are, in fact, a century old in design. In summer they overheat like greenhouses driving huge energy usage and costs[30] and HVAC system replacement costs as building loads exceed mechanical capacity and need complete replacement. Proposed solutions include adding mass to facades to reduce heat gain, but such refurbishments can be very expensive, and not all buildings are suitable conversion. The sheer scale of the HSBC tower makes it an unlikely candidate for housing as, with many offices, the space in the centre of the deep plan floor plates is unusable in housing. Light-wells and atria can be cut into their centres for a very high cost. Ventilation is a major factor as many office buildings have sealed facades relying on central ventilation, heating and cooling systems that re-circulate air between spaces, a problem during any future pandemics where issues of cross infections between spaces arose. Complete overhauls of the HVAC systems are needed to convert centrally serviced deep plan towers into a number of individually serviced housing units with access to daylight on multiple floors. Planners have no training in the energy or comfort mechanics of how buildings work or their impacts on the mental and physical well-being of occupants. This must change. Perhaps plans for the demolition of a building should be submitted during its original planning process and if the burden will fall on the public purse the building should not be given planning permission at all to avoid areas of cities becoming wastelands of empty, insecure between the comma and the word and unusable buildings.

The Well-Being Industry and Issues of Trust

The corporate world now has its own certification scheme for well-being 2014 by the team who developed the LEED rating scheme and, in 2014, started the International WELL Building Institute (IWBI). All four entities sell products and hold events that promote high-performing buildings, from offices and multi-storey residences to educational complexes and kitchens.[31]

The IWBI offers an international WELL standard certification scheme that for a full service across the full scope of project design, construction and operations, with an enrolment fee of \$3,000 and a programme fee of \$0.16/ft^2 with total fees starting at \$8,000 and capped at \$98,000. Like the LEED rating scheme that originally did not award their Platinum certification to buildings without a full central air conditioning system,[32] the well-being certification only caters for conditions in highly serviced buildings. It has seven measures, of which air, water, comfort, and light are all largely delivered by the HVAC systems of the building. Three others include nourishment that can be provided by a nice canteen or free lunch, fitness by cycle racks, showers and a gym locally or on-site, and the mind issues revolving around working practices, furnishings, meetings points and protocols, 'fun' opportunities, and possibly the size of the salaries paid to individual workers in high-spec offices. No doubt the well-being industry will evolve in light of WFH trends, just as the green certification industry had to in response to changing societal priorities, but in a world of flat lining economies and the stampede out of certain office types it will be interesting to see where it goes.

Peace of Mind

What a relief it is to enter a cool, air-conditioned building on a hot day outdoors. What a relief to open a window or turn on a fan in a hot room and sit in a cooling breeze or have a cool shower after an outdoor run or workout in the gym. The knowledge that one can do so can bring peace of mind to the building user. However, the growing mortality figures show that it is the poor and disengaged who die first in the heat because they do not have such life-saving opportunities at hand, or often the ability to use them. Not everyone can afford to install or run mechanical conditioning systems, even in rich countries. Most can, however, afford to run the types of low-energy fans and local personal heating and cooling that can bring peace of mind during extreme weather, as the following case study shows.

Climate-Safe Rooms: The Geelong Study

During the January 2014 heatwave in South Australia, 136 heat-related hospital admissions were recorded in Adelaide; most were elderly. In response, Roaf wrote a short magazine article asking readers to 'buy your gran a cool corner: [a] low-cost climate refuge',[33] promoting the idea of just fitting out the coolest room in a house to keep occupants safe during extreme hot or cold events. Governments have produced pie-in-the-sky reports on the total cost of refurbishing their building stocks for eye-watering sums, but everyone who knows this will never happen. A public–private partnership to share the costs of creating at least one climate-safe room could be affordable for many and keep people out or hospital as a consequence.

Concerned by cold related deaths, the City of Greater Geelong in Victoria, Australia, partnered with CSIRO and industry to retrofit one room to create a safe winter weather haven in the homes of 26 individuals who were very vulnerable to the health impacts of climate extremes. Based on initial evaluations of each house, upgrades were planned to optimise thermal performance within the allocated budget in the most frequently used room in the house, typically the living or dining room. Perhaps the most remarkable finding of the project was that participants reported increased comfort, health, and happiness. Participants reported being more comfortable during winter and in hot weather with major positive effects on participants' health and well-being. Reported feelings of depression, anxiety, pain, and discomfort were all reduced while activity levels increased and higher self-care ability and fewer visits to doctors were reported.

Fears around paying energy bills reduced, with electricity bills plummeting by an average of more than 45% in the summer months, when installed solar PV generation peaked. Participants saved approximately $1,462 Australian on health and energy combined while experiencing reduced exposure to unhealthy temperatures. The Climate Safe Rooms project was so successful with its knock-on economic and health benefits for the public health system, with an overall happier, healthier, more equitable community that was rolled out to another hundred vulnerable households in Victoria.

Anxiety around paying bills and fear of possible health consequences indoors from extreme heat or cold were all but eliminated in this project, where comfort, health, wealth, and well-being are intimately connected.[34]

Conclusions: Mental Health in the Post-COVID World

In an unstable world with an unpredictable future, one mantra from the last tumultuous years stands out: *No one is safe until everyone is safe*. The status quo must change in our increasingly unequal societies, in which the very rich get unimaginably richer in their multiple luxurious abodes while the world's poor struggle to eat adequately, let alone remain thermally safe in their own homes.

While top-down political policies to enhance comfort often push clumsily for heavily lobbied-for, often high-cost, machines and services, the growing realisation must be that it is the buildings themselves in which people can stay safe in the more extreme climates of the future or die in. The rich in their soaring towers and gated communities are also vulnerable – when the grid fails, their options are few.

The first step to instilling citizen confidence in the desire or ability of systems around them to keep them safe in the face of the growing climate challenges is to understand the problems at hand. In our increasingly unequal world, the importance of understanding the simple idea that enabling people to find both mental and physical comfort and peace of mind in the buildings and lives they occupy is a keystone to maintaining political stability and system resilience across continents, climates, and cultures. That drive to survive in the evolving extremes of climate ahead requires clear, credible, and trustworthy policies and investment aimed at managing the anxieties, fears, and emotions of populations about the changing world around them.

Notes

1 Porges, S.W. (2017). *The Polyvagal Theory: The Transformative Power of Feeling Safe*, New York and London: W.W. Norton and Company.
2 Steimer, T. (2002). The biology of fear- and anxiety-related behaviors. *Dialogues in Clinical Neuroscience*, 4(3)(September), pp. 231–249. https://doi.org/10.31887/DCNS.2002.4.3/tsteimer. PMID: 22033741; PMCID: PMC3181681.
3 Klinenberg, E. (2015). *Heat Wave a Social Autopsy of Disaster in Chicago*, 2nd Edition, University of Chicago Press.
4 Mayer, L., M. Meuli, U. Lips, et al. (2006). The silent epidemic of falls from buildings: analysis of risk factors. *Pediatric Surgery International*, 22, pp. 743–748. https://doi.org/10.1007/s00383-006-1731-7
5 City of London Corporation (2022). *Preventing Suicides in High Rise Buildings and Structures Planning Advice Note*, Published by the City of London Corporation Environment Department and Public Health Team, Accessed on 10/06/2024 at: https://democracy.cityoflondon.gov.uk/documents/s168370/Preventing%20suicides%20in%20high%20rise%20buildings%20and%20structures%20PT%2026.04.22.pdf
6 Haque, M., M. Sartelli, J. McKimm and M. Abu Bakar (2018). Health care-associated infections – an overview. *Infection and Drug Resistance*, 11(November 15), pp. 2321–2333. https://doi.org/10.2147/IDR.S177247. PMID: 30532565; PMCID: PMC6245375.
7 Roaf, S. (2021). Minimising hospital acquired infections using good design: Future trends, *Proceedings of the 42nd AIVC Conference "Ventilation Challenges in a Changing World" Rotterdam*, Netherlands, 5–6 October, pp. 195–215, Rotterdam: AIVC, Accessed on 26/03/2023 at: https://www.aivc.org/resource/ventilation-challenges-changing-world-book-proceedings
8 Nissen, K., J. Krambrich, D. Akaberi, et al. (2020). Long-distance airborne dispersal of SARS-CoV-2 in COVID-19 wards, *Scientific Reports*, 10, p. 19589. https://doi.org/10.1038/s41598-020-76442-2
9 Dancer, S.J., K. Cormack, M. Loh, et al. (2022). Healthcare-acquired clusters of COVID-19 across multiple wards in a Scottish health board, *Journal of Hospital Infection*, 120, pp. 23–30.
10 Hobday, R.A., and S.J. Dancer (2013). Roles of sunlight and natural ventilation for controlling infection: Historical and current perspectives, *Journal of Hospital Infection*, 84, pp. 271–282. https://doi.org/10.1016/j.jhin.2013.04.011
11 Heschong, L. (2021). *Visual Delight in Architecture: Daylight, Vision and View*, Chapter 15, London: Routledge.
12 Roaf, S. (2021). Building preparedness for pandemics: Recommendations and summary for policy makers, in *AMR and the Environment: A Global & One Health Security Issue*, Edited by G.F. Upham, Geneva: Published by the AMR Think Do Foundation, pp. 62–77, ISSN: 2673-9232. https://www.amrthinkdotank.org/book
13 Roaf, S. (2022). COVID-19: Trust, windows and the psychology of resilience, in *The Routledge Handbook of Resilient Thermal Comfort*, Edited by F. Nicol, H. Bahadur Rijal and S. Roaf, London: Routledge, ISBN: 978-1-032-15597-5

14 Guo, Y., X. Li, S. Luby and G. Jiang (2022). Vertical outbreak of COVID-19 in high-rise buildings: The role of sewer stacks and prevention measures. *Current Opinion Environmental Science & Health*, 29(October), p. 100379. https://doi.org/10.1016/j.coesh.2022.100379. Epub 2022 Jul 14. PMID: 35856009; PMCID: PMC9279164.
15 Pease, L.L., N. Wang, T.I. Salsbury, et al. (2021). Investigation of potential aerosol transmission and infectivity of SARS-CoV-2 through central ventilation systems, *Building and Environment*, 197, p. 107633, ISSN: 0360-1323. https://doi.org/10.1016/j.buildenv.2021.107633
16 Rhett Power (2023). How leaders are enticing workers back into the office, *Forbes*, Accessed on 11/06/2024 at: https://www.forbes.com/sites/rhettpower/2023/04/30/how-leaders-are-enticing-workers-back-into-the-office/
17 Rabb, W. (2025). The new approach to hurricane-strong homes? Concrete walls but affordable, *Insurance Journal*, (832362), Accessed on 21/07/2025 at: https://www.insurancejournal.com/news/southeast/2025/07/21/832362.htm
18 Valliant, J. (2024). *Fire Weather: A True Story from a Hotter World*, London: Hodder and Stoughton Ltd., ISBN: 978-1-39-972023-6
19 Kenny, G.P., J. Yardley, C. Brown, R.J. Sigal, O. Jay (2010). Heat stress in older individuals and patients with common chronic diseases, *CMAJ*, 182(10)(July 13), pp. 1053–1060. https://doi.org/10.1503/cmaj.081050. Epub 2009 Aug 24. PMID: 19703915; PMCID: PMC2900329.
20 Ballester, J., M. Quijal-Zamorano, R.F. Méndez Turrubiates, et al. (2023). Heat-related mortality in Europe during the summer of 2022. *Nature Medicine*, 29, pp. 1857–1866. https://doi.org/10.1038/s41591-023-02419-z
21 Gasparrini, A., P. Masselot, M. Scortichini, et al. (2022). Small-area assessment of temperature-related mortality risks in England and Wales: a case time series analysis, *The Lancet Planetary Health*, 6(7), pp. e557–e564, ISSN: 2542-5196. DOI: 10.1016/S2542-5196(22)00138-3
22 Soules, M. (2021). *Icebergs, Zombies, and the Ultra Thin: Architecture and Capitalism in the Twenty-First Century*, Princeton University Press. ISBN: 978-1-61-689946-2
23 Soules, M. (2024). Zombies and ghosts, *Places Journal*, Accesses on 12/06/2024 at: https://doi.org/10.22269/210521
24 Ballester, J., M. Quijal-Zamorano, R.F. Méndez Turrubiates, et al. (2023). Heat-related mortality in Europe during the summer of 2022. *Nature Medicine*, 29, pp. 1857–1866. https://doi.org/10.1038/s41591-023-02419-z
25 https://en.wikipedia.org/wiki/Crown_Sydney
26 Accessed on 12/06/2024 at: https://www.theguardian.com/artanddesign/architecture-design-blog/2015/sep/02/walkie-talkie-london-wins-carbuncle-cup-worst-building-of-year
27 Campbell, D. (2024). Hybrid working makes employees happier, healthier and more productive, study shows, *The Guardian*, 16/06/2024, p. 18.
28 Yang, L., D. Holtz, S. Jaffe, et al. (2022). The effects of remote work on collaboration among information workers, *Nature Human Behaviour*, 6, pp. 43–54. https://doi.org/10.1038/s41562-021-01196-4
29 Newton, L., P. Scott and V. Barnes (2023). Why banks once flocked to Canary Wharf's high-tech superstructures, but are now starting to return to the City, *The Conversation*, Accessed on 11/06/202 at: https://theconversation.com/why-banks-once-flocked-to-canary-wharfs-high-tech-superstructures-but-are-now-starting-to-return-to-the-city-208888
30 Chandler-Wilde, H. (2023). What to do with a 45-Story Skyscraper and No tenants, *Blomberg.com*, Accessed on 11/06/2024 at: https://www.bloomberg.com/news/features/2023-08-29/hsbc-8-canary-wharf-tower-will-it-become-flats-how-can-qataris-upcycle-it
31 https://resources.wellcertified.com/people/leadership/rick-fedrizzi/
32 Shaviv, E. (2008). Passive and Low Energy Architecture (PLEA) vs green architecture (LEED), in *Paper 371 in Proceedings of the PLEA 2008 Conference*, Dublin.
33 Roaf, S. (2016). Buy your gran a cool corner: A low-cost climate refuge, *ReNew: Technology for a Sustainable Future*, 134, pp. 70–73. http://www.jstor.org/stable/renetechsustfutu.134.70
34 Geelong Sustainability (2023). *Climate Safe Rooms Findings and Final Report*, Published by Geelong Sustainability, 53 Pages, Accessed on 10/06/2024 at: https://www.geelongsustainability.org.au/wp-content/uploads/2023/11/Climate-Safe-Rooms-Final-Report-Sep-2023.pdf

26

EMOTIONAL COMFORT AND SOCIAL WELL-BEING

What Are Emotions?

Emotions are complicated but hugely important to our feelings of being well and comfortable. Whether good or bad, beneficial or harmful, our emotions can trigger both virtuous and vicious cycles in our well-being as our mental activities stimulate physiological responses that can make us feel warmer or cooler regardless of local ambient temperatures. Being depressed and isolated can result in the lowering of body temperatures, reinforcing feelings of ill-health. Being hot and bothered or angry can exacerbate the circumstances that triggered those feelings, and so the cycles continue. Conversely, well-designed environments can have a calming effect on the mind. A good designer will incorporate features that alleviate mental stress, not least because emotions can have such a direct effect on how comfortable people feel and, in turn, their well-being, ability to cope, and productivity.

Philosophers over millennia have deliberated on the relationship between feelings, inclinations, and actions. In his A Human History of Emotion, Richard Firth-Godbehere explained:[1]

> We've had *temperaments* (the way people's feelings make them behave), *passions* (feelings felt first in the body that affect the soul), and *sentiments* (the feelings you get when you see something beautiful or someone acting immorally). We've left most of these historical ideas behind, replacing them with a single catchall term that describes a certain type of feeling processed in the brain: *emotion*.

He points out that there are almost as many definitions of *emotions* as people studying them. Modern neuroscience provides new tools for exploring how emotions affect the body, but many ideas are almost as old as the hills. For instance, we have looked at how fear can affect building occupants, owners, and society at large. In his book *Rhetoric*, Aristotle details certain emotions in depth. He puts emotions in pairs like fear/confidence and love/hate. Aristotle concluded that

DOI: 10.1201/9781315645070-34

there are two primary emotions from which all others emanate, which might also be thought of as key drivers of comfort:

Pleasure. 'A movement by which the soul as a whole is brought into its normal state of being', otherwise called an *unperturbed soul*. Rather like the physiological sensation of alliestheisa that gives pleasure as the body returns to its normal, or comfort, state.

Pain. The opposite of pleasure, associated with the conditions that caused the pain in the first place. Thermal discomfort and stress can certainly *perturb the soul*.

Aristotle described fear as a reaction to potential suffering that makes people 'inclined to deliberation', that is, makes people dwell on suffering with a view to avoiding it.[2] Modern science has discovered that fear arises in groups of cells in the brain, the largest of which is the amygdala, which some phycologists used to call the lizard brain that reacts so quickly to threats that the conscious brain is unaware of the response till after the urge to fight, flee, or freeze kicks in. But there are a whole range of mental drivers and behaviours associated with fear, like phobias and anxiety disorders, that can be learnt through culture, upbringing, and education, which we will touch on more in the next chapter.

What concerns us here is how emotions can affect how comfortable we feel, how we can use that knowledge to design better buildings. To start with, can we measure emotions?

Emotions and Mental Health

Lauri Nummenmaa in Finland has done some ground-breaking work on mapping emotions in the body illustrated in Figure 26.1. Emotions are often felt in the body via its *somatosensory* feedback systems. In physiology touch is called the somatosensory system, a network of neural structures linking the brain and body to produce the perception of touch, temperature, and pain and where it occurs in the body. The touch sensation enables the body to feel the environment via the skin using specialised receptors in the skin that send signals to the brain indicating soft and harder pressure, heat and cold, body position, and pain. It is a subset of the nervous system that along with the sight, sound, smell, taste, balance and spatial orientation systems enables us to sense the world around us.

His team developed a method to map bodily sensations associated with different emotions using a unique topographical self-reporting method.[3] In five experiments, participants were shown two silhouettes of bodies alongside emotional words, stories, movies, or facial expressions. They were asked to colour the bodily regions whose activity they felt increasing or decreasing while viewing each stimulus. Different emotions were consistently associated with statistically separable bodily sensation maps across experiments involving West European and East Asian subjects, leading them to propose that the emotions represented in their somatosensory system may be common to populations around the world. They posited that the ways emotions were sub-consciously felt triggered bodily changes that may play a key role in generating consciously felt emotions.

Sex, size, weight, and socio-demographic variables affected body maps only weakly, but with age, the intensity of the emotional responses did diminish. The elderly were shown to be more liable to feel negative emotional sensations, like depression and sadness, and did not report so often being prone to feelings of happiness, as do younger cohort.[4]

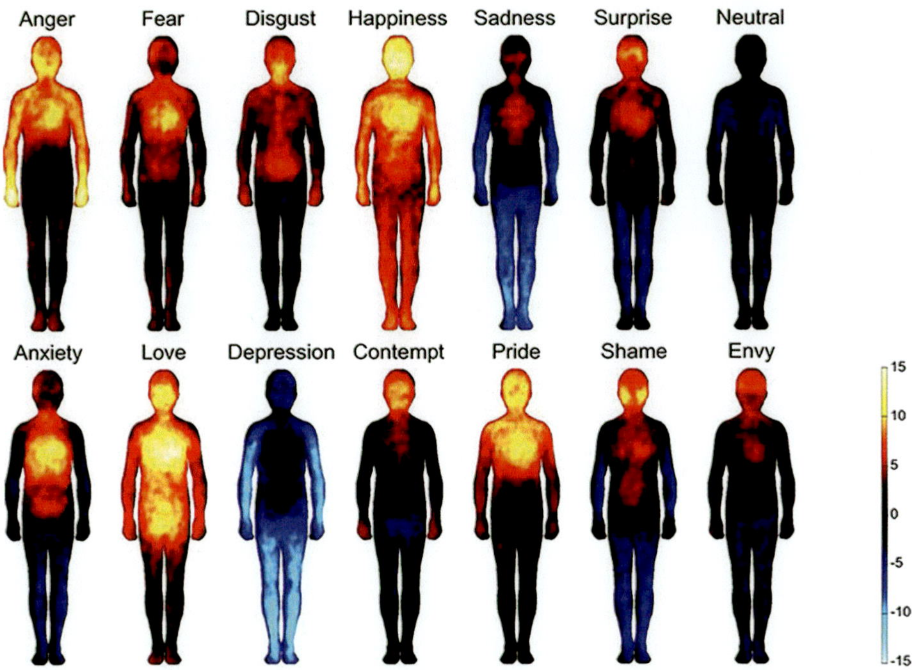

FIGURE 26.1 Maps made of bodily sensations associated with different emotions using a unique topographical self-report method developed by Nummenmaa et al. (2013). The colours denote regions of the body where the subjects report experiencing increased (red) and decreased (blue) bodily sensations, not the temperatures in those areas.

Emotions, Temperature, and Colour

Nummenmaa's emotion maps report levels of sensation and are not thermal maps. We all know from experience that emotions do manifest themselves thermally in and on our bodies. We can see blush with embarrassment or flush with anger. We often describe people in terms of heat and colour: boiling mad (angry), blue (depressed), hot tempered, cold (unempathetic), etc. Temperature is one of the most common indicators used by doctors to get a quick idea of how ill a patient is, and in buildings, how sick or well an indoor environment is. Can we use temperature to gauge the comfort or well-being of a person? Emotions play an important role in the way in which humans relate to each other, and reading those emotions in visual signals may help us make use of those interactions and, in extremis, survive.

A team in Mexico has been experimenting with taking thermal images of faces to see if we can do just that. Their methodology is illustrated in Figure 26.2. Using infrared thermography (IRT) which non-invasively measures the radiation of energy (heat) that a body emits to diagnose the emotions they are feeling during the experiment. Using forty-four subjects were recorded while experiencing joy, disgust, anger, fear, and sadness. The analysis of the thermograms, differentiating between different regions of interest on the face they were able to identify biomarkers that indicated when one of the five basic emotions (joy, disgust, anger, fear, and sadness) were taking place with an almost 90% agreement.[5] What matters here is that people occupying buildings can feel the whole gamut of human emotions and their body temperatures change as a result, so just

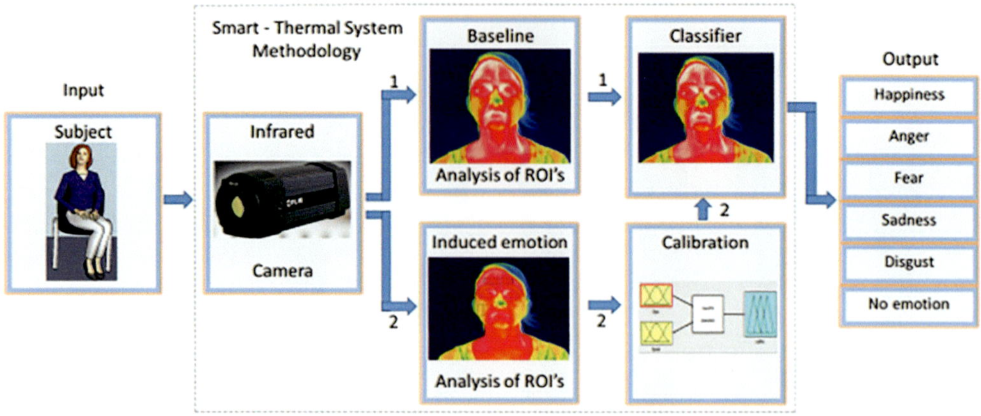

FIGURE 26.2 Methodology used for the smart thermal system to record emotions.

Source: Cruz-Albarran et al. (2017).

as there is no typical person in any one building's type or location, each individual will perceive their environmental temperatures differently according to the emotions they are experiencing at any one time as a result of interactions in the real or imagined world. During menopause, menstruation, or pregnancy, ladies can experience hot flashes that can fluctuate temperatures around the body and skin by up to 3°C. We are thermally victims of our own bodies as well as emotions.

Old, Cold, and Lonely

Age is a significant factor in how people register emotional and thermal feelings. This can be compounded if they are isolated and lonely. Researchers have shown that for people being depressed can actually make them blue with cold as vasodilation is triggered, drawing blood vessels further into the body to conserve heat.[6]

Much work is now going on around the idea of how body temperature and thermoregulatory systems are affected by, and respond to, the proximity of others within a group, or isolation from them.[7] The warm glow of feeling hygge (Chapter 25) is not only a result of the visible fire or candles and cosiness but also due to interactions between the people sharing that experience in the group.

Emotions can affect our thermal states and our productivity at work. When measuring the thermal performance of a tent designed for the Antarctic in a meat storage facility in Hull during a heatwave in August 2018, Roaf found it surprising to hear how emotions thermally affected four large and tough looking workers in the huge cold stores that are kept at minus 20°C.[8] Despite being always dressed the same, on different days they all admitted that they could withstand the cold sometimes for only half an hour when they were feeling low, but on other days they could tolerate the freezing air for an hour and a half when feeling upbeat. It surprised them, but they all agreed this was the case, that their ability to cope with cold was dependant on their emotions on the day. It must have been very difficult for Scott's team in the Antarctic not to suffer from the same thermo-emotional ups and downs, with fatal consequences.

Social Isolation and Anger

People often use metaphors such as 'she is cold and aloof' or 'she is a warm person'. Temperature metaphors conveniently establish and explain what is meant and often show a remarkable similarity in imagery across languages. Perceptions of physical warmth play a central role in how people are perceived and interact in groups.

People's *social temperature* is central to the perception of relationships. But emotional connection can influence their body temperatures. In one experiment, people who were socially excluded showed a drop in finger temperature in research, but warming the fingers through holding a warm cup of tea after a period of social exclusion was also shown to alleviate negative affects typically experienced after social suffering.

Robust social and emotional, supportive relationships can significantly improve health and decrease the risk of mortality in comparison to those who have low quantity or quality of social relationships. Social isolation has been identified as an independent major risk factor for all-cause mortality.[9]

Anger has the opposite effect, and people's finger temperatures have been shown to increase as a function of being angry.[10] Hot temperatures increase people's feelings of hostility, their understanding of hostile environments, so setting the scene for 'excitation transfer effects', that is, make them more likely to be aggressive in certain social situations.[11]

Community and Comfort

The physical and social environments people live in obviously profoundly affect their physical health and mental and emotional well-being. The nature and quality of the personal relationships people have to each other, and the ease or difficulty with which they are formed and retained matters. Human contact can improve thermal comfort and so encouraging social connections, both indoors and out, within communities is important.

The search for ideal outdoor temperatures may waste considerable time. People vary so much. Acceptable outdoor temperatures vary with climates and cultures. Scandinavians in safe communities regularly put babies outside in freezing temperatures for a daily nap to make them hardier. In Sweden, *there is no bad weather, only bad clothing*. Such routines are customary and safe as parents and nursery staff understand the thermal limits of the practice, bringing babies in word nap inside as temperatures fall below −15°C (5°F).

The elderly who have free time on their hands often congregate outdoors to undertake group activities, playing games in tree-shaded squares in the Mediterranean or taking part in group exercises in open areas of Chinese suburbs. People instinctively understand when they need to retreat indoors when temperatures get too high or low temperatures for them, and such communal activities tail off in more extreme weather for health reasons. This fight-or-flight response is acquired and, in part, physiologically triggered, as one Chinese study found that people with cardiovascular diseases were more likely to decrease physical activity in extreme cold, while those with hypertension were less likely to decrease physical activity in extreme heat for instance.[12]

Outdoor areas between buildings are the spatial glue that sticks people together in communities. People adapt to the outdoor climates they occupy and pass through on their everyday journeys by adjusting their clothing, activity times and locations and pathways through available spaces in different seasons. In winter, they may sit on a sunny bench or a shaded one in summer and their routes through spaces may change over the seasons as they seek to minimise discomfort en route,[13] that is, if and when they feel safe to do so.

Light, Colour, and Emotions

Goethe, in 1810, published his extraordinary book *The Theory of Colour*.[14] Two volumes cover the phenomena associated with light and colour, which end with a section on the *moral associations of colour*. Because of the elementary role played by light and colour in our understanding of the world around us, he points out that

> *we shall not be surprised to find that its effects are at all times decided and significant, and that they are immediately associated with the emotions of the mind.*

Goethe explained his view of what colours mean emotionally, including more detailed descriptions with the impact of different colours:

- The colours on the *plus* side are yellow, red-yellow (orange), yellow-red (minium, cinnabar). The feelings they excite are quick, lively, aspiring.
- The colours on the *minus* side are blue, red-blue, and blue-red. They produce a restless, susceptible, anxious impression.

Nice ideas, but very difficult to prove. Despite much research since Geothe's day, the complexity of the relationships between light, colour, and emotions work is still at an early stage. Andrew Elliot pointed out in 2015, in his review of theoretical and empirical work on colour,[15] that 'much writing on colour and psychological functioning has been completely atheoretical', focused exclusively on finding answers to applied questions like 'What wall colour facilitates worker alertness and productivity?' or 'What colour improves recovery times in hospitals?', as noted in Table 26.1.

Such findings can be provocative and media-friendly, but results like these apply to specific research contexts only,[16] not yet to a comprehensive theoretical colour framework. The subject is also made more complex because of the effect of light and ambient temperatures on how colour and spaces are perceived.[17] How differently does one feel about a colour in warm and dim, or cool and bright, light?

TABLE 26.1 Examples of research on colour and psychological functioning

Area of research	*Central finding*
Colour and selective attention	Red stimuli have been shown to receive an attentional advantage.
Colour and alertness	Blue light has been shown to increase subjective alertness and performance on attention-based tasks.
Colour and intellectual performance	Viewing red prior to a challenging cognitive task (intelligence test) undermined performance.
Colour and aggressiveness/dominance evaluation	Participants rated males wearing red as more dominant.
Colour and attraction	Viewing red on or near a female has been shown to increase attraction in heterosexual males.
Colour and store/company evaluation	Blue stores/logos have been shown to increase quality and trustworthiness appraisals.

Source: Elliot (2015).

Emotional Interiors

Colours are often used to evoke emotions in the interiors of buildings to help people in them function better for a given purpose. In the Google Headquarters in Madrid designed by Jump Studios, colour is used dynamically to set very different tones and moods for meeting, social, and working spaces. Figures 26.3a and b show how immediate an impact colour might have on the emotions of occupants where the sharp vibrant yellows clearly aim to stimulate activity and ideas while the quiet green space provide a softer, more thoughtful ambience for thinking and discussions. In homes or workplaces colour can be used to promote personal well-being and peace of mind with softer colour hues while other interiors are designed to benefit from the energy generated in occupants by surrounding them with stimulating, colourful, visual environments. More sophisticated designers and organisations may exploit a range of colour themes to appeal to the range of emotional states expected or desired in a single building.

A designer of interiors will often trust their own instincts and experiences to inform colour choices. There are some experts in this field, but there is no definitive rule book to follow, and common sense is a useful design tool here. In a 2006 study of nearly a thousand occupants in four buildings in different latitudes by Küller et al., they showed that the light and colour in the

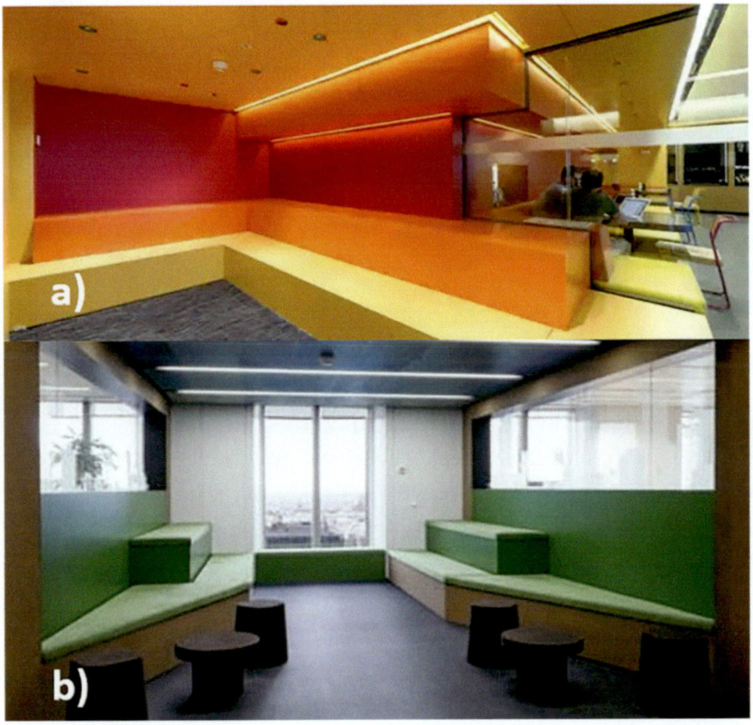

FIGURE 26.3 (a) Stimulating colours in the meeting spaces at the Google Headquarters buildings in Madrid, with interiors designed by Jump Studios in London, and (b) quiet restful green creates a more contemplative ambience in meeting areas in Google Headquarters buildings.

Source: Shaun Fernandes

workplace itself had a real influence on the subjective mood of employees. Mood was lowest when lighting was much too dark, it improved and reached its highest level when the lighting was reported as 'just right', but when it became too bright, the mood declined again. On the other hand, illuminance, as measured in simple objective terms, showed no significant impact on mood at any time of the year, but good design is really not that simple. Lisa Heschong writes in the *Visual Delight in Architecture* of the many ways that our lives are enriched by the presence of natural daylight illumination and window views within our buildings and shows how light can be used to trigger delight.[18]

The colour of the measured workspaces was shown to be important in the Küller et al. study. Their index of emotional status was higher throughout the year for those who had the most colourful work environment. Thus, it may seem that the brighter the colour, the better it is for those who work there. However, this should be seen in light of the fact that the majority of the environments were subdued or neutral. Only 5% were experienced as very colourful. In some laboratory studies of colour and mood, very strong colours were applied, with results that seemed undesirable. The authors suggested that a moderate increase in the use of good colour design will serve to improve the overall mood of the working staff, which seems sensible.[19]

On the 6th floor of Daikin's Technology Centre in Osaka,[20] they have been experimenting with providing different indoor climates at different parts of the Future Labs zones, so their highly world-leading scientists can choose optimal local climates for themselves. They may want to be in a warmer or cooler environment for meetings, which may well depend on the time of day or year, but the key point is that they can choose. This is such a simple idea – to allow people thermal flexibility to work is optimal conditions for themselves. Colours can also be used to enhance the available range of thermal experiences with warmer colours used in the warmer zones and cooler colour ranges in cooler zones. This servicing of occupant's emotional preferences for different indoor climates can be very rewarding especially in offices where long hours are spent on computers for the online gaming industry and even architectural design practices.

Love Conquers All

In his seminal 1958 essay *The Nature of Love* H. F. Harlow's study of infant monkeys reared with the aid of laboratory-constructed mother-substitutes made it obvious that contact comfort is overwhelmingly important in the development of affectional responses.[21] He wrote that love is a wondrous state, deep, tender, and rewarding, and suggested that more attention should be paid to what leads to love or affection and how it develops.

Warmth is a fundamental feature of people's perceptual systems and is used to interpret the social environment.[22] Not just interpersonal warmth, and many of us will know that a good hug can go a long way in improving emotional well-being and repairing negative feelings of exclusion,[23] but also the warmth afforded by interactions in a communal area or focal warm feature in a space.

In an age when commercial buildings are so often over-cooled, the importance of emotional and physical warmth can be used to make people fonder of the spaces they occupy and the people they work with. Emotions can be strongly affected by the behaviours of people around us; for instance, being ostracised or excluded, even briefly and by strangers, is painful and threatens fundamental emotional needs. Studies show that excluded individuals perceive a room as cooler and that they compensate for related emotions, for instance, with a desire for warmer drinks.[24] There is so much research still to do it is impossible to be certain of the impact of design actions;

for example, in a hotter space, does the demand for cooler drinks increase and feelings of anxiety arise if that demand cannot be met?

Emotions Over Matter

Emotions are complex and can colour how we feel about going to work or being at home. A huge range of factors can impact on emotions, including a person's sex or weight or physiognomy and life history, which can all relate to the emotional state of a person. The *thermal disposition* of individuals is a different thing. First discussed in 1979 by Howell and Kennedy, who coined the term *perceived cold-naturedness*, there was much interest in the contribution of psychological variables to the prediction of thermal comfort judgments in real-world settings. Howell and Stramler went on to develop their findings, showing that cognitive variables have a substantial impact on thermal sensation judgements.[25]

Healey and Webster-Mannison in 2012, looking at the qualitative influences on office worker in Brisbane, then developed the idea of individuals having their own *thermal disposition*[26] as they couched their findings in the notion of thermal comfort as a sociocultural achievement, as opposed to a universal and quantifiable condition. They suggested that an individual's feelings of discomfort are related to who they are, how they interact with the group in the office space, what opportunities they have at hand to adapt their environment, and how much access they had to 'thermal delight'. Here we have a mélange of factors being drawn upon to explain differences in comfort response, ranging possibly from the physical attributes of the person to their emotional ability to deal with those around them and then to the state of their own spiritual well-being.

FIGURE 26.4 Peace of mind and mental solace can be found in simple things. A monk's room in Onsenji Temple in Kinosaki.

Source: Susan Roaf.

More recently, Rupp et al. defined *thermal disposition* as a person's self-assessment of their own thermal sensitivity, that is, being more sensitive to heat, cold, both, or neither, in relation to their perception of the thermal sensitivity of their co-workers.[27] They highlight the potential benefits of categorising occupants according to their thermal dispositions to improve their thermal comfort experiences by placing them in the local micro-climate that best suited to their preferences, which might relate to light levels, access to natural ventilation and local comfort technologies like fans, heaters, etc., a strategy that can result also in very significant energy savings.[28]

Given the huge range of emotional impacts thermal conditions can have on people, it must be clear that trying to find a single comfort temperature for a large cohort of moody people in a space, building, city, region, or world is actually impossible. Combine that with the fourth *spiritual well-being and beliefs* wall of the Māori *Whare* and it provides even more confounding variables to the challenge of creating comfortable working environments for busy office workers or peacefully meditating monks in the Japanese temple shown in Figure 26.4.

Notes

1 Firth-Godbehere, R. (2021). *A Human History of Emotion: How the Way We Fell Built the World We Know*, London: 4th Estate.

2 Gross, D.M. (2006). *The Secret History of Emotion: From Aristotle's Rhetoric to Modern Brain Science*, Chicago: University of Chicago Press.

3 Nummenmaa, L., E. Glerean, R. Hari and J.K. Hietanen (2013). Bodily maps of emotions, *Proceedings of the national Academy of Sciences of the USA*, First Published December 30. https://doi.org/10.1073/pnas.13216641113

4 Volynets, S., E. Glerean, J.K. Hietanen, R. Hari and L. Nummenmaa (2019). Bodily maps of emotions are culturally universal, *Journal of the American Psychological Association*. DOI: 10.1037/emo0000624

5 Cruz-Albarran, I.A., J.P. Benitez-Rangel, R.A. Osornio-Rios and L.A. Morales-Hernandez (2017). Human emotions detection based on a smart-thermal system of thermographic images, *Infrared Physics & Technology*, 81, pp. 250–261, ISSN: 1350-4495. https://doi.org/10.1016/j.infrared.2017.01.002

6 Ijzerman, H., M. Gallucci, W.T. Pouw, S.C. Weibetagerber, N.J. Van Doesum and K.D. Williams (2012). Cold-blooded loneliness: social exclusion leads to lower skin temperatures, *Acta Psychologica*, 140(3), pp. 283–288.

7 Ijzerman, H., A. Szymkow and M. Parzuchowski (in press). Warmer hearts, and warmer, but noisier rooms: Communality does elicit warmth, but only for those in colder ambient temperatures, *Journal of Experimental Social Psychology*, 67, pp. 88–90.

Ijzerman, H., J.A. Coan, F.A.M. Wagemans, M.A. Missler, I. Van Beest, S.M. Lindenberg and M. Tops (2015). A theory of social thermoregulation in human primates, *Frontiers in Psychology* (April 21). https://doi.org/10.3389/fpsyg.2015.00464

8 Roaf, S. (2019). Extreme design: Lessons from antarctica, *Proceedings of the Comfort at the Extremes Conference*, Dubai, 10–11 April, pp. 22–30, pp. 806–816, Accessed on 13/05/2019 at: www.comfortattheextremes.com

9 Reblin, M., and B.N. Uchino (2008). Social and emotional support and its implication for health, *Current Opinion in Psychiatry*, 21(2), pp. 201–205. https://doi.org/10.1097/YCO.0b013e3282f3ad89. PMID: 18332671; PMCID: PMC2729718.

10 Ekman, P., R.W. Levenson and W.V. Friesen (1983). Autonomic nervous system activity distinguishes among emotions, *Science*, 221(4616), pp. 1208–1210. https://doi.org/10.1126/science.6612338

11 Anderson, C.A., W.E. Deuser and K.M. DeNeve (1995). Hot temperatures, hostile affect, hostile cognition, and arousal: Tests of a general model of affective aggression, *Personality and Social Psychology Bulletin*, 21(5), pp. 434–448.

12 Ho, J.Y., H.Y.C. Lam, Z. Huang, et al. (2023). Factors affecting outdoor physical activity in extreme temperatures in a sub-tropical Chinese urban population: An exploratory telephone survey, *BMC Public Health*, 23(101). https://doi.org/10.1186/s12889-022-14788-0

13 Li, K., Y. Zhang and L. Zhao (2016). Outdoor thermal comfort and activities in the urban residential community in a humid subtropical area of China, *Energy and Buildings*, 133, pp. 498–511, ISSN: 0378-7788. https://doi.org/10.1016/j.enbuild.2016.10.013

14 Goethe (1810). *Theory of Colours. German Translated by Charles Locke Eastlake (1840)*, London: Published by John Murray.

15 Elliot, A.J. (2015). Color and psychological functioning: A review of theoretical and empirical work, *Frontiers in Psychology*, 6(April 2), p. 368. https://doi.org/10.3389/fpsyg.2015.00368. PMID: 25883578; PMCID: PMC4383146.

16 Elliot, A.J., and M.A. Maier. (2012). Color-in-context theory. *Advances in Experimental Social Psychology*, 45, pp. 61–125. DOI: 10.1016/B978-0-12-394286-9.00002-0

17 Kompier, M., K. Smolders and Y. de Kort (2022). Effects of light and ambient temperature on visual and thermal appraisals, Chapter 21, in *The Routledge Handbook of Resilient Thermal Comfort*, Edited by F. Nicol, H. Bahadur Rijal and S. Roaf, pp. 347–362. London: Routledge, ISBN: 978-1-032-15597-5

18 Heschong, L. (2021). *Visual Delight in Architecture: Daylight, Vision, and View*, London: Taylor & Francis Ltd. ISBN: 978-0-36-756323-3

19 Küller, R., B. Seifeddin, T. Laike, B. Mikellides and G. Tonello (2006). The impact of light and colour on psychological mood: A cross-cultural study of indoor work environments, *Ergonomics*, 49, pp. 1496–507. DOI: 10.1080/00140130600858142

20 https://www.daikin.com/about/corporate/tic

21 Harlow, H.F. (1958). The nature of love, *American Psychologist*, 13(12), pp. 673–685. https://doi.org/10.1037/h0047884

22 Ijzerman, H., and S.L. Koole (2011). From perceptual rags to metaphoric riches – Bodily, social, and cultural constraints on sociocognitive metaphors: Comment on Landau, Meier, and Keefer (2010), *Psychological Bulletin*, 137(2), pp. 355–361. https://doi.org/10.1037/a0022373

23 Ijzerman, H., M. Gallucci, W.T.J.L. Pouw, S.C. Weiβgerber, N.J. van Doesum and D.W. Kipling (2012). Cold-blooded loneliness: Social exclusion leads to lower skin temperatures, *Acta Psychologica*, 140(3), pp. 283–288. https://doi.org/10.1016/j.actpsy.2012.05.002

24 Zhong, C.B., G.J. Leonardelli (2008). Cold and lonely: Does social exclusion literally feel cold?, *Psychological Sciences*, 19(9)(September), pp. 838–842. https://doi.org/10.1111/j.1467-9280.2008.02165.x. PMID: 18947346.

25 Howell, W.C., and C.S. Stramler (1981). *Contribution of Psychological Variables to the Prediction of Thermal Comfort Judgments in Real World Settings*, United States: N. p., Web.

26 Healey, K., and M. Webster-Mannison (2012). Exploring the influence of qualitative factors on the thermal comfort of office occupants, *Architectural Science Review*, 55(3), pp. 169–175. https://doi.org/10.1080/00038628.2012.688014

27 Rupp, R.F., J. Toftum and E. Ghisi (2022). Thermal comfort and occupant disposition in mixed-mode offices in a Brazilian subtropical climate, Chapter 18, in *The Routledge Handbook of Resilient Thermal Comfort*, Edited by F. Nicol, H. Bahadur Rijal and S. Roaf. London: Routledge, pp. 300–314, ISBN: 978-1-032-15597-5

28 Malkawi, A., B. Yan, Y. Chen and Z. Tong (2016). Predicting thermal and energy performance of mixed-mode ventilation using an integrated simulation approach, in *Building Simulation*, Vol. 9, Tsinghua University Press, pp. 335–346.

27

SPIRITUAL COMFORT AND BELIEFS

Introduction: Mind Over Matter

In his classic book *Ways of Seeing*, John Berger posited that seeing things comes before words and can never be fully covered by them. Sight gives access to part of the world, and then touching situates a person in relation to it. The way we interpret what we see is also *affected by what we know or believe*. A medieval man seeing a fire may have triggered thoughts of hell,[1] and that might cause him to draw away from it. To people on a cold winter's night, a fire may mean comfort, succour, and security and be drawn towards it.

Traditionally, philosophers tried to silo aspects of the thermal environment to better understand it. Gaston Bachelard, in his book *The Psychoanalysis of Fire* in 1938, tried to meld poetry and science by understanding, individually, the many different ways in which fire manifests itself in a person's psychology and in the real world. He wrote in relation to fire of respect, reverie, purity, love, sex, history, and chemistry,[2] separating the psychological and physical sciences into discrete chapters, but always stressing the interdependences at the heart of 'comfort'. Bachelard quotes Bernardin de Saint-Pierre in his 1791 book *Voyage en Arcadie*,[3] taking two sticks of different wood and rubbing them together until his unconscious rediscovers the first fire to be lit in mutual love.

The analogy is extended to assert that *eurhythmy* of an active rubbing motion, on condition that it be sufficiently gentle and prolonged, brings about *euphoria*. He claims that this joy cannot be explained objectively. Then he extends the allusion, saying you cannot equate this joy to that of rubbing, cleaning, furbishing, and polishing with meticulous care, as done by certain housewives. He quotes the author Honoré de Balzac in his 1830 novella *Gobseck*,[4] who noted that the 'cold houses' of old maids had some of the shiniest furniture. Bachelard built on this image to note that psychoanalytically speaking, cleanliness is really a form of uncleanliness, giving an insight into his own mental interpretation of the quote. Read that as you may, but this example links physical cold possibly with poverty (unable to afford a warm house), age, behaviours to cope with cold (vigorous rubbing), and sexual allusions and gives some insight into the complexity of trying to understand mind and matter issues around the subject of comfort.[5]

DOI: 10.1201/9781315645070-35

The Ecology and Anthropology of Perception

The psychologist James Gibson developed a more unifying theory to explain how we perceive world in *The ecological approach to visual perception* in 1979:[6]

> The environment of animals and men is what they perceive. The environment is not the same as the physical world, if one means by that the world described by physics. The observer and his environment are complementary.

Tim Ingold developed this idea in his book *The perception of the environment* in 2000.[7] He argues that two people, placed in the same environment, may perceive it differently, not because they have different models in their heads for processing what is fundamentally the same input of sensory data, but because their perceptual systems (of vision, hearing, touch, etc.) are 'tuned', by way of previous experience, to attend to different aspects of what the environment affords; they become part of their environment.

People who create models of the world built from data inputs of physical attributes of a place such as measures of light, sound, temperature, pressure on the skin, etc. and then organise it into a model of the supposed experience of an individual may capture the biophysical aspects of a place but cannot touch on how:

- Those physical factors will impact psychologically on the propensities, expectations, and beliefs of a particular person
- People will interact with their environments as a consequence of those thought processes

Ingold holds that the physical and the sociocultural worlds are not held together with a film of psychological cement as much mainstream thinking would have us believe, but rather, that we should find a way of talking about human life that eliminates the need to slice it up into these different layers.

How well this unifying idea fits into the reality of the *thermal comfort systems* in which people co-evolve with their environments and are mutually dependant in the resulting interactions, for survival, the elimination of pain, the promotion of pleasure, and finally for the maintenance of homeostasis in the body. We do not occupy environments; we are a part of the various environments we occupy, a series of natural units encountered on our passage through a day and a life. Our psychological personas can vary considerably within each ecological niche and cultural setting. We are different people at home, at work, or at play because, as we adapt physically to them, we may well adopt different values, principles, beliefs, attitudes, and ethical stances that colour the way we perceive and shape the different places. The ability to adapt to varied environments involves the use of behavioural *skills* or capabilities, including the ability to access different physical tools, and mental attitudes to how the surrounding world is perceived. Ingold maintains that these *skills* are not genetic but are learnt again, or abandoned, in each successive generation, who will build their own *modus operandi* in the sequence of environments they co-evolve with.

Think of chameleons, a highly specialised sub-group of the Old World lizards family (Figure 27.1). The ability to colour-shift provides camouflage in the very different habitats of the African rain forests, where they are green, or the deserts of Africa, where they are brown. Some range between reds, yellows, greens, or blues to blend into their backgrounds. They are not in

FIGURE 27.1 Chameleons colour-shift their external skin to blend into their backgrounds.

Source: MathKnight. Own work, based on, CC BY-SA 4.0, https://commons.wikimedia.org/w/index.php?curid=11349030.

the environment; they become part of the environment, as we humans do, too, with the various ways in which humans interact with the world around them. Not all are as simple as changing clothes to maintain homeostasis indoors, and the following vignettes show how complex and important are spiritual world views and beliefs to shaping the thermal environments we occupy.

Mind Over Matter – Wim Hof

Wim Hof, the Dutch Iceman, is famous for both being able to withstand extreme cold and as a storyteller. Attracted to cold in his youth, one day he broke the ice on a pond in Vondelpark in Amsterdam, and the rest, one might say, is history captured in stories about his relationship with his warm friend – cold. He has survived various extreme cold endurance feats from staying in a tub filled with ice cubes up to his neck for almost two hours, running a half marathon barefoot in the snow, and swimming under the polar ice between ice holes. Bolstered by such high visibility exploits he developed and marketed the Wim Hof Method (WHM) combining cold exposure, meditation, and breathing techniques to 'strengthen your inner nature and prevent disease'. In his best-selling book on the WHM, in the first paragraphs he asks readers if they would like more energy, less stress, and a stronger immune system, to sleep better, improve your athletic performance and sex life, boost your mood, lose weight, and alleviate anxiety? He then explains how to do that with his three pillars of daily practice:[8]

1. Breathing practice linked to Tummo meditation that involves vigorous respiratory muscle activity that has been found to increase metabolic activity that, to some extent, delays non-shivering thermogenesis.
2. Mindset programming with repetition of reinforcing mantras, like 'I feel happy, healthy and strong', and willpower training by mentally setting and extending personal cold endurance targets and aligning body and mind to achieve them.
3. Cold emersion and acclimation exercises, starting with cold shower, taking a warm one, and moving on for some people to full emersion in cold ice tubs and water.

Ice bathing is not new and has been practised for millennia in Scandinavia and Russia for physical and mental health and well-being. The Vikings followed hot sauna sessions with a plunge into icy waters, to enhance their 'livskraft' (life force) and mental clarity, a practice continued to this day across northern Europe. Hof has been able to monetise his WHM to make millions from his workshops and retreats, courses and programmes, books and speaking engagements. His example does shed light on issues of mind over matter when adapting to more extreme temperatures.

Wouter van Marken Lichtenbelt was able to study Wim Hof's physiological responses to cold exposure in his laboratory in Maastricht, and for details of those experiments, read his 2017 paper *Who is the Iceman*?[9] He ultimately concluded that the Iceman endures extreme levels of cold exposure

> through a combination of several factors: an increased heat production (non-shivering thermogenesis, brown fat activation, contraction respiratory muscles), efficiently increasing body tissue insulation (vasoconstriction) thereby reducing heat loss and conserving the body core heat content, and finally, possibly not unimportantly, he used his well-trained mental ability to endure the cold (change of mindset, as he calls it).

Many of Wim Hof's pupils have benefitted greatly from his teachings experiencing their bodily feelings during almost bioenergic hyperventilation sessions that they never experienced before. They are also pushing themselves to complete extreme body challenges and afterwards feel relaxed and a sense of personal achievement, often associated with improving mental health and well-being, just as many feel after a hot sauna or a cold-water swim. His method trains people's minds to control their bodies to do this. As soon as Hof stepped out of the ice in Lichtenbelt's laboratory, he started shivering just as everybody else would, due to the redistribution of the cooled blood from his limbs to the body core – also known as the 'afterdrop' effect.

Warnings about cold therapy have been issued by the American Heart Association and the British Heart Foundation, advising people to see a doctor before attempting to follow his extreme emersion techniques, noting that there is poor scientific evidence supporting the practice. As of March 2024 there were 32 reports of people dying in relation to the Wim Hof Method reinforcing the important message that matter-over-mind practices have their limits.[10]

Meditation Over Matter: Monks

Monks and hermits have gone into the woods or mountains for millennia in search of Enlightenment. Ascetics from many religions practice abstinence from worldly pleasures in pursuit of spiritual goals, living frugal lifestyles and renouncing material possessions and physical pleasures in search of an expanding consciousness and the ultimate goal of connecting to the infinite universal consciousness. Sensibly, many hermits in inclement climates choose caves to live in that offer stable micro-climates and if remote enough, solitude.

In 2023, Jay Dhariwal et al. published a study in Jain religious communities in Northern India in which monks and nuns reported being comfortable in very high temperatures. In the original thermal comfort survey of 20 Jain monks and nuns during hot, dry heatwave conditions in May 2023, the ascetics expressed acceptability of the thermal conditions with the indoor operative temperatures (*top*) varying between 35°C and 40°C.[11] A later study of 65 ascetics from Svetambara Jain sects in hot and dry as well as warm and humid weather conditions in Delhi, Jodhpur, Siriyari, and Ahmedabad reported that 90% of them found the thermal conditions to be acceptable when the UTCI was in the strong to very strong heat stress range. UTCI stands for the Universal Thermal Climate Index, a method of assessment for the thermophysiological effects of the atmospheric environment, used in human biometeorology.[12] Neither the adaptive thermal comfort nor the PMV methods of assessing whether a person will be comfortable at a proscribed temperature really works for locally adapted populations in hot climates, as becomes clear when checking high temperatures with either rating system as is easily done using the ASHRAE comfort tool developed at the Centre for the Built Environment at the University of California, Berkeley.[13] However, what Figure 27.2 does show is that up to core temperature of around 37°C (98°F), the majority of participants were comfortable, even in the humid climate, but above that it was those in the hot, dry regions who reported being neutral to warm in temperatures up to 40°C (104°F). It is clear that above core temperature in the humid areas, people adapt their surroundings to keep temperatures below that critical level, where heat loss by evaporation is hampered by high levels of water in the atmosphere.

Just as Win Hof trained his mind to conquer cold, so have this *mindful* Jain community been practising methods of psychological adaptation to high temperatures from their founding over 25 years ago. Community members vowed to live a life in sync with nature ever since they attained *Diksha*, divine grace, bestowed on them by a guru at their consecration ceremonies. Many came from modern households with cars and air conditioners before committing to their new life. Becoming a Jain monk or a nun is not an instantaneous act; rather, they go through a phased adaptation that can last for years, in which they incrementally reduce their dependency on external worldly goods and habits, like the use of ACs for comfort. Like other living beings, such as animals, birds, and insects, they try to adapt to the changing climate without any external means of thermal comfort by mentally committing to doing so and adopting daily and seasonal routines and mind-sets and occupying physical and biotic infrastructures in harmony with natural energy flows, like wind, sun, and rain. One monk pointed out that many of Indian villages still had no access to electricity decades ago, and they try to learn from older generations about how they managed their lives in eco-friendly ways in earlier times.

As the planet heats, existing approaches may not be enough to prevent some getting heat stroke. Recognising this, some Jainism sects have recently let their visitors use fans for thermal

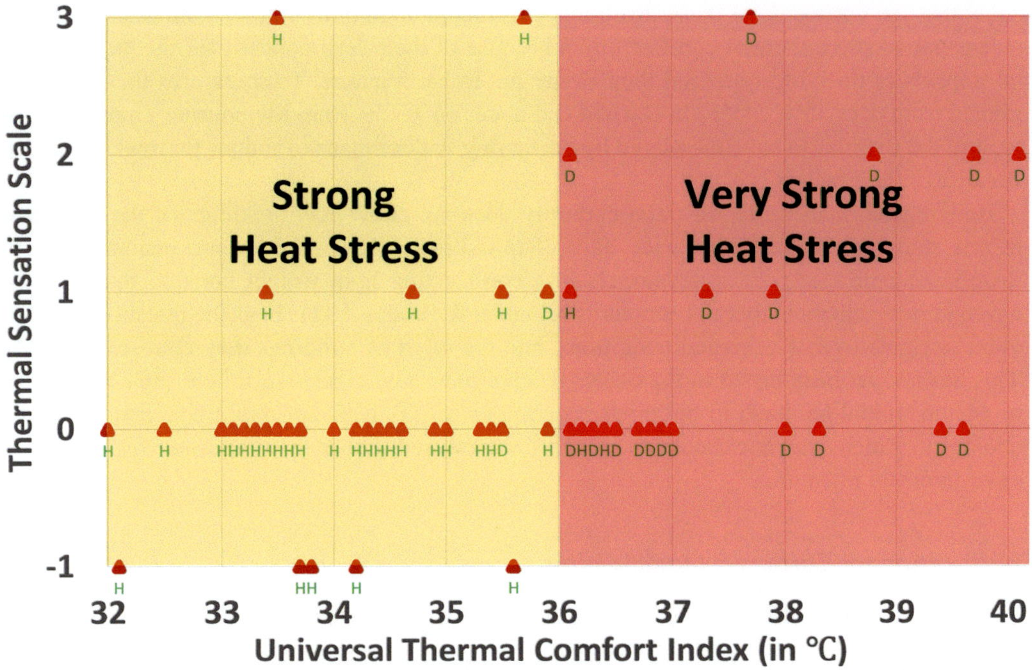

FIGURE 27.2 Thermal sensation responses of Jain monks and nuns in northern India. H = *humid* and D = *dry* conditions. *Note:* −1 = slightly cool, 0 = neutral, +1 = slightly warm, +2 = warm, +3 = hot.

Source: Jay Dhariwal.

comfort due to increasing temperatures and concerns for their health, while some ascetics have started wearing slippers when they go out in the scorching sun on the hot asphalt roads. Others argue the soles of their feet become dead through walking barefoot and they like slippers so negating the need for shoes. Some monks refused to participate in the surveys because the equipment used by the team was responsible for energy consumption, which they had pledged not to use. They perceptively suggested the researchers stayed with them for a few days to observe their routine instead of recording the temperatures to understand how they lived with high temperatures just as Michael Humphreys suggested to understand how to design a comfortable building for hobbits one would just need to follow them around and find out how they did it.

Environmental Movements and Thermal Mavericks

There are many communities of Thermal Mavericks around the world who choose to occupy temperatures that most people might find uncomfortable, for reasons including cost, opportunity or personal convictions. A twelve-month thermal comfort study by Lyrian Daniels in non-standard Australian households in two very different climates, cool temperate Melbourne and hot humid Darwin, studied the relationships between environmental attitudes and comfort behaviours. The results show that the occupants of these dwellings considered conditions comfortable often well outside usual adaptive thermal comfort limits because of their deep concern for the environmental impacts of their lifestyles and their desire not to harm nature. Temperatures they occupied ranged from over 35°C (95°F) in Darwin and under 15°C (59°F) in Melbourne. They preferred to, and did their bit to save the planet by occupying wider ranges of indoor thermal conditions because of their beliefs.[14]

Building Regulations in Australia currently does not allow new buildings of the sort some in this study occupied, for instance built of mud-brick, under its domestic residential certification scheme that pushes developers into constructing light weight housing that must be air-conditioned to meet thermal standards set under the codes.[15] This begs the questions of individual's rights over the thermal conditions, and the types of buildings they choose to occupy. This has not yet been tested in the courts. People have a deep need to follow their beliefs and provision should be made in buildings regulations for non-standard buildings, and life styles choices in which 'maverick' cohorts can feel empowered, rather than punished for their good environmental intentions.

Justification for Corporeal Mortification

Perhaps a full Thermal Satisfaction scale should run from the extremely desired state of *ecstasy* to the undesired (by most) state of *pain* end via measures of heat, neutrality and cold. Some religions at different points in their history adopt different approaches to test or reward followers including practices involving self-mortification and pain. Corporal mortification is believed, for instance by the Opus Dei sect of Catholicism to help people resist their natural drive toward personal comfort which prevents them from answering the Christian call to love God and serve others for the love of God. In short, pain provides a potent inducement not to stray from the straight and narrow of what is required by a particular religion at a particular point in their history. Opus Dei members regularly use the cilice (a spiked chain worn around the thigh), flagellation, comfortless living conditions and sleeping on cold floors to remind them of their duty to God. Some Shia Muslims also ritually practice flagellation to commemorate and share the suffering of the profit Husayn ibn Ali.

Milder Suffering for Religious Beliefs

Closer to the centre of the Thermal Satisfaction scale, in mid-19th-century England denying oneself of the pleasure of a comfortable seat in church might provide visible evidence of a degree of fleshy mortification. When Protestantism was fracturing into new churches in the early 1800s the use, or avoidance, of pews types was seen as a test of the high or low character of a Protestant church. Describing a mid-19th century liturgical conflict between Henry Edward manning and Archdeacon Hare, Lytton Strachey remarked that '*Manning had been removing the high pews from the church in Brighton, and putting in open benches in their place. Everyone knew what that meant; everyone knew that the high pew was one of the bulwarks of Protestantism, and that an open bench had upon it the taint of Rome*'.[16] High pews shielded the worshipper from the draughts in the body of the church while the pews exposed their backs to the often biting cold. People believed that suffering can be a form of spiritual purification. According to

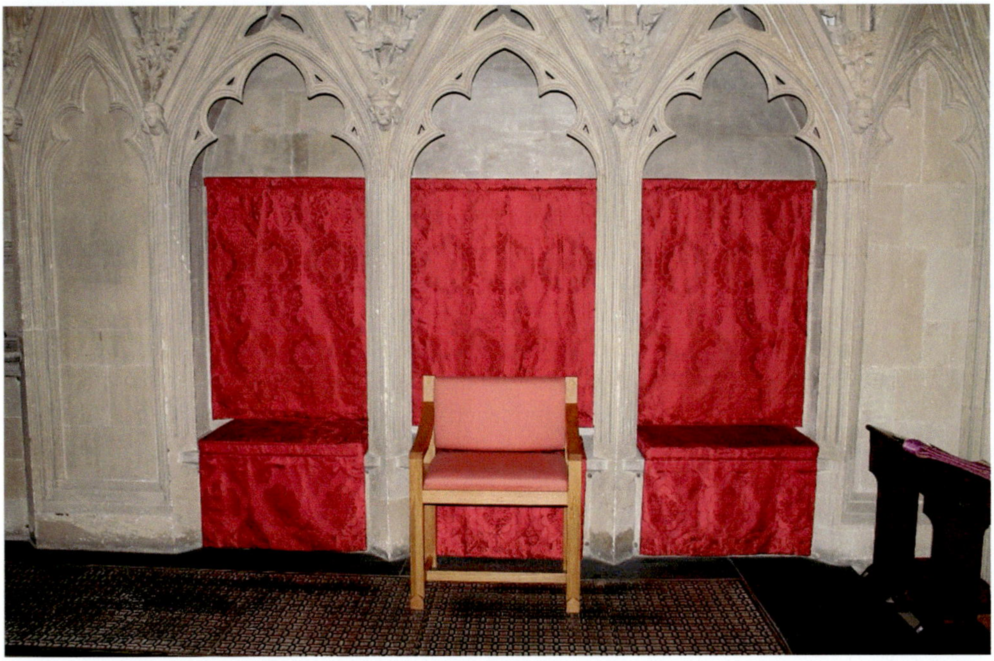

FIGURE 27.3 The sedilia, the group of three stone seats for clergy in the south chancel wall of Merton College Chapel, Oxford, used liturgically about four or five times a year. The cushioned seats and wall hanging does not provide adequate comfort for the modern clergy who choose a modern chair in front rather than sit next to the stone walls

Source: Roaf.

the book of Job, the reason the righteous suffer is to test their faith in God, to make them more like Him, and to bring glory to Him.

The fervour to suffer for one's beliefs, while still comfortably surviving sermons that could last over hours lacked some appeal to many more of the English decadent wealthy who attended church faithfully every Sunday. Comfortable timber cubicles were provided to protect families from the cold and enable them to pray, and or snooze in private. Those most often in the church like the clergy, choirs, organists and the privileged prayers often sat protected by timber panelling, screens or curtains from the cold of adjacent stone walls and draughts, as shown in Figure 27.3, just as the nobility furnished the cold walls of their chilly castles and manor houses with wall hangings or arrases.

Belonging and Belief

Religions can provide comfort in difficult times, connect us to communities, offer a sense of purpose, and give lives meaning. Many seek spiritual solace through communion with spiritual powers greater than themselves, often in buildings like churches, chapels or mosques. For modern populations often unused to discomfort at home very hot or cold conditions when praying can take a focus away from imagined spiritual dialogues and the succour sought from it. European churches are often large stone spaces in which efforts are made to alleviate the thermal discomfort of congregations.

Oxford Colleges all have chapels that were traditionally freezing to visit in winter. Many students find themselves confused and lost during their often-turbulent years at university and need, and seek, solace in their chapels. Many young students far from their homes in very different climates and cultures feel more or less lost in their new lives. To pray in a chapel, many entering it will sit on a wooden pew or bench, often using a seat cushion for comfort, and a fabric kneeler to make praying on the cold floor bearable. Both cushions are often embroidered with a unique allegorical motif or pattern. The very act of using these comfort-enhancing cushions on a chair or the floor can 'ground' the worshipper in that particular place, not just physically, but also spiritually, in time and space, connecting the insignificant human to the vastness of history embedded in the fabric of the chapel and to the occupying deity within its walls, floors, and roof. The profound sense of belonging to, being part of, a belief system and its community that can be engendered in such ritualised habits as praying is such an important feature of many religions. Insignificant as they seem, those small comfort cushions do encourage people to feel centred and belonging to that space.

In Islam, the use of the prayer mat in a mosque has the same effect, enabling the worshipper to feel they have their own unique connection to that holy place, and in belonging they are freed up to be open to the most important communion of all, with Allah. They become part of the place they are in, linked by that comforting connection on mat or cushion.

Financial and Physiological Benefits of Spiritual Comfort

Making worshippers comfortable can pay dividends both financially and by enhancing the quality of life and well-being of worshippers. A study of thermal comfort in mosques in Malaysia has shown that worshipers are more drawn to pray in air-conditioned mosques than in unconditioned warmer buildings.[17] Just as a church or mosque will gain more income from services by popular speakers, so both places of worship gain more financial support from hosting many worshippers, and having a comfortable indoor climate attracts income.

Islam evolved early on in the very hot climates of the Middle East. For those working on the streets, in the fields or market places of such regions there is an overwhelming advantage to take a break regularly during the heat of the day to rest physically and mentally in a cool prayer hall, to drain the accumulated heat from the body and prepare them for their next exposure to high temperatures. Muslims are obligated to pray five times a day – Fajr, Zuhr, Asr, Maghrib, and Isha, answering to the call for prayer, the Salah is prayed in the congregation every day at dawn, lunch time, tea time, in the evening, and at night. In terms of maintaining homeostasis this makes perfect sense, and for a labourer coming out of temperatures of 40°C+ (104°F) in hot dry Kuwait into the coolth of a mosque at c. 25°C (77°F)[18] it must seem like a heaven sent gift. Even for the wealthy in Kuwait there may be thermal benefits because it has been found that in residences often have their air conditioners set to American standards of around 23°C (73.4°F), while locals actually express preferred temperatures a couple of degrees higher as were found in local measured mosque temperatures. The massive investment in large solid religious buildings may also provide important climate refuges for communities during extreme cold and hot spells in our unpredictable future, where the grid fails, people cannot afford to heat or cool their own buildings or the weather is so extreme that conventional conditioning systems are not able to deal with the extremes and for instance widespread over heating occurs indoors.

Belief Baffles Brains: The Over-Cooling Cult

Why do people over-cool buildings? From shopping malls, restaurants, hotels, homes, offices, hospital and many in the USA are over-cooled, at a huge cost to buildings users and to the country as a whole. It has been estimated that in 2012 alone over-cooling cost the US 103,929 GWh of waster energy, an unnecessary environmental cost of 57,125 kt CO2e, and a financial cost of around $10 Billion US.[19] Reasons for over-cooling include the beliefs that:

- Comfort temperatures included the international Steady State Standards are deemed to be correct when in reality they promote unnecessarily low temperatures despite having been widely discredited for decades.[20]
- Cold is associated with higher status.
- Cold people are more productive, despite the fact there are no academically credible studies that show a relationship between temperature and office work performance. This does not mean it does not exist, but that it is not been proved.[21]
- Cooling energy costs are insignificant in relation to worker's wages. This is no longer the case as energy costs have soared, the climate destroying impacts of excessive energy use are now understood, and many workers now work from home.

The madness of wearing over-coats to work, hotels, hospitals, colleges, conferences, cinemas etc. is recognised now that many building managers understand that the fastest and least invasive way to reduce energy use in buildings is to raise or lower the thermostat settings within reason with little impact on the perceived comfort of occupants. Beliefs change as circumstances around them require them to and a fundamental lesson for building operators is that people adapt to those conditions they normally occupy. Change the thermostat a degree or so and often few will notice. Stop buildings being too cold and many occupants will be grateful. Despite these facts people still over-cool buildings.

Comfort and Consumerism

Our everyday lives are organised around a routine of ordinary habits and actions built on patterns of inconspicuous consumption or energy and products.[22] From the cleaning products in our homes, to high end expenditure for special occasions our spending expectations and practices are shaped by what we believe to be due to us and achievable even if, from different perspectives such purchases and their contexts appear incomprehensible.

Anyone who has visited Cartier, Chanel or other high end shops in Hong Kong shopping malls may wonder why in that hot, humid climate luxury retailers keep their shops at freezing temperatures. A Time Out report on air conditioning abuse in Hong Kong found that the luxurious IFC Mall was kept at a frigid 13°C (55.4°F), while the local grocery stores were as cold as 12°C (54°F). As an economic superpower Chinese consumers of the 'Bling Dynasty' account for more than a third of the global luxury market which has become heavily dependent on the Chinese sales globally.[23] If the Asian super-rich believe that low temperatures equate to luxury,

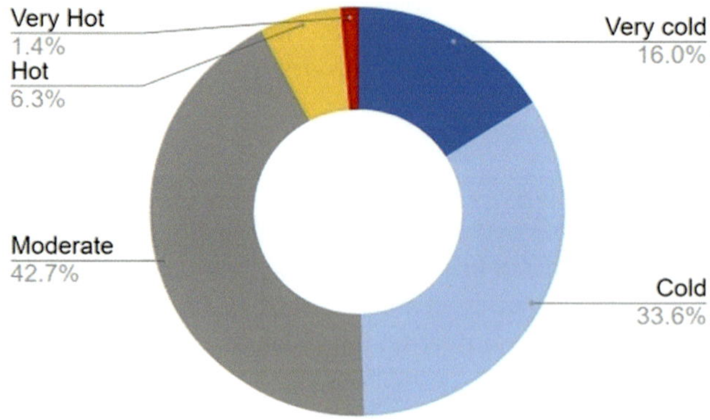

FIGURE 27.4 Responses when office workers in Hong Kong were asked whether the temperature in the office is too hot or too cold

Source: Green Sense, 2019.

then luxury shops around the world will lower their temperatures to cater for them. Money talks and turns down thermostats.

Cold air machines, as they are known in Cantonese, are ubiquitous in the city, where many people carry an 'indoor' jacket to put on in public buildings and transport.[24] In 2019, commercial buildings in Hong Kong accounted for 66% of total electricity consumption, of which 30% was used in air conditioning, rising to 55% in offices.[25] In one survey, 40% of workers could not adjust their office temperature, despite the fact that over 50% of workers were cold or uncomfortably cold. Attitudes to their discomfort included people accepting they cannot change their thermal conditions so they must adapt to them, and if it means they are at work, such cold was good for them.

In the face of energy shortages and soaring costs the Hong Kong government proposes to mandate that air conditioner temperatures be set to 25.5°C, a temperature some 47% of respondents of one comfort survey considers to be in their comfort range of 24–26°C (74°–79°F; Figure 27.4). Many shifts in comfort temperature narratives can managed in reality as happened in Japan in 2011.

Catastrophes Change Beliefs

The Great East Japan Earthquake in 2011 caused enormous damage, including the devastation of the Fukushima nuclear power plant and the subsequent shutting of all other nuclear plants in Japan. Immediately, peak power on the grid was reduced by 15%, and to compensate and keep the lights on, energy conservation measures, including that all government offices had to set air conditioners to come on only when the indoor temperature reached 28°C (82.4°F). Although Japanese citizens believed that it was right to reduce energy demand, the temperature was found in a subsequent study to be too warm. In a study, it was recommended that it be reduced to 27°C (82.4°F) or less, to maintain worker productivity levels. In extremis, the belief of the Japanese citizens that they were working for the good of the country helped them accept the overnight step change in comfort temperatures imposed.[26] The lesson here is that in adversity, what is enormously important is to mould acceptable narratives to change beliefs to maintain social stability. People are capable of great things if they are psychologically resilient.

The Power of Personal Beliefs

One study of the lives of African American women in 'high-disorder' neighbourhoods found that despite poor social conditions, the studied African American women displayed signs of considerable adaptability and resilience. Through self-belief they were able to overcome the damaging effects of the impoverished and crime-ridden neighbourhoods they lived in, protected and buffered by strong positive, optimistic outlooks, personal resources (religious beliefs, positive affectivity, and physical health) and good interpersonal relationships. Even in their high-disorder neighbourhoods they made the places safe for them and theirs. Researchers noted that religion was a key factor, not least in reducing the distress associated with hopelessness. Hope for them lay in a belief that the future would be better than the present. Beliefs were demonstrated to be independent of current circumstances and the beliefs and behaviours of those women offered their communities hope, improved mental, emotional and physical health, enhanced well-being and emotional comfort.[27] To build a comfortable community necessitates building a strong community, and that can be assisted by ensuring the existence of the places and spaces needed for people to interact with each other, safely.

Mindfulness and Well-Being

Even in less challenging environments the importance of self-belief can be key to achieving peace of mind. Valuing workers self-worthy and spiritual well-being has too often ignored in the commercial world. One reason why so many people are unhappy working in open plan offices is that they offer so few opportunities to think clearly or deeply in their work spaces,[28] too little breathing time and space to be *Mindful* in. being able to work with a lack of distractions like noise or discomfort can not only enable people to be present in their own thoughts and feelings and the world around them, it can enhance their ability to concentrate on ideas and questions to hand, to solve problems where solutions might be obscured by such distractions. *Mindful* moments of heightened awareness may appear to be irrelevant for many in the course of a working day, while it is prioritised more for others in the thinking industries like academia and high-tech firms. However, most people will benefit from time during a day to find some peace of mind, to forget their worries stresses and be 'in the moment with thoughts, sounds and sensations'. In Britain the National Institute for Health and Care Excellence (NICE) recommends that employers make mindfulness available to all employees, to support mental well-being at work. In some cultures, this might be in a church or mosque, while in some modern offices it may be in a meditation rooms, peaceful places or a garden available for staff to book into, or just be in. In reality for others it may just be a cool back step in an alley behind the hot kitchen of a busy restaurant. Mental comfort and well-being across a working day may be enormously enhanced when it involves access to periods of contemplation within a person's inner life, rather than a total concentration on the physical conditions around them and the work at hand. Just as people need to top up their thermal batteries they as need to address and replenish their spiritual needs and beliefs.

The Tragedy of the Commons

Comfort micro-cultures, whether secular or religious, can often be defined by how their values, principles, beliefs, attitudes, and ethics manifest themselves in the temperatures occupied by their adherents. At time when humanity is manifestly overpowering many natural environments around the world some groups act with impunity to accelerate the destruction of nature. Other groups consciously behave to minimise their own, often willing to sacrifice comfort and convenience to protect local and global ecologies. People are shaping the physical and thermal worlds driven by what they believe is due to them, what they can afford and what they believe to be right. The fact lies at the core of the dilemma at the heart of the how we should choose to move forwards in our quest for ease, comfort and peace of mind. Do we opt for the American Dream and live in air-conditioned bubbles that protect us from increasingly extreme weather outside? Do we radically re-order the world to ensure that our own ecological niches adapt with the evolving environments around them, like the Chameleon, not only to minimise our impact on nature but also to protect ourselves from attack by it.

The idea of the *Tragedy of the Commons* suggests that humanity will likely succumb to the lures of worldly ease over the solace of spiritual comfort and peace of mind, if they can afford to. The Tragedy of the Commons is a term used to describe the dilemma that arises from the situation in which multiple individuals, acting independently and rationally consulting their own self-interest, will ultimately deplete a shared limited resource, even when it is clear that it is not in anyone's long-term interest for this to happen. Garrett Hardin developed first in a 1968 article that was taken up by many attempting to rationalise why, in the face of our growing knowledge base, decisions were and are being made that appear to push progress into socially, economically and environmentally damaging directions. His concern then was for the impacts of uncontrolled populations growth.[29] It applies to so much more, and as global populations expand with rising expectations for living standards and comfort combined with ever more efficient systems for plundering earth's resources ecosystems and economies are rapidly being pushed beyond their capacities to function fairly, or in the worst cases, function at all.

Harding started his paper with the statement that the population problem has no technical solution: It requires a fundamental extension in morality. The preceding narratives have shown that the mental and spiritual strengths and the beliefs of individuals and communities can have great power to provide the skills and strengths that will be needed to co-evolve with nature through the traumatic decades ahead. The tragedy of the commons indicates otherwise, that fundamentally our species is motivated by the desire to promote the interests of the individual and those around them, regardless of the consequences. Unfortunately, no man is an island.

No Man Is an Island

No man is an island,
Entire of itself;
Every man is a piece of the continent,
A part of the main.

If a clod be washed away by the sea,
Europe is the less,
As well as if a promontory were:
As well as if a manor of thy friend's
Or of thine own were.

Any man's death diminishes me,
Because I am involved in mankind.
And therefore never send to know for whom the bell tolls;
It tolls for thee.

John Donne (1624)

Notes

1 Berger, J. (1972). *Ways of Seeing*, London: Penguin Books Ltd.

2 Bachelard, G. (1964). *The Psychoanalysis of Fire*, Boston: Beacon Press. 1st Published in French in Paris in 1938.

3 Saint-Pierre, J.-B.D. (1791). *Etudes de la Nature*, 4th Edition, Vol. 4, Paris: Published by De l'Imprimerie de Monsieur, p. 34. Studies of nature, translated from the French of J. H. B. de St. Pierre, carefully abridged. With a copious index. By L. T. Rede in 1798, Accessed on 24/07/2024 at: https://archive.org/details/bim_eighteenth-century_etudes-de-la-nature-en_saint-pierre-bernardin-_1798/page/n13/mode/2up

4 Balzac, H.D. (1980). Gobsek, published by Dyalpha, Cleveland, Ohio. 1st published in 1830.

5 Balzac, H.D. (1830). Gobsek, Accessed on 22/11/2025 at: https://www.gutenberg.org/files/1389/1389-h/1389-h.htm

6 Gibson, J.J. (2014). *The Ecological Approach to Visual Perception: Classic Edition*, 1st Edition, Psychology Press. https://doi.org/10.4324/9781315740218

7 Ingold, T. (2011). The perception of the environment: Essays on livelihood, dwelling and skill, 2nd Edition (1st in 2000), Abingdon, Oxon: Routledge.

8 Hof, W. (2020). *The Wim Hof Method: Own Your Mind, Master Your Biology, and Activate Your Full Human Potential*, Published by Sounds True, ISBN: 978-1-68-364409-5.

9 van Marken Lichtenbelt, W. (2017). Who is the iceman?, *Temperature*, 4(3)(July 11), pp. 202–205. https://doi.org/10.1080/23328940.2017.1329001. PMC 5605164. PMID 28944263.

10 https://en.wikipedia.org/wiki/Wim_Hof

11 Dhariwal, J., S. Gangrade and P. Dhariwal (2024). Learnings from the extreme thermal comfort adaptation of Jain ascetics during the summer and the monsoon months in India, *Proceedings of the 2023 Comfort at the Extremes Conference*, Ahmedabad | India, pp. 79–89, Accessed on 24/07/2024 at: CATE2023_Book_of_Proceedings.pdf (carbse.org).

12 https://www.utci.org/

13 Tartarini, F., S. Schiavon, T. Cheung and T. Hoyt (2020). CBE thermal comfort tool: Online tool for thermal comfort calculations and visualizations, *SoftwareX*, 12, p. 100563. https://doi.org/10.1016/j.softx.2020.100563, Accessed on 26/07/2024 at: https://comfort.cbe.berkeley.edu/

14 Daniel, L., T. Williamson, V. Soebarto and D. Chen (2015). Learning from thermal mavericks in Australia: Comfort studies in Melbourne and Darwin, *Architectural Science Review*, 58(1), pp. 57–66. https://doi.org/10.1080/00038628.2014.976537

15 Hickson, P. (2019). Australian energy efficiency policy is killing passive low energy earth building, *Proceedings of the 2019 Comfort at the Extremes Conference*, Dubai, pp. 174–188, Accessed on 26/07/2024 at: https://windsorconference.com/wp-content/uploads/2019/04/CATE2019_Proceedings_v3_web.pdf

16 Strachey, L. (1918). *Eminent Victorians*, 1979 Folio Society Edition, p. 42. https://en.wikipedia.org/wiki/Pew

17 Hussin, A., E. Salleh, H.Y. Chan and S. Mat (2015). The reliability of predicted mean vote model predictions in an air-conditioned mosque during daily prayer times in Malaysia, *Architectural Science Review*, 58(1), pp. 67–76. https://doi.org/10.1080/00038628.2014.976538

Al-ajmi, F. and D. Loveday (2010). Indoor thermal conditions and thermal comfort in air-conditioned domestic buildings in the dry-desert climate of Kuwait, *Building and Environment*, 45, pp. 704–710. DOI: 10.1016/j.buildenv.2009.08.018

18 Al-ajmi, F.F. (2010). Thermal comfort in air-conditioned mosques in the dry desert climate, *Building and Environment*, 45(11), pp. 2407–2413.

19 Derrible, S., and M. Reeder (2015). The cost of over-cooling commercial buildings in the United States, *Energy and Buildings*, 108, pp. 304–306, ISSN: 0378-7788. https://doi.org/10.1016/j.enbuild.2015.09.022

20 Humphreys, M., F. Nicol and S. Roaf (2016). *Adaptive Thermal Comfort: Foundations and Analysis*, London: Earthscan and Routledge, March 2012, ISBN: 978-0-41-569161-1

21 Porras-Salazar, J.A., S. Schiavon, P. Wargocki, et al. (2021). Meta-analysis of 35 studies examining the effect of indoor temperature on office work performance, *Building and Environment*, 203, p. 108037, ISSN: 0360-1323. https://doi.org/10.1016/j.buildenv.2021.108037

22 Shove, E. (2003). *Comfort, Cleanliness and Convenience: The social organization of Normality*, Oxford: Berg, ISSN: 1472-2895.

23 Rambourg, E. (2014). *The Bling Dynasty: Why the Reign of Chinese Luxury Shoppers Has Only Just Begun*, Published by Wiley Finance.

24 Cummins, A. (2016). Hong Kong's air con abuse: It's not cool, *Time Out Magazine, Hong Kong*, 20th May, Accessed on 28/07/2024 at: https://www.timeout.com/hong-kong/blog/hong-kongs-air-con-abuse-its-not-cool-052016
25 Greensense (2019). *Green Sense Air Conditioning Survey 2019*, Accessed on 29/07/2024 at: https://greensense.org.hk/en/2019/09/%E7%84%A1%E5%86%B7%E6%B0%A3%E5%A4%9C2019%EF%BC%9A%E5%86%B7%E6%B0%A3%E8%AA%BF%E6%9F%A5%E7%B5%90%E6%9E%9C/
26 Tanabe, S.I., Y. Iwahashi, S. Tsushima and N. Nishihara (2013). Thermal comfort and productivity in offices under mandatory electricity savings after the Great East Japan earthquake, *Architectural Science Review*, 56(1), pp. 4–16. https://doi.org/10.1080/00038628.2012.744296
27 Cutrona, C.E., D.W. Russell, R.M. Hessling, et al. (2000). Direct and moderating effects of community context on the psychological well-being of African American women, *Journal of Personality and Social Psychology*, 79(6), pp. 1088–1101.
28 Musser George (2023). Fixing the hated open plan office, *Scientific American*, April 1, Accessed on 24/06/2024 at: https://www.scientificamerican.com/article/fixing-the-hated-open-design-office/
29 Harding, G. (1968). The tragedy of the commons, *Science*, 162(3859), pp. 1243–1248.

Conclusions

28

ADAPTABLE BUILDINGS NEED ADAPTIVE COMFORT

Introduction – New Thinking on Comfort

This book is a compendium of information on comfort and an exploration of the many different ways in which experiences of thermal comfort impact our lives and well-being. We started writing this book over ten years ago, but every year since, new catastrophic events and trends have made it necessary to update the text, and our thinking. The standout comfort experience of those years was the COVID pandemic from 2020 to 2022. That actually changed all our lives and the whole narrative around indoor comfort. Many of us found out then that what actually mattered when working from home was simply keeping a person affordably comfortable where they sat, not conditioning the whole building. The pandemic was set against an even bleaker backdrop of records of temperatures, precipitation, flooding, and wind speeds being regularly broken around the world and the emergence of new and ever more extreme weather phenomena.

It is not surprising that in the face of such rapid environmental change, the glacially slow, regulation-heavy process of mandating comfort standards could not keep up the pressures for social, economic, and political change that are driving the urgent need for better buildings in our ever-hotter world. We are pleased to have been able, eventually, to pull together this wide-reaching overview of some of the comfort challenges we face individually and as populations in different climates. We have offered across these chapters a range of possible types of solutions for making buildings safer in a changing climate, not in prescriptive straightjacketed design mantras or detailed inflexible rating systems or computer programmes that do not reflect the infinite variety of factors and forces that shape the built environment. People will have to think for themselves on how to best integrate the comfort and design ideas presented here for their own unique clients and sites.

The Big Message – Be Prepared

Humanity is patently unprepared for the scale of adaptation needed to survive in buildings during extremes of heat and weather, even now. Policy- and decision-makers, designers, and the general public need to urgently wake up to that fact and tool up to be better prepared in

DOI: 10.1201/9781315645070-37

FIGURE 28.1 A 2025 development to the north of Oxford where the structure with glass walls on all sides with no internal or external shading creates an energy hog and will result in radiant overheating for sedentary workers exposed to solar gain through the glazing regardless of the internal air temperature.

Source: Susan Roaf.

improved buildings and more cohesive communities to stay thermally safe in the future. From Bauhaus to Passivhaus, the 19th-century building reliance on local sun, wind, biomass, and shade to make buildings comfortable gradually transformed over the course of the 20th century to be increasingly reliant on mechanical conditioning to remain habitable. The need to design for 'well-behaved' indoor climates stands at odds with the growing love affair between architects with glazing, engineers with machines, and developers with cheaper, lightweight construction. The result is un-resilient buildings that often fail disastrously in heatwaves and blizzards, particularly when their grids fail.

The question now must be, for those who have read this book and in light of what it contains: How and when can the necessary changes happen to both mandate and create genuinely low-energy buildings in time to improve all (species) of our chances of surviving comfortably?

Changing Mind-Sets

Change must begin with ourselves before we help others transition to new thinking. The first step of all is to recognise the nature and scale of the problems faced, then to act upon that understanding to adapt in time. True, many environmentalists have already grasped this fact and are acting upon it. However, entrenched thought patterns that were developed in a different past can hold us all back. It is important not to be overwhelmed by the ecological implications and impacts of climate change on buildings, but to evolve workable solutions to 'bounce forwards' to safer states, it is necessary to have a clear, future-facing world view that both gives directions forward and can accommodate change.

How and what a designer creates is strongly influenced by their own *backstory* that colours their own unique world view, their *normal,* their *imaginary*.[1] Anthropologists use the word *imaginary* to describe a set of cultural beliefs, the ethos of a culture, or a group of people who share an understanding of the way the world is. Today, easy access to a huge range of information sources, many generated by 'hidden persuaders', has resulted in the many varied and often conflicting and entrenched world views at play today. Family members can inhabit very different world views, one being a committed Republicans or another a Democrats in the USA, or being a Brexiteer or anti-Brexiteer in the UK. One architect may be convinced by the need for light weight, low embodied energy buildings with lots of insulation and a heat recovery system to reduce emissions, while another swears by heavyweight, naturally ventilated buildings to keep us affordably, thermally safe indoors. A HVAC engineer may think that any building just needs very efficient mechanical systems with good back up energy supplies to survive through heatwaves. All viewpoints contain truths reinforced by personal experiences. One architect working in a big city on office towers in an international practice may have design opinions that are poles apart from her brother designing small houses on a Scottish island. Every reader will have their own views on what we have written in this book. Interestingly, the same person may inhabit more than one imaginary, one perhaps for the office selling high-tech smart buildings, and another at home, where the simplest, low-cost building solutions are preferred. However, all of our imaginaries will inevitably have to *adapt* in the face of changes ahead.

Quoting Maynard Keynes, J. K. Galbraith wrote:

> *We are ruled by ideas and very little else. . . . But the rule of ideas is only powerful in a world that does not change. Ideas are inherently conservative. They yield not to attack of other ideas . . . but to the massive onslaught of circumstances with which they cannot contend.*[2]

We live in a world now in which circumstances are catastrophically changing around us. Those people, buildings, and systems that are best able to adapt to a different future will have to be *flexible* in the face of such changes. There are very few places left on this planet where business-as-usual in the built environment is a safe future option going forwards. The path to transitioning also requires us to find a way through the tsunami of messaging we are hit by every day in the media. With so little trust existing between groups in society at large, a first step is to build a clear and credible picture of what future climates and conditions lie ahead that we can trust in ourselves and aim towards with our actions. Backcasting is one way of creating that future view.

Backcasting

We typically use weather *forecasts* based on historical and current climate data to predict future climates using hugely complex simulations. A *backcasting* study produces alternative images of the future that have been thoroughly analysed as to their feasibility and consequences.[3] Future climate predictions are added to other data forecasts from different fields, including the political, social, and pure sciences, to build a picture of what a preferred future might look like in, say, 2050 or 2100. Experts from different disciplines then negotiate and reshape that future by trying to understand the impacts of the interaction between their fields and the whole. The outcome may be one or a few well-informed and convincing pictures of different futures from which the most likely to succeed can be selected to build upon. The second part of the exercise involves working backwards from that ideal future to today to identify policy and action steps that need to be put in over time, to make that preferred future happen. The fundamentals of the method were first outlined by John B. Robinson in 1990, who kick-started the method by asking the simple question 'If we want to attain a certain goal, what actions must be taken to get there?'[4]

People who have gone through backcasting exercises often produce what to others may appear to be counter-intuitive answers to simple questions about the future. This is because backcasters have worked hard to reimagine, in depth, an often-detailed future world view that others have no access to so cannot understand the logic behind their responses.

We have all probably undertaken backcasting exercises ourselves when asking ourselves questions that might never have occurred to us 10 or 20 years ago. Is that new house we want to buy on the floodplain? Will that west-facing glass wall overheat the house? Is there any cross-ventilation in this flat to help cool us and it down on hot days? Is that house on the cliff too exposed to extreme winds? Such questions reflect our already-altered visions of future climates and our own evolving imaginaries. People are also altered by being exposed to temperature extremes themselves. Such experiences do actually affect our minds, brains, and bodies through altered behaviours, environmental interactions, mental health, and moods. Clayton Page Aldern pointed out in his book *The Weight of Nature* in 2025 that it is not just neurological effects causing those alterations but a growing oneness and empathy with, and connection to, nature triggered perhaps by the louder sounds of rain, wind, or thunder around a house, or the hotter feel

FIGURE 28.2 Michael Humphreys (second from left), discussing the measurement of temperatures with SBSE's Alison Kwok and Bruce Haglund and Fergus Nicol (*right*) and Susan Roaf.

Source: Tisha Egashira.

of sun on the skin.[5] We instinctively feel this and know that the climate is changing, prompting the impetus to act. That is why it has been suggested that the best way to prepare to design for hotter futures is to go and live and work in hotter places for a while, to get a feel for them and the way buildings work well in them.

Moving people beyond their personal experiences of temperatures can change minds about the potential impacts of weather extremes. British born and bred Michael Humphreys shown in Figure 28.2 was astounded by the fact he could feel very comfortable at 32°C (89.6°F) during fieldwork in Pakistan, altering his understanding of comfort at warmer temperatures forever.

Marketing Imaginaries – The American Dream

At the heart of most imaginaries are dreams that can be exploited to sell ideas. Political policies and commercial endeavours can fail because they misunderstand those dreams or do well by correctly targeting the aspirations of a generation to pursue them. The air conditioning industry, for instance, did well by weaving the growth of their products into the mid-20th-century American dreams of a burgeoning middle class. Three excellent books by Cooper (1998),[6] Ackerman (2002),[7] and Cox (2010)[8] cover the history of how the US air conditioning industry shaped the mind-set of the American people, using narratives of personal success involving attractive claims of improved health, commercial productivity, and quality of life. A national, and later international group-think assumption that air conditioning was eventually hard-baked into the *American dream* of corporate success and carefree futures. The 20th-century air-conditioned imaginary painted a world of effortless comfort and ease that was dependent on mechanically serviced buildings that would be affordable in booming post-war markets. Sales pitches rarely include the dark sides of the offering to customers.

Long before the environmental impacts of the HVAC technologies were understood, the downside then, as now, was the cost of the energy needed to run the machines. Fortunately, the emergence of the air conditioning dream coincided with the *nuclear dream* launched spectacularly in 1945 with the atomic bombs that cemented the USA at the pinnacle of global civilisation with its superiority in the discipline of physics, putting America at the heart of the control centre of 'colonial modernity'. To gloss over the tragic image of the deaths in Japan, on 8 December 1953, President Dwight Eisenhower launched the project to develop civil nuclear power worldwide with his *Atoms for Peace* speech to the United Nations.[9] Eisenhower was trying to sell nuclear energy to the American public, so it was claimed to be of 'great benefit to all mankind' and of 'universal, efficient and economical' energy. In 1954, Lewis Strauss, President of the American Atomic Commission, promised the public an energy 'too cheap to meter'.

Nuclear narratives varied. The Japanese concentrated on selling nuclear power to citizens as the main route to energy security and self-sufficiency in a country without coal, so essential for technological progress for national revitalisation after the war.[10] In the UK, nuclear energy was sold to people as being the miracle energy technology that would power the post-war economic recovery. In 2003, France claimed that nuclear energy would be the only source of decarbonised energy massively available in the world at that time. More recently, marketing strap-lines from the nuclear industry have claimed that carbon neutrality by 2050 cannot be achieved without nuclear (fission) power. Billions have also been spent over the last 70 years by governments seeking that even more dreamlike technological utopia where everything from cars to toasters and vacuum cleaners will be powered by the 'atoms for dreams' miracle with nuclear fusion technologies. Nuclear power is still being promoted heavily by vested interests to global governments, even in Australia, which was badly traumatised by British testing of H-bombs on Aboriginal lands in the 1950s. As far back as 1974, the pioneering Australian nuclear scientist, Mark Oliphant, said in a talk to Adelaide University:

> Speaking as a cynic I would expect that any decisions to build nuclear power plants in Australia in the near future would be made, not because they are necessary to meet our energy needs, but because they would enable us to make nuclear weapons.

Every country has its own nuclear narrative, and why does this matter to the issue of comfort in buildings and the architects who design them? Like the ecological cycles that power life on Earth, the narrative cycles that power political landscapes are also deeply connected. By 1980, the German government had already decided that *nuclear power is a potentially catastrophic threat that we cannot control. Therefore, we need a nuclear phase-out as fast as technically possible.*[11] Germany pioneered the use of solar and wind technologies in Europe as well as better building construction in the early 1990s and 2000s, and by 2024, nearly 60% of all German energy came from renewables. Denmark also adopted a strong anti-nuclear stance with its mantra of 'Atomkraft Nein Danke'. *Nuclear Power, No, Thank You*. Renewables now supply over 80% of Denmark's power and aim to get to 100% of all energy by 2050.

The invisible link now between all national market narratives is the need to drastically reduce fossil fuel energy use while still keeping citizens thermally safe and comfortable in buildings. Denmark and Germany invested heavily in improving the thermal performance of actual buildings over time rather than just pushing more electricity-hungry technologies into their building markets. France, the main nuclear economy in Europe, was able to provide cheap electricity to voters using heavy state subsidies until the catastrophic circumstances around climate change struck at the heart of the nuclear industry. Rivers used to cool the plants dried up or overheated, and with surging air conditioning demand in the French south, the northern nuclear plants could not meet the soaring summertime peak loads and widespread grid failures resulted. The recent move in France to promote solar and battery technologies to serve those peak summertime loads is tardy but hopefully not too late. The way different countries provide the power needed to keep populations comfortable is intimately intertwined with the values and interests of those in power nationally and their friends in the governments of the day.

Nowhere is this clearer than in the United States; even after the major climate catastrophic hurricanes of Katrina in 2005 and Sandy in 2012 and the ever-growing number of hurricanes, tornados, fires, blizzards, floods, heat domes, and droughts, the Trump incumbency continued to pretend that nothing untoward was occurring with the climate. Using accusations of 'fake news' to push ahead promoting the interests of their own funders with narratives like 'Drill, Baby, Drill', the comfort and well-being of most Americans on so many levels were, and are, being seriously threatened by such narratives.

Comfort and the Net-Zero Narrative

Narratives really matter. Good marketing often presents well-constructed narratives as single issue, simple truths that are designed to appeal to populations while underplaying or marginalising other cost issues. With large, all-glass super-polluting buildings without opening windows, the construction and running costs, pandemic vulnerability, heat island impacts, and degraded local environments are avoided topics. The narrative glass towers are sold into commercial markets that include their status as big, shiny symbols of corporate success and that they are also very 'energy-efficient' or 'sustainable' or enhance occupant 'well-being' according to government-sanctioned, often almost mendacious rating systems. Those high-end buildings do not actually have to be sold to the general public, whose views on such developments are completely ignored anyway. Building funders, typically now foreign investors in the UK, don't worry themselves or their friends in government over the implications of local unpopularity. For domestic markets, many rating schemes are in place that promote developers' interests for lightweight construction, unshaded glazing powered by 'efficient' machines, and result to buildings that overheat and increasingly require air conditioning to keep them habitable.

Moving populations into investing in a *net-zero* economy has been challenging for the UK, not least because of their unpopular sales pitches. Take for example the push for the uptake of electric heat pumps in the UK to replace the alternative gas boilers that cost far less to install and run. Britain has, in the last decade, hiked the electricity unit price to pay for the buildings of an expanded national electricity grid and 'big' energy projects, resulting in the fact that in 2025 the cost in Britain of a unit of electricity was 60% more than any other European. Rather tin-eared politicians railed against the public for their low uptake of the heat pumps that most could not afford or saw relatively little benefit in. Heat pump sales did eventually grow with the support of generous subsidies and dedicated environmentalists. Not a good look, though.

Other *net-zero earworms* that permeated political spheres and the design industries included the developer's favourite, which is the promotion of low-cost, lightweight (cheap) construction under the banner of it being 'better' because it has *low embodied energy*. This narrative sounded convincing and struck a chord with politicians, who fulsomely backed it. In reality, what matters in a design is the *total or whole life energy costs* of a building, including both embodied and operational energy costs, over the building's lifetime. You could live in a really low-embodied-energy cardboard home but need lots of HVAC, energy, and income to remain comfortable in it. In ever-more extreme weather, the resilience of a structure is critically important. Gomes et al. pointed out in 2022 that dynamic life cycle analysis shows that 'it does not necessarily cost more energy and emissions now to build resilient buildings, than they save over their lifetimes'. They argue that decisions are seldom made with a proper understanding of the whole life cycle efficiency of the tested alternatives of their future consequences. Their advice is that buildings should now be designed for self-sufficiency by equipping buildings to run on local renewable energy from the sun, wind, and in thermal mass storage, even during the most extreme events.[12] Clarifying all sides of the issues around narratives clearly to the public is critical to informing their own decision-making on climate-safe design in a heating world. A further complication comes from the news media, who often air the views of plausible 'independent' pundits they know who are expert in the marketing discourses that they themselves find most useful. There is not much real honesty out there in a world that rises and falls on funding from industry and developers. Ignorance impedes change.

Regulating for Comfort

Thermal comfort in all environments is achieved by the building occupants adapting to the conditions they find or create. These conditions may be very constant and predictable in buildings with air conditioning or quite variable but also may be predictable in buildings with less mechanical control.

In the case of constant indoor conditions, building occupants will know from experience what to expect, but the provision of constant indoor temperatures will inevitably be costly in energy terms. In the case of the variable indoor environments in naturally conditioned and ventilated buildings, there may be more need for building occupants to know how to anticipate what environment they will encounter and what action will be most likely to enable them to become comfortable. As we have seen, the amplitude of the thermal heartbeat of the building will depend on its design, the amount of thermal mass it has, and how extreme the weather outside the buildings really is. Simple responses to temperature changes indoors, like opening a window or changing clothing, will likely be familiar to occupants and easy to control. The use of energy, for instance, by switching on a ceiling fan can require relatively little energy. The success of an *adaptive design* will also be measurable in the amount of energy expended by the occupants to effect the necessary changes in their environments.

Regulating the acceptable range of indoor thermal environments was historically undertaken to name a temperature within the range available which was least likely to give rise to discomfort. Cynics might suggest that comfort standards as they have evolved are more about locking buildings into requiring comfort provided by evermore HVAC systems, than about encouraging the design of buildings that enable their occupants to remain affordably comfortable for as much of the year as possible by adapting themselves or their environments.

The 20th-century narrative of the importance of an absolutely constant temperature range needs to be urgently reset to promote not adaptive thermal comfort standards but *adaptable, thermally comfortable buildings.* Mixed-mode buildings may conform to the regulated comfort standards when operating in air-conditioned mode. When in *free-running* mode, without heating or cooling, a wider range of environments and temperatures are possible and often preferred in them despite a slightly higher chance of discomfort, but with significant savings in energy and cost typically seen as adequate pay-off. That begs the question: Is the reset to mandating mixed-mode buildings even possible where committees overseeing building regulations are composed largely of HVAC engineers?

The Adaptive Thermal Comfort Narrative

Governments must get more savvy about understanding the real barriers and incentives that the critical role comfort has to play in the language of net zero for both reducing GHG emissions from the built environment and also in improving the living conditions and resilience of their own populations. Translating policy goals and actions into issues relevant to stakeholders on the ground is important not only for their understanding but also for the acceptance of the changes described in the policies. Narratives that attract people are those that are likely to be seen to directly improve their own interests and those of their nearest and dearest.

Ultimately, the only people most of us actually care about are family and close friends, and for them, people will act.[13] How many local focus groups after a heatwave said they need more heat pumps? Did they talk about the distress of not being able to open or shade their windows or afford to buy or run a fan or coolers? Did they say they had to move downstairs to be cool enough to get a good night's sleep? Who is listening to them? In winter, at least people can put on an extra jumper or warmer socks or turn on a local heater or retreat to bed with a hot water bottle to stay warm enough. In summer, there are only so many clothes one can take off. Promoting genuinely low-cost comfort for people in their own homes would really land well with voters.

Politicians are learning that putting solar systems on the roofs of the poor so they can generate their own energy is a sure way of helping to lift them out of fuel poverty, with huge benefits in terms of their health, education, finances, and social opportunities.[14] Why not promote personal comfort technologies, like small heaters, coolers, or fans to heat the person, not the building? Why not mandate all residences to have easily accessible and secure openable windows so homes can be run for as much of the year as possible in natural ventilation mode? Why not promote the use of external shading rather than using planning laws to ban it? Why not start programmes to help create climate refuges in homes, particularly of the vulnerable, to protect them during extreme weather? Why not start education programmes in schools around the new skills needed for living with more heat?

A very clear spectrum of useful actions to stay thermally safe in buildings could be very popular with the public. Policies should be expressed in simple, commonsense terms, like 'sensibly sized windows', 'shaded where exposed', 'good cross-ventilation and some thermal mass to make spaces more comfortable', and 'store energy for later use when needed'. Extra financial help for the vulnerable in cold winter or hot summer fuel payments are incredibly popular with voters, who like the flexibility of being able to use the cash to buy a small fan or hot water bottle, as well as for help with paying energy bills. Selling policies to the public by mandating black box rating products sold by the incumbency discourages the public from wanting to act and the ability to act. Trust was lost between some governments and the public during COVID that has faltered in retuning.

There is a tone-deafness to many of the current public discussions around thermal comfort that could be generously attributed to a degree of ignorance on the subject amongst the politicians, developers, financiers, designers, planners, and employers, the people who have the greatest influence on the development of modern buildings and cities. Some of the reasons for this became apparent in the final report on the Grenfell Tower fire in London in 2017 provided clear evidence that the construction industry is riddled with staggering levels of self-serving players, many of whom demonstrably promoted the vested interests of their powerful friends over concerns for the well-being of tower block's occupants. Many died as a result of the exposed systemic corruption in play, with huge subsequent knock-on consequences for thousands of other tower block occupants and owners.[15] That is not a good narrative either, and again, many of the circumstances that were involved concerned decisions around how to keep people comfortable in their own homes, decisions made by people who were gaming their understanding of the risks involved.

Risk and Uncertainty

> *Risk is where we know what the possible outcomes are and can estimate their probabilities. Uncertainty is about where we are unsure of the probabilities of particular outcomes.*[16]

The way we see the future is influenced by how happy we are in taking risks with it. People's appetites for risk vary enormously, as does their confidence that they understand the probabilities involved. Unfortunately, in an unpredictable future, possibilities evaporate in the face of unanticipated uncertainties, as we all found out with the COVID pandemic.[17] David Spiegelhalter concludes in his book *The Art of Uncertainty* that established ways of acknowledging uncertainty and formal statistical models are useful, but we should not be deluded into thinking they represent reality.[18] He posits that modelled predictions of short- or long-term futures can be effective tools underpinning decision-making, but they will not be able to deal with deep uncertainties. There is also the problem of the inevitable biases in the world views and resulting behaviours of the individuals and organisations involved in attempts to plan the future. One obvious solution for governments is to ensure that committees are peopled by a range of experts, not just one interest group, and they should certainly not be chaired by individuals who have vested interests in the outcomes of the committee's work.

Why gamble with the future of humanity by playing with people using a loaded dice?

Misplaced Faith in Simple and Complex Models

The ecological resilience expert Robert Ayres pointed out that simple quantifiable models will not be adequate to identify timings, and other attributes of turning points, but that 'naive intelligence and intuition may be the best tool for coping with a very complex and non-deterministic future'.[19]

Faith in quantifiable models, even highly sophisticated one, can be misleading, as exemplified by a climate scientist in April 2022. At the Comfort at the Extreme Conference in London, Alan Kennedy-Asser claimed that by 2050 Britain could expect temperatures above 40°C. By July 2022, several

FIGURE 28.3 Visitors to the herdsman's tent in Tuva, South Siberia, come well wrapped up. The tent-wife is perfectly happy in light clothing, sorting out her yard. Many might claim 'she cannot be comfortable in near zero temperatures', but evidence of our eyes shows she really is pretty happy in her T-shirt and jeans.

Source: Dolaana Khovalyg.

UK locations recorded temperatures over 40°C. He told a subsequent conference in September 2022 that one should never trust an optimist and that designers would be sensible to design for maximum temperatures of 43–45°C in the UK from now on.[20]

Building design simulation models generally still use 20th-century steady-state engineering comfort zones to inform their predicted energy demands. They are not quite as bad as the old days, when there were two settings, one for winter and one for summer, switched over on the same days every spring and autumn, but some not far off. By the year 2000, ASHRAE could no longer maintain the stance that its comfort standards developed for mechanically conditioned buildings represented the only conditions that people can be comfortable in. A major improvement occurred when the adaptive model was integrated into US Building Codes through the efforts and analyses of global field study data by Gail Brager in California and Richard de Dear in Sydney. In 2000, ASHRAE accepted revisions on thermal environmental conditions for human occupancy into their *ASHRAE Standard 55 – Adaptive Comfort Standard* that allows for warmer indoor temperatures to be deemed acceptable indoors for naturally ventilated buildings during summer and in warmer climate zones. Prior to that, naturally ventilated buildings had been slated as unacceptable in terms of their comfort provision under previous ASHRAE design codes.[21] In reality, adaptive standards like *ASHRAE Standard 55* can encompass comfort conditions in all building types, both naturally ventilated and air-conditioned, but as we have learnt in this book, they really do not describe the full gamut of conditions that people around the world find comfortable in buildings from the Tuvan nomads in Figure 28.3 to those who dwell in tropical climates or the hot deserts of the world.

Next-Generation Buildings: Energy Sufficiency

The most basic function of a building is to provide safe, acceptable, and affordable indoor conditions for building occupants to live and work in without discomfort, with adequate space that is fit for its purposes.[22] All the rest might be seen as window dressing. In the last decades of the 20th century, the mantra for providing those functions in an environmentally benign way was to make the building as 'energy-efficient' as possible. The IPCC 6th Assessment Working Party III Report in 2022 explained that *building efficiency* is not what we should be aiming for as it is about continuous short-term marginal technological improvements. They contrasted that with *building sufficiency*, which is about long-term actions driven by non-technological solutions, which consume less energy in absolute terms and have minimal impacts on the global climate. Look at the buildings in Figure 28.1 and 28.4 and guess which one has been sold to clients with the narrative of *efficiency* and which one operates in a *sufficient* manner, as a thermally well-behaved building.

Designing for sufficiency means working out how to minimise the whole life energy costs of buildings, including embodied and operational energy, in addition to maintenance, replacement, and demolition costs. The IPCC authors encourage new thinking for sufficiency using ideas around bioclimatic design, density issues, the use of nature-based solutions, multifunctionality of spaces, sizing of buildings to the evolving needs of households, the circular use of materials, and repurposing unused existing buildings, optimisation of the use of buildings through lifestyle changes, use of the thermal mass of buildings to reduce thermal needs and the sharing of appliances. They estimate that at a global level, up to 17% of the mitigation potential in the buildings sector could be captured by 2050 through sufficiency interventions. Sufficiency thinking opens

FIGURE 28.4 Kampung Admiralty building in Singapore by WOHA Architects, not only providing a mixed-use urban building but also cooling it and the locality with natural ventilation, planting, and internal and external ceiling fans, as well as some air conditioning.

Source: Patrick Bingham-Hall.

doors to genuine innovation and avoids the *energy efficiency fudge* that so many people are still being duped by:

> *It is possible for a building, vehicle or system to be highly energy-efficient while at the same time being highly energy consuming.*[23]

We hope this book might be a good introductory text for designing energy-sufficient buildings.

Cool Cities – Singapore

It is fitting to end the book with a really hopeful case study. In May 2025, the Singapore government and Singapore officially launched *'Go 25'* to encourage sustainable cooling in homes and offices and workplaces by setting indoor air conditioning at, or above, 25°C (77°F). They were following the example of the Japanese government's 'Coolbiz', which, following the Fukushima meltdown, discouraged temperatures below 28°C in air-conditioned offices to save energy. Singapore is a low-lying island state almost on the equator, giving it a very stable climate with mean temperatures around 27°C (80°F)–28°C (82.5°F) and absolute maxima around 35°C (95°F), with relative humidity around 80% all year. Temperatures in the centre of the city can be up to 7°C (12.6°F) hotter than in rural areas, and temperatures across the island are rising faster than in many other locations across the globe.[24] Buildings account for over 20% of Singapore's carbon emissions, with air conditioning being responsible for around 60% of energy consumption. The historic problem of overcooling buildings in the state is increasing as temperatures rise, leading to higher electricity usage and also driving up GHG gasses. A vicious circle is formed with emissions accelerating climate change and exacerbating the urban heat island effect that then further heats the city, in turn pushing up demand for air conditioning.

Singapore is a world leader in acting to reduce the long-term impacts of air conditioning on the environment. It helps that it is a small state with well-connected actors across the built environment. Although it has a low tax regime, since independence from Malaysia in 1965, it has developed high-value industries, and in 2024, Singaporeans earned over US$53,000, just below Canada, and ranking 19th globally for mean annual salaries. Its own vulnerability to climate change as a low-lying island and fairly strict and benign political regime has resulted in significant investment in future-proofing the state against higher temperatures. Regular backcasting exercises are undertaken using their advanced meteorological and urban heat island models[25] and a Virtual Singapore Twin 3D simulation of the whole city and state that was launched in 2014 by national government agencies.[26] Working with the nation's academics, the government has funded and acted upon studies showing energy savings of up to 12% for each degree rise in air conditioning temperature from 23°C (73.4°F) to 25°C (77°F) in offices. With no reported increases in occupant discomfort, additional energy savings were achievable when air conditioning temperatures were set above 25°C.

Singapore is heavily promoting hybrid cooling solutions, where higher air conditioning is used in parallel or sequence with natural ventilation and cooling by fans, resulting in even greater energy savings. Such advances are possible because the architects and engineers working in Singapore are actually capable of designing buildings that can run using either air conditioning, natural ventilation, or fans, where appropriate. Many residences in Singapore still run for much of the day and year in natural ventilation mode, and their design schools also teach the necessary skills to ensure students leave confident of being competent mixed-mode architects and engineers. In a brilliant move, the new wing of the School of Architecture at the National University of Singapore was designed to be a net-zero building and is run multi-modally in order to serve as a living laboratory to inspire staff and students to explore innovative ideas and push the boundaries of sustainable design.[27] This co-development of architectural, engineering, and urban design solutions has flourished, enabled by the strong ties between government and academia. These ties also helped in the government's hugely successful urban greening programme to reduce the heat island impacts that are closely monitored and reviewed by academic colleagues. The ties between designers and the regulatory wings of government are key to enabling change to happen and has also fostered the growth of internationally acclaimed architects, like WOHA, whose Kampung Admiralty is shown in Figure 28.4, who are pioneering the reconnection of tropical architecture back to nature, even in larger buildings.[28] Next-generation adaptive buildings are only

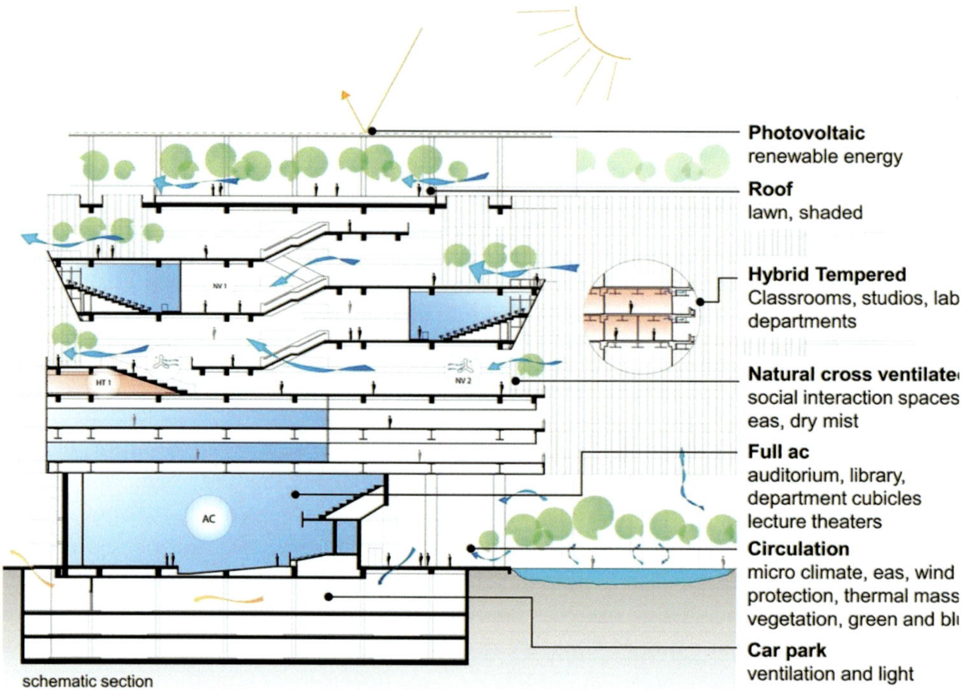

FIGURE 28.5 Annotated cross section of the BRAC (Building Resources Across Communities) University campus in Dhaka showing a range of comfort strategies at play in the hybrid building complex.

Source: Transsolar TSE.

possible when designed with first-class, 'next-generation' engineers who are willing, and able, to push design boundaries to create more resilient comfort conditions.

This next-generation design thinking has then been pushed out across continents to new buildings in Dubai, India, and Bangladesh. Figure 28.5 shows a cross section of the BRAC (Building Resources Across Communities) University campus in Dhaka, led by Mun Summ Wong of WOHA and engineered by Transsolar.[29] Many of the spaces are naturally ventilated, or hybrid, while others are air-conditioned. The complex houses 20,000 occupants and has transformed what was a polluted swamp area into an internationally leading-edge example of a genuinely low-energy building that enhances the lives of not only its occupants but also the city around it. The future of natural energy design is already being nurtured by such next-generation thinking in places like China, India, Singapore, Vietnam, and Bangladesh. One dream must be that *reverse comfort colonialism* will bring the genius of the East and the South back into the sterile deserts of high-energy building clusters in European and American cities, with new examples of such wise, resilient, hybrid buildings. A new and hopeful imaginary is blossoming internationally of how to design climate-proof and climate-ready buildings for a different future. Singapore was lucky to have the right mix of politics, economy, and education to grow some of those early seeds of change that are now spreading around the world.

Adaptive Thermal Comfort Needs Adaptable Buildings

We started this book looking at ideas around adaptive thermal comfort. Over the course of 28 chapters, our eyes and minds have been opened up to a myriad of multidimensional issues around what actually constitutes thermal comfort. Along the way, the emphasis has shifted from the engineer's viewpoint, which has dominated thinking on the subject for a century, back to the realms of architecture, because what really matters in designing for comfort at the extremes, for a different future, are the form and construction of the building itself and the adaptive opportunities it offers to people to alter themselves, their behaviours, and their environments to remain comfortable. But that is not the whole story.

We have learnt that there is no such thing as a single comfort temperature or comfort zone. Comfort is achieved in many different ways in different climates, cultures, and circumstances. Some people simply use machines to deliver it; others simply open windows or shade a wall to make themselves more comfortable. Comfort hours and months can be extended with the clever use of energy stored in the building fabric, site, or landscape charged by cycles of the sun and wind. Some people in extreme climates need to stay warm with little fuel so only use heat when exactly needed at controlled points in their life cycles. Some nomads in their flimsy tents move their dwellings from winter to summer quarters to remain thermally safe and even delightfully comfortable using little or no fuel. Many people practice intra-mural migration, occupying warmer rooms in winter and cooler rooms in summer. Some people choose to adapt their lifestyles or even perceive themselves to be comfortable in more extreme conditions. Far-sighted planners use planting to cool their sites, cities, and states. Coolth can be harvested from distant mountains and heat from the depths of the earth beneath buildings. We have seen examples of all these strategies in this book and will see many more in the future as the ecological buildings of tomorrow are increasingly run on local natural energy through choice or necessity.

FIGURE 28.6 *Open the windows.* Adaptable buildings work with nature – not work to destroy it.

Source: Susan Roaf.

Global and local climates are becoming rapidly more extreme. Humanity must now prepare for different future climates. The issues of how decision-makers and designers can keep people and populations affordably and reliably thermally comfortable, safe, and alive in those climates is becoming one of the hottest political topics for governments globally in our rapidly heating world. Figure 28.6 is included as a final reminder that at the heart of any future-facing solutions for the built environment must be the principles that underpin adaptive thermal comfort thinking that enable windows to be opened again and energy sufficient buildings to be run for as long as possible on local natural energy. That is the only way forward that can enhance the thermal resilience, comfort, health, and well-being of occupants while minimising the damage done to the global climate by buildings.

Notes

1 Strauss, C. (2006). The imaginary, *Anthropological Theory*, 6(3), pp. 322–344. https://doi.org/10.1177/1463499606066891

2 Galbraith, J.K. (1958). *The Affluent Society*, 1999 Edition, London: Published by Penguin Books, p. 17.

3 Dreborg, K.H. (1996). Essence of backcasting, *Futures*, 28(9), pp. 813–828, ISSN: 0016-3287. https://doi.org/10.1016/S0016-3287(96)00044-4

4 Robinson, J.B. (1990). Futures under glass: A recipe for people who hate to predict, *Futures*, 22(8), pp. 820–842.

5 Aldern, C.P. (2025). *The Weight of Nature: How a Changing Climate Changes Our Minds, Brains and Bodies*, London: Penguine Books.

6 Cooper, G. (1998). *Air-Conditioning America: Engineers and the Controlled Environment, 1900–1960*. Baltimore: Johns Hopkins University Press.

7 Ackermann, M.E. (2002). *Cool Comfort: America's Romance with Air-Conditioning*, Washington, DC: Smithsonian Institution Press.

8 Cox, S. (2010). *Losing Our Cool: Uncomfortable Truths about Our Air-Conditioned World*, New York: The New Press.

9 Marignac, Y. (2023). Dreams of a nuclear revival overtaken by reality, *Alternatives économiques*, Accessed on 21/05/2025 at: https://www.alternatives-economiques.fr/reves-dune-relance-nucleaire-rattrapes-realite/00109026

10 Valentine, S. and B. Sovacool (2019). Energy transitions and mass publics: Manipulating public perception and ideological entrenchment in Japanese nuclear power policy, *Renewable and Sustainable Energy Reviews*, (101), pp. 295–304. https://doi.org/10.1016/j.rser.2018.11.008

11 Hermwille, L. (2016). The role of narratives in socio-technical transitions – Fukushima and the energy regimes of Japan, Germany, and the United Kingdom, *Energy Research & Social Science*, 11, pp. 237–246, ISSN: 2214-6296. https://doi.org/10.1016/j.erss.2015.11.001

12 Gomes, V., M.R.M. Saade, L.O. Neves, I. Loche, L.M. Pulgrossi and M.G. Silva (2022). Sight beyond reach: Dynamic life cycle assessment to support resilient retrofit decision-making in a changing climate, in *The Routledge Handbook of Resilient Thermal Comfort*, Edited by F. Nicol, H. Bahadur Rijal and S. Roaf, London: Routledge.

13 Larson, S. (2010). Understanding barriers to social adaptation: Are we targeting the right concerns, in *Transforming Markets in the Built Environment: Adapting for Climate Change*, Edited by S. Roaf, Earthscan Publications Ltd., April.

14 Andreadis, G., S. Roaf and T. Mallick (2012). Tackling fuel poverty with building-integrated solar technologies: the case of the city of Dundee in Scotland, *Energy and Buildings* (2010). https://doi.org/10.1016/j.enbuild.2012.11.032

15 Roaf, S. (2024). *Overhaul the Building Regulations: The Role of Research, Buildings and Cities: Commentary on the Challenges Ahead: Research Has a Role to Protect the Public Interest and Inhabitants*, Accesses on 16/01/2025 at: https://www.buildingsandcities.org/insights/commentaries/building-regulations-research.html

16 Scoones, I., and A. Stirling (2000). *The Politics of Uncertainty Challenges of Transformation*, London: Routledge.

17 Koffman, J., J. Gross, S.N. Etkind and L. Selman (2020). Uncertainty and COVID-19: How are we to respond? *Journal of the Royal Society of Medicine*, 113(6), pp. 211–216. https://doi: 10.1177/0141076820930665
18 Spiegelhalter, D. (2024). *The Art of Uncertainty: How to Navigate Chance, Ignorance, Risk and Lunch*, London: Pelican Books.
19 Ayres, R.U. (1998). *Turning Point: The End of the Growth Paradigm*, London: Earthscan.
20 Kenedy, A. (2022). *Video on His Two Lectures in 2022*, April. https://youtu.be/cqQ1zQHGHTk [Accessed:] and [Missing reference]
21 Parkinson, T., R. de Dear and G. Brager (2020). Nudging the adaptive thermal comfort model, *Energy and Buildings*, 206, p. 109559. https://doi.org/10.1016/j.enbuild.2019.109559
22 Santos, H., P. Samani and E. de Oliveira Fernandes (2018). Energy sufficiency in buildings, a synonym for passive and low energy architecture (PLEA), *Architectural Science Review*, 61(5), pp. 292–297. https://doi.org/10.1080/00038628.2018.1505332
23 Darby, S., and T. Fawcett (2018). *Energy Sufficiency – An Introduction: A Concept Paper for ECEEE* (European Council for an Energy-Efficient Economy). https://doi.org/10.13140/RG.2.2.31198.08006
24 Singapore Ministry of Sustainability and the Environment (2025). *Launch of the GO25 National Movement*, Accessed on 26/05/2025 at: https://www.mse.gov.sg/latest-news/launch-of-go25-national-movement
25 Mughal, M.O., X.-X. Li and L.K. Norford (2020). Urban heat island mitigation in Singapore: Evaluation using WRF/multilayer urban canopy model and local climate zones, *Urban Climate*, 34, p. 100714, ISSN: 2212-0955. https://doi.org/10.1016/j.uclim.2020.100714
26 https://en.wikipedia.org/wiki/Virtual_Singapore
27 https://cde.nus.edu.sg/arch/facilities/net-zero-energy-building-sde-4/
28 Bingham-Hall, P. (2016). *Garden City – Mega City: Rethinking Cities for the Age of Global Warming*, Singapore: Pesaro Publishing.
29 https://transsolar.com/de/projects/dhaka-brac-university

APPENDICES

APPENDIX 1

The Admittance Method – The Basis!

J. Harrington-Lynn

Introduction

The admittance method has been in use in the UK for over 60 years. It provides a basic method for predicting the thermal performance of buildings and, as such, has been adopted and promoted by the CIBSE.

Although the development of powerful personal computers has allowed the use of sophisticated numerical methods to predict the thermal performance of buildings, there is still a demand from the industry for a simple numerical method to provide:

- A quick method of estimating the thermal requirements for simple buildings
- The likely performance of more complex buildings
- A teaching tool to assist in the understanding of the basics of thermal transfer with buildings

Basic Theory

The energy requirements of a building can be described by a series of relatively simple equations linking the radiant and convective energy transfers through the external and internal spaces and the conduction and storage within the building material.

Despite the fact that these equations have been well-known and understood for many years, and despite the growth of computers, no exact method of thermally modelling a building is generally available. The reasons for this are:

- The number of energy transfer paths required to describe a building (A simple cubical room requires at least 21 equations to describe the radiant and convective transfers between the surfaces.)
- The lack of precise information on the properties of building materials, climatological data, building usage patterns, and the operating characteristics of the plant
- The problems associated with building information management, BIM

As noted earlier the admittance method was developed over many years at BRE to provide a solution to the above problems. Within the admittance method, the following assumptions are adopted:

- The use of an index temperature to simplify the radiant and convective transfers with a room space – this led to the development of environmental temperature.
- The use of a numerical solution based on periodic inputs to the diffusion equation.
- The direct analogy between the thermal transfers in a building and a simple electrical network consisting of resistance and capacitors only.
- The development of a solar gain and alternating solar gain factor to simplify the solar gains through glazing.

These assumptions are explored further in the following pages.

The Need for an Index Temperature

One of the first uses of an index temperature was by Mackey and Wright (1943) in the development of the sol-air temperature to model the convective, solar radiant, and long-wave radiant exchanges at the outer surfaces of a building. The sol-air temperature, or sol-air excess temperature, provided a simple solution to the complex heat transfer between the sun, sky vault, surrounding buildings, and external building surface.

Within a building the transfers are more complex. Energy is transferred between the internal surfaces of a room or space by longwave radiation and from these surfaces to the air in the space by convection.

In a simple cubical room, there are 15 radiant transfer paths between the surfaces and six convective transfer paths between the surfaces and room air. These 21 paths are the minimum, which need to be considered (Figure A1.1). They assume that each surface is isothermal and that the air temperature is constant throughout the space. Such isothermal conditions rarely occur in a single room in a building. Apart from the disruptions caused by corners and junctions and the insertion of windows and doors to the radiant transfers, stratification and ventilation all cause departures from these assumed conditions and add to the number of transfers. The presence of furniture, occupants, and equipment (including building services equipment) all add to the number of transfers and the complexity of the problem.

This ignores the additional complexity caused by the fact that both the radiant and convective transfers are non-linear. What is therefore required to aid thermal modelling is a single temperature, an index temperature, and a set of transfer coefficients, through which all these transfers can be assumed to occur

Prior to the development of the environmental temperature model, it was assumed that the complex network of interchanges was a simple star network, with each arm of the star representing the combined radiant and convective energy flows between the room surface and an undefined room 'space' or index temperature.

Although such a simplified model can provide acceptable answers when sizing heating plant to meet the maximum design conditions, its limitations quickly become apparent when the calculation of annual energy consumption or the behaviour of a passive solar building is to be modelled. What is required is a method of defining more accurately what 'the space' temperature represents and inputting energy gains at the appropriate points in the calculation. The environmental temperature is a solution to this problem.

The Environmental Temperature

The environmental temperature is used within the admittance method to simplify the calculation of the rate of heat flow by convection and radiation from surfaces and other radiant surfaces within a building.

Over the last 30 years, there has been much argument and discussion on the need for and the development of the environmental temperature as a suitable index temperature to ease the calculation of heat gains and losses within a space. Muncey has argued that the use of such a temperature is an unnecessary complication, and Davies that the original development of the environmental temperature by Danter and as used by the CIBSE guide was flawed.

The most important point to remember is that the *environmental temperature* is a mathematical convenience to aid in resolving the complex problem of heat transfer within a space. It is not a measurable temperature or a comfort temperature, although it has been seen and used as a proxy for a comfort temperature. It is analogous to the radiant/star transpose used by electrical engineers to resolve electrical network problems.

In modelling the heat transfers within a space, we also have to remember that both the radiant and convective transfers are non-linear.

The stages in developing the environmental temperature were to:

- Derive linearised functions for the convective transfers
- Ditto for the radiant transfers
- Transpose the radiant transfers between surfaces to a simple radiant star network
- Combine the resultant convective and radiant networks

Convective Transfers

The convective transfer, Qc, between any internal surface of a room and the room air is governed by an equation of the form:

Equation 1: Equation for Convective Heat Transfer, Non-Linear

$$\mathbf{q_c = f.\,(t_{si} - t_{ai})^{1/4}.}$$

Where: f = a factor depending on the boundary conditions and direction of heat flow at the surface
q_c = convective transfer coefficient
t_{ai} = air temperature adjacent to the surface
t_{si} = temperature of the surface

The preceding equation can be written in a linearised form as:

Equation 2: Linearised Equation for Convective Heat Transfer

$$\mathbf{q_c = h_c\,(t_{ai} - t_{si})}$$

Where: ***hc*** = a convective transfer coefficient

The convective transfer coefficient is a function of the air and surface temperature, direction of heat flow, roughness of surface, velocity of the air movement at the surface, etc.

The CIBSE, in their guide, in common with other similar guides, has assumed that a constant value can be assigned to the convective transfer coefficient, *hc*, depending on:

- The direction of horizontal heat flow, into or from the surface
- The exposure
- The surface roughness
- The air velocity at the surface
- The temperature difference between the surface and the surface temperature

Radiant Transfers

The radiant transfers occur between the surfaces. As has been noted earlier, in a simple cubic-shaped room, with each surface assumed to be isothermal, there are 15 (factorial 5) of these exchanges. Each of these transfers can be expressed by an equation of the form of Equation 3.

Equation 3: Equation for Radiant Transfer Between Two Surfaces

$$q_{r12} = \sigma\ e1\ e2\ fe.F_{12}\ (t_{s1}^{\ 4} - t_{s2}^{\ 4})$$

Where: q_{r12} = radiant transfer between surfaces 1 and 2
σ = Stephan Boltzman constant
e_1 and e_2 = emissivities of surfaces 1 and 2
f_e = factor representing the interreflection of radiation between surfaces
F_{12} = view factor to allow for spatial relationship between surfaces 1 and 2.
t_{s1} and t_{s2} = surface temperatures of surfaces 1 and 2 (K)

This equation can be expressed in a linearised form (see Equation 4):

Equation 4: Linearised Version of Radiant Transfer Equation

$$q_{r12} = h_r\ E_f.F_{12}\ (t_{s1} - t_{s2})$$

Where: h_r = radiant coefficient.
Ef = emissivity factor

(Note: ***hr*** is a variable with temperature, but it is normally assumed constant over the temperature range occurring in buildings.)

For each of the individual surfaces in a room, the total radiant transfer from the surface to the other surfaces may be expressed as in Equation 5.

Equation 5: Equation Representing the Sum of Radiant Transfer Between a Surface and the Other Surfaces

$$\mathbf{q_r = h_r\, \Sigma\, E_f\, F_{in}\, (t_{si} - t_{sn})}$$

Where: $\mathbf{q_r}$ = equivalent radiant transfer between a surface and other surfaces in a room
Σ = sum of the radiant transfer between a surface and other surfaces

The next step is to transform this delta network of linear radiant transfers representing between the surfaces to an equivalent star network of six exchanges in a cubical room terminating in a common radiant index or node, equal to the mean surface temperature of the space.

This can be expressed as in Equation 6.

Equation 6: Equation Representing Simple Star Network of Radiant Exchanges

$$\mathbf{q_r = h_r\, E_X (t_x - t_{sm})}$$

Where: E_X = emissivity factor with respect to the radiant transfer node and the room surfaces
t_x = temperature of radiant transfer node
tsm = mean surface temperature of room surface
= $\mathbf{\Sigma\, An\, tsn/\Sigma An}$

In a cubical room, the emissivity factor, *Ex*, approximates to 6/5. In a non-cubical room or a room where the surfaces may have been subdivided because they contain elements at differing surface temperatures, for example, windows or embedded radiators, *Ex* will depart from 6/5. However, in the majority of cases, the variation from 6/5 factor is small and can safely be ignored.

This transform of the 15 elements of the delta network to a 6-element star network is similar to the traditional transformation of the 3-element delta network to the 3-element star network used in electrical resistance network theory and is equally valid. In both cases, the resulting centre of the network is a hypothetical point, which is used to simplify further analysis of the network.

Combined Transfers

The preceding equations (2 and 6) for convective and radiant transfer from the surfaces can be combined together to provide a single equation for the heat transfer from a surface within a room.

Equation 7: Equation for the Combined Radiant and Convective Transfers With Respect to Environmental Temperature

$$\mathbf{q = q_c + q_r = h_c\, (t_{sm} - t_{ai}) + h_r\, E_X(t_{sm} - t_x) = (h_c + h_r\, E_X)\, (t_{sm} - t_{ei})}$$

Where $\mathbf{t_{ei}} = (6/5\ h_r\, E_X\, t_{si} + h_c\, t_{ai})/(6/5\ h_r\, E_X + h_c)$
= environmental or index temperature

For the majority of typical room shapes encountered in buildings and the values of *hc* and *hr* used in the UK, the environmental temperature, *tei*, has been shown to approximate arithmetically to 1/3 air temperature and 2/3 the mean surface temperature. The inclusion of embedded or even free-standing heaters, furnishings or even occupants can usually be included in the calculation without unduly disturbing this ratio between air and surface temperature.

The Environmental Temperature and Comfort

As noted in the earlier section, the environmental temperature is an index temperature, a mathematical convenience, which simplifies the calculation of the rate of energy transfer to or from the internal surfaces of a space by radiation or convection. The environmental temperature is typically assumed to be 1/3 air temperature plus 2/3 the mean surface temperature. It is not a physical quantity that exists outside of the calculation. However, since it is a function of mean surface temperature and air temperature, it can be derived from the direct measurement of the surface temperatures and air temperature.[1]

In the late 60s and early 70s, many users of the admittance method made a connection between the environmental temperature and comfort temperature since both were a mixture of air temperature and radiant or surface temperature. They began to explore the use of the method to predict the likely comfort conditions in their design. Initially, they ignored the difference in the 2/3 to 1/3 makeup of the environmental temperature and the 1/2 and 1/2 makeup of the dry resultant temperature. More importantly, many ignored or did not emphasise the difference between radiant and surface temperature in non–black body situations in their use of the method. This led to confusion and criticism of the basis of the method, which distracted effort from the development and use of the method. This need to make allowance for comfort conditions encouraged the reworking of the equations to allow the use of an index temperature based on the dry resultant temperature.

The Solution to the Diffusion Equation for Heat Transfer

Dynamic or non-steady-state solutions to the diffusion equation for heat transfer through a building element vary with complexity. The majority of these require the use of a computer to provide a solution.

The admittance method is one of the simpler of the available methods and can be applied by manual calculations It assumes that the energy inputs to the building is cyclic, which are repeated over three or more 24-hour periods. The solution is obtained by matrix analysis of the diffusion equation and is based on the work of Pipes.

The method uses three factors to define the dynamic characteristics of the building element:

- The admittance
- The decrement factor
- The surface factor

In addition to these factors, the thermal transmittance, or U-value, is used to define the steady-state characteristic.

In deriving a solution to the diffusion equation, the following assumptions are made:

- The energy transfers to or from the internal surface of the elements are transfers to or from the environmental temperature or index temperature.

- The energy transfer through the building element is unidirectional and can be considered as steady – cyclic, that is, repeated in a predictable fashion over three to five days.
- The solution based on the fundamental daily frequency (i.e. 24 hour) is sufficiently accurate.
- The energy transfer can be considered in two parts, namely, the steady-state or daily mean transfer and the transfer due to the cyclic swing about the steady-state transfer through the element.

For both the mean and the swing, the same rules apply to the input of energy to the element. Radiant energy, from sunshine, lighting, or other shortwave sources, is absorbed at the room surface and readmitted to the room, or energy from radiators or other hot bodies can be emitted directly to the environmental temperature, and convective energy from convective energy can be emitted to the room air. Although in practice energy can occur simultaneously at all these three possible input points, the rules of mathematical superimposition apply, and we can treat each input separately in deriving an overall solution.

The dynamic factors are functions of the thickness, density, and specific heat of the materials making up the element of construction. The dynamic factors are complex numbers, that is, with a real and imaginary part, that can be expressed as amplitude with associated time lead/lag.

General Solution

Derivation of the Admittance

The admittance factor, which gives its name to the method, is the most important of the three dynamic factors. It is the dynamic equivalent to the thermal transmittance and is a measure of the combined amount of energy being stored in and transmitted through the element. In electrical terms, it is directly equivalent to the reciprocal of the impedance of an RC network.

It is normally applied to the energy transmitted from an enclosure/room via the environmental temperature point to either the outside air or to other enclosures within the building and includes an allowance for the energy stored within the structure.

Equation 8: Definition of Admittance

$$\mathbf{Y = q_i/t_i}$$

Where: Y = admittance
q_i = cyclic energy input at index point i
t_i = cyclic variation in temperature at index point i

As noted earlier, the *admittance* is a function of the thickness, density, and specific heat of the materials making up the element. In the case of homogeneous elements, for slabs less than 50 mm thick of any density, the admittance approximates to the thermal transmittance of the slab, and the associated time lead is short. With increasing thickness and density, the admittance tends to a constant value equal to the reciprocal of the surface resistance, that is, 8 W/m2 K (1/0.123), and time leads tending to three hours (Figures A1.1 and A1.2).

This characteristic of increasing admittance with density leads to the use of the term 'lightweight' or 'heavyweight' when referring to the thermal characteristics of building elements.

However, caution is required when referring to multilayered constructions, since the admittance is determined mainly by the characteristics of the first layer of the element and the order of the materials making up a structure. Lining a dense and thick concrete or masonry construction with 25 mm of insulation or more will reduce the admittance and cause the combined element to behave as a 'lightweight' rather than an 'heavyweight' construction. Conversely placing insulation behind a covering of plaster will enhance the effectiveness of the thermal capacity of the plaster and make the element behave as a 'heavyweight' construction. Therefore, the position of the individual materials in a multilayered structure is important. There are many examples of low-energy buildings, which, because of the position and thickness of the insulation with respect to the inner lining, have behaved in a totally unexpected fashion. Although they have a low energy demand, they have behaved as extremely 'heavyweight' building exhibiting all the characteristics of a Norman castle, slow to warm up and cool, with corresponding problems in designing a heating system.

The Decrement Factor

The decrement factor, the second of the three thermal factors, was in use several years prior to the development of the admittance method. It was first defined by Mackey and Wright and introduced to the UK by Dander in 1960.

The decrement factor is a measure of the fraction of cyclic energy transmitted through a building element from one side to another. It differs from the admittance in that it does not include the effect of energy stored in the structure. It is normally applied only to the energy transmitted from the outside through the structural elements to the internal spaces. It is expressed as the ratio of cyclic transmittance to the steady state/mean transmittance, as given in Equation 2.2. As with the admittance, the value of the decrement factor decreases and its associated time lag increases with increasing thickness, density, and thermal capacity of the elements of the structure.

Unlike the admittance, the decrement factor is, however, independent of the order of the elements in a structure.

Equation 9: Definition of Decrement Factor

$$\mathbf{f} = \tilde{U} / \mathrm{U} = \tilde{Q}_{\mathbf{o}} / \tilde{t}_{\mathbf{i}}\mathbf{U}$$

Where: f = decrement factor

$\tilde{U}$ = alternating transmittance for energy flow from outside to inside

U = steady-state transmittance

$\tilde{Q}_o$ = cyclic energy gain to inner space due to cyclic energy flow through structure

$\tilde{t}_i$ = cyclic temperature of inner space

Surface Factor

The third factor, the surface factor, is the fraction of heat gain by radiation falling on an internal surface that is absorbed by the element prior to readmission to the environmental temperature. Note that it is not a measure of energy reflected from a surface. Within the context of the admittance method, it is applied to shortwave radiation from solar gains directly transmitted through glazing or emitted within the space as visible light from lighting systems.

The value of the surface factor decreases with increasing thermal capacity and, as with the admittance, is influenced by the order of the materials in a multilayered construction. Therefore, for structural elements where the order of materials is not symmetrical, that is, internal floors and ceilings, two values of the surface factor have to be calculated.

Internal Partitions: Special Case

Admittance

The calculation of the thermal behaviour of internal partitions separating spaces with identical thermal conditions can be simplified by including within the calculated admittance an allowance for the cyclic transmission of energy through portions the see Equation 10a.

This precludes the need to include within the overall energy balance equations a term equivalent to the cyclic transmittance through the external walls due to the temperature fluctuations in the internal spaces.

The net effect is to reduce the value of the admittance by an amount equivalent to the U-value times the decrement factor. For thin internal partitions, the admittance tends to zero. For thick massive partitions, the change is negligible.

Equation 10: Derivation of Admittance of Internal Partitions

$$\tilde{q} = (\mathbf{Y} - \mathbf{Uf})\tilde{t}$$

Where: $\tilde{q}$ = cyclic energy transmitted through partition
$\boldsymbol{Y}$ = admittance unadjusted for cyclic transmission
$\boldsymbol{Uf}$ = cyclic transmittance
$\tilde{t}$ = cyclic variation of temperature of enclosure

Surface Factor

The surface factor for internal partitions can be modified in a similar fashion, assuming that each surface of an internal element is exposed to a similar amount of radiation.

For internal floors and ceilings, where these are not symmetrical in construction about their mid-plane since both the admittance and surface factor are influenced by the order of the materials in the construction, two sets of admittance and surface factors are required. Also since the direction of flow the cyclic energy transfers will change during the cycle a common value of the surface heat transfer coefficient is used, rather than calculating separate values indicative of the direction of energy flow.

The Basic Energy Balance Equations

The combination of the preceding thermal factors and the environmental temperature enables the energy balance equations for an enclosure to be expressed by a pair of equations. These represent the steady-state or mean temperature balance and the dynamic or cyclic temperature balance for an enclosure with one or more external surfaces.

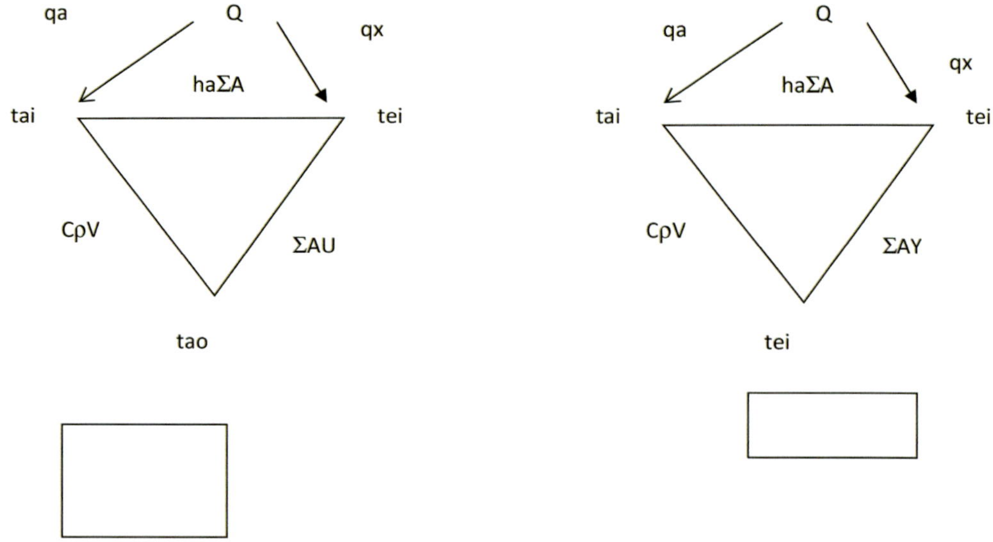

FIGURE A1.1 Steady-state transfers.

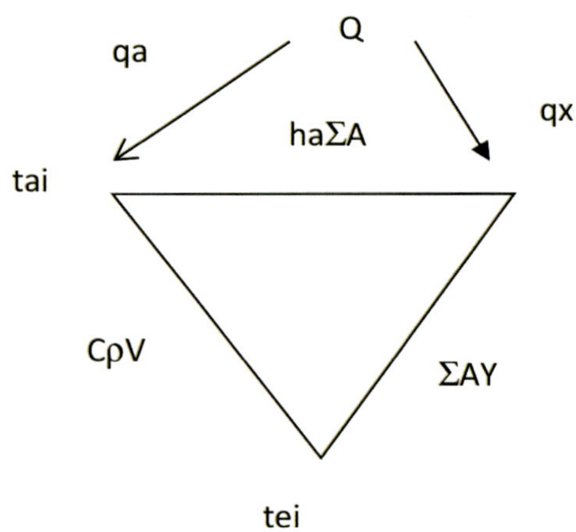

FIGURE A1.2 Cyclic transfers.

To achieve this, it is necessary to generate a hypothetical conductance, *Aha*, linking the internal air temperature, *tai*, and the environmental temperature, *tei*. See Figures A1.1 and A1.2.

The energy transfer between a surface and air is given by Equation 11.

Equation 11: Energy Transfer Between Surface and Air

$$Qa = A_s h_c (t_s - t_{ai})$$

Where: ***Qa*** = energy transfer between surface and air
A_s = area of surface

h_c = convective coefficient
t_s = surface temperature, K
t_{ai} = air temperature, K

In an enclosure, assuming the convective coefficient, h_c, is the same for all surfaces and directions of energy flow, this can be rewritten as Equation 12:

$$\mathbf{Q_t = \Sigma A_s\, h_c\, (t_{ms} - t_{ai})} \qquad (12)$$

Where:

Equation 12: Total Energy Transfer Between All Room Surfaces and Air

Q_t = total energy transfer between all room surfaces and room air
ΣA_s = sum of area of all room surfaces
t_{ms} = mean surface temperature of all room surfaces

The mean surface temperature, $\mathbf{t_{ms}}$, can be expressed in terms of environmental temperature, $\mathbf{t_{ei}}$, as follows:

$$\mathbf{t_{ei} = 1/3\, t_{ai} + 2/3\, t_{ms}}$$
$$\mathbf{t_{ms} = 3/2\, t_{ei} - 1/2\, t_{ai}}$$

Substitution in **Equation 12**:

$$\mathbf{Q = 3/2\ hc\ \Sigma As\ (tei - tai) = ha\ \Sigma As\ (tei - tai)}$$
$$\mathbf{Q_t = 3/2\, \Sigma\, A_s\, h_c\, (t_{ei} - t_{ai})}$$
$$= h_a\, \Sigma\, A_s\, (t_{ei} - t_{ai})$$

Where:
h_a = hypothetical conductance between air and environmental temperature

Energy Balance With Respect to Dry Resultant Temperature

The adoption of the dry resultant temperature (1/2 tai + 1/2 tei) requires the derivation of two further hypothetical conductances, one between air temperature and the dry resultant temperature, *hac*, and the other between environmental temperature and the dry resultant temperature, *hec*.

This leads to the following energy balance equations for a space.

Steady state:

$$\mathbf{(Fu\Sigma AU + Fv.CrV)\ (mtc - mtao) = \Sigma\ (FuQmx + Fv\ Qma)} \qquad \mathbf{4.1}$$

Dynamic/cyclic:

$$\mathbf{(Fy\Sigma AY + Fv\ CrV)\ (tc - mtc) = \Sigma\ (Fy\ (Qx - Qxm) + Fv\ (Qa - Qma))} \qquad \mathbf{4.2}$$

Where ΣAU = sum of the product of surface area and corresponding thermal transmittance of all external surfaces

CrV = ventilation conductance

ΣAY = sum of the product of surface area and corresponding thermal admittance

Fv = ventilation conductance correction factor with respect to dry resultant temperature = **ΣAhc/(ΣAhac + CrV)**

Fu = steady-state transmission conductance correction factor with respect to dry resultant temperature = ΣAhec/(ΣAhec + ΣAU)

Fy = dynamic transmission conductance correction factor with respect to dry resultant temperature = ΣAhec/(ΣAhec + ΣAY)

Qx = energy input at environmental temperature point (2/3 radiant + 1/3 convective)

Qmx = mean energy input at environmental temperature point over 24 hours (2/3 radiant + 1/3 convective)

Qa = energy input at air point (100% convective)

Qma = mean energy input at air point over 24 hours (100% convective)

tc = dry resultant temperature

mt = mean dry resultant temperature, over 24 hours

$mtao$ = mean external air temperature, over 24 hours

$mtai$ = mean internal air temperature

Energy Inputs

In generating the preceding sets of energy balance equations, two possibilities exist for the input of energy inputs.

The first choice assumes that energy can be input as mixture of a 100% convective input at the air temperature node or as a 100% radiant input, which is absorbed at the enclosure surfaces before being input at the environmental temperature. In this choice, the surface factor reduces the radiant input.

The second assumes that energy can be input to the system as a mixture of three sources, namely:

- A 100% convective input at the air temperature node
- A 'hot-bodied' input at the environmental temperature node
- A 100% radiant input which is absorbed at the enclosure surfaces before being input at the environmental temperature

Where the 'hot body' is assumed to be 1/3 convective plus 2/3 radiant.

Although the first option is apparently more straight forward, the second approach is recommended since it allows for the influence the surface temperature of the 'hot-bodied' energy source has on the environmental temperature of the space. For example, a radiator system, apart from providing an energy input to the system, also provides a direct temperature input to the space. Therefore, treating it as a direct, hot-bodied input makes allowances for both these inputs.

Internal Gains

The preceding approach can be applied to all the common energy sources occurring within buildings.

TABLE X Approximate radiant and convective output for a range of energy sources

Energy source	*Actual makeup*		*Environmental makeup*		
	% convective	*% radiant*	*% environmental*	*% convective*	*% radiant*
Heated surface	33	67	100	0	0
Human body	33	67	100	0	0
Warm air heater	67	33	50	50	0
Heavy curtains	50	50	67	33	0
Venetian blind	67	33	50	50	0
Induction unit	67	33	50	50	0
Air handling unit (heating or cooling)	100	0	0	100	0
Radiators	35	75	100	0	0
Chilled beams	35	75	100		
Fluorescent fitting	5	85	15	0	85
Radiant heaters	10	90	30	0	70

With lighting systems the direct visible light, in the form of shortwave radiation, is first absorbed at the room surface before being readmitted to the environmental temperature. The surface factor comes into play at this stage.

Table X lists the approximate convective and radiant makeup for a range of common internal energy sources and the consequential environmental, convective, and radiant breakdown.

Although in practice the outputs of most energy sources will vary as a result of changes in operating conditions the suggested distribution between environmental, convective, and radiant inputs is of sufficient accuracy for use within the admittance method.

External Gains

Solar gains, either through the glazing systems or the opaque parts of the building structure, can also be dealt with in a similar manner to the preceding.

Solar Gains Through the Opaque Fabric

The sol-air temperature method of dealing with the resulting energy gains to the inside of a building due to solar radiation falling on the opaque surfaces of a building was first proposed by Mackey and Wright. The method, in its approach, was the precursor of the admittance method in that it divided the solar gains into a daily mean and a swing and solved the diffusion equation for the 24-hour fundamental harmonic. It forms an integral part of the current admittance method.

The first stage in the method is to convert the solar radiation falling on an external surface into an apparent temperature, the sol-air temperature. This is achieved by an equation of the form:

$$\mathbf{teo = \{(\alpha rad\ Gr + L\ b)/hc) + tao}$$

Where: *teo* = sol-air temperature

αrad = shortwave surface absorptance

Gr = global solar irradiance (Wm2)

L b = longwave irradiance balance

hc = convective heat transfer coefficient due to wind
tao = outside air temperature

Traditionally, this provided two equations to define the energy gain to a space due to the solar gain through an opaque surface:
Mean gain:

$$\mathbf{mQSo = \Sigma\{(\Sigma AU)\ (mteo - mtao)\}}$$

Swing in gain:

$$\mathbf{QSo - mQSo = \Sigma\{(\delta\Sigma AU)\ (teo - mteo\}}$$

Since the global solar radiation incident on the surface varies with the slope and orientation of the surface this requires a different sol-air temperature and different set of equations for each exposed opaque surface of a building.

To incorporate this solar gain within the basic equations used in the admittance method the easiest approach is to use the concept of sol-air excess rather than the sol-air temperature to define these gains. This is achieved by dividing the sol-air temperature into the two components due to the solar gain and the external temperature, where:

$$\mathbf{\Delta teo = teo - tao}$$

The preceding equations for the gains can then be expressed as:
Mean sol-air excess:

$$\mathbf{\Sigma\{(\Sigma AU)\ (m\Delta teo + (mtei - tao))\} = mQso}$$

Swing in sol-air excess.

$$\mathbf{\Sigma\{(\delta\ \Sigma AU)\ ((\Delta teo - m\Delta teo) + (tao - mtao))\} = Qso - mQSo}$$

The advantage of this split can best be seen when we include the solar gains through the opaque surfaces in the general energy balance equations.
Steady state:

$$\mathbf{(\Sigma AU + Fv.CrV)\ (mtei - mtao) = mQso + mQei = \Sigma\{(\Sigma AU)\ (mteo - mtei) + m(Qx + Fv.\ Qa)}$$

Dynamic/cyclic:

$$\mathbf{(\Sigma AY + Fv\ CrV)\ (tei - mtei) = sQso + sQei =}$$
$$\mathbf{\Sigma\{(\delta\ \Sigma AU)\ (\Delta teo - m\Delta teo) + \Sigma\{(\delta\ \Sigma AU)\ (tao - mtao)\} + S(Qx + Fv.\ Qa)}$$

Note the term, Σ{(ΣAU) (mtei − tao)}, which represents the mean conduction gain, now appears on the LHS of the general equation.

Solar Gains Through Glazing

The calculation of the energy gains to a space due to solar radiation falling on glazing or window system is complex.

To calculate the solar gains, we have to take account of:

- The variation of solar intensity throughout the year as a function of the solar geometry and cloud cover
- The physical characteristics of the glass and blinds making up the window system
- The position of the window within the building geometry
- The shadowing effects of surrounding buildings, etc.

Within the admittance method this complexity was originally overcome by the use of solar gain factors, mean and alternating, which are applied to the solar intensity falling on the window systems and used to calculate the peak temperatures in summer.

These factors are defined as the ratio of the solar gains to the room space divided by the solar intensity, viz.:

S = Qex/I . . . . . .

The calculation of these solar gain factors is complicated and requires knowledge of:

- The intensities of the direct and diffuse solar radiation falling on a surface, particularly at the times of peak load.
- The transmission, absorption and reflection values, the TAR values, for the window system, that is, glass and blinds, being modelled.
- The distribution of the resulting radiation within the room space.

With the use of computers more complex and accurate methods can be used to predict the solar gains based on the TAR values of the window system and the solar incidence falling on the window system.

A BRE paper by Professor J. F. Nicol provides guidance on the methodology to be used.

Where internal blinds are in use, the gains will be split between the air – point and the environmental point.

Where external blinds are used the gain all the gains appear at the environmental point.

Application

The method as described above is normally applied only to one specific space in a building, that is, a single room or open planned floor. There is however no reason to extend the method to analyse a complete building by cross linking the energy balance equations for a number of rooms using matrix mathematics.

Conclusion

The preceding summary is a brief description of the admittance method (or cyclic method), which has been developed over the past 50 years, initially by the BRE, and in later years by task groups of the CIBSE and others. The method, if used properly, provides a good picture of the potential thermal performance of buildings under a variety of weather conditions.

John Harrington-Lynn

Note

1 Unfortunately, the similarity between environmental temperature and comfort temperature, in that both are functions of the room air temperature and surface temperature, has in many cases led to confusion in the use and application of the admittance method. This confusion has been enhanced by a certain casualness in terminology by drafting committees of the CIBSE and authors of papers on the method when referring to surface and radiant temperatures.

This confusion in terminology has led to criticism of the method by M. G. Davies and others. However, it is the view of the author and the CIBSE that this criticism does not justify the scrapping of the method. In fact, the CIBSE, in their Guide A, has adopted dry resultant temperature, the average of air and mean radiant temperature. Users are reminded of the following:

- The mean surface temperature in a room is independent of an observer's position within the room.
- The mean radiant temperature varies with the observer's position.
- It is only in the geometrical centre of a cubical room that the mean radiant and mean surface temperatures are equal.
- The environmental temperature is a function of the mean surface temperature and air temperature.
- The comfort temperature is a function of radiant temperature, air temperature, air movement, etc.

APPENDIX 2

Thermal Mass and Time Lags: Thermal Comfort in the Havelis of Jaisalmer

Jane Matthews

People have lived for thousands of years in some of the hottest regions in the world. By studying an example of a highly evolved large house type in the state of *Rajasthan* in north-west India, Jane Matthews was able to understand the role and importance of thermal mass in enabling people to stay comfortable in that hot desert climate. She was awarded her PhD in 2000 for her study of the thermal performance of a *haveli*, a historic mansion house in Jaisalmer.[1] Two critical factors in the building's performance are its thick sandstone and limestone construction and the co-location and interactions of spaces both with each other and with the earth and the sky at different times and places within the overall building form. In the extreme heat of north Indian summers, the building operated to dampen temperature swings in many spaces occupied in the summer. It was designed to utilise the mass in summer to dissipate the heat of the day from the structure to the cooler night sky and air. Crucially, the lower, heavy mass walls shown in Figure A2.1and the earth beneath the building acted to slow down the rate at which the external heat of the day penetrated the spaces. Extended time lags between outdoor and indoor peak temperatures existed in some lower, high-mass rooms of the building, as can be seen in Figure A2.2 The basement in particular responded at times to outdoor temperatures that had occurred up to one month previously.

The use of building fabric as a thermal buffer to both shave and shift temperature swings is becoming increasingly important in a heating climate characterised by more extreme weather events. However, designers often only consider the effects of mass in a notional fashion because of the time and degree of specialist knowledge required to comprehensively model the impact of exposed building mass. Conventionally, time lag is discussed and calculated with regard to a single building element, for example, a particular wall or floor. Calculations are complex and often require input based on assumptions. Matthews's study adopted an empirical approach to the determination of time lag, not of individual building elements, but of indoor temperature in relation to outdoor temperature.

It is possible to estimate both time lag and the dampening of indoor room temperature of existing buildings by examining graphs of indoor and outdoor temperature plotted against time. The time lag (tL) in Figure A2.2 is given by the difference between the occurrence of the peak temperature outdoors and the corresponding peak temperature indoors. One important

FIGURE A2.1 (a) The north-facing street façade of the haveli and (b) the floor of the inner courtyard.

Source: Matthews (2000).

FIGURE A2.1 (Continued)

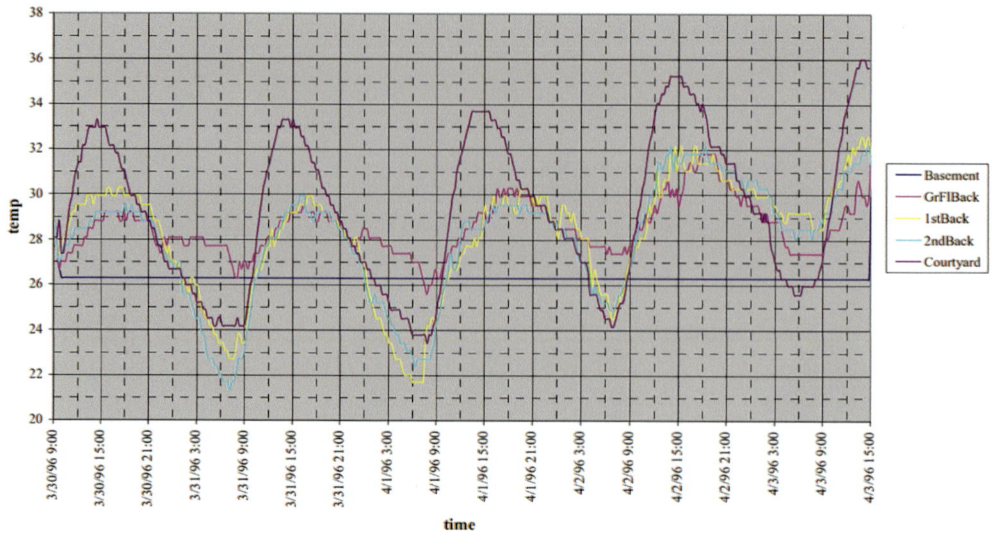

FIGURE A2.2 Graph of recorded temperatures in March 1996 in different rooms around the haveli showing the huge range of temperatures in the exposed, aboveground courtyard, and the virtually unchanged temperature recorded in the basement.

observation here is that the daily maximum (or minimum) cannot be shifted by more than six hours (a quarter of the cycle frequency) through the introduction of mass alone. Once the outdoor temperature drops below the indoor temperature, the room and mass will start to cool (without other forces in play). The dampening of the series is quantified by means of a decrement factor. This is the ratio of the amplitude of internal temperature wave to the amplitude of external temperature wave.

Preliminary analysis of the data collected in Jaisalmer over 12 months indicated that the indoor temperature time lag and decrement factor were not constant from season to season, or even from day to day. Two methods were developed and evaluated to accurately determine the average time lag between indoor and outdoor monitored temperature of different rooms for both winter and summer seasons. The detailed methodology, results, and discussions in Matthew's PhD are freely available online.[1] The following observations were made:

- In the summer, the majority of indoor spaces have cooler daily average temperatures than outdoors, while in the winter the majority of indoor spaces have warmer daily average temperatures than outdoors.
- Daily and annual temperature ranges increased, while time lag decreased from the bottom (high mass) to the top (low mass) of the haveli at all times of year, giving a variety of maximum and minimum indoor temperatures at any one time.
- More consistent time lag results are achieved with data sets recorded over periods of transient daily average temperature (e.g. spring or autumn), particularly when examining high-mass zones. Thermal performance studies have, in the past, tended to concentrate on periods of extreme climate, when the outdoor daily average temperatures are relatively stable.
- Some zones within the haveli have sufficient available mass to shift the annual as well as the daily temperature cycle.

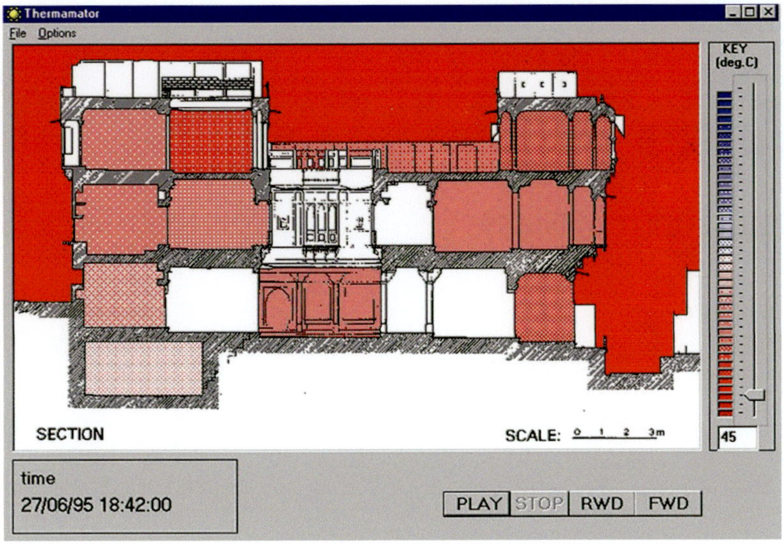

FIGURE A2.3 Section through the haveli showing different temperatures at one time of day and year from Comfator, a thermal design animation tool developed for her PhD by Jane Matthews.

Source: Jane Matthews.

- When considering building zones where it is anticipated that the available mass is sufficient to substantially shift the annual temperature cycle, daily average temperature data yields the best results. In all other cases, it was found that a recording interval in the region of one hour would be more than adequate. In addition, outdoor temperature data for a minimum of one month previous to the beginning of the monitoring period would be advantageous, particularly when analysing high-mass zones.
- Long-term logging, with monitoring periods spanning, ideally, a minimum of 12 months, is important, particularly in high-mass buildings.

The climatic success of the haveli is largely due to the wide choice of indoor climates it provides for its inhabitants. Considerable differences in temperature can be experienced by horizontal, and even more so by vertical, migration through the building. Only a few rooms in a haveli were traditionally designed for a specific purpose. Instead, larger rooms were used as season, need, and mood dictated. These climatic differences are influenced by the orientation of the room, exposure to direct and indirect radiation, and ventilation rates, but the results of this study suggest that the range of temperature is predominantly a product of the varying amounts, particularly floor to floor, of available mass within the building. These findings were visualised with the Comfator tool tracking temperatures in every space across a whole year in the haveli, an image of which is shown in Figure A2.3.

Note

1 Matthews, J. (2000). *Thermal Comfort in the Havelis of Jaisalmer*, Thesis University of East London, Accessed on 11/02/2025 at: https://www.researchgate.net/publication/33776485_Thermal_comfort_in_havelis_of_Jaisalmer

APPENDIX 3

Modelling Thermal Mass

Murray Milne

By the early 1970s, Murray Milne had realised there was no easy-to-use, intuitive tool that would allow students to design a building and show them hourly annual performance throughout the year, correctly accounting for thermal mass under actual weather and varying internal loads. He addressed that gap by writing Solar 5, which later became HEED (*Home Energy Efficient Design*[1]). He wanted it to be user-friendly enough so that students could see useful results within the first hour and, thus, instantly make design changes to compare the performance of multiple iterations of it, the way designers used to use yellow tracing paper to make small changes on each subsequent layer. The computational problem of correctly simulating the performance of thermal mass with varying internal load schedules and varying external weather was really difficult. The next problem was how to automatically create an accurate rotatable 3D graphic representation of the building, its windows and sunshades, and its surrounding landscape and other shade-casting structures under an accurate moving sun. The result was a bar chart showing how each scheme compared to a code-compliant building (scheme 1). When the designer got the last bar down to zero, it was a zero net energy building.

Among the concepts HEED teaches is how to correctly use, or not use, thermal mass to help designers understand the assets and liabilities of their unique climates. One can lecture on the principles, but until students have to fight through getting the performance bars down to zero for their own designs, they do not internalise the challenges. That is, a team was built to write *Climate Consultant*[2] to also be intuitive to use, and to graphically present complex data in an easy-to-comprehend form. Students needed still more guidance because each climate type needs different design responses in terms of thermal mass, ventilation, etc. Building, then, off Baruch Givoni's work, boosted by Don Watson, *Climate Consultant* plots out an hourly annual psychometric chart, and based on various comfort models, it calculates the number of hours each design strategy can yield indoor comfort using passive design. It then creates a ranked list of design guidelines for a design that is climate-specific, based on research by Donald Watson, Vivian Loftness, and many others.

The last step was to make the psychometric chart understandable, so the *Psychometric Chart Tutorial* was produced.[3] All these tools have to be free because students could not afford to pay their true cost. None of this would have been possible without the hard work of Carlos Gomez,

Don Leeper, Robin Liggett, and many other graduate students, with support from DOE, California utilities, the California Energy Commission, and ultimately, the Californian ratepayers.

These are useful American tools for those seeking to appropriately use thermal mass to create zero net buildings. All these design tools plus a technical bibliography are freely available on https://energy-design-tools.sbse.org/.

Notes

1 https://www.sbse.org/resources/heed
2 https://www.sbse.org/resources/climate-consultant
3 https://www.youtube.com/watch?v=OsWm8dfhP_U

APPENDIX 4

The LT Method – A Strategic Design Tool

Nick Baker

Origins

Modern design tools like REVIT see the performance of buildings through the lens of performance metrics of the HVAC industry. But 40 years ago innovative thermal design thinkers were evolving ways of seeing buildings in more the architectural terms of their thermal, ventilation and lighting landscapes. There is much to learn about ways forward for the THERMAL DESIGN of buildings from the 'old ways of modelling' of building performance in the early design stages, as evidenced by this case study of the relatively short life of the LT Method.

The LT method was born in the snowy wastes of Alaska in the late 1980s. David Hoch, from the Scott Polar Institute, Cambridge, was researching the energy costs of remote off-grid rural schools, where energy supplies, in the form of diesel oil for heating and electricity generation, had to be flown in at huge cost. This expense focused attention to the pros and cons of glazing – admitting daylight and reducing the dependence on expensively generated electricity for lighting but, at the same time, causing huge heat loss, only partially offset by solar gain (Figure A4.1). Nick Baker at the Cambridge University Department of Architecture used a simple daylight factor algorithm and the familiar U-value calculation (both responding to glazing area) and applied them to a spreadsheet that, at that time, was an innovative step in itself. For climatic data, monthly average daytime temperatures were used, and hourly sky illuminance was calculated from solar radiation totals and a solar geometry algorithm.

To calculate useful solar gain, a solar/load ratio method was used. For the lighting energy, a monthly average hourly model had to be employed to cover the variation of sky luminance during the occupancy period. In this original study, the ground reflectance was of special interest, for during the spring, when snow was on the ground, ground-reflected light had a major influence. That study concluded that windows were by far the best strategy, and that the overall energy performance could be greatly improved by insulating shutters.

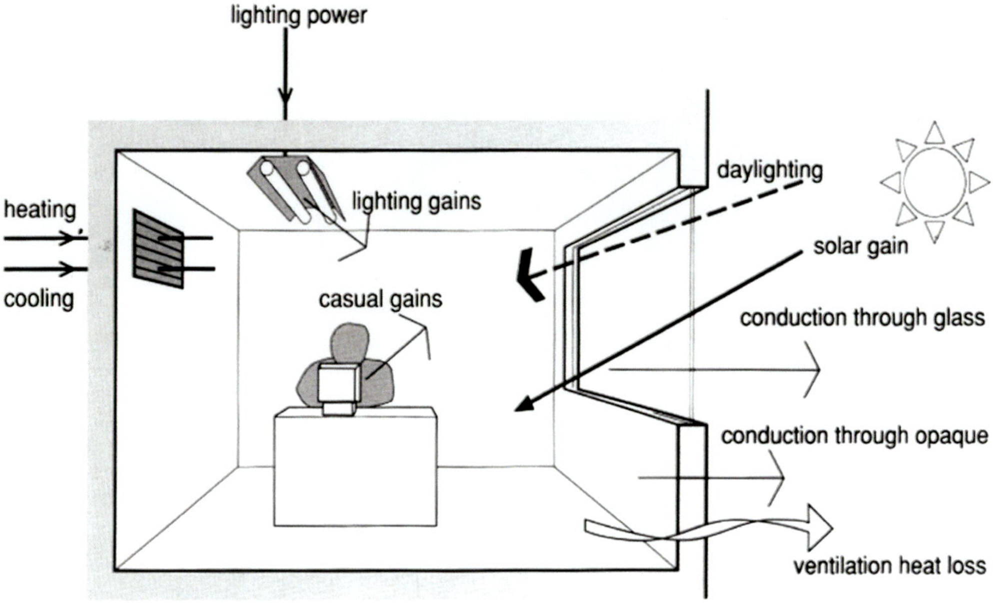

FIGURE A4.1 Heat gains and losses in an office space.

The Simplified Design Tool

The next application of LT was as a design tool supporting the third European Architecture Competition 'Working in the City' in 1991.[1] This competition addressed non-domestic buildings set in urban sites in a number of European cities to reduce energy use in them in the LT method by increasing their passive performance (Figure A4.2). The tool was to be distributed to all entrants with the competition documentation, and a minimal use of the tool to analyse one or more design variants was mandatory. It followed that the tool had to respond to parameters available early in the design – for example, local climate, building form and orientation, glazing area and type.

The non-domestic category made the energy balance relating to heating, cooling, and lighting particularly relevant and demanded two more features. Firstly, the impact of an atrium (then fast becoming a 'must have' for energy-conscious urban buildings), and secondly, the impact of over-shading from adjacent buildings on daylight availability and solar gain.

The atrium (conservatory or sunspace) had to be integrated with the basic model in such a way that did show not only a beneficial effect of reducing heat losses through the separating wall but also a negative effect of reducing the available daylight to the adjoining spaces. This was achieved by calculating a monthly 'atrium thermal climate' and using this to calculate both the conductive and ventilation heat loss via the separating wall. Furthermore, by selecting one of four ventilation modes, the effect of *ventilation pre-heating* from the atrium could be evaluated. To model the detrimental effect on daylighting, LT reduced the depth of the daylit zone for atrium-adjacent spaces.

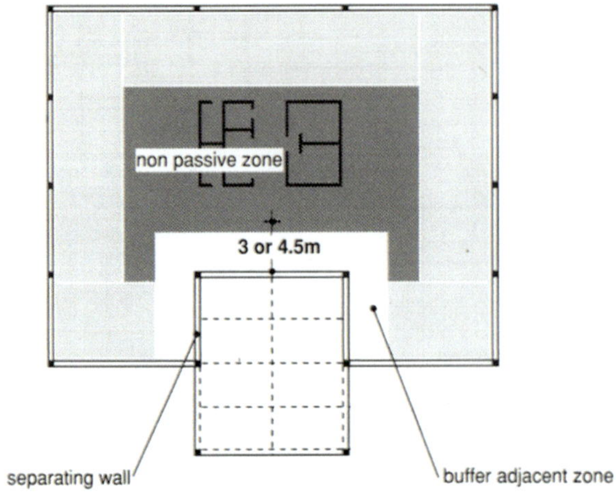

FIGURE A4.2 Plan of an office block showing passive and non-passive zones.

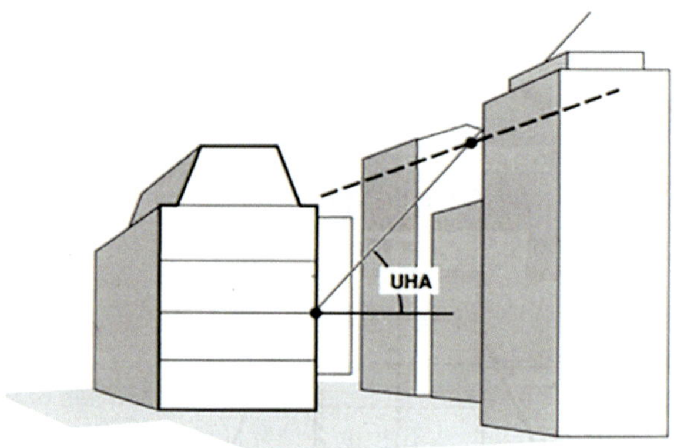

FIGURE A4.3 Diagram of formula used to calculate overshading in the method.

Overshading had a detrimental effect on daylight and useful solar gains in winter but could reduce cooling load in the summer so was included in the tool (Figure A4.3). It also had an impact on lighting energy. It is quite a complex issue since the amount of impact is sensitive to glazing ratio and orientation, as well as the urban horizon angle (UHA).

Access to computers was still the preserve of researchers and only a very few practising architects. We decided to prepare a set of curves like those shown in Figure A4.4 which showed the annual energy consumption per square metre, for heating, cooling, and lighting, as a function of glazing ratio (of the window wall). Sets of curves were generated for different orientations and five different European climatic zones, and this immediately exposed the impact of different combinations of temperature, solar radiation, and day length on appropriate low-energy strategies when integrated (quantitatively) over the year. We began to realise that the curves had an intrinsic value themselves, apart from being a source of data which could be used to estimate the energy consumption of the building as was done for Table A4.1.

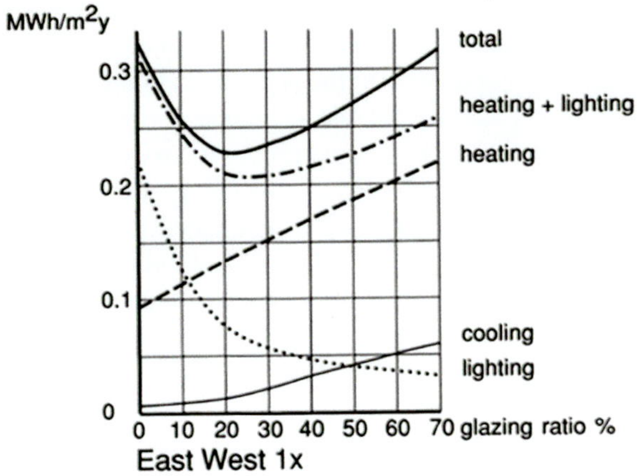

FIGURE A4.4 Graph showing curves for annual energy consumption per square metre, for heating, cooling, and lighting, as a function of glazing ratio of the window wall.

TABLE A4.1 Table showing the buffer space thermal savings (BTS), in MWh/m², from an atrium and urban overshading

	A		B		C		D	
	x1	x2	x1	x2	x1	x2	x1	x2
1 N	0.62	0.71	0.71	0.77	0.65	0.72	1.27	1.39
2 N	0.62	0.84	0.82	1.01	0.69	0.84	1.24	1.50
3 N	0.47	0.64	0.60	0.75	0.57	0.71	1.02	1.27
4 N	0.39	0.60	0.60	0.80	0.49	0.62	1.01	1.34
5 N	0.35	0.49	0.47	0.67	0.41	0.54	0.60	0.89
6 N	0.62	0.95	0.84	1.22	0.71	0.99	1.09	1.57
7 N	0.56	0.75	0.67	0.90	0.65	0.81	1.01	1.34

The effect of an atrium and urban overshading was also pre-computed and presented in a tabular form. Again, when observing users of the LT method, we saw that the tables had an educational value in themselves.

Energy and CO_2 Emissions

In providing data of energy consumption for different end uses – heating, cooling, lighting – it had to be decided how to present them. If there was a pay-off between the uses, how could this be shown with a common unit? In the late 1980s, it was not uncommon to see building performance with *total energy consumption* calculated by adding electrical energy to energy delivered as fuel for heating, such as oil or gas. This was in spite of the fact that electrical energy cost at least three times as much as heating energy and, more to the point, at that time caused more than three times as much environmental impact (such as acid rain and global warming). But these were early days, and the use of CO_2 equivalent as a common currency of environmental impact was not widespread. So *primary energy* was used, representing the energy value of the fuel at source, before refining, transportation, and final combustion, or more importantly, being converted to electricity. This had the effect of greatly enhancing the value of daylighting and escalating the cost of air conditioning. In later versions of LT, we used CO_2 equivalent emissions.

Passive and Non-Passive Ones

The *passive zones*, as shown in Figure A4.5, were an essential part of how the design tool applied the energy values from the curves, which were *annual primary energy consumption per square metre*, to the building being considered. The *passive zone* was the part of the floor plan which was within two floor-to-ceiling heights of the perimeter wall, or on the top floor. The *passive zone* could be daylit and naturally ventilated and make use of solar gains in winter. However, it suffered from heat loss through the envelope in winter and unwanted solar gains in summer (which contributed to cooling load). The remaining *non-passive zone* had to be artificially lit and mechanically ventilated all the time and cooled when internal gains raised the temperature above the cooling set point. These modes, together with occupancy gains, and reference to the appropriate LT curve, yielded a specific energy consumption that could be applied to a corresponding floor area. Different passive and non-passive zones were then summed for the whole building.

Although the passive zone concept was developed as part of the LT calculation, it began to have life of its own. Some years before, Dean Hawkes had proposed the terms *selective* and *exclusive* (referring to the building's degree of coupling with the outside climate), in this case to apply to the whole building. The passive/non-passive concept was really applying the same distinction to parts of the same building. In looking at real buildings, the *passive zone ratio* (PZR)

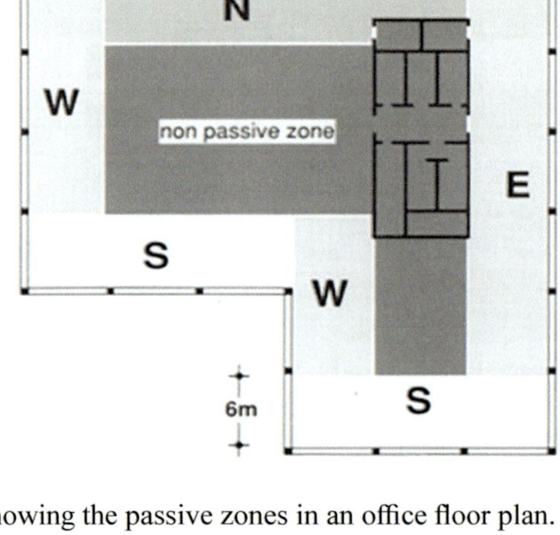

FIGURE A4.5 Diagram showing the passive zones in an office floor plan.

was found to be a useful parameter to judge the potential of a building to be daylit, be naturally ventilated, and make use of solar gains – all of which would have a strong influence on energy consumption. Skip ahead two or three decades and the same parameters are found to be relevant to the current issues around the health and well-being of occupants, in relation to exposure to daylight and views of outside.

Synthesis

The data from the graphs and the building were brought together on the LT worksheet, which guided the user through the steps to the bottom line – *the total primary energy consumption* shown in Figure A4.6. But several subtotals were visible en route, and the effect of an atrium or buffer space (positive or negative), together with other parameters, such as the solarity of the atrium, and the PZR. Explanatory notes were referenced alongside data entry boxes, and again the opportunity was taken to explain the principles behind the data.

This final procedure gave the user an awareness of the workings of the building, quite unlike the simple result that is typical of a modern simulation programme.

The LT method was used for two more European Architecture Competition. LT3 (1994) was a version developed for Zephyr, an ideas competition promoting passive cooling, and focused on sites in southern Europe. Since shading was the number one strategy to prevent overheating, we had to devise a way of accounting for its effect on both cooling load and lighting. LT4 was

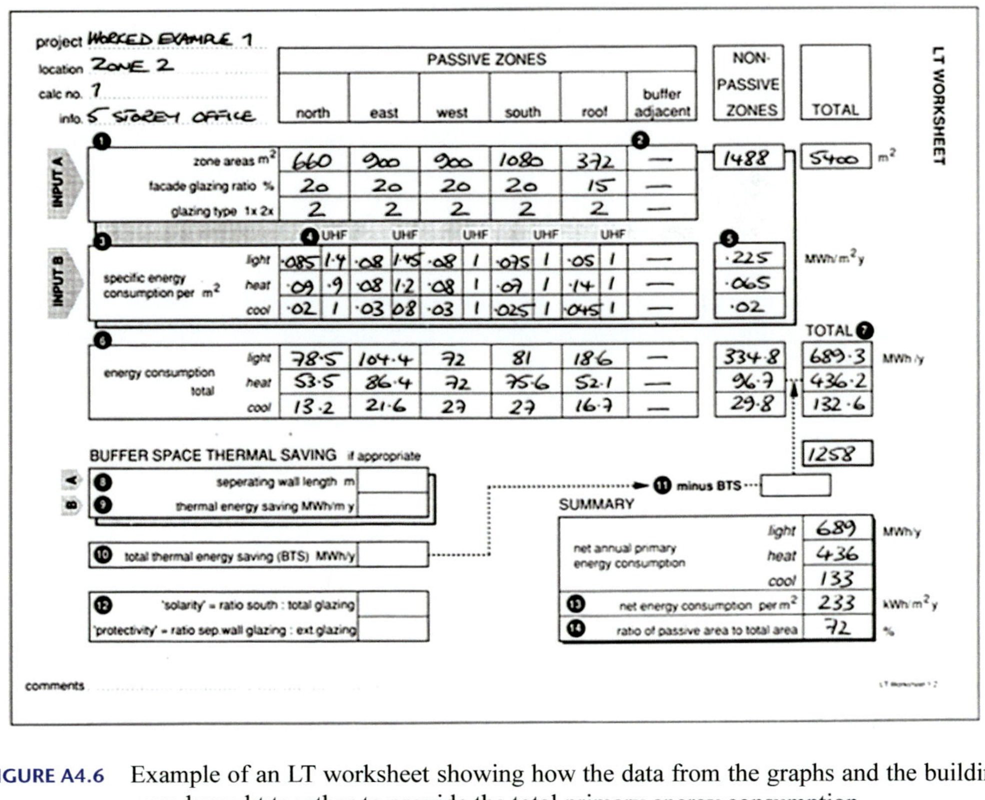

FIGURE A4.6 Example of an LT worksheet showing how the data from the graphs and the building were brought together to provide the total primary energy consumption.

used for the final European Architecture Competition, 'Living in the City', and was modified to suit the retrofit of multiple occupancy blocks in eastern Europe.

LT Method: The Software-Based Tool

Rapid developments in computing power soon made the on-paper graphical form obsolete, and various iterations of software-based LT were developed, ranging from a computer-based version using pre-computed data to the later versions which had a drawing tool data input, an internal solver, and graphical output. The last development of LT (known as Climate Lite) was a joint project between the Building Research Establishment, Cambridge University, and a software company, South Facing. The better resolution of local micro-climate, including noise and pollution. The facility to look at multiple buildings was also developed, by using a copy tool and editing, which allowed developments of multiple buildings to be assessed in relation to the local micro-climate and to mutual overshadowing. Sadly, due to limitations of resources, this version was never completed. Computing capabilities have expanded exponentially since the method was developed but have tended not to present buildings at the design stage as having some spaces that can be passively, naturally light and ventilated, which is an important factor to emphasise as the move back to mixed-mode buildings and away from deep-plan fully mechanically service buildings is gathering speed.

Conclusions

The great strength of the LT method was that it provided a transparent energy design tool that used environmental parameters available at the earliest stages of a design to instil a generic understanding of how the main energy sources, flows, and uses interact in response to the design, an understanding that could be carried through to future design exercises. The LT had an educative and strategic value, rather than just a tactical tool to test specific design iterations.

Student designs are now inevitably accompanied by the results of simulations in energy use and carbon emissions, comfort, daylight renderings, even CFD simulations of airflows, all presented with stunning graphics. But it is rare to find evidence that all this analysis has had any real influence on the development of the design. It is possible to conclude that modern simulation tools are not design tools in the sense that they help understand the process of design, but design-proving tools. They do not provide students with a basic understanding of the physics on which the simulations are based. This leaves them very vulnerable to the built-in assumptions, default values, and artefacts of the computation models. It follows, then, that even when the buildings meet the target in the simulated world, they may well fail in real life.

In education, proof-testing simply isn't what is needed. Maybe it is time to revisit the world of LT again, using all that today's IT has to offer and seeing its main objective as being to develop a generic understanding and expertise in the user, rather than a simple performance figure produced from a black box model for a hypothetical building.

Note

1 Ratti, C., D. Robinson, N. Baker and K. Steemers (2000). LT Urban: The energy modelling of urban form, in *Architecture + City + Environment, Proceedings of the 17th International Passive and Low Energy Architecture Conference*, Edited by K. Steemers and S. Yannas, Cambridge, 2–5 July, pp. 660–665, London: James + James Publishers, ISBN 1-902916-16-6

INDEX

Note: page numbers in *italics* indicate figures, and page numbers in **bold** indicate tables on the corresponding page. References following 'n' refer to the notes.

For Product Safety Concerns and Information please contact our EU representative GPSR@taylorandfrancis.com Taylor & Francis Verlag GmbH, Kaufingerstraße 24, 80331 München, Germany

Batch number: 10733627

Printed by Printforce, the Netherlands